COLD ANTIMATTER PLASMAS AND APPLICATION TO FUNDAMENTAL PHYSICS

To learn more about AIP Conference Proceedings, including the Conference Proceedings Series, please visit the webpage
http://proceedings.aip.org/proceedings

COLD ANTIMATTER PLASMAS AND APPLICATION TO FUNDAMENTAL PHYSICS

Okinawa, Japan 20 – 22 February 2008

EDITOR

Yasuyuki Kanai
RIKEN, Japan

Yasunori Yamazaki
University of Tokyo and RIKEN, Japan

SPONSORING ORGANIZATIONS
RIKEN
University of Tokyo
ITAMP

AMERICAN INSTITUTE OF PHYSICS

Melville, New York, 2008
AIP CONFERENCE PROCEEDINGS ■ 1037

Editors:

Yasuyuki Kanai

Atomic Physics Laboratory, RIKEN
2-1 Hirosawa, Wako, Saitama, 351-0198
Japan
E-mail: kanaiyasuyuki@riken.jp

Yasunori Yamazaki

College of Arts and Sciences
University of Tokyo
3-8-1 Komaba, Meguro, Tokyo, 153-8902
Japan
E-mail: yasunori@phys.c.u-tokyo.ac.jp

and

Atomic Physics Laboratory, RIKEN
2-1 Hirosawa, Wako, Saitama, 351-0198
Japan
E-mail: yasunori@riken.jp

L.C. Catalog Card No. 2008932705

ISBN 978-0-7354-0561-5
ISSN 0094-243X

Printed in the United States of America

CONTENTS

Preface

Recent progresses of trapping, cooling, and manipulating antiprotons and positrons accelerate antimatter plasma physics as well as fundamental physics applying such new techniques. The Workshop "Cold Antimatter Plasmas and Application to Fundamental Physics" was held intending to provide an occasion for scientists in the related fields to gather together, to exchange ideas, and to brain-storm new research subjects.

In studying fundamental symmetry of nature, the straightforward strategy has been to prepare energetic particles and to make hard collisions to study deeper and deeper inside. Apparently, there exists or will exist technical as well as financial limits in constructing very large facilities. Instead of pursuing higher energies, an alternative scenario is to apply high precision measurements to cooled exotic particles, which also fits well with the idea of "environmentally friendly". From this viewpoint, antiprotons and positrons have exceptional advantages over various exotic particles, i.e., they have infinite lifetimes. In other words, they can be cooled using macroscopic time, and accordingly they can really be cold, as cold as a few K or even lower although they are initially produced at relativistic energy regions.

This book contains the latest knowledge of the field presented/discussed at the workshop. The subjects covered are from fundamental questions about CPT symmetry and gravitation, to atomic collisions and atomic physics including antihydrogen as well as non-neutral plasma physics involving antimatter. Technological aspects are also included, which is expected to develop hand in hand with the research itself particularly when the field is new and active.

It was the pleasure of the organizing committee that the whole period of the Workshop was filled with active and vivid discussions.

This workshop was sponsored by RIKEN, University of Tokyo, ITAMP, and The Society for Atomic Collision Research, and was held in Naha, Okinawa, Japan from February 20th to 22nd, 2008 having about 60 participants from nine different countries.

On behalf of the Organizers: Y. Yamazaki

Sponsors

RIKEN
University of Tokyo
ITAMP
The Society for Atomic Collision Research

Committee Members

Yasunori Yamazaki, University of Tokyo and RIKEN
Yasuyuki Kanai, RIKEN
H. R. Sadeghpour, ITAMP

Antimatter Factory

Yoshio Yamaguchi

Theoretical Physics Laboratory, RIKEN, 2-1 Hirosawa, Wako, Saitama 351-0198, Japan

Abstract. Following the proposed $\bar{d}$ production by $\bar{p}$ ($\bar{p}$, π^-) $\bar{d}$, (light) antinuclei synthesis can be made by successive use of one and two antinucleon transfer reactions, ($\bar{d}$, $\bar{n}$), ($\bar{d}$, $\bar{p}$), ($\bar{t}$, $\bar{p}$), ($\bar{t}$, $\bar{n}$), ...

Keywords: antiparticle, antiproton, antineutron, $\bar{d}$, $\bar{t}$, $\bar{\alpha}$
PACS: 25.43.+t,

INTRODUCTION

1908, 100 yrs ago, is the celebrated year for the atomism. The existence of molecules was definitely established by Perrin in this year. Before 1908, atoms, molecules and atomic theory are merely interesting, but mysteriously workable hypothesis, which was the firm view-point shared by the main stream of chemistry and physics.

After this event, physics has made revolutionary development, relativity, quantum physics, nuclear and particle physics, astro-physics, during the past 100 years. The 20th century was indeed the Century of Physics.

Antiparticles stemmed from the Dirac equation of relativistic electron. The Dirac equation gave the negative energy solutions. Dirac introduced the new notion of the vacuum filled with negative energy electron states, a hole in which behaves like a positively charged particle. Paul Dirac, Herman Weyl and others considered this positively charged particle to be a proton. Soon Dirac changed his mind, a hole is positively charged electron, called later positron, whose mass equals that of the electron. This is the birth of the notion, antiparticle. Theoretically, soon, electron-positron pair produced by energetic photons and charged particles in matter has theoretically been studied.

Meanwhile, cosmic-ray physicists utilizing cloud chambers saw events like below:

CP1037, *Cold Antimatter Plasmas and Application to Fundamental Physics,* edited by Y. Kanai and Y. Yamazaki

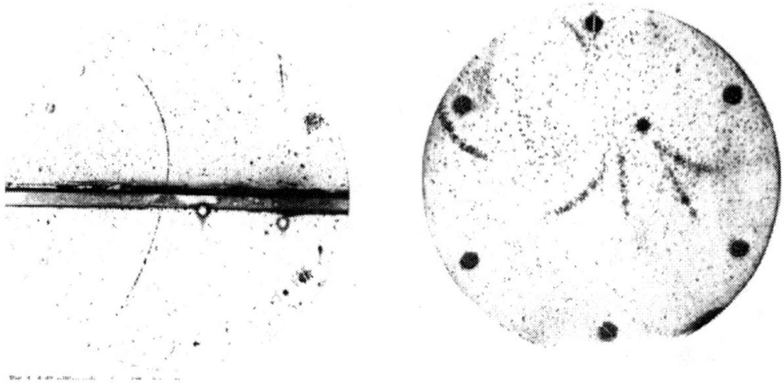

FIGURE 1. Taken from Fig.1 (p. 492), Fig. 3 (p.493) in the paper by C. D. Anderson, Phys. Rev. **43**(1933)491.

One of cloud crews said the electron likes to make back-scattering. It is C.D. Anderson who first realized they are the experimental verification of the positron (1932).

Later, Anderson — Neddermeyer(1936)and C.F. Powell et al (1949) discovered μ^{+}, μ^{-} and π^{+}, π^{-}, charged μ, π and their antiparticles. Antis of photon and π^{0} are themselves.

The Antiproton and the antineutron were discovered at 1955 and 1957, respectively by the Bevatron. Before these discoveries, physicists gossiped who shall win the Nobel Prize, either accelerator builders or experimentalists who first detected the antiproton? The Bevatron was set its highest energy 6.2 GeV above the antiproton production threshold at p-p collisions.

Antinuclei with A= - 2 and - 3, produced by high energy proton-synchrotron were detected much later,

$^{2}\overline{\mathbf{H}}$ in 1965,
$^{3}\overline{\mathbf{He}}$ in 1970,
$^{3}\overline{\mathbf{H}}$ in 1974.

Antialphas and heavier antinuclei are not yet detected, even though contemporary high energy accelerators / colliders can "energetically" and in principle produce them. This is because antinucleus production rate there are exponentially decreasing with increasing (anti) mass numbers.

$\bar{d}/\bar{p}$ production ratio at high energy hadron machines is typically $\sim 10^{-3.}$ One may need to synthesis even $\bar{d}$ and other light antinuclei in order to make precession experiments on properties of anits.

ANTINUCLEI, ANTIATOMS, ---

A particle and its antiparticle have symmetry as stated by the CPT theorem. We know about the asymmetry expressed by CP violation and T violation. The gravity acts(theoretically) equally on particles and antiparticles, though direct experimental demonstrations are meager, e.g., direct free fall experiments on e+, $\overline{\mathrm{p}}$ and $\overline{\mathrm{n}}$ are not yet performed.

The CPT theorem holds for the special relativistic local quantized field theories. New situations, e.g., strong gravity may offer something unexpected.

We live in the world where symmetries are spontaneously broken, in spite of the fast that the fundamental law might (or should) be symmetrical. Special and general relativity deny the existence of the inertial system. However, there is the cosmic background radiation (CBR) in our universe (our universe in the special solution of the ultimate law of the Nature, in which many symmetries are broken). And the local rest frame of CBR offers the inertial system for us. We, living in the universe with the broken symmetries, are searching more symmetrical law of the Nature.

We are looking for something new beyond the Standard Model of particle physics and cosmology. One of such important trials must be to see whether there is any asymmetry between particles and their antis besides both CP and T violation.

ANTIMATTER FACTORY

To investigate seriously the difference, if any, among matter (nucles, atoms, ---) and antimatter, we may wish to have various anticulei($A \geq 4$) and antiatoms. High energy accelerators/colliders at presently running are not able to produce $^4\overline{\mathrm{He}}$ and heavier antis. Then it may be better to try to cook heavy antinuclei from accumulated antiprotons. In a way our synthesis looks similar to primordial nucleo synthesis just after the Big Ban and nuclear synthesis at stellar central zones of main sequence stars.

However there exist several differences. We can not store antineutrons $\overline{\mathrm{n}}$. We can not rely on weak and radiative processes, such as $\overline{\mathrm{p}} + \overline{\mathrm{p}} \rightarrow \overline{\mathrm{d}} + e^{-} + \overline{\nu}_e$, $(\overline{\mathrm{p}}, \overline{\gamma})$ reactions.

Production ration $\overline{\mathrm{d}}/\overline{\mathrm{p}}$ at high energies is small $\sim 10^{-3.}$ Therefore, it seems we must first mass-produce $\overline{\mathrm{d}}$ from $\overline{\mathrm{p}}$. Such a production method has already been proposed [1, 2, 3], utilizing the reaction $\overline{\mathrm{p}} + \overline{\mathrm{p}} \rightarrow \overline{\mathrm{N}} + \overline{\Delta} \rightarrow \overline{\mathrm{d}} + \overline{\pi}^{-}$ with the energetically asymmetric $\overline{\mathrm{p}} - \overline{\mathrm{p}}$ collider [2]. After enough accumulation of $\overline{\mathrm{d}}$, $\overline{\mathrm{d}} + \overline{\mathrm{d}}$ shall produce $^3\overline{\mathrm{H}}$ and $^3\overline{\mathrm{He}}$. $\overline{\mathrm{d}} + {}^3\overline{\mathrm{H}}$ and $\overline{\mathrm{d}} + {}^3\overline{\mathrm{He}}$ will produce $^4\overline{\mathrm{He}}$. Repeated use of pick-up reactions, $(\overline{\mathrm{d}}, \overline{\mathrm{p}})$, $(\overline{\mathrm{d}}, \overline{\mathrm{n}})$ and $({}^3\overline{\mathrm{H}}, \overline{\mathrm{n}})$ etc., we may cook heavier nuclei.

To realize these nuclear reactions we need storage rings (and colliding zone) of ~1GeV/c and relevant bending magnets.

Such facility, consisting of set of asymmetric colliders (a few GeV/c and beam transports, beam cooling parts), if constructed, may also be useful to measure cross sections of many nuclear reactions (including radiative capture processes) relevant to nucleo synthesis of astrophysical importance. Such measurement may also serve to the Fusion Projects for energy production.

Furthermore, μ^+ catalyzed fusion may be tried to produce very light antinucle.

This facility may cost substantially. But there must be many (used part of) colliders and detectors, already shut down, in the world. Economical use of such parts is the way to cost down.

Eventually I wish to see the synthesis of nuclei (atoms) up to the heaviest. Very accurate test of properties of antinuclei (atoms) would uncover something new different from nuclei (atoms), beside the accurate check of *CPT* theorem, etc.

REFERENCES

1. D. Möhl *et al.*, CERN-EP/82-11. 23 Jan 1982.
2. M. Abd El Bary *et al.*, Antideuteron Breeding in a Double Rings Collider, Letter of Intent for the Facility for Antiproton and Ion Research, April 16, 2004.
3. D. Möhl *et al.*, Nucl. Instrum. Nethods **202**(1982)427.

Formation Of A Cold Antihydrogen Beam in AEGIS For Gravity Measurements

G. Testera[a], A.S. Belov[b] , G. Bonomi[c], I. Boscolo[d], N. Brambilla[d], R. S. Brusa[e], V.M. Byakov[f], L. Cabaret[g], C. Canali[h], C. Carraro[a], F. Castelli[d], S. Cialdi[d], M. de Combarieu[i], D. Comparat[g], G. Consolati[j], N. Djourelov[l], M. Doser[m], G. Drobychev[w], A. Dupasquier[j], D. Fabris[v], R. Ferragut[j], G. Ferrari[o], A. Fischer[h], A. Fontana[c], P. Forget[g], L. Formaro[d], M. Lunardon[v], A. Gervasini[d], M.G. Giammarchi[d], S.N. Gninenko[b], G. Gribakin[p], R. Heyne[h], S.D. Hogan[q], A. Kellerbauer[h], D. Krasnicky[u], V. Lagomarsino[a], G. Manuzio[a], S. Mariazzi[e], V.A. Matveev[b], F. Merkt[q], S. Moretto[v], C. Morhard[h], G. Nebbia[v], P. Nedelec[n], M.K. Oberthaler[r], P. Pari[i], V. Petracek[u], M. Prevedelli[o], I. Y. Al-Qaradawi[z] , F. Quasso[j], O. Rohne[t], S. Pesente[v], A. Rotondi[c], S. Stapnes[t], D. Sillou[n], S.V. Stepanov[g], H. H. Stroke[s], G. Tino [o], A. Vairo[d], G. Viesti[v] , H. Walters[p], U. Warring[h], S. Zavatarelli[a] , A. Zenoni[c], D.S. Zvezhinskij[g] (AEGIS Proto-Collaboration)

[a] *Istituto Nazionale di Fisica Nucleare and Department of Physics, University of Genoa, via Dodecaneso 33, 16146 Genova, Italy*

[b] *Institute for Nuclear Research of the Russian Academy of Sciences, Moscow 117312, Russia*

[c] *Istituto Nazionale di Fisica Nucleare and Department of Physics, University of Pavia, via Bassi 6, 27100 Pavia, Italy*

[d] *Istituto Nazionale di Fisica Nucleare and Department of Physics, University of Milan, via Celoria 16, 20133 Milano, Italy*

[e] *Department of Physics, University of Trento, via Sommarive 14, 38100 Povo, Italy*

[f] *Institute for Theoretical and Experimental Physics, Moscow 117218, Russia*

[g] *Laboratoire Aimé Cotton, CNRS, Univ Paris-Sud, Bât. 505, 91405 Orsay, France*

[h] *Max Planck Institute for Nuclear Physics, Postfach 103980, 69029 Heidelberg, Germany*

[i] *Laboratoire de Basse Tempe'rature du SPEC/DRECAM/DSM, CEA Saclay, 91191 Gif-sur-Yvette Cedex-France*

[l] *Institute for Nuclear Research and Nuclear Energy, 72 Tzarigradsko Chaussee, 1784 Sofia, Bulgaria*

[m] *Department of Physics, CERN, 1211 Gene've 23, Switzerland*

[n] *Lab. d'Annecy-le-Vieux de Phys. des Particules, 9 Chemin de Bellevue, B.p. 110, 74941 Annecy Cedex, France*

[j] *Department of Physics, Politecnico di Milano, Piazza L. da Vinci 32, 20133 Milano, Italy.*

[o] *Istituto Nazionale di Fisica Nucleare and Department of Physics, University of Florence and LENS, CNR-INFM, via Sansone 1, 50019 Firenze, Italy*

[p] *Department of Applied Mathematics and Theoretical Physics, Queen's University, University Road, Belfast BT7 1NN, United Kingdom*

[q] *Laboratorium für Physikalische Chemie, ETH Zürich, Zürich 8093, Switzerland*

[r] *Kirchhoff Institute of Physics, University of Heidelberg, Im Neuenheimer Feld 227, 69120 Heidelberg, Germany*

[s] *Department of Physics, New York University, 4 Washington Place, New York, NY 10003, USA*

CP1037, *Cold Antimatter Plasmas and Application to Fundamental Physics,* edited by Y. Kanai and Y. Yamazaki

[i]University of Oslo, Department of Physics, Oslo, Norway
[ii] Czech Technical University in Prague, Faculty of Nuclear Sciences and Physical Engineering
[v]Istituto Nazionale di Fisica Nucleare and Department of Physics, University of Padova, Via Marzolo 8, 35131 Padova - Italy
[z] Qatar University, Doha, Qatar P. O. Box 2713, Doha, Qatar
[w]Institute for Nuclear Problems of the Belarus State University, 11, Bobruiskaya Str., Minsk 220030, Belarus

Abstract. The formation of the antihydrogen beam in the AEGIS experiment through the use of inhomogeneous electric fields is discussed and simulation results including the geometry of the apparatus and realistic hypothesis about the antihydrogen initial conditions are shown. The resulting velocity distribution matches the requirements of the gravity experiment. In particular it is shown that the inhomogeneous electric fields provide radial cooling of the beam during the acceleration.

Keywords: Antihydrogen, Equivalence Principle, Rydberg atoms.
PACS: 04.80.Cc, 32.80.Ee, 34.70.+e, 36.10.-k, 37.10.Ty

INTRODUCTION

The equivalence principle is a foundation of General Relativity and a large experimental effort is placed in testing its consequences in all possible fields: this research activity includes tests about the equality of the inertial and gravitational mass, the universality of the free fall, the search for non Newtonian corrections to the gravitational law, the measurement of the gravitational red shift, the search for time variation of the fundamental constants. Measurements about the equality of the inertial and gravitational mass of different bodies began well before the general relativity (they started with Galileo and Newton) and then continued over the years with larger and larger precisions: sensitivity of the order of 10^{-15} and 10^{-18} are expected from modern space experiments [1] while laboratory torsion balance setups have already reached an accuracy of 1 part in 10^{13} [2].

While all these tests concern macroscopic bodies, a sensitivity to the gravitational acceleration g of the order of 10^{-10} has been obtained with cold atoms in atomic fountains [3].

The important point is that all these measurements (with macroscopic bodies or cold atoms) are performed on matter system: there are no direct measurements about the validity of the principle of equivalence for antimatter. The validity of the equivalence principle for antimatter is extrapolated from the matter results or it is inferred using indirect arguments (several of them are controversial and model dependent [4]). Particularly interesting is that some quantum gravity models leave room for possible violations of the equivalence principle for antimatter [4].

The AEGIS experiment has been proposed [5,6] at the CERN Antiproton Decelerator to directly measure the Earths gravitational acceleration g on a beam of cold antihydrogen. The basic setup of the experiment should allow one to reach an accuracy of 1% and the experiment is designed to allow higher precision measurements through radial cooling of the beam.

The gravitational acceleration g will be obtained by detecting the vertical deflection of the antihydrogen beam flying horizontally with a velocity of a few 100 m/s for a path length of about 1 meter. The small deflection (few tens of μm) will be measured using two material gratings coupled to a position sensitive detector working as Moiré deflectometer in the classical regime. A device of this type with an additional grating in place of the position sensitive detector has been already operated with atoms [7] by members of the collaboration.

While details about the antihydrogen formation in AEGIS and the Moiré device are reported elsewhere [5], here we will focus on the formation of the antihydrogen beam.

PULSED COLD ANTIHYDROGEN FORMATION

The antihydrogen beam formation will take places in AEGIS by a two step process: the first one is the production of very cold antihydrogen in Rydberg states and the second one is the acceleration in one direction (the horizontal one) of the Rydberg antihydrogen using inhomogeneous electric fields. A key point is that the antihydrogen formation will happen within a short time (of the order of 1 μs). This "pulsed production scheme" strongly differs from the existing schemes where the antihydrogen is produced during long time intervals (seconds in the ATHENA experiment) [8].

Cold antihydrogen in Rydberg states will be produced by the charge exchange reaction [5] [9]

$$Ps^* + \overline{p} \rightarrow \overline{H^*} + e^- \qquad (1)$$

between Rydberg positronium Ps* and cold (100 mK or below) antiprotons. Rydberg positronium (with principal quantum number n of the order of 20-30) will be formed by laser exciting the ground state positronium emitted by a porous material hit by a positron bunch (10^8 positrons with a time length of 10-20 ns).

The production of antihydrogen happens when the Rydberg positronium traverses the cold antiproton cloud. Taking into account that the velocity of the Rydberg positronium must be of the order of 10^4-10^5 m/s to optimize the cross section for the reaction of Eq. 1 and that the antiproton cloud dimensions are of the order of a few mm, therefore the production time is defined within about 1 μs.

This pulsed production scheme of antihydrogen offers the possibility to easily measure the resulting antihydrogen temperature (by time of flight); it also provides a t=0 time for gravity measurement and it allows to form the antihydrogen beam as described below.

The trapped antiproton temperature is the main factor determining the resulting antihydrogen temperature provided that it is greater than about 100 mK: below this value kinematics effects in the charge exchange reaction are not completely negligible. FIGURE 1 (not to scale) shows the scheme of the region of the experiment where the antihydrogen will be formed, accelerated and sent through the grating system. Not shown in the drawing are the regions of the apparatus devoted to catch and cool antiprotons from the CERN Antiproton Decelerator and to accumulate positrons.

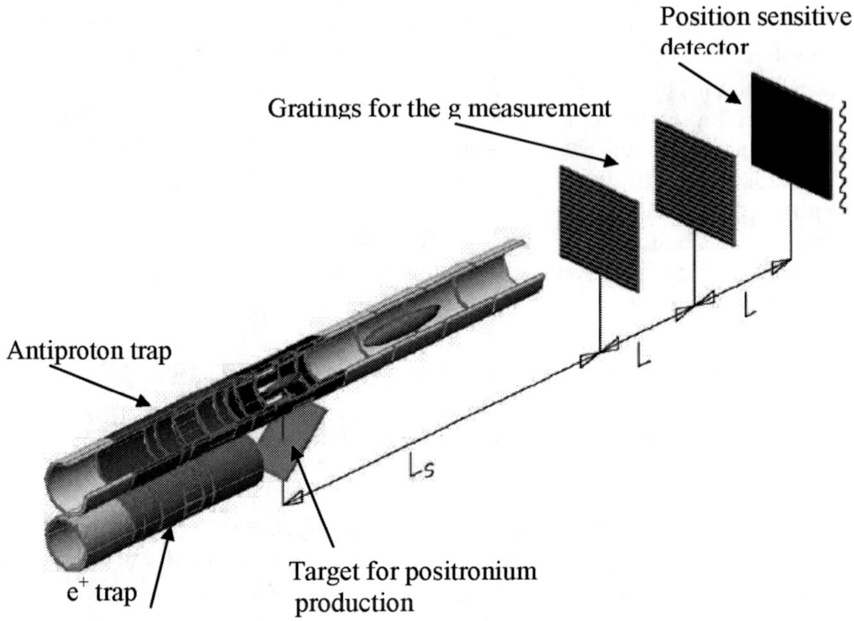

FIGURE 1. The figure (not to scale) shows the two parallel Penning-Malmberg traps that will be used in AEGIS to manipulate the antiprotons and positrons and to form and accelerate antihydrogen. They will be mounted inside a 100 mK cryostat in 1 Tesla magnetic field. The upper trap is devoted to antiprotons. Cold antiprotons wait for positronium in the black region. The lower trap is devoted to positrons: they will be sent to to the porous target mounted in front of the antiproton trap to produce positronium. Laser pulses will excite the positronium to selected Rydberg states having n in the range 20-30. The yellow region of the upper trap shows the bunch of antihydrogen after the acceleration (as described in the text). Finally the two material gratings followed by a position sensitive detector for the gravity measurement are shown. L_s is 30 cm, L=40 cm. The trap radius is 8 mm. The gratings and detector dimensions are 20x20 cm^2.

THE PENNING TRAP RYDBERG ACCELARATOR

The use of inhomogeneous electric fields to accelerate and manipulate Rydberg atoms has been proposed long time ago [10] but only recently this idea has become an experimental reality. In particular experimental demonstration of the acceleration, deceleration and even trapping has been achieved by members of AEGIS with hydrogen atoms in Rydberg states with quantum numbers very similar to those to be used in AEGIS [11].

To first approximation the energy levels E of the (anti)hydrogen atom in an external, homogeneous electric field of magnitude F are given, in atomic units, by

$$E = -\frac{1}{2n^2} + \frac{3}{2}nkF \tag{2}$$

where k is a quantum number which runs from $-(n-1-|m|)$ to $(n-1-|m|)$ in steps of two and m is the azimuthal quantum number. Here 1 a.u. = 27.211 eV in the case of the energy and 5.14×10^9 Vcm^{-1} for the electric field strength.
If the excited atoms are moving in a region where the amplitude of the electric field is changing then their internal energy changes according to Eq. 2 and thus, to conserve the total energy, they are accelerated or decelerated; the change of kinetic energy is 3/2 nk ΔF. This is described as a force acting on the atom by taking the space derivative of Eq. 1.

The experiments with hydrogen have shown that using time varying electric fields changes of velocity of the several hundreds m/s can be achieved in a few mm space using electric fields only limited by field ionization.
The formation of the beam of antihydrogen in AEGIS will be obtained by switching the voltages applied to the antiproton trap electrodes (immediately after the antihydrogen formation) from the usual Penning trap configuration to a new configuration that we call the "Rydberg accelerator".
To achieve this it is required to generate an electric field having an amplitude decreasing (or increasing) with z. Antihydrogen with positive (negative) k will be accelerated toward the system of gratings . The accelerating electric field will stay on for a selected time interval.
The gravity measurement procedure sets the requirements about the beam. The gratings of the Moiré device select specific trajectories of the atoms that can arrive on the detector. The distribution of the number of atoms arriving on the detector as a function of the vertical coordinate shows a periodical pattern due to the gratings. The gravity force causes a vertical shift of this pattern which depends on the time of flight T between the two gratings. Then to get g it is necessary to reconstruct, in addition to the vertical position on the detector, the horizontal velocity of the particles while they travel through the gratings. For this reason, it is important to have a start time well defined and a limited axial spread of the beam at the time when the accelerating electric field is switched off.
The transverse velocity of the beam does not influence the measurement result but it affects of course the number of particles arriving on the detector and the size of the gratings and detector itself. It is then mandatory to implement an acceleration procedure that does not heat the beam in a significant way in the radial direction.
We have simulated the expected velocity distribution in AEGIS taking into account the electric field obtained with the true geometry of the trap electrodes and realistic hypothesis about the initial antihydrogen spatial extension (few mm radius and about 1 cm axial length) and the expected state population.
The antihydrogen state population has been calculated by a dedicated CTMC code simulating the charge-exchange process in the AEGIS regime [5]. The Rydberg antihydrogen atoms will be produced not in a single quantum state but they will populate a distribution of states. The mean value of the principal quantum number n_H of antihydrogen is related by the positronium one by $n_H \approx \sqrt{2} n_{Ps}$, but a distribution having a rms deviation of few units is expected.

As a reference the results here reported refer to a Gaussian distribution of n_H having mean value 30 and a rms deviation equal to 4. For every n all the possible k have been included. In these conditions the product nk ranges from zero to about 1000.
The main consequence is that the various atoms will acquire different velocities in the same electric field and a beam with a large velocity spread will be obtained. To partially reduce the velocity spread it is important to use a time-varying electric field as shown in FIGURE 2.

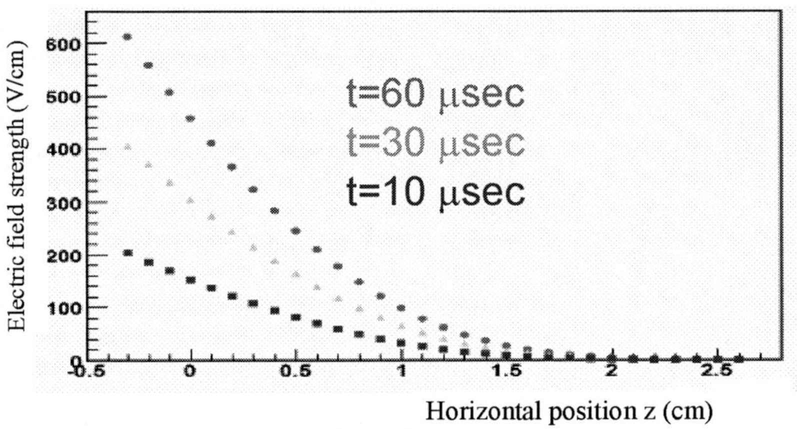

FIGURE 2. The figure shows the amplitude of the accelerating electric field on the trap axis versus z after 10, 30, 60 μs from the beginning of the acceleration process. The electric field stays on for 80 μsec in these simulations. It is obtained by applying voltages to radial sectors of the cylindrical trap electrodes. The accelerating electrodes are constituted by those of the antiproton trap shown in figure 1 together with some neighboring electrodes. These electrodes are split into four sectors (in the radial plane) with 135 degrees and 45 degrees angular extension. A voltage V_0 is applied to one of the big sectors while a voltage $-V_0$ is applied to the one opposite to it. The smaller sectors are grounded. The required distribution of electric field strength along the propagation direction of the beam is generated by applying a time-dependent potential to the relevant electrodes. The resulting electric field direction is almost perpendicular to the magnetic field. The antihydrogen cloud after production is centered at z=0.

FIGURE 3 shows the axial velocity before and after the acceleration for a cloud of antihydrogen produced with antiprotons cooled to 100 mK and distributed uniformly within 3 mm radius (that is within a large fraction of the 8 mm trap radius) and 8 mm length. Only 10% of the particles are lost on the trap electrodes. The corresponding distribution of the axial positions immediately after the accelerating electric field is switched off is shown in Fig 4: its spread is consistent with the requirements about the gravity measurement.

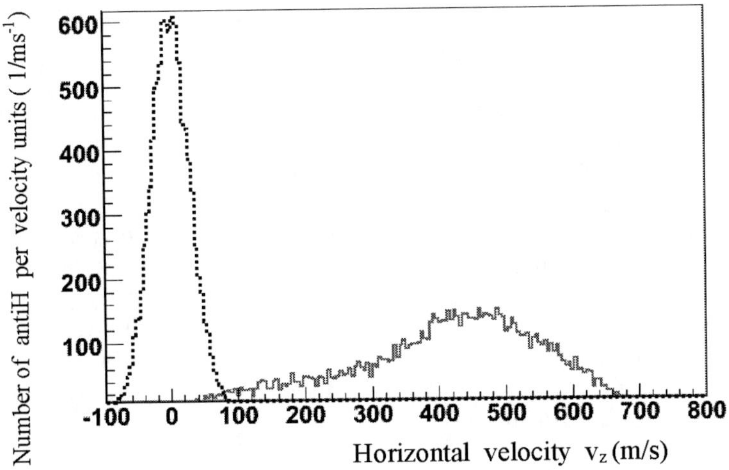

FIGURE 3. The red continuous curve is the distribution dN/dv_z of the horizontal velocity at the time when the accelerating electric field is switched off . For comparison the dotted black is the initial velocity distribution (Maxwell-Boltzmann distribution at a temperature of 100 mK). The mean value of the red curve can be tuned by playing with the field intensity.

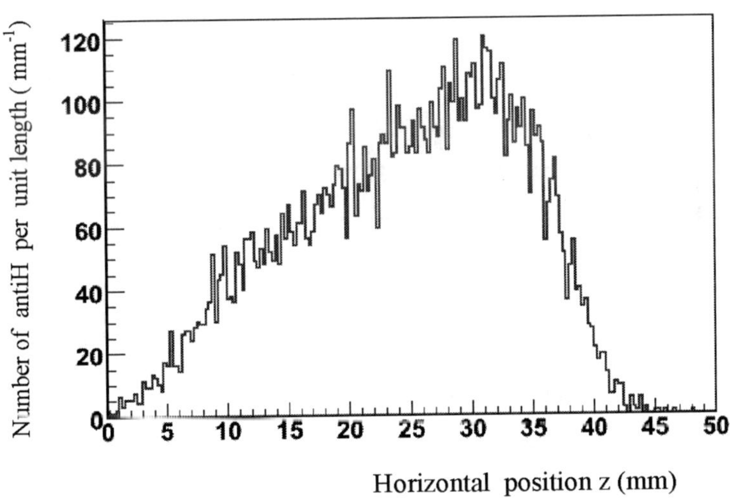

FIGURE 4. Distribution of the horizontal position of antihydrogen after acceleration. The plot refers to the same conditions as figure 3.

TRANSVERSE VELOCITY DISTRIBUTION OF THE ANTIHYDROGEN BEAM

The amplitude of the radial electric field increases with the distance from the trap center thus providing an average deceleration force resulting in a reduction of the mean radial velocity. This is particular relevant for antihydrogen produced within 1-2 mm from the trap center. In the conditions of figure 2 (100 mK antihydrogen distributed within 3 mm radius) the distributions of the transverse velocity before and after the acceleration are practically identical. This is already a positive result because it indicates that, on average, the acceleration procedure is not destroying the efforts placed in producing ultracold antihydrogen. Particularly interesting if the comparison between the transverse velocity before and after the acceleration for antihydrogen initially produced within a smaller radius: 1.5 mm. In this case the net effect is radial cooling as shown in FIGURE 5.

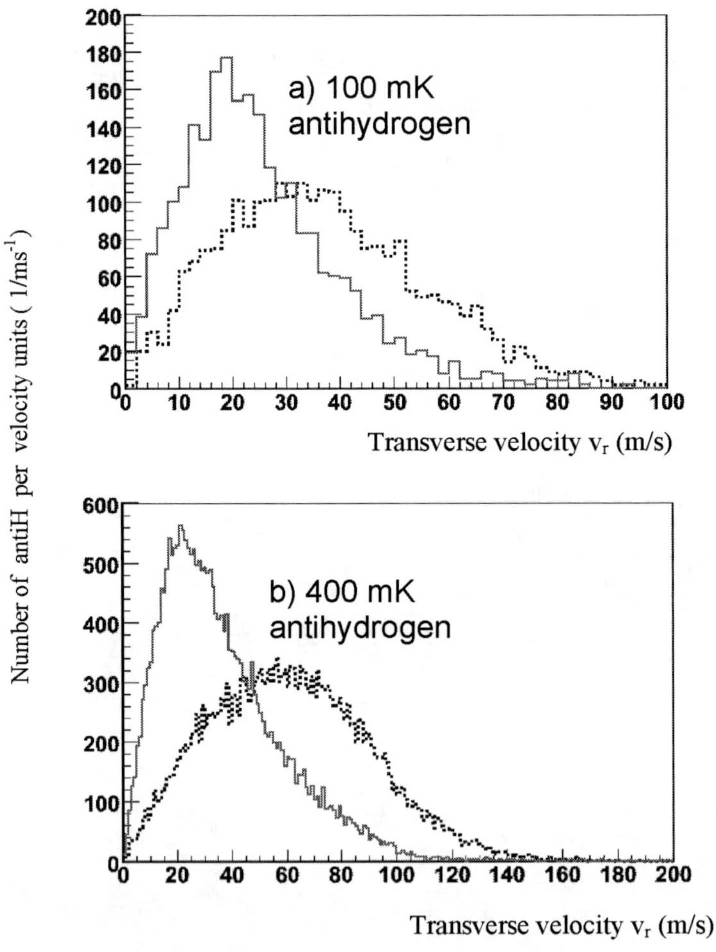

FIGURE 5. These two figures compare the transverse velocity distribution dN/dv_r as a function of v_r of antihydrogen before (black dotted plot) and after (red continuous plot) the acceleration. It is assumed that the antihydrogen is initially produced within a cylinder having 1.5 mm radius and 8 mm length. The initial velocity, before the acceleration, corresponds to a Maxwell distribution with 100 mK in figure a) and 400 mK in figure b). The cooling effect is evident in both cases.

EFFECT OF THE MAGNETIC FIELD

The Stark acceleration of antihydrogen in AEGIS will happen in presence of the uniform magnetic field of the Penning-Malmberg traps. Under completely general conditions, the dynamics of Rydberg hydrogen atoms in electric and magnetic fields with arbitrary mutual orientation are rich and complex [12]. Several regimes are possible depending on the range of parameters (values of the electric and magnetic fields and principal quantum number values). The magnetic field in the antihydrogen formation region is chosen following a compromise between the requirement of the charged particle trap (that demands high magnetic fields) and the need to softly perturb the Rydberg antihydrogen. The choice of B=1 Tesla realizes this compromise for Rydberg antihydrogen having n not higher than about n=35. In this regime n is still a good quantum number and the diamagnetic correction is small (but not completely negligible) compared to the Zeeman one in the Hamiltonian and the classical dynamics is regular [13]. For larger n and higher magnetic field the classical dynamics could become chaotic and the quantum levels can show large avoided crossing [11]. These effects could influence the efficiency of the acceleration procedure. Detailed quantum calculations to exactly quantify these effects are in progress. The presence of the magnetic field is here accounted modifying Eq. 2 with the one giving the energy levels to the first order in the perturbation theory for perpendicular electric and magnetic fields [14]

$$E = -\frac{1}{2n^2} + n'\sqrt{\gamma^2 + 9n^2F^2} \quad . \tag{3}$$

Here γ is the magnetic field in atomic units and n' is a quantum number assuming the values $-\left(\frac{n-1}{2}\right), -\left(\frac{n-3}{2}\right), \ldots .. \left(\frac{n-1}{2}\right)$.

Figure 6 compares the simulated horizontal velocity distribution of figure 3 with the one obtained using the same electric field and the same initial conditions of the antihydrogen cloud but including the magnetic field. The effect is a reduction of the final velocity values which is tolerable. The radial cooling effect discussed in the previous paragraph is maintained in presence of magnetic field.

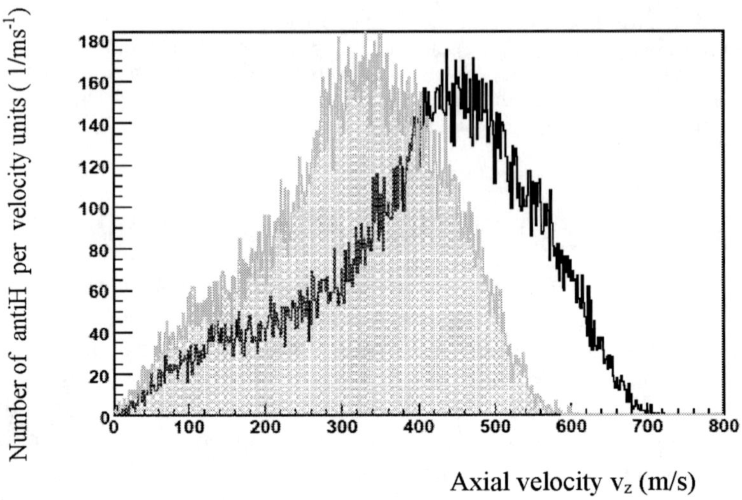

FIGURE 6. Simulated distributions of the horizontal antihydrogen velocity with (filled green plot) 1 Tesla magnetic field and without (black curve). The black curve is the one of figure 3. The initial antihydrogen cloud are the ones used to obtain figure 3 (a cloud of 100 mK temperature with 3 mm radial size and 8 mm length and having a mean value of n equal 30 with rms=4). For every n all possible values of n' with uniform distribution are considered.

CONCLUSIONS

The recent developments about the manipulation of Rydberg hydrogen by using inhomogeneous electric fields together with the experience gained at the CERN AD to form antihydrogen are merged together in the design of AEGIS to form a beam of cold antihydrogen. The beam will be used to perform the first gravity measurement on antimatter. In the long term physics program of the collaboration several improvements are planned: here we mention the use of negative ions to better cool the antiprotons [15] and of quasi-continuous Lyman-α laser for cooling the formed antihydrogen beam [5,16]. Finally the possibility of accelerating (and decelerating) and thus transporting antihydrogen atoms opens the way to investigate novel antihydrogen trapping strategies in which the production and the trapping regions are spatially separated [5].

REFERENCES

1. STEP Website: http://einstein.stanford.edu/STEP/step2.html
   μSCOPE Website: http://www.onera.fr/dmph/accelerometre/index.html
2. Baeble S. et al.,Phys. Rev. Lett. 83 (1999) 3585
3. A. Peters et al, Nature 400 (1999) 849
4. M. Nieto and T. Goldman, Phys. Rep. 205, 5 221-281,(1992)
5. http://doc.cern.ch/archive/electronic/cern/preprints/spsc/public/spsc-2007-017.pdf)
6. A. Kellerbauer et al. (AEGIS Proto-Collaboration) NIM B 266 (2008) 351
7. M. Oberthaler et al., Phys Rev. A 54 4 1996
8. M. Amoretti et al, (ATHENA Collaboration) Phys. Lett. B 578, 23 (2004)
9. E. A. Hessels et al., Phys. Rev. A 57 3 1668 (1998)
10. T. Breeden, J. Metcalf PRL 47 (1726) 1981
11. E. Vliegen et al., Phys. Rev. A 023405 (2007); E. Vliegen, F. Merkt J. Phys. B 39 (2006) L241
12. J. Main et al., Phys. Rev. A 57, 1149 (1998)
13. J.B. Delos et al., Phys. Rev. A 30 3 1208 (1984)
14. F. Penent et al., Phys. Rev. A 87 4707 (1988)
15. A. Kellerbauer and J. Walz, New J. Phys. 8 (2006) 45.
16. L. Cabaret et al., IEEE J. of Quantum Electronics 36, 323 (2000).

Baryogenesis and its implications to fundamental physics

M. Yoshimura

Center of Quantum Universe and Department of Physics, Okayama University Tsushima-naka 3-1-1 Okayama Japan 700-8530

Abstract. In this talk I shall explain some basic concepts of baryogenesis and leptogenesis theory, and a new idea of experimental method of verification of fundamental ingredients of leptogenesis theory; the Majorana nature and the absolute magnitude of neutrino masses. Both of these are important to the quest of physics beyond the standard theory, and have far reaching implications irrespective of any particular medel of leptogenesis. If this new method works ideally, there is even a further possibility of detecting relic neutrinos.

Keywords: Baryogenesis, Leptogenesis, Neutrino, Majorana particle, Neutrino mixing
PACS: 12.10.-g,11.30.Fs,13.15.-v

INTRODUCTION

Before going to explain basic ideas of baryogenesis theory, let me mention the situation of particle physics when the theory became a hot subject of discussion.

The year 1978 is characterized by a wide recognition of the standard model as acceptable theory of particle forces and fundamental constituents, quarks and leptons. By this year the structure of neutral current weak interaction, the new prediction of electroweak unification based on SU(2) $\times$ U(1) gauge group, has been confirmed to a high precision, and all weak phenomena have been explained by a single parameter of the weak mixing angle. Moreover, quantum chromodynamics (QCD) of the strong interaction has emerged as the leading candidate of explaining data of deep inelastic electron and neutrino scattering, along with a possible hint towards the color confinement by the lattice gauge theory.

As is well known, the structure of electroweak theory is chiral, namely the left-right (LR) symmetry is broken at a fundamental level, and this appeared to explain both the V - A current $\times$ current interaction of ordinary weak interaction, and the 2 component massless neutrino.

However, it was suspected by many [1], already in mid 70's, that the standard theory is not the end of story. According to proponents of these grand unification ideas, the standard theory is incomplete in essential points; (1) it lacks understanding of exact relation $1/3$ of electric charges of quarks and charged leptons, (2) the asymptotic freedom of strong interaction hints towards a unification of three gauge coupling constants of SU(3) $\times$ SU(2) $\times$ U(1) in the standard theory (this was later confirmed experimentally by the aid of supersymmetry [2]), (3) all conserved quantum numbers should ideally be associated with the gauge force, yet there exist non-gauge quantum numbers empirically conserved such as baryon and lepton numbers.

CP1037, *Cold Antimatter Plasmas and Application to Fundamental Physics,* edited by Y. Kanai and Y. Yamazaki

To many, these seemed to suggest a further truly unified picture of all particle forces, bridging quarks and leptons as an identical entity in the symmetry limit. A natural unification seemed a left-right symmetric unification such as SO(10) theory. In this grand unified theory (GUT) the fundamental unit of one generation of quarks and leptons has 16 components (16 spinor of SO(10) group);

$$\begin{pmatrix} u_R & u_G & u_B & \nu \\ d_R & d_G & d_B & e \end{pmatrix}_L + \begin{pmatrix} \overline{u_R} & \overline{u_G} & \overline{u_B} & \bar{N} \\ \overline{d_R} & \overline{d_G} & \overline{d_B} & \bar{e} \end{pmatrix}_L . \tag{1}$$

Chirality $L,R = (1 \mp \gamma_5)/2$ is flipped in the second entity here. Thus $(\bar{f})_L = \overline{f_R}$, and this theory is LR symmetric.

Existence of a chiral partner $N_R = \overline{N_L}$ makes this model quite distinct from the famous SU(5) GUT in which the fundamental unit is reducible representation of $\underline{5} + \underline{10}$. The SU(5) theory has been studied extensively, but it now seems to be excluded by negative search of proton decay. The new chiral partner N_R is distinct from the ordinary neutrino ν_L, and opens a possibility of understanding very small neutrino masses and futher suggests leptogenesis theory, as will be further explained. The ordinary neutrino mass is generated by a 2×2 matrix diagonalization of two Majorana fermions,

$$\begin{pmatrix} 0 & h\Lambda_{EW} \\ h\Lambda_{EW} & M \end{pmatrix} \rightarrow m_\nu = \frac{h^2 \Lambda_{EW}^2}{M} , \tag{2}$$

where the usual Dirac mass term $h\Lambda_{EW}\overline{\nu_L}\nu_R$ acts as a mixing of two 2-components spinors of ν_L and N_R, with M a large Majorana mass of N_R (h is a Higgs-Yukawa coupling). As a result, both diagonalized ν_L and N_R are massive Majorana fermions; they are 2-component massive fermions. The lepton number is violated by neutrino mass terms. The small neutrino mass $m_\nu \propto 1/M$ is a reflection of large mass scale $M \geq 10^{10}$GeV far beyond the standard theory, much larger than the electroweak scale $\Lambda_{EW} \approx 250$GeV. This is the celebrated formula of the seesaw mechanism [3]. For instance with $h\Lambda_{EW} = 100$GeV and $M = 10^{15}$GeV, the neutrino mass is of order 10meV, roughly a right order of magnitude to explain the neutrino oscillation.

Baryon number non-conservation that exists in these GUT has been most extensively studied experimantally, presumably by the suggestion [4] that this violation combined with CP violation may lead to a quantitative understanding of the baryon asymmetry of our universe. The initial phase of experimental searches of proton decay has however failed.

BASIC OF BARYOGENESIS

There is no observation that hints for existence of anti-matter in our vicinity of the universe. Moreover, the big bang cosmology, confirmed by many facts including the very impressive COBE and WMAP microwave obervations, galaxy and supernovae surveys, suggests strongly that our present universe is dominated by matter, and this can be traced back to the epoch of anti-matter annihilation. The emerging picture is then a tiny asymmetry of $O[10^{-10}]$ between matter and anti-matter existing prior to this epoch.

The theory of baryogenesis proposed around 1978 [4], [5] attemps to explain this baryon asymmetry using GUT. It is a quantitative idea supported by numerical computation [6]. The earlier suggestion [7] has not been noticed until much later, presumably due to a lack of strong motivation at that time. Many review papers have appeared since then, including my own [8].

The basic concept may be summarized by three key issues needed to explain a generation of the baryon asymmetry; (1) need for baryon (B) non-conservation (this is later replaced by B − L non-conservation, opening a possibility for leptogenesis), (2) CP and C violation, (3) departure from thermal and chemical equilibrium. All these seem obviously necessary conditions, but their interplay is rather intricate. Let me explain two points with this regard.

The third condition implies a high mass scale for the epoch of baryogenesis. If a process giving a finite baryon number is balanced by its inverse process that simultaneously occurs, a net generation necessarily vanishes. The only possible way that this is evaded is existence of the arrow of time by cosmic expansion. Thus, an effective condition for a net baryon generation is as follows; the rate of inverse process is smaller than the expansion rate. In the early hot universe of temperature T the expansion rate is given by $\approx \sqrt{N_F}T^2/m_{pl}$ where $m_{pl} = 1/\sqrt{G}$ with G the gravitational constant, and N_F is an effective number of species contributing to the energy density of the universe. Let us take a heavy boson X decay that mediates baryon number non-conservation. The rate of inverse decay is then of order, $\alpha m_X e^{-m_X/T}$, with α a coupling constant factor of X-boson to quarks and leptons. The condition is then an inequality that must be obeyed around $T \sim m_X$; $m_X \geq O[0.01]\alpha m_{pl}$. This condition is made consistent with GUT parameters, if $m_X \geq O[10^{15}]$ GeV.

Next, let us explain the second condition. Effects of CP violation show up in a subtle way. In particular, it is caused by a quantum mechnaical interference. To illustrate this, take an example of X-boson decay, and consider for simplicity that only two decay modes exist; $X \to \bar{q}\bar{q}$ (two anti-quarks) and $X \to ql$. These two decay channesl have different baryon number change, the difference in the two is $b_q - b_l = -1$. Considering the decay rate for these processes, the rate of baryon number generation by X-boson decay is then

$$-\gamma_X \frac{2}{3} + \gamma_l \frac{1}{3} + \overline{\gamma_X}\frac{2}{3} - \overline{\gamma_l}\frac{1}{3} = \overline{\gamma_X} - \gamma_X \,. \tag{3}$$

The second relation follows from the equality of decay rates of particle and anti-particle; $\overline{\gamma_X} + \overline{\gamma_l} = \gamma_X + \gamma_l$. The non-vanishing of this quantity requires CP violation; $\overline{\gamma_X} \neq \gamma_X$.

The rate of the net baryon number generation is then computed in perturbation theory, which typically proceeds

$$\overline{\gamma_X} - \gamma_X = \sum_i \left(|\overline{f}_i|^2 - |f_i|^2\right) , \tag{4}$$

where i symbolically implies both the kinematic phase space, decay channels, and contributing Feynmann amplitudes. The most important contribution arises from the tree (given by $g_1 f_1$ when coupling factors are separated) and the one-loop (given by $g_2 f_2$) diagrams of Feynmann amplitudes. Hence, the asymmetry rate is of order,

$-4\Im(g_1 g_2^*)\Im(f_1 f_2^*)$. The inteference effect makes the generated baryon number necessarily a small quantity.

LEPTOGENESIS

The three conditions that have been listed for baryon asymmetry generation are actually satisfied in the standard theory of particle physics. The problem with the baryon number generation in the standard model is however that baryon number is violated by tunneling effect [9], and its rate is suppressed by a factor like $O[e^{-1/\alpha}]$ with $\alpha \sim 1/137$. This rate is too small.

But it was pointed out in [10] that the effect might be enhanced at the early universe of finite temperatures, where the tunneling effect is replaced by a barrier crossing. Necessary ingredients for leptogenesis have been developed gradually since then, by works of [11], [12], [13]. An important point is B and L violation, B−L being kept conserved. This leads to strong damping of preexisting B and L numbers, preserving B − L, which occurs at finite temperatures in the standard model. It is however important to note that a generation of a net asymmetry from symmetric condtion (zero baryon number) is difficult to achieve in the standard model.

A standard leptogenesis model is what is called thermal leptogenesis [14], [15]. This model has many similarities to baryogenesis in GUT context discussed above, the difference being the lepton asymmetry generation instead of the baryon asymmetry generation. The lepton asymmetry thus generated is converted by baryon number asymmetry via electroweak B and L violation at finite temperatures.

Leptogenesis has been boosted by the discovery of finite neutrino mass in oscillation experiments. Parameters of neutrino mass matrix are indirectly related to parameters in leptogenesis. There is an interesting interplay of two kinds of parameters, as extensively discussed in [15]. It is desirable to push these connections, and in particular to experimentally verify key issues related to leptogenesis.

HOW TO EXPERIMENTALLY VERIFY INGREDIENTS OF LEPTOGENESIS

Three important ingredients for success of leptogenesis are (1) the Majorana nature of neutrino masses, along with its implied lepton number non-conservation, (2) the small absolute magnitude of neutrino masses $\leq 0.1\,\mathrm{eV}$ for instance, (3) CP violation in the leptonic sector,

Our new experimental proposal [16] aims to verify the first two items in the initial experimental phase. Let us first clarify the fundamental difference between the Dirac and the Majorana fermions. This is most straightforwardly done by using independent, two component spinors. In the case of Dirac equation two independent spinors denoted by φ and χ are related to each other by the Dirac mass term $\propto m$;

$$(i\partial_0 - i\vec{\sigma}\cdot\vec{\nabla})\varphi = m\chi\,, \quad (i\partial_0 + i\vec{\sigma}\cdot\vec{\nabla})\chi = m\varphi\,. \tag{5}$$

On the other hand, the Majorana equation is described by

$$(i\partial_0 - i\vec{\sigma}\cdot\vec{\nabla})\varphi = im\sigma_2\varphi^* . \tag{6}$$

Put it differently, the Majorana equation holds with the identification of Dirac χ-spinor by conjugate of φ, $\chi = i\sigma_2\varphi^*$. As is well known, this identification is possible from the point of Lorentz covariance. Physically, this makes anti-partilce indistinguishable from their own particle, much like a self-conjugate photon of an integer spin. With zero mass, two descriptions of Dirac and Majorana equations are equivalent since χ and φ in the Dirac equation are decpoupled, but with a finite mass distinction is present.

Our proposal [16] uses transitions of metastable excited atoms to another metastable or ground state, $|i\rangle \to |f\rangle$ emitting two neutrinos; for the Dirac case a neutrino-antineutrino pair, $\nu_i\bar{\nu}_j$ and for the Majorana case a pair of two neutrinos, $\nu_i\nu_j$. Only in the case of Majorana pair two identical particles are involved, and the Pauli exclusion principle operates. The best way to directly distinguish the Majorana pair emission from the Dirac pair emission is to detect the interference effect of two identical fermions by measuring rates as a function of available energy. The effect is the greatest near the mass threshold of $\nu_i\nu_i$ where the difference in rates is $\propto m_i^2$. There are altogether six thresholds for linear combinations $\nu_i\nu_j, ij = 1,2,3$. At these six thresholds, rates also involve different combinations of mixing angles, thus leading to the neutrino mass spectroscopy including mixing angles.

A great merit of atomic pair emission is the closeness of neutrino masses to atomic level difference. Indeeed, there are almost infinite varieties of atomic transitons with varying energy level differences. A demerit is the small rate; for a pure pair emission the rate $\approx 10^{-33}(\Delta/eV)^5$ sec^{-1}, scaling with the 5th power of energy diffence Δ^5. Obviously one needs an enhancement mechanism. Another problem is how to prove the weak nature of transition without detecting neutrinos.

The best experimental method [17] towards enhancement we have thought of so far is to utilize the coherent target effect in much the same way as in superradiance. The neutrino pair emission has an advantage over radiative superradiance because the spatial coherence region is much larger due to effect of finite neutrino masses. What is involved in coherence is the momentum magnitude, and this is independent of the atomic energy difference for a particle emission of a finite mass. Indeed, at the neutrino mass threshold the neutrino momenta can become arbitrarily small. On the other hand, the photon momentum is always fixed by its energy, and its coherence region is limited by its wavelength in superradiance phenomena [18], [19].

In order to maximally extract information on neutrino masses, we propose to use radiative pair emission, $|i\rangle \to |f\rangle + \gamma + \nu_i\nu_j$. The three body nature of decay kinematics also helps much in the coherent emission rate. The threshold photon energy for the neutrino pair emission $\nu_i\nu_j, i,j = 1,2,3$ appears at

$$\omega_{ij} = \frac{\Delta}{2} - \frac{(m_i+m_j)^2}{2\Delta}, \tag{7}$$

where $\Delta = E_i - E_f$ is the energy difference of initial and final atoms. Needless to say, there is no way to distinguish the radiative pair emission from a misidentifed radiative background process on the event by event basis. But the distinction is possible statistically.

There is a background peak at $\Delta/2$ coming from two photon process $|i\rangle \rightarrow |f\rangle + 2\gamma$, but this peak is very sharp, its width being inversely proportional to the size of a coherent target region. Moreover, the two photon pair emission has unusual kinematic features; the pair is emitted back to back and the emitted direction of the pair is collimated along the cylinder axis of target region. Hence, one can hope to reject the background experimentally.

We shall quantify a few key points. A more detailed explanation is given in our forthcoming paper [17]. One of the best processes for performing neutrino mass spectroscopy is radiative neutrino pair emisson from alkali-earth atoms such as Ba. The level structure of low-lying excited Ba contains two electron system of multiplet, ${}^1D_2, {}^3D_{2,1,0}, {}^1S_0$, with energy hierarchy, $E({}^1D_2) > E({}^3D_{2,1,0}) > E({}^1S_0)$. Radiative pair emission from ${}^1D_2$ may have a rate of $O[10^{-2}]\,\mathrm{sec}^{-1}$, when the coherent rate $\propto N^3$, with N the number of target atoms. We have assumed $N = 10^{15}$ and the target volume $V = 10^3\mathrm{cm}^3$. Presumably this choice of target parameters is too optimistic, but the rate seems promissing.

METHOD OF RELIC NEUTRINO DETECTION

The principle of detection [20] of relic neutrinos of abundance $\approx 100\mathrm{cm}^{-3}$ for each species of neutrinos is simple; in the presence of neutrino sea the pair emission rate is reduced by the Pauli blocking. The effect is maximal at the neutrino mass threshold, since the Fermi-Dirac distribution function gives

$$f(p) \rightarrow \frac{1}{e^{m_\nu/(k_B T_d)} + 1} \sim \frac{1}{2}, \tag{8}$$

with T_d the decoupling temperature of $O[0.1]\,\mathrm{MeV}$. The pair emission thus gives $1/4$ reduction (product of two $1-f$'s) of the rate compared with the case of absence of the relic sea. The important experimental question is how wide this reduction region is in terms of the momentum or the equivalent photon energy as in the case of radiative neutrino pair emission. In [20] it is shown that this region is rather wide for a range of smallest neutrino masses $\leq$ a few meV.

REFERENCES

1. H. Georgi and S.L. Glashow, *Phys. Rev. Lett.***32**(1974)438; J. C. Pati and A. Salam, *Phys. Rev.***D10**(1974)275.
2. U. Amaldi et al.,*Phys. Lett.***B260**(1991)447; P. Langacker and M. Luo, *Phys. Rev.***D44**(1991)817.
3. T. Yanagida, in Proc. Workshop on the Unified Theory and the Baryon Number in the Universe, Tsukuba, Ibaraki, Japan, edited by A.Sawada and A.Sugamoto (KEK Report No.KEK-79-18,1979). M. GellMann, P. Ramond and R. Slansky, in Supergravity, edited by D.Z.Freedman and P.Van Niewenhuizen (North-Holland, Amsterdam, 1979).
4. M. Yoshimura,*Phys.Rev.Lett.***41**(1978)281;**42**(1979)746(E); *Phys.Lett.***88B**(1979)294.
5. D. Toussaint, S.B. Treiman, F. Wilczek and A. Zee,*Phys.Rev.* **D19** (1979)1036; S. Weinberg,*Phys.Rev.Lett.***42**(1979)850.

6. E.W. Kolb and S. Wolfram, *Nucl. Phys.***B172**(1980)224. J.N. Fry, K.A. Olive, and M.S. Turner, *Phys. Rev.***D22**(1980) 2953, 2977.
7. A.D. Sakharov, *Pis'ma Zh. Eksp. Teor. Fiz.* **5**(1967)32; [JETP Lett. **5**(1967) 24].
8. M. Yoshimura, *Cosmological Baryon Production and Related Topics*, Proc. Fourth Kyoto Summer Institute on Grand Unified Theories and Related Topics, ed. by M. Konuma and T. Maskawa (World Scientific Pub., Singapore. 1981).
9. G. t' Hooft, *Phys. Rev. Lett.***37**(1976)8.
10. S. Dimopoulos and L. Susskind, *Phys.Rev.***D18**(1978)4500.
11. F.R. Klinkhamer and N.S. Manton, *Phys. Rev.* **D30**(1984)2212.
12. N. Christ, *Phys.Rev.***D21**(1980)1591.
P.Arnold and L.McLerran, *Phys.Rev.***D36**(1987)581.
M. Dine, O. Lechtenfeld, B. Sakita, W. Fischler and J. Polchinski, *Nucl.Phys.***B342**(1990)381.
13. V.A. Kuzmin, V.A. Rubakov and M.E. Shaposhnikov, *Phys.Lett.* **155B**(1985)36.
14. M. Fukugita and T. Yanagida, *Phys.Lett.***174B**(1986) 45.
15. W. Buchmuller and M. Plumacher, *Int. J. Mod. Phys.***A15**(2000) 5047. G. F. Giudice, A. Notari, M. Raidal, A. Riotto, and A. Strumia, *Nucl. Phys.***B685**(2004) 89.
16. M. Yoshimura, *Phys. Rev.***D75**(2007) 113007, and hep-ph/0611362 (2006).
17. M. Yosimura, A.Fukumi, K. Nakajima, I. Nakano, H. Nanjo, and N. Sasao, *Coherent collinear photon pair and radiative neutrino pair emission*, in preparation.
18. R.H. Dicke, *Phys. Rev.***93**, 99(1954).
19. For a theoretical overview of superradiance, M. Gross and S. Haroche, *Phys.Rep.***93**, 301(1982).
20. T. Takahashi and M. Yoshimura, *Effect of Relic Neutrino on Neutrino Pair Emission from Metastable Atoms*, hep-0703019.

Accelerating universe, WEP violation and antihydrogen atoms

Yasunori Fujii

Advanced Research Center for Science and Engineering, Waseda University, Tokyo, 169-8555, Japan

Abstract.
Apart from the suspected violation of the CPT invariance, we might expect if the measurements of antihydrogen atoms provide testing Weak Equivalence Principle (WEP) in the gravitational phenomena. We start with how its violation can be related to the expected idea of unification of particle physics and gravitation, an attempt beyond the standard theories, including general relativity of Einstein. A particular emphasis will be placed on the issue of an accelerating universe, a rather recent development since nearly 10 years ago, suggesting a strong motivation toward attempts beyond the conventional concepts of the traditional cosmology. We face today's version of the cosmological constant problem. A candidate of the new ingredient appears to be provided by a scalar field, sometimes under the names like quintessence or dark-energy. In this article, we discuss the subject from a point of view of a more theoretical approach based on the scalar-tensor theory of gravitation. By exploiting the concepts of conformal transformations and conformal frames unique to this type of approach, we show that a successful understanding of the observed cosmological acceleration entails an unexpected outcome of breakdown of WEP, which may show up as a distinct behavior between hydrogen and antihydrogen in the gravitational influence, from a further wider perspective including a vector field as well. We intended to introduce most of the new concepts as plainly and briefly as possible, according to the nature of the talk. On the other hand, however, we find it also useful to go into some more details, depending on the interests of individuals among the audience. For a compromise, we decided to add a number of footnotes which were not delivered during the presentation, but might be helpful in the proceedings published later.

Keywords: Weak equivalence principle, Antihydrogen, Accelerating universe, Scalar-tensor theory
PACS: 03.75.Be, 04.50.Kd, 04.80.Cc, 04.90.+e

INTRODUCTION

From the point of view of general relativity (GR), we have no distinction between particles and antiparticles in any of the gravitational phenomena. But on the other hand, there has been a strong quest for unifying particle physics and gravity. In this approach, we might be allowed to make the distinction mentioned above. In fact in almost any of the theoretical models of unification we find a number of fields which are yet to be discovered, but supposed to couple to the ordinary type of matter nearly as weakly as the ordinary gravity does. For this reason we call them Non Geometric Gravitational Fields (NGGF), partially under the name GF. At the same time, however, we find that these fields belong to a category of the matter in Einstein's sense, instead of another category that accommodates the metric field. In order to remind the audience of this non-geometric nature, we decided to put another attribute NG. Due to this feature, these fields or forces will couple to the matter depending on some specific properties of individual object. Consequently, matter objects will fall freely with acceleration different

CP1037, *Cold Antimatter Plasmas and Application to Fundamental Physics,* edited by Y. Kanai and Y. Yamazaki

from object to object, hence violating the law of universal free-fall (UFF).[1] This is also called a violation of Weak Equivalence Principle (WEP). What do we mean by the term "weak"? We offer a very brief reply.

WEP represents a phenomenological aspect of the more general law expressed in the mathematical term; a tangential spacetime attached to any point on the Riemannian manifold is Minkowskian. The message is that, even with violation of WEP, we can still maintain Einstein's central idea of dynamics of spacetime geometry unspoiled.[2] The only thing is that what we call "gravitational phenomena" might be extended to be contaminated partially by NGGF,[3] hopefully as one of the barely visible traces of the unification mechanism, supposed to be realized fully only at extremely high energies. There can be many different kinds of NGGF. In practice we may focus upon scalar or vector fields supposed to have a very small masses, sometimes also called non-Newtonian gravity [2], or the fifth force [3].

WHY WEP VIOLATION IN COSMOLOGY?

Accelerating universe

Let us now tell how the cosmology comes to suggesting WEP violation. As usual, we assume a homogeneous and isotropic universe, with $\rho(t)$, the matter density depending only on the cosmic time t. Typical inter-galactic separations grow in proportion to the scale factor $a(t)$, which obeys the Einstein equation,

$$3H^2 = \rho, \quad \text{with} \quad H = \frac{\dot{a}}{a}. \tag{1}$$

In front of the density ρ we usually have the coefficient like $8\pi G/c^2$. But we use the (reduced) Planckian unit system with

$$c = \hbar = M_{\rm P}(= (8\pi G/(c\hbar))^{-1/2}) = 1. \tag{2}$$

The units of length, time and energy are re-expressed as follows in the traditional units,

$$8.10 \times 10^{-33}\text{cm},\ 2.70 \times 10^{-43}\text{s};\ 2.44 \times 10^{18}\text{GeV}, \quad (t_0 = 13.7\text{Gy} \approx 10^{60.2}). \tag{3}$$

Also the Planck mass $M_{\rm P}$ defined in (2) in terms of Newton's constant G, or the gravitational constant, is supposed to provide the basic energy scale of the unification processes. Particularly interesting is t_0, today's age, now believed to be 13.7 Gy. In units of the Planck time, as shown in the last portion of (3), this is $\sim 10^{60}$, a number worthwhile to be remembered, as we are going to show shortly.

---

[1] Basically Galilei's observation in the Leaning Tower in Pisa.

[2] Comma-goes-to-semicolon rule remains true [1], for example.

[3] In the same way as the ordinary "matter" fields, like the electromagnetic or the strong-interaction field, do not respect the universal coupling to other fields, like the electrons and nucleons, NGGFs do not observe the simple results as implied by UFF.

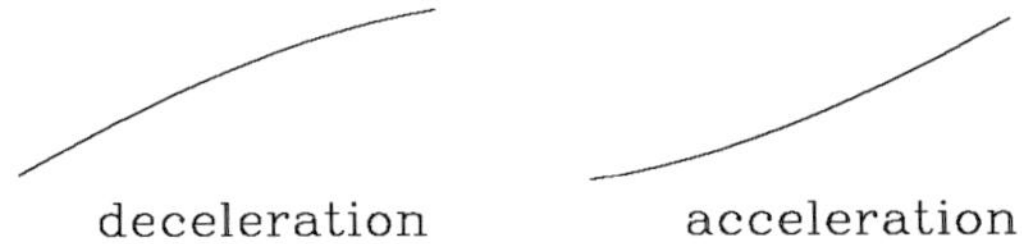

FIGURE 1. Two typical behaviors of $a(t)$.

Before 1998, we had taken it for granted that the universe was decelerated, because under virtually any of the reasonable assumptions, we always ended up with $a(t)$ which behaves in the left diagram in Fig. 1, with $a''(t) < 0$, thus decelerating. In 1998, however, two groups of astronomers reported their detailed observation of the very distant objects, with a startling result; an accelerating universe with $a''(t) > 0$, as shown by the right diagram [4].

They also showed that they fitted their observation by adding a positive constant $\Lambda > 0$ on the right-hand side of the Einstein equation;

$$3H^2 = \rho + \Lambda. \tag{4}$$

We add an explanation based on Fig. 2 which was prepared for the sake of illustration, in which we conveniently introduced the "critical density" ρ_{cr} defined by

$$\rho_{cr} \equiv 3H_0^2, \tag{5}$$

implying, with the subscript 0 for the present time, the entire right-hand side of (4).

As we find, ρ or ρ/ρ_{cr} falls off generically like t^{-2}, while Λ or $\Omega_\Lambda \equiv \Lambda/\rho_{cr}$ stays constant. They cross each other. Before this crossing, a is decelerating, while after the crossing, a is accelerating, though slightly, roughly speaking. The observation by these groups determined $\Omega_\Lambda \sim 0.7$. This implies that so far unknown component Λ makes up as much as 70% of the entire cosmic energy. In the Planckian unit system, ρ_{cr} is given basically like t_0^{-2}. As was shown in (3), we thus find

$$\Lambda_{obs} \approx \rho_{cr} \approx (10^{60})^{-2} \approx 10^{-120}, \tag{6}$$

where we used the result $\Omega_\Lambda \approx 1$ for the "observed" value Λ_{obs}.

On the other hand, nearly any of the theoretical models of unification showed it unavoidable to have Λ of the order of $\sim M_P^4$ or ~ 1 in the reduced Planckian unit system. This implies $\Lambda_{th} \sim 1$, which we compare with the observed value. This is certainly the worst record of disagreement between theory and observation. This is today's version of the cosmological constant problem, which consists of the two questions.

First, what a powerful theory is it to make it possible to fine-tune the parameters to the incredible accuracy of 120 orders of magnitude? This is a fine-tuning problem. Suppose

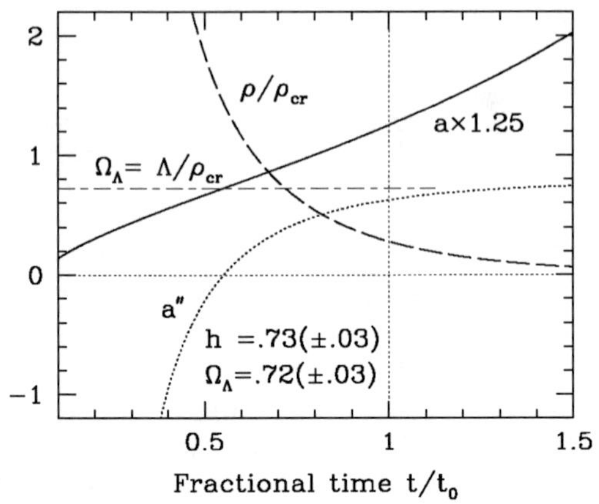

FIGURE 2. Illustrative plots of $\rho(t), a(t), a''(t), \Lambda$ against t/t_0.

we may nevertheless assume a truly constant Λ of this small. As we showed in Fig. 2, it crosses ρ, hence causing an acceleration only once throughout the whole history of the universe, which we are now watching in some sense. What a coincidence? This is the coincidence problem.

From these considerations we come to wondering if Λ is not a true constant, but is of some dynamical nature. In this connection we notice that the Λ term on the right-hand side of (4) can also be described by an ideal fluid with the equation of state $p = -\Lambda < 0$. It is this negative pressure that provides an extra acceleration to the universe. Also this negative pressure is precisely what a scalar field distributed uniformly everywhere in the universe, as described by $\phi(t)$, can provide with. This is sometimes called a dark energy different from the dark matter.

Some people wish to see the scalar field as a major player in the universe under the name of quintessence [5], but following quite a cautious and phenomenological approach. We are going to say, however, that a more theoretical approach could be not only promising but also rewarding.

Scalar-tensor theory

As one of such theoretical approaches, we are going to talk about the scalar-tensor theory, invented in 1955 by Jordan [6] with the basic Lagrangian

$$\mathcal{L}_{\rm ST\Lambda} = \sqrt{-g}\left(\frac{1}{2}\xi\phi^2 R - \varepsilon\frac{1}{2}g^{\mu\nu}\partial_\mu\phi\,\partial_\nu\phi + L_{\rm matter} - \Lambda\right), \tag{7}$$

where ϕ is the scalar field, with its kinetic energy term, the second term. We use the notation slightly different from Jordan's ω, but nothing is serious about this.[4] We use again the Planckian unit system. Unlike in Jordan's original theory, we have added Λ in order to show the relevance to the cosmological constant problem.

The first term on the right-hand side of (7) is called a nonminimal coupling (NM) term, which is truly unique to this theory, accommodating ϕ in a nontrivial manner. We compare this with the traditional Einstein-Hilbert (EH) term in GR;

$$\mathscr{L}_{\rm EH} = \sqrt{-g}\frac{1}{16\pi G}R = \sqrt{-g}\frac{1}{2}R, \tag{8}$$

showing that this theory has no truly constant G. Instead it has an effective $G_{\rm eff}$ defined by

$$\frac{1}{8\pi G_{\rm eff}} = \xi\phi^2(x), \tag{9}$$

in terms of the spacetime function $\phi(x)$. One of Jordan's motivations was to propose a consistent theory which allows a time-dependent gravitational "constant," as had been suggested earlier by Dirac [7].

Sometime later Brans and Dicke added an important feature [8]. They required that ϕ is decoupled from $L_{\rm matter}$, or ϕ does not enter $L_{\rm matter}$. By doing so they save the idea of WEP. The derivation in this "Brans-Dicke model" is a little complicated. For more technical details, we would like to ask the reader to refer to our book [9], or the related paper [10].

We also face another complication related to what is known as a conformal transformation

$$g_{\mu\nu} \rightarrow g_{*\mu\nu} = \Omega^2(x)g_{\mu\nu}, \tag{10}$$

with an arbitrary spacetime function $\Omega(x)$. This is a local scale transformation of the metric. The Lagrangian in (7), for example, is a specific functional of $g_{\mu\nu}$. We substitute the inverse of (10) into (7), hence re-expressing the same Lagrangian as a different functional of the new metric $g_{*\mu\nu}$. In a sense the same Lagrangian looks different. This might be termed as different Lagrangescapes [10].

Pauli pointed out that the importance of a special choice

$$\Omega = \xi^{1/2}\phi, \tag{11}$$

for which the re-expressed Lagrangian is put into

$$\mathscr{L}_{\rm ST\Lambda} = \sqrt{-g_*}\left(\frac{1}{2}R_* - \frac{1}{2}g_*^{\mu\nu}\partial_\mu\sigma\partial_\nu\sigma + L_{*\rm matter} - V(\sigma)\right), \tag{12}$$

with the EH term, instead of the NM term. For this reason we say that by this transformation we have moved to the Einstein conformal frame (ECF) from the Jordan conformal

---

[4] The exact relations to Jordan's original notations $\phi_{\rm J}$ and ω are given by $16\pi\phi_{\rm J} = (\xi/2)\phi^2$, $\varepsilon = {\rm Sign}(\omega)$, $\xi^{-1} = 4|\omega|$.

frame (JCF). Because Ω is arbitrary, we have infinitely many different CFs. Out of them the above two, JCF and ECF, are of distinctive importance among others.

We notice that the "canonical" scalar field in (12) is no longer the original ϕ but is σ newly defined by

$$\phi = \xi^{-1/2} e^{\zeta\sigma}, \tag{13}$$

where ζ is given by $\zeta^{-2} = 6 + \varepsilon\xi^{-1} = 6 + 4\omega$. Also important is to note that the Λ term in (7) has been converted to the potential of the scalar field σ,

$$V = \Lambda e^{-4\zeta\sigma}. \tag{14}$$

Now G in (7) is variable while the gravitational constant, which might be denoted by G_*, in (12) is a constant. In other words, G is constant or variable depending on the choice of the CF. The same is true not only for G, but also for many other so-called constants, like particle masses, and the coupling constants. It is then crucially important to determine what or which CF we are now in, or what or which CF a "physical" CF is. In order to reply this question, we have to go through a long series of detailed discussions. Skipping all the details, we will show what our conclusion is like.[5]

- JCF is a theoretical CF with a large $\Lambda \sim 1$ expected from unification,
- ECF is a physical CF with a small observed Λ_{eff} as small as 10^{-120}.

We add a few comments. For Λ in JCF, we assume the "large" value $\sim M_{\text{P}}^4 \sim 1$ in order to keep a contact with the unification idea. Sometimes this CF is called a string CF. On the other hand, we have "small" and observed Λ in ECF. Let us explain why.

Consider the energy density ρ_σ of the scalar field σ in ECF;

$$\rho_\sigma = \frac{1}{2}\dot{\sigma}^2 + V(\sigma), \quad V = \Lambda e^{-4\zeta\sigma}, \tag{15}$$

where (14) has been used, while the dot implies a differentiation with respect to the cosmic time t_* in ECF. We may interpret this ρ_σ as an effective cosmological constant, or the dark energy.

According to the solution of the Einstein equation in ECF, we find,[6]

$$\rho_\sigma \sim t_*^{-2}. \tag{16}$$

At the present time this is

$$\sim t_{*0}^{-2} \sim (10^{-60})^2 = 10^{-120}. \tag{17}$$

This is precisely the number in (6) we found for Λ_{obs} to fit the data of acceleration. This is nice. This relation can be called a "scenario of a decaying cosmological constant"

---

[5] Chapter 3, also in Chapters 2.6 and 4.4, in [9] have been devoted to discussing various aspects of the concepts of conformal transformations and conformal frames. The content and its consequences are restated, sometimes even extended, also in our recent paper [10].

[6] See (4.106) of [9]. The same result had been obtained [11], however, based on a somewhat complicated model.

which provides us with a simple and natural derivation of the mysterious number 10^{-120}. According to this derivation, today's Λ is this small only because our universe is this old, not due to any of the fine-tuning of parameters, thus solving the fine-tuning problem as emphasized in Section 2.1. We would say that this is a major success of the scalar-tensor theory, not shared by the more phenomenology-oriented approaches, including the quintessence approach.[7]

However, we still have a problem. We started off with the matter Lagrangian according to the Brans-Dicke model in JCF. Naturally the electron field, for example, as well as the electron mass term are present in $L_{\rm matter}$. The absence of ϕ as required by WEP implies no time-dependence of the electron mass, $m_{\rm e}$. Now a constant $m_{\rm e}$ in JCF is likely to be time-dependent in ECF, supposed to be a physical CF. According to a more detailed analysis we do find a behavior;[8]

$$m_* = m{t_*}^{-1/2} \sim {t_*}^{-1/2}. \tag{18}$$

This is too much time-dependent, as we are going to argue in what follows.

Suppose we use atomic clocks in our measurements in physics, or in astronomy. Our time-unit is provided basically by the electron mass. Also suppose the electron mass changes with time, like (18), for example. Our time unit also does. How can we detect that? We can do it by comparing our clock with other kind of clock of a different principle. Putting the other way around, we have no way to detect it, as long as we insist to use the atomic clock. This might be stated as "Own-unit insensitivity principle" [10]. This might be translated into saying that $m_{\rm e}$ is a true constant in the physical CF which corresponds to the use of atomic clocks, whose time unit is provided by the same $m_{\rm e}$. In other words, the physical CF should be such that it leaves $m_{\rm e}$ constant. More generally speaking, choosing a physical CF depends on what clock we use. From this point of view, the time-dependence like (18) is unacceptable.

We may readily extend the same analysis to more realistic situation of measuring redshift of the light coming from far-away objects, hence estimating how long ago the light was emitted. Eventually we come again to appreciating the constancy of the electron mass, in the same way as in using atomic clocks.[9]

We emphasize again that choosing ECF provides with many features that appear to qualify ECF to be a physical CF. One of such features is the scenario of a decaying cosmological constant as mentioned above.

This argument above is a consequence of a long series of details, but is truly unique to the presence of Λ. We would say that this is something hard to refute. Then we have no other choice but to revise one of the initial assumptions that particle masses are constant

[7] Strictly speaking, the simple falling-off time-dependence shown in (16) is short of the role of a constant-like behavior which we should expect as a substitute of the cosmological constant causing an extra acceleration of the universe. This issue demands a departure from a simple-minded scalar-tensor theory. As discussed in detail in Chapter 5.4.2 of [9], we introduced another scalar field χ to reproduce the occurrence of a temporarily constant-like behavior, whose size is basically given by (17), inheriting the scenario of a decaying cosmological constant implemented before χ is introduced. In this way we come also to replying the coincidence problem, at least by lessening its gravity.

[8] See (4.121) of [9] and (3.13) of [10].

[9] See Section 3.3 of [10].

in the JCF. We have to revise the Brans-Dicke model. We in fact implemented the desired modification, resulting in the constant mass in ECF rather than in JCF.[10] Obviously in the modified model back in JCF, ϕ enters the matter Lagrangian, against Brans-Dicke's premise, hence breaking WEP. This is the way how we come to suspecting the validity of WEP,[11] in general situations with the universe accelerating.[12]

POSSIBLE MEASUREMENTS ON $\bar{\mathrm{H}}$

We have specifically discussed the cosmological implications of the scalar field, which, we still believe, has an origin in the unification physics. From this wider view, there is no reason why we confine ourselves only to the scalar field. We may think of a vector field as well. This choice is particularly interesting because it does allow a distinction between H and $\bar{\mathrm{H}}$.[13] In what follows we will try two phenomenological attempts on the basis of the scalar and vector fields.

Level-spacings of H and $\bar{\mathrm{H}}$

We first try to see how much these fields affect the spectroscopy of H and $\bar{\mathrm{H}}$, like the energy levels between $1S$ and $2S$ with hyper-fine splittings included. Let us begin with preparing a theoretical and artificial source of NGGF; a sphere with the radius R filled with the substance of $\rho = 10$gm/c.c., as an example. We choose $R = 10$m smaller than the force-range λ, supposed to be roughly around 100m.[14] We place the atoms near the surface with the z axis chosen as shown in Fig. 3, with the origin at the center of p or $\bar{p}$.

We are interested only in the order-of-magnitude estimate of the level energies. Then we may ignore p or $\bar{p}$ at the center. In the non-relativistic approximation, we may also ignore the possible differences between e and $\bar{e}$, and then between the scalar and vector fields. In this context, the scalar field and the zeroth component of the vector field may be represented collectively by a single scalar field Φ, which is produced by the artificial source as described above. We also expand as $\Phi(z) \approx \Phi(0) + z\Phi'(0)$, with the coefficients evaluated as

$$\begin{aligned} \Phi(0) &\sim \zeta \int d^3r \frac{\rho e^{-r/\lambda}}{4\pi r} \sim \zeta \frac{M}{4\pi R} \sim \frac{1}{3}\zeta \rho R^2, \\ \Phi'(0) &\sim -\Phi(0)/R. \end{aligned} \tag{19}$$

---

[10] See Chapter 4.4.3 of [9].

[11] The similar conclusion of WEP violation has been reached from string theory, but in a less compelling manner.

[12] We admit that this conclusion derives from the scalar-tensor theory, part of which is too intricate to allow a naive interpretation. It is this theory, however, that, in spite of possible loose ends [12], offers a theoretical implementation of the relation $\Lambda_{\rm obs} \sim 10^{-120} \sim (10^{60})^{-2} \sim t_0^{-2}$, hardly to be dismissed as a mere numerical coincidence.

[13] In the language of the geodesic equation, NGGF contributes to the "right-hand side."

[14] See (6.140) of [9].

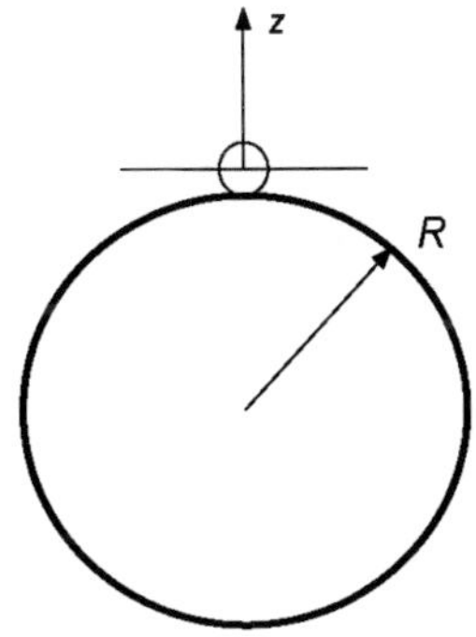

FIGURE 3. An artificial source of NGGF. $R = 10\text{m} < \lambda \sim 100\text{m}$, $\rho = 10\text{gr/cc}$.

We try to import ζ from the diagrams in Fig. 4, which had been obtained from various different sources. In this connection, however, we point out that the value of ζ, representing the coupling strength of NGGF to the matter, is model-dependent.[15] They can be different for the scalar and the vector fields, for leptons and nucleons, and so on. At present, we must be satisfied only by the crude and averaged estimates. As such we tentatively choose $\zeta^2 \lesssim 10^{-4}$.

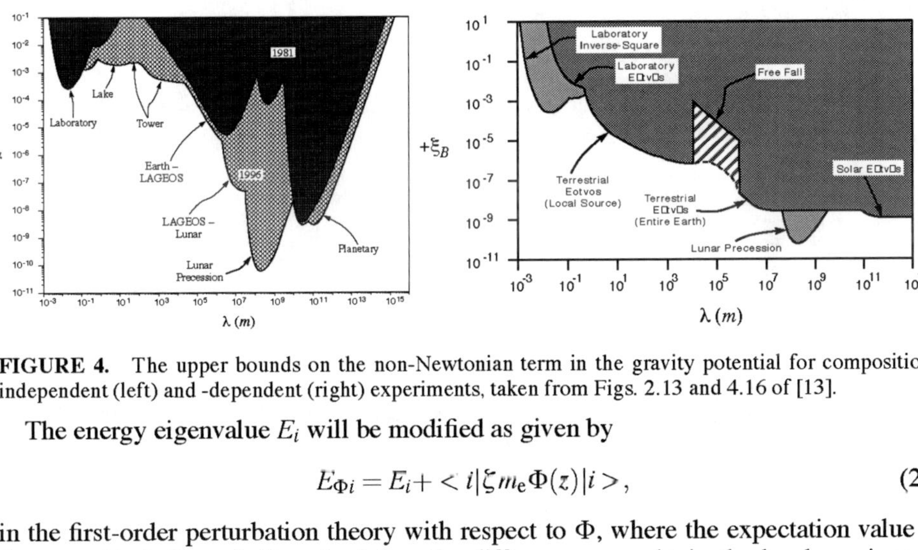

FIGURE 4. The upper bounds on the non-Newtonian term in the gravity potential for composition-independent (left) and -dependent (right) experiments, taken from Figs. 2.13 and 4.16 of [13].

The energy eigenvalue E_i will be modified as given by

$$E_{\Phi i} = E_i + < i|\zeta m_e \Phi(z)|i >, \tag{20}$$

in the first-order perturbation theory with respect to Φ, where the expectation value in the state i is indicated. By calculating the differences we obtain the level-spacings as

---

[15] A simple relation $\zeta^2 = \alpha_5/2$ can be derived only for the composition-independent experiments, i.e. probing possible deviation from the pure $1/r$ behavior, for the assumed form $V(r) = -(Gm_1m_2)(1 + \alpha_5 e^{-r/\lambda})$ based on the Brans-Dicke model for the scalar field. More uncertainties are unavoidable for any other kinds of estimates.

given by

$$\Delta E_\Phi = \Delta E + \delta\Delta E, \quad \text{with} \quad \delta\Delta E = \zeta m_e \Phi'(0) a_0 K \sim 10^{-53}\zeta^2, \tag{21}$$

where $a_0 = (m_e\alpha)^{-1}$ while K is of the order unity, including zero. As the ratio to the typical level-spacings $\Delta E \sim \alpha^2 m_e$, we finally arrive at

$$\frac{\delta\Delta E}{\Delta E} \sim 10^{-30}, \tag{22}$$

for $\zeta^2 \sim 10^{-4}$. In this way we find that the level-spacings of H or $\bar{\mathrm{H}}$ remain virtually unaffected by the NGGF. This will simplify the analysis in the next subsection.

Gravitational redshift from a falling emitter

A very small number obtained above is partly due to a quantum mechanical analysis applied to the very small system. We now discuss the free-fall of $\bar{\mathrm{H}}$ or H, basically a classical phenomenon, but combined with the Pound-Rebka type measurement of the gravitational redshift, partly from our own theoretical intrigue.

Let us consider an apparently simplified situation in which an $\bar{\mathrm{H}}$ or H falls off freely with the acceleration $\bar{g} = g + \delta g$, where g has an origin in the metric, while δg is due to NGGF, so that

$$z_{\mathrm{P}} = -\frac{1}{2}\bar{g}t_{\mathrm{P}}^2, \tag{23}$$

as shown in Fig. 5, where the subscript P refers to the world point of the falling atom. Note that the size of δg produced by the artificial source mass described before is given by

$$\left|\frac{\delta g}{g}\right| \sim 10^{-5}\zeta^2, \tag{24}$$

which is $\sim 10^{-9}$ if $\zeta^2 \sim 10^{-4}$. We assume that the acceleration is constant, just to simplify the analysis. The falling $\bar{\mathrm{H}}$ emits radiation which is received by a receiver R fixed at the height $z = 0$ at the world point P$'$. We are going to apply basically the same analysis used in the Pound-Rebka experiment [1].

We must be careful, however, because this freely-falling emitter with the acceleration $\bar{g}$ does not agree with the local inertial frame in which the frequencies of radiation, emitted and absorbed, are determined on the basis of quantum mechanics in flat spacetime.[16] In Einstein's sense, the local inertial frame should be defined purely in terms of g without δg.[17]

[16] This point was emphasized also in [14].

[17] As another detail in this connection, a light-ray propagates through the spacetime endowed with the metric g without affected by NGGF. The candidate of a vector field does not couple to the photon due to the charge-conjugation invariance, while the possible coupling to the scalar field has no effect in the geometrical-optics limit [15].

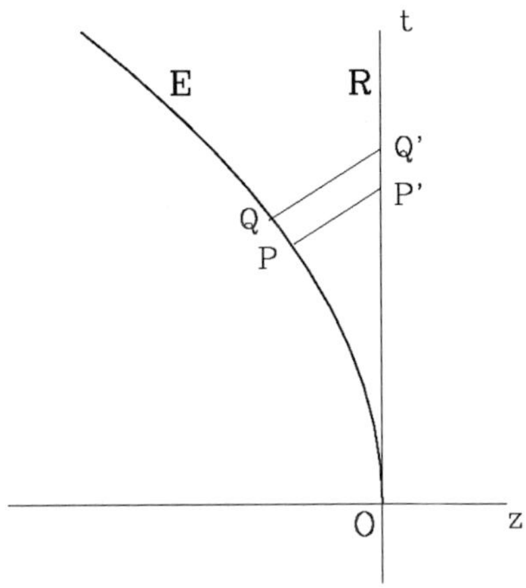

FIGURE 5. E stands for the trajectory of a falling atom emitting radiation at the world point P, received at P$'$ by the receiver R fixed at $z = 0$.

Taking this dual nature properly into account, we finally obtain

$$\frac{\nu_{\rm R}}{\nu_{\rm E}} \approx \frac{1+gz_{\rm P}}{1-\bar{v}} \left(1 - \frac{1}{2}\bar{v}^2 + \cdots \right), \tag{25}$$

where with $c = 1$, $\nu_{\rm R}$ is the frequency measured at the receiver, while $\nu_{\rm E}$ stands for the frequency at the emitter given by one of the level-spacings, for which we exploit the near independence on NGGF, as verified in the preceding subsection.

The term $1 + gz_{\rm P}$ in (25) is the same as in the ordinary Pound-Rebka result, while the term $(1 - \bar{v})^{-1}$ represents the Doppler shift due to the falling of E, where $\bar{v}$ is the downward velocity $\bar{v} = -\bar{g}t_{\rm P}$ given in terms of $\bar{g}$. We have another factor which we do not consider any more at this moment.

Roughly speaking, the Doppler shift may determine $\bar{g}$ at relatively earlier times, while the Pound-Rebka type observation determines g at later times. Suppose we determine or put an upper-bound on $\delta g/g$ to the accuracy of 10^{-9}. In view of (24) we may then have reached nearly the same level of $\zeta^2 \sim 10^{-4}$, as has been derived from other types of estimates in Fig. 4. But as we mentioned before, those other results might have suffered from considerable uncertainties. We may also say that the proposed experiment is probably the first to deal with such a simple system like H or $\bar{\rm H}$. From these points of view, even less accurate results, say $\delta g/g \sim 10^{-7}$ or $\zeta^2 \sim 10^{-2}$, might be still nontrivial at all. But to go ahead further, more scrutiny is obviously needed both on the theoretical as well as the experimental fronts.

REFERENCES

1. C.W. Misner, K.S. Thorn and J.A. Wheeler, *Gravitation*, W.H. Freeman, 1973.
2. Y. Fujii, Nature Phys. Sci. **234** (1971), 5; Y. Fujii, Ann. Phys. **69** (1972), 494; Phys. Rev. **D9** (1974), 874.
3. E. Fischbach, D. Sudarsky, A. Szafer, C. Talmadge and S.H. Aronson, Phys. Rev. Lett. **56** (1986), 1427.
4. A. G. Riess et al., Astron. J. **116** (1998), 1009; S. Perlmutter et al., Nature **391** (1998), 51; S. Perlmutter et al., Astrophys. J. **517** (1999), 565.
5. R. R. Caldwell, R. Dave and P. J. Steinhardt, Phys. Rev. Lett. **80** (1998), 1582; L. Wang, R. R. Caldwell, J. P. Ostriker and P. J. Steinhardt, Astrophys. J. **530** (2000), 17.
6. P. Jordan, *Schwerkraft und Weltall*, Friedrich Vieweg und Sohn, Braunschweig, 1955.
7. P.A.M. Dirac, Nature **139** (1937), 323.
8. C. Brans and R.H. Dicke, Phys. Rev. **124** (1961), 925.
9. Y. Fujii and K. Maeda, *The scalar-tensor theory of gravitation*, Cambridge University Press, 2003.
10. Y. Fujii, Prog. Theor. Phys. **118** (2007), 983.
11. Y. Fujii, Phys. Rev. **D26** (1982), 2580.
12. See, for example, Y. Fujii, Prog. Theor. Phys. **110** (2003), 433.
13. E. Fischbach and C. Talmadge, *The search for non-Newtonian Gravity*, AIP Press/Springer, 1999.
14. Y. Fujii, Delivered at The International Workshop on JHF Science (JHF98), KEK, Tsukuba, Japan, March 4-7, 1998; arXiv:gr-qc/9806092.
15. Y. Fujii and M. Sasaki, Phys. Rev. **D75** (2007), 064028.

A Scheme To Produce The Antihydrogen Ion $\bar{H}^+$ For Gravity Measurements

P. Pérez[a], L. Liszkay[a,*], M.F. Barthe[b], J-P. Boilot[c], M. Carty[a], C. Corbel[c], P. Crivelli[a,f], A. Curtoni[a], O. Delferrierre[a], P. Desgardin[b], M. Etienne[d], U. Gendotti[f], M. Hassan[g], A. Mohri[g], T. Ohdaira[e], J-M Rey[a], A. Rubbia[f], H. Saitoh[g], Y. Sauce[a], R. Suzuki[e], A. Walcarius[d], Y. Yamazaki[g]

[a]*DSM/Irfu, CEA/Saclay, F-91191 Gif-sur-Yvette Cedex, France*
[b]*Conditions Extremes et Matériaux : Haute Température et Irradiation, UPR3079, 3A rue de la Férollerie, 45071 Orléans Cedex 2, France*
[c]*Laboratoire des Solides Irradiés, École Polytechnique, 91128 Palaiseau Cedex, France*
[d]*Laboratoire de Chimie Physique et Microbiologie pour l'Environnement, UMR7564, CNRS-Nancy-Université, 405 rue de Vandoeuvre, F-54600 Villers-lès-Nancy, France*
[e] *Institute of Advanced Industrial Science and Technology (AIST), Tsukuba, Ibaraki 305-8568, Japan*
[f]*Institut für Teilchenphysik, ETHZ, CH-8093 Zürich, Switzerland*
[g]*Atomic Physics Laboratory, RIKEN, Wako 351-01, Japan*

Abstract. We propose to use the charge exchange reaction of antiprotons with positronium atoms in order to produce antihydrogen atoms, $\bar{H}$, and $\bar{H}^+$ ions. The ions can be cooled down to μK temperatures and then ionized to recover an ultra slow neutral $\bar{H}$ atom. Its acceleration is then measured by time of flight. Results on the conversion of slow positrons into positronium are presented. This is a first step towards the creation of a dense cloud of positronium atoms to be used as a target for the antiprotons. The source of positrons is based on a 6 MeV industrial electron linac with 0.2 mA average current to be installed in CEA-Saclay. Equipped with a tungsten target and a moderator, it is aimed at producing rates of order 10^8 s^{-1} slow positrons.

Keywords: Antimatter, Gravity, Positron beam, Antihydrogen, Positronium.
PACS: 04.80.Cc, 07.77.Ka, 29.25.-t, 41.75.Fr.

INTRODUCTION

The possibility that antimatter may behave differently from matter in the earth's gravity field is a long-standing issue. Extended supergravity theories were indeed found to potentially contain gravivector fields that would lead to an antigravity component [1]. Although indirect experimental constraints lead to severe limits on such forces [2], it is worthwhile to perform a direct measurement.

The antihydrogen atoms currently produced at CERN [3] are obtained mainly via a three-body interaction of an antiproton in a dense cloud of positrons. This process creates highly excited atoms emitted isotropically at high velocities, thus with characteristics not adequate for a free fall measurement. Walz and Hänsch proposed to

* on leave from: KFKI Research Institute for Nuclear and Particle Physics, P.O.Box 49, H-1525 Budapest, Hungary

CP1037, *Cold Antimatter Plasmas and Application to Fundamental Physics,* edited by Y. Kanai and Y. Yamazaki

use instead the $\overline{H}^+$ ion in order to collect ultra cold atoms [4]. This ion can be cooled down to μK temperatures (i.e. m/s velocities). The excess positron can then be laser detached in order to recover the neutral $\overline{H}$ atom. This process can be set up to minimize momentum transfer in the vertical direction. The temperature achieved in cooling of the $\overline{H}^+$ ion gives the main systematic error. Collecting 1000 events should provide a 2% measurement error on g.

We proposed a way to prepare such ions using interactions of the antiprotons with a dense target of positronium atoms (Ps) [5, 6]. The charge exchange process $\bar{p} + Ps \rightarrow \overline{H} + e^-$ is followed by $\overline{H} + Ps \rightarrow \overline{H}^+ + e^-$. The matter counterpart of the first process has been measured [7] in agreement with calculations with cross sections of order 10^{-15} cm^{-2} for antiproton energies around 6 keV [8]. The calculated cross section for the second process is 10^{-16} cm^{-2} at the same energy [9]. Thus if 10^7 antiprotons interact with a density of 10^{12} cm^{-3} Ps atoms, 1 $\overline{H}^+$ ion is produced, together with 10^4 $\overline{H}$ atoms.

The project would comprise an electron accelerator to produce MeV positrons via pair production on a target. The positrons are then moderated to be stored in a Penning trap and then dumped onto a material to convert them into positronium. The antiprotons are injected in the Ps cloud thus formed to produce the antihydrogen atoms and ions according to the above reactions. A synoptic view is shown in figure 1.

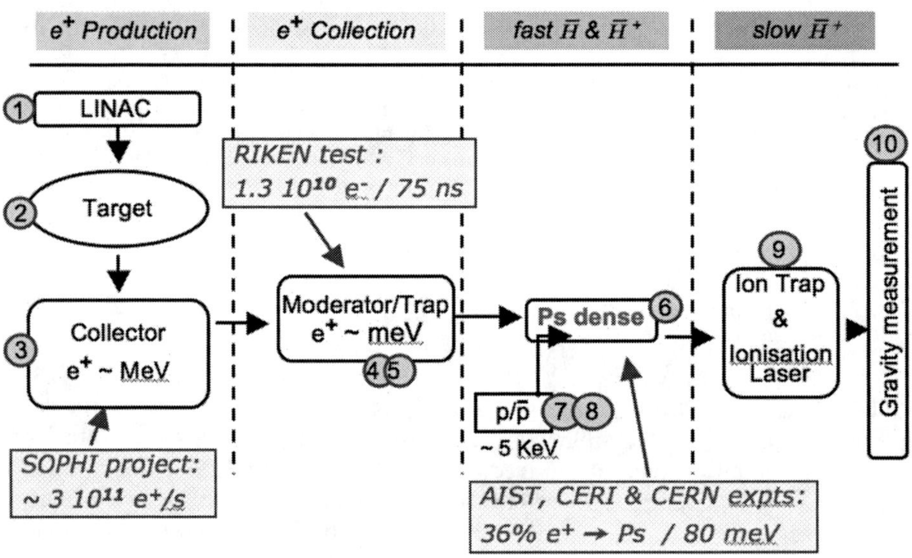

FIGURE 1. Overall scheme

ANTIPRON BEAM

The best option to prepare the $\bar{p}$ beam is to fill and store the antiprotons in a Penning-Malmberg trap. The MUSASHI multi ring trap and RFQD setup that is a part of the ASACUSA collaboration apparatus would certainly be optimal [10]. Routinely, over 10^6 $\bar{p}$ are stored, radially compressed and extracted. Increasing this number to 10^7 was reached by accumulating AD shots over a 20 minute period [11]. Development will be needed to compress the antiprotons in a bunch shorter than 100 ns and accelerate it at an energy of 6 keV with less than 1 mm transverse beam size at the location of the Ps target.

POSITRON PRODUCTION

In [6] we proposed the idea to use a 10 MeV / 2 mA electron accelerator in order to achieve the high positron rates needed. The low energy induces low positron production rates that can be overcome with a high average current. Recently we have had the opportunity to purchase a small industrial linac from Linac Technologies [12]. Its nominal energy and average current are 6 MeV and 0.2 mA. The repetition rate is 220 Hz with bunch length of 2 μs. The magnetron peak power is 1.9 MW. The total power consumption is 35 kVA. The acceleration length is 21 cm after which the beam diameter is 1 mm. The overall dimensions are roughly 1 m x 0.8 m x 0.8 m. The low energy and current, compared to our initial aim, are due to financial issues. However, the average positron energy is lowered to 800 keV, which is advantageous concerning moderation efficiency [13].

Simulations, performed with the GEANT3 software [14], predict rates of 5 10^{11} e^+ s^{-1} when the beam hits a 200 μm thick target at 5 degree incidence angle. At such low incident energies the positrons are emitted almost isotropically from the target, with 2/5 of these positrons emitted backwards from the incident beam direction.

A tungsten moderator of 10^{-4} efficiency would thus produce 3 10^7 slow e^+ s^{-1}. We are planning to install a set of transport coils and a dipole in order to separate the electrons from the positrons, thus opening the possibility to use a solid gas moderator with at least 10 times higher efficiency after including transport losses due to the large beam size.

The overall dimensions of this project [13], including the concrete shielding against X rays, is 6 m x 5 m x 3 m, making it a compact setup compared to higher energy accelerators or nuclear reactors. First results should be obtained by the end of 2008.

Such a source may be adapted to the needs of materials science research by transforming the pulse time structure for instance with a trap and additional buncher to obtain sub ns bunches each with few positrons, a variable bunch spacing and a high repetition rate [15].

POSITRON STORAGE

Given the antiproton accumulation time of 20 minutes, a slow positron flux of 10^8 s^{-1} allows the stacking of up to 10^{11} e^+ in an accumulation trap. The well depth of the present RIKEN 5 Tesla Multi Ring Trap (MRT) allows storing 10^{10} positrons [16]. It is also foreseeable to improve the isolation in such a trap in order to obtain a 10 kV deep potential well, and thus accumulate 10^{11} positrons.

The positrons must be extracted from the trap in a time less than the lifetime of *ortho*-positronium (*o*-Ps), i.e. less than ~100 ns. We have performed a test with the RIKEN MRT filled with electrons. A static harmonic potential well allows storing the charged particles. In order to extract the electrons we have modified the potential distribution with a fast switch enabling to quickly lower the potential on one side of the well (figure 2a). The current at the trap exit was measured with a phosphor screen. A CCD camera provided a transverse view of the extracted beam. We obtained 75 ns FWHM wide pulses containing 1.3 10^{10} positrons (figure 2b). The beam diameter inside the trap is as low as 0.7 mm. It extends to 1 cm at 80 cm from the trap center due to the divergent fringe field. When rotating wall compression is applied the diameter inside reduces to 100 μm so that we may expect ~ 1 mm at exit. Further bunch length compression is possible as shown for instance at lower intensity by Cassidy et al. [17].

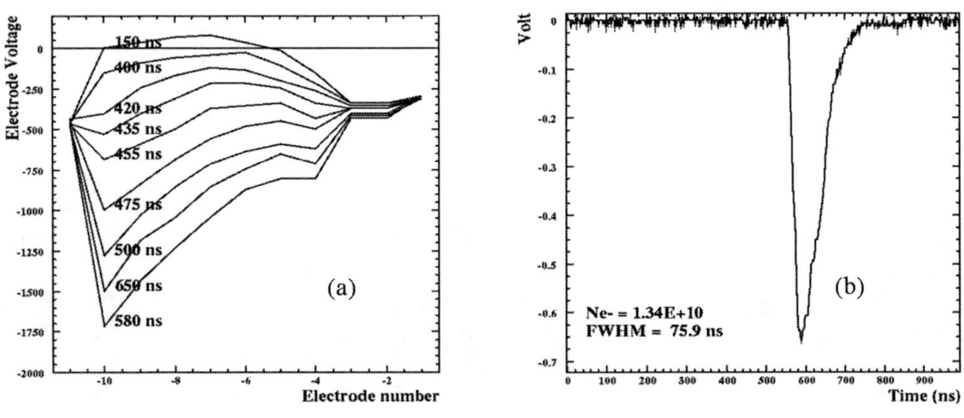

FIGURE 2. (a) Trap potential as a function of electrode number and time; (b) Pulse output of 1.3 10^{10} e- .

POSITRONIUM TARGET

The Ps target is formed by the interaction of the positron beam, extracted from the storage trap, onto a porous SiO_2 film. We have prepared films that have a high e^+ to Ps conversion efficiency, and reemit a large fraction of the produced oPs atoms into vacuum at a rate depending on the structure of the nanopores. The composition of these layers is detailed in [18]. In order to detect Ps produced in the films, we used first a 0.5-25 keV positron beam at CNRS/CEMHTI in Orléans, France [19].

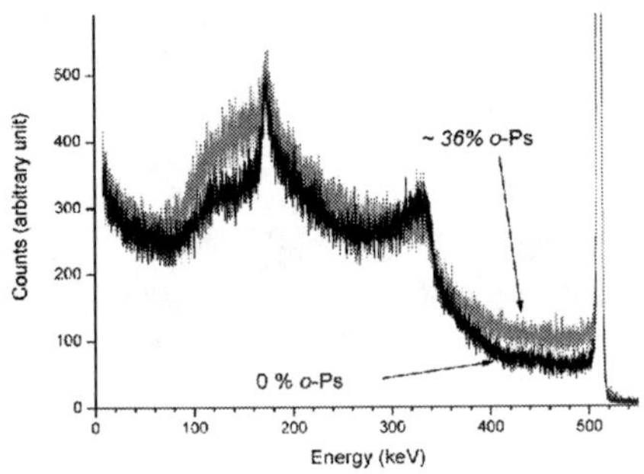

FIGURE 3. Measured γ energy

The 3 γ annihilation fraction was measured, based on the difference of the energy distribution of the annihilation photons in events yielding two and three photons (figure 3) [18]. A high purity germanium detector, placed perpendicularly to the beam direction detected the annihilation gamma photons. The fraction of impinging positrons annihilating through 3 γ photons has been determined from the measured positron spectra, and calibrated using the observed change in the intensity of the 2 γ annihilation peak. Conversion efficiencies into *o*-Ps above 30% have been obtained for implantation energies up to 4 keV, when the porogen content is high. We interpret this high efficiency as being due to the interconnectivity of the pores leading to emission into vacuum.

A second set of measurements was performed with the slow positron beam installed at CERN [20], to which we added a set of four large BGO crystals to measure the lifetime of the produced oPs, and a ^{22}Na source of 400 Mbq. The BGO scintillation γ detectors make a wall of 200 x 220 mm^2 parallel to the incoming e^+ beam, covering a region 55 mm beyond and 165 mm before the sample. The presence of the long 142 ns component in the lifetime distribution, for the samples at high porogen content gives a first indication that a large fraction of the oPs is emitted into vacuum (figure 4). This component is depleted when the porogen content is lowered (figure 5b

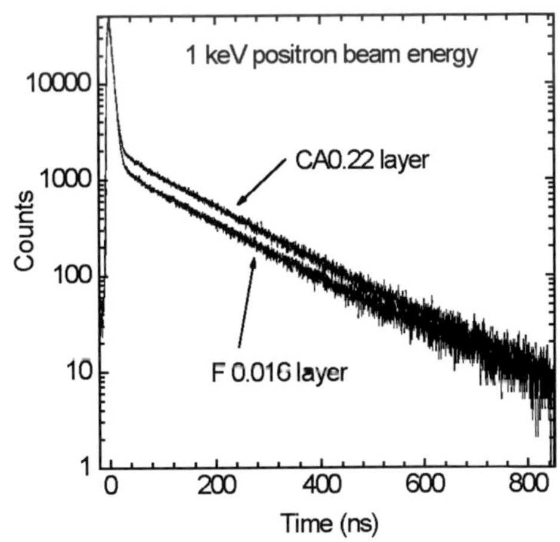

FIGURE 4. Lifetime spectra

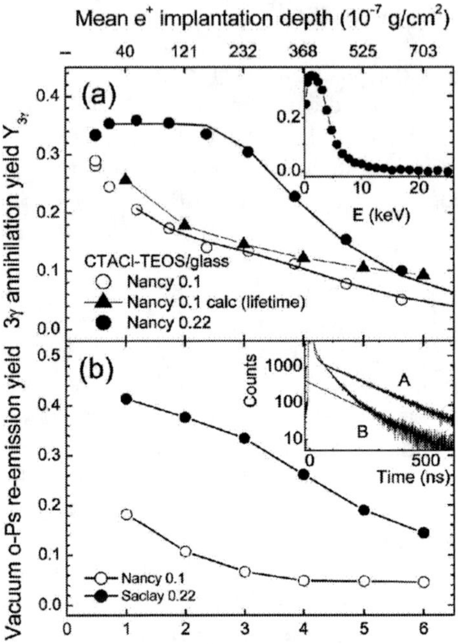

FIGURE 5. from [23]
(a) 3 γ *o*-Ps annihilation measured yield or calculated from vacuum lifetime component (triangles) ;
(b) vacuum *o*-Ps reemission yield and lifetime spectra for 3 keV *e+* (insert).

insert). The 3 γ oPs yield calculated from the long lifetime component matches the yield obtained from the energy distribution (figure 5a).

o-Ps time-of-flight measurements have been performed at the AIST Linac-based slow positron beam in Tsukuba, Japan [21]. We also installed a similar setup on the CERN beam. Figure 5 shows typical time of flight distributions for implantation energies between 0.5 and 4 keV. At low implantation energy, the reemitted ground state *o*-Ps peak corresponds to 1 eV kinetic energy. It goes down to ~80 meV for 4 keV implantation energy. This low value improves over the usual emission from metals with 2-3 eV work function because the density of the Ps target cloud, for the $\overline{H}$ experiment, is inversely proportional to its velocity. Gidley et al. claim that Ps may even be thermalized in the material before emission into vacuum [22]. Our apparatus at CERN includes a sample holder that can be cooled in order to study this possibility. The *o*-Ps emitted from our present samples did not reach thermalization energy. The time-of-flight peak did not move when cooling them down to 20K. The layers already tested are sufficient for the *o*-Ps target and are easily reproducible.

Further experiments are underway to adapt these materials for the making of a Ps target suitable for the antihydrogen and $\overline{H}^+$ project.

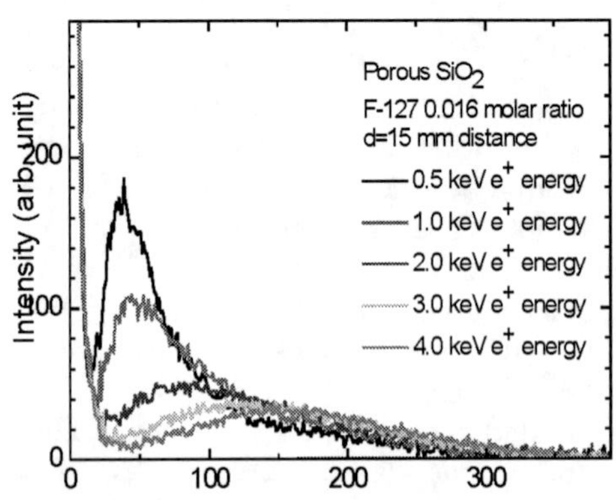

FIGURE 6. Time of flight spectrum

The Ps cloud may also be illuminated by a laser system to excite these atoms in order to benefit from the increase in the $\bar{p}$ + Ps* → $\overline{H}$* cross-section, expected to be proportional to the principal quantum number to the fourth power. A large fraction of the antiproton bunch would then be converted into excited antihydrogen atoms. Part of the $\overline{H}$* atoms may further interact with other Ps atoms to produce $\overline{H}^+$. The n = 3 level of Ps* has an energy of 0.75 eV close to the binding energy of $\overline{H}^+$.

The antiproton beam should be collimated in order to match the transverse dimensions of the cavity holding the Ps cloud, and delivered in a short bunch at energies, depending on the excitation state of the Ps atoms, up to 10 keV. We believe the best way to achieve these goals is to store the antiprotons in a trap such as the one developed in the ASACUSA collaboration [10].

SUMMARY

In order to demonstrate the feasibility of an experiment to produce the $\overline{H}^+$ ion, we have tested porous SiO2 plates that convert positrons into orhopositronium with up to 40% efficiency, and emit *o*-Ps into vacuum at a kinetic energy around 80 meV. Extraction of ~ 10^{10} electrons from a Penning trap has been obtained with a pulse width of 75 ns, adequate for further *o*-Ps production in the SiO2 target. A project to build an intense source of slow positrons, based on a small electron linac, is underway at CEA-Saclay. It is aimed at producing up to 10^8 slow positrons per second.

REFERENCES

1. J. Scherk, *Phys. Lett.* **B 88**, 265-267 (1979).
2. S. Bellucci and V.Faraoni, *Phys. Lett.* **B 377**, 55-59 (1996).
3. G. Andresen et al., *Phys. Rev. Lett.* **98**, 023402 (2007); G. Gabrielse et al., Phys. Rev. Lett. **98**, 113002 (2007).
4. J.Walz and T. Hänsch, *General Relativity and Gravitation* **36** 561-570 (2004).
5. P.Pérez and A. Rosowsky, *Nucl. Inst. Meth.* **A 545**, 20-30 (2005).
6. P.Pérez and A. Rosowsky, *Nucl. Inst. Meth.* **A 532**, 523-532 (2004).
7. J.W. Humberston, M. Charlton, F.M. Jacobsen and B.I. Deutch, J. Phys. **B 20,** L25-29 (1987) ; J. P. Merrison et al., Phys. Rev. Lett. **78**, 2728 (1997); J.P. Merrison et al., Hyperfine Interact. **109,** 313321 (1997).
8. M. Yamanaka, Y. Kino, Phys. Rev. **A 65,** 062709 (2002) and references therein; A. Igarashi, N. Toshima, T. Shirai, J. Phys. **B 27,** L497 (1994); J. Mitroy, G. Ryzhikh, J. Phys. **B 30,** L371 (1997); K. Ratnavelu, J. Mitroy, J. Phys. **B 28,** 287 (1995).
9. H.R.J. Walters and C. Starett, *Phys. Stat. Sol.* **C**, 1-8 (2007).
10. N. Kuroda et al., Phys. Rev. Lett. **94**, 023401 (2005).
11. ASACUSA Status Report, CERN-SPSC-2008-002, SPSC-SR-027, January 2008.
12. Linac Technologies S.A., 16 rue N. Appert, 91400 Orsay, France;
13. P. Pérez, L. Liszkay, J-M Rey, O. Delferrierre, V. Blideanu, M. Carty, A. Curtoni, N. Ruiz, Y. Sauce, contribution to the SLOPOS 11 workshop, preprint DAPNIA-07-99, accepted for publication in *Appl. Surf. Sci.*
14. GEANT3.21, CERN Program Library W5013 (1994).
15. R. Suzuki et al., *Jap. J. Appl. Phys.* **30**, 532-534 (1991); D. Segers et al., *Nucl. Instr. Meth.* **A 337** 246-252 (1994).

16. N. Oshima, T.M. Kojima, M. Niigaki, A. Mohri, K. Komaki, and Y. Yamazaki, *Phys. Rev.Lett.* **95,** 195001 (2004); A. Mohri, H. Higaki, H. Tanaka, Y. Yamazawa, M. Aoyagi, T. Yuyama and T. Michishita, *Jpn. J. Appl. Phys* **37,** 664-670 (1998).
17. D.B Cassidy, S.H.M Deng, R.G. Greaves, A.P. Mills Jr., *Rev. Sci. Instrum.* **77**, 073106 (2006).
18. L. Liszkay, M.-F. Barthe,C. Corbel, P. Crivelli, P. Desgardin, M. Etienne, T. Ohdaira, P. Pérez, R. Suzuki, V. Valtchev, A. Walcarius, contribution to the SLOPOS 11 workshop, preprint DAPNIA-07-125, accepted for publication in *Appl. Surf. Sci.*
19. P. Desgardin, L. Liszkay, M.-F. Barthe, L. Henry, J. Briaud, M. Saillard, L. Lepolotec, C. Corbel, G. Blondiaux, A. Colder, P. Marie, M. Levalois, *Mater. Sci. Forum* **523,** 363-365, (2001).
20. N. Alberola et al., *Nucl. Instr. Meth.* **A 560**, 524 (2006).
21. R.S. Yu, Y. Kobayashi, T. Ohdaira, R. Suzuki, K. Ito, K. Hirata and K. Sato, *Mater. Sci. Forum* **361**, 445-446 (2004);
22. R.S. Vallery, P.W. Zitzewitz and D.W. Gidley, *Phys. Rev. Lett.* **90**, 203402-1 (2003).
23. L. Liszkay et al., Appl. Phys. Lett. **92,** 063114 (2008).

Precision measurements of positronium decay rate and energy level

S.Asai, Y.Kataoka, T.Kobayashi, T.Namba, T.Suehara, G.Akimoto, A.Ishida, M.M.Hashimoto*, H.Saito†, T.Idehara** and M.Yoshida‡

*ICEPP, Faculty of Science, University of Tokyo, 7-3-1 Hongo, Bunkyo-ku, Tokyo 113-0033
†Graduate school of Arts and Sciences, University of Tokyo, 3-8-1 Komaba, Meguro-ku, Tokyo, 153-8902
**FIR Center, University of Fukui, 3-9-1 Bunkyo, Fukui, 910-8507
‡High Energy Accelerator Research Organization (KEK) 1-1 Oho, Tsukuba,Ibaraki, 305-0801

Abstract.
Positronium is an ideal system for the research of the bound state QED. New precise measurement of orthopositronium decay rate has been performed with an accuracy of 150 ppm, and the result combined with the last three is $7.0401 \pm 0.0007 \mu s^{-1}$. It is the first result to validate the 2nd order correction. The Hyper Fine Splitting of positronium is sensitive to the higher order corrections of the QED prediction and also to the new physics beyond Standard Model via the quantum oscillation into virtual photon. The discrepancy of 3.5σ is found recently between the measured values and the QED prediction ($O(\alpha^3)$). It might be due to the contribution of the new physics or the systematic problems in the previous measurements: (non-thermalized Ps and non-uniformity of the magnetic field). We propose new methods to measure HFS precisely without the these uncertainties.

Keywords: Positronium, QED, decay rate, HFS
PACS: PACS numbers

INTRODUCTION

Positronium (Ps), the bound state of an electron and a positron, is a purely leptonic system and the triplet (1^3S_1) state of Ps, orthopositronium(o-Ps), decays slowly into three photons. It is good sample to measure decay rate directly and precisely. The o-Ps has an energy higher than the single state (1^1S_0) of Ps, parapositronium(p-Ps). The difference of the energy level is called as Hyper Fine Splitting (HFS) and is significantly larger(about 203 GHz) than the hydrogen-atom (1.4 GHz). Since o-Ps has the same quantum number of photon, the quantum oscillation o-Ps $\rightarrow \gamma^* \rightarrow$ o-Ps contributes the HFS (87 GHz), and the oscillation has the good sensitivity to the new physics beyond Standard Model.

Precise measurements of the decay rate of o-Ps and the HFS give us direct information about quantum electrodynamics(QED) in bound state. The higher order calculations in the bound state have be developed recently, and the we have a good chance to compare the experimental results with the QED predictions directly.

There are two topics in this note. We report the first test of the 2nd order calculation of the orthopositronium decay rate in the first part. We perform the accurate measurement (error $<$150ppm) of the o-Ps decay rate, and the 2nd order calculation is directly validated for the first time. In the second part, we propose the new methods to measure

CP1037, *Cold Antimatter Plasmas and Application to Fundamental Physics,* edited by Y. Kanai and Y. Yamazaki

the HFS. There is discrepancy(3.5σ) between the prediction($O(\alpha^3)$) and the precise measured values. We point out the possible two systematic errors in the previous measurements and propose the new methods without these systematic uncertainties.

ORTHOPOSITRONIUM DECAY RATE

History and current status

Three precise measurements[1, 2, 3] of the o-PS decay rate were performed, in which reported decay rate values much larger, i.e., 5.2 – 9.1 experimental standard deviations, than a QED prediction[4] (7.039934(10) μs^{-1}) corrected up to $O(\alpha^2)$. This discrepancy has been referred as 'orthopositronium lifetime puzzle', and was long-standing problem.

As some fraction of o-Ps inevitably results in 'pick-off' annihilation due to collisions with atomic electrons of the target material, the observed o-Ps decay rate λ_{obs} is a sum of the intrinsic o-Ps decay rate $\lambda_{\text{o-Ps}}$ and the pick-off annihilation rate into 2γ's, λ_{pick}, i.e.,

$$\lambda_{obs} = \lambda_{3\gamma} + \lambda_{pick}. \tag{1}$$

In the old measurements[1, 2, 3], λ_{obs}'s were measured by varying the densities of the target materials, size of the cavities and also the entrance aperture of the cavities. The extrapolation to zero density or aperture was expected to yield the decay rate in a vacuum, $\lambda_{3\gamma}$, under the assumption of the constant o-Ps velocity. We pointed out[5, 6] that this assumption was the common/serious systematic uncertainties in these results, since it takes much time that Ps is well thermalized.

We have proposed the following entirely new method[6], which is free from above-mentioned systematic error. The energy distribution of photons emitted from the 3-body decay is continuous below the steep edge at 511 keV, whereas the pick-off annihilation is 2-body which produces a 511 keV monochromatic peak. Energy and timing information are simultaneously measured with high-energy resolution germanium detectors such that $\lambda_{pick}(t)/\lambda_{3\gamma}$ can be determined from the energy spectrum of the emitted photon. Once a precise thermalization function is obtained, $\lambda_{pick}(t)$ will contain all information about the process. The population of o-Ps at time t, $N(t)$ can be expressed as

$$N(t) = N_0' \exp\left(-\lambda_{3\gamma}\int_0^t \left(1+\frac{\lambda_{pick}(t')}{\lambda_{3\gamma}}\right) dt'\right). \tag{2}$$

Providing the ratio is determined as a function of time, the intrinsic decay rate of o-Ps, $\lambda_{3\gamma}$, can be directly obtained by fitting the observed time spectrum.

We obtained decay rate of 7.0398(29) μ[6], are consistent with the first order QED calculation, and quite differ from the previous results as shown in Fig. 1. This result are confirmed by two different and more accurate (200ppm) measurements[7, 8] and the 'lifetime puzzle' has been solved. Now the interesting is focused on the validation of the higher order correction itself.

Non relativistic QED approximation has been developed recently, and it is useful to calculate higher oder correction of the bound state. The second order correction[4],

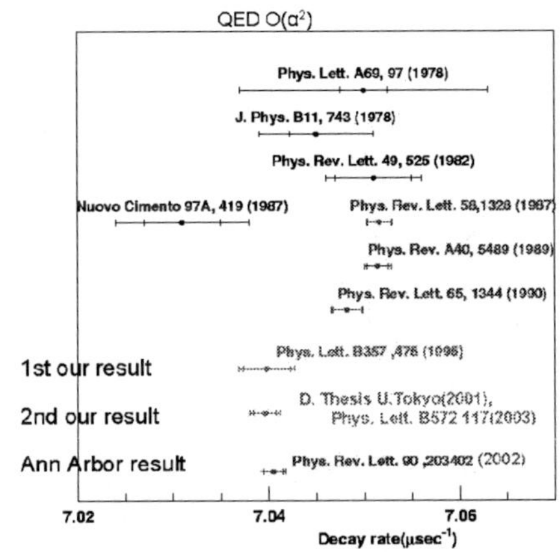

FIGURE 1. Historical plots of the measured o-Ps decay rate

whose contribution is about 160ppm, has been performed in 2002, and the more accurate measurement is necessary to examine the 2nd order correction.

Experimental setup

Figure 2 shows a diagram of the experimental setup[9]; A $^{68}Ge-Ga$ positron source (dia., 10 *mm*) with the strength of $0.3\mu Ci$, being sandwiched between two sheets of plastic scintillators(NE102 thickness=200μm) and held by a cone made of aluminized mylar. The cone was situated at the center of a cylindrical vacuum container made of 1 *mm*-thick plastic scintillators and glass, being filled with SiO_2 aerogel (RUN-I) or powder (RUN-II), and evacuated down to 1×10^{-2} Torr. Density of the aerogel and powder is 0.03g/cc both, and the surface of the primary grain are replaced into hydrophobic in oder to remove the electric dipole of the OH-. The sizes of the primary grain are 10 and 16 nm for the aerogel and powder, respectively. Thus the mean free path of the collision between positronium and SiO_2 are different, and it makes difference in the thermalization process and the pickoff probability.

The emitted positrons pass through thin scintillator, transmitting a light pulse to a photomultiplier(Trigger PMT) and forming Ps when stopped in the silica powder or aerogel. Since the kinetic energy of the positron is high (E_{end}=1.9MeV: $^{68}Ge-Ga$) about 40% of the positron pass through the silica and annihilate at the vacuum container. If the positron deposits energy(>100 keV) in the 1 *mm*-thick plastic scintillators, inside wall of the vacuum container, the produced trigger signal is vetoed to remove these annihilate events in the wall.

Three high-purity coaxial germanium detectors (Ortec GEM38195) precisely measured the thermalization process. Energy resolutions were measured using several line γ sources, with typical resultant values of 0.5 keV at 514 keV. Four YAP($YAlO_3$: Ce doped: size is 50×50×33 mm) scintillators simultaneously measured the time and en-

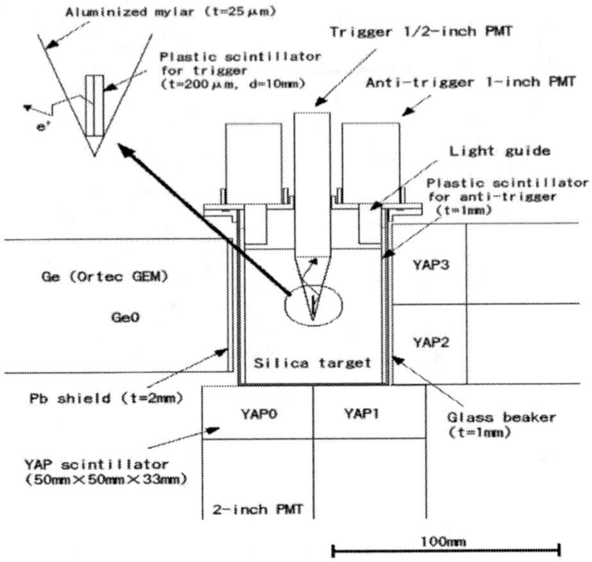

FIGURE 2. Schematic diagram of apparatus.[9]

ergy information from each decay. The good time resolution of (1 nsec) is obtained for $E_\gamma >$150 keV. Since the decay constant of the YAP is fast(30 nsec) and the slow components of the scintillation is negligible small[9], the event pileup probability is small and the time walk correction can be performed perfectly as function of the deposited energy.

Analysis

The ratio $\lambda_{pick}(t)/\lambda_{3\gamma}$ is determined using the energy spectrum measured by the Ge detectors. The energy spectrum of the o-Ps decay sample, referred to as the *o-Ps spectrum*, is obtained by subtracting accidental contributions from the measured spectrum. The 3γ-decay continuum spectrum is estimated using Monte Carlo simulation in which the geometry and various material distributions are reproduced in detail[9]. For every simulated event, three photons are generated according to an order-α-corrected energy spectrum. Successive photoelectric, Compton, or Rayleigh scattering interactions of every photon are then followed through the materials until all photon energy is either deposited or escapes from the detectors. The response function of the detectors is determined based on the measured spectrum of monochromatic γ-rays emitted from ^{152}Eu, ^{85}Sr, and ^{137}Cs, with this function being used in the simulation[9]. These material and detector effects are taken into account precisely, and the *3γ-spectrum* is obtained. These are continuous distributions, and is normalized to the observed o-Ps spectrum with the ratio of event numbers within the region (480-505 keV).

Figure 3(a) shows the o-Ps spectrum and 3γ spectrum observed in RUN-II (powder). The pick-off 2γ spectrum, which is obtained from the o-Ps spectrum after subtracting the

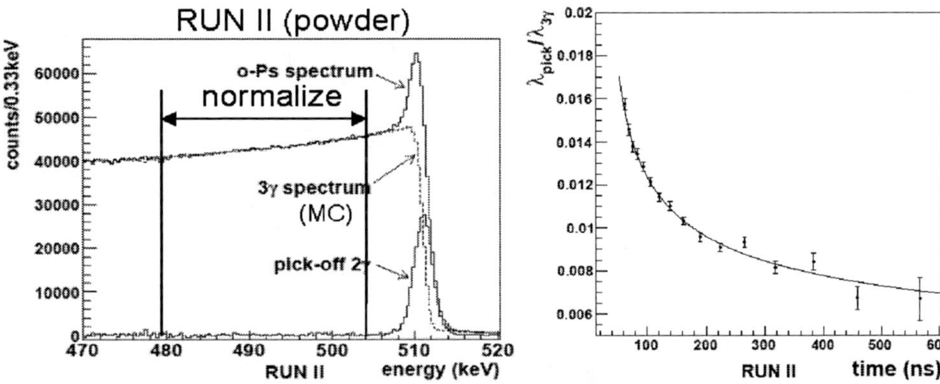

FIGURE 3. (a) Energy spectrum of o-Ps decay γ's obtained by Ge detectors. The solid line represent data points in a time window of $60-700$ *nsec*, and the dotted line shows the 3γ-decay spectrum calculated by the Monte Carlo simulation. Pick-off spectrum obtained after subtracting the 3γ contribution from the o-Ps spectrum is superimposed. (b) The ratio $\lambda_{pick}(t)/\lambda_{3\gamma}$ are plotted as a function of time for RUN-II(powder). Only the statistical errors are shown and the solid lines represent best fit results obtained.

3γ-spectrum, is also superimposed in the same figure. The $\lambda_{pick}/\lambda_{3\gamma}$ can be calculated directly from the ratio of the pick-off 2γ and 3γ spectrum, the ratio does not depend on the absolute values of the detection efficiencies. The calculations of the $\lambda_{pick}/\lambda_{3\gamma}$ are performed for the various time windows, and its time dependence is shown in Fig.3(b). Horizontal axis of the figure is the time between the positrons emission and decay, this slop show the thermalization process of the positronium. It takes much time than the lifetime to well thermalized, as we have already shown in the previous measurements[6, 7].

Figure 4 shows the time spectrum of the YAP scintillators with an energy window above 150 keV, in which a sharp peak of the prompt annihilation is followed by the exponential decay curve of o-Ps and then the constant accidental spectrum. Time resolution is 1.2 nsec is obtained and the o-Ps curve is widely observed over 1.0 μs. We fit resultant time spectrum using the least square method, i.e.,

$$N_{obs}(t) = \exp(-R_{stop}t)\left[\left(1+\frac{\varepsilon_{pick}}{\varepsilon_{3\gamma}}\frac{\lambda_{pick}(t)}{\lambda_{3\gamma}}\right)N(t)+C\right], \tag{3}$$

where ε_{pick} and $\varepsilon_{3\gamma}$ are the detection efficiencies for pick-off annihilation and 3γ decays, and R_{stop} is an experimental random counting rate representing the fact that time interval measurement always accept the first γ as a stop signal. $\lambda_{pick}/\lambda_{3\gamma}$ is about 1% due to the low-density of the SiO_2 powder or aerogel, i.e., the ratio of error propagation to decay rate is suppressed by a factor of 100.

The obtained decay rates[9] are $\lambda_{3\gamma} = 7.03876 \pm 0.0009(stat.)\ \mu s^{-1}$ for RUN-I and $7.04136 \pm 0.0009(stat.)\ \mu s^{-1}$ for RUN-II, which are consistent with each other.

Estimates of various systematic errors are summarized in Table 1 and the details are shown in reference[9]. We just mention here about errors related to simulation and physics source(Stark effect and 3γ annihilation in the pick-off collision).

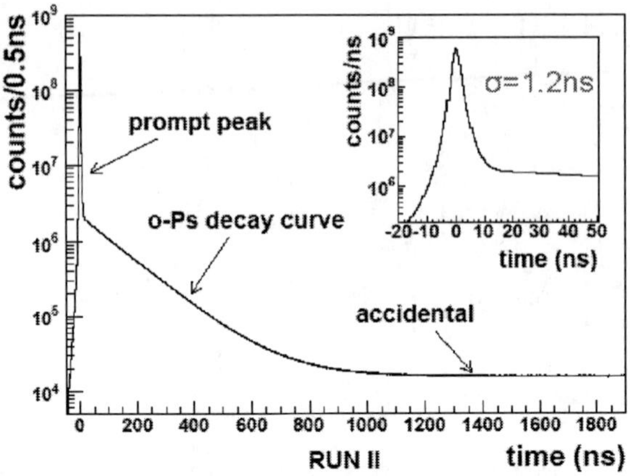

FIGURE 4. Timing spectrum measured by the YAP scintillators within an energy window above 150 keV for RUN-II. The figure in the right corner shows the magnified view of the prompt peak.

TABLE 1. Summary of the systematic errors

Source	RUN-I(ppm)	RUN-II(ppm)
TDC non linearity	$< \pm 15$	$< \pm 15$
Pile up	$< +10$	$< +10$
Pickoff estimation		
(1) 3γ subtraction	$< \pm 89$	$< \pm 91$
(2) Ge efficiency	$< \pm 33$	$< \pm 28$
(3) YAP efficiency	$< \pm 64$	$< \pm 19$
Physics source		
(1) Zeeman effect	< -5	< -5
(2) Stark effect	$< +3$	$< +4$
(3) 3γ annihilation	< -91	< -33
Total	-147 and + 115	-104 and +98

The predominant contribution to total systematic error is produced by uncertain normalization of the 3γ spectrum. That is, the number of pick-off events are determined by subtracting the normalized 3-γ spectrum of Monte Carlo simulation from the o-Ps spectrum, where changing the normalization factor affects the $\lambda_{pick}(t)/\lambda_{3\gamma}$ values and eventually propagates to the final result. Since the sharp fall-off of the 3γ-spectrum at 511 keV is solely produced by the good Ge energy resolution of $\sigma = 0.5$ keV, this subtraction only affects the lower side of the pick-off spectrum such that improper subtraction results in asymmetry of the pick-off spectrum shape. Comparison of the asymmetries of the pick-off peak shape and the prompt peak annihilation spectrum is a good parameter for estimating this systematic error. The errors are about 90 ppm for both measurements.

The Stark shift stretches the lifetime of Ps atoms, i.e., a probative calculation shows that the shift is proportional to a square of the effective electric field E such that $\Delta\lambda_{3\gamma}/\lambda_{3\gamma} = 248\cdot(E/E_0)^2$, where $E_0 = m_e^2 e^5/\hbar^4 \approx 5.14\times 10^9\ V/cm$. Silanol functional groups on the surface of the grain behave as an electrical dipole moment creating an effective field around the grains. Average densities of Silanol are measured to be $0.44nm^{-2}$ and the effective field can be analytically calculated such that the contribution to the o-Ps decay rate is determined to be $-5ppm$. These estimations were confirmed using results from precise hyper-fine-structure (HFS) interval measurements of ground state Ps in silica powder[10], where the interval is proportional to the size of Stark effect. Considering the difference in powder densities used, the HFS results are consistent with our estimation.

Other sources of systematic errors: Error contribution due to the Zeeman effect is estimated using the measured absolute magnetic field around the positronium assembly ($-5\ ppm$). Since the 3-γ pick-off process can only occur at a certain ratio, the calculated relative frequency $\sigma_{3\gamma}/\sigma_{2\gamma} \sim 1/378$ is consistent with previous measurements[11], being $-91\ ppm$ for RUN-I and $-33\ ppm$ for RUN-II.

The above discussed systematic errors are regarded as independent contributions such that the total systematic error can be calculated as their quadratic sum, resulting in $-147\ ppm$, $+115\ ppm$ for RUN-I and $-104ppm$, $+98ppm$ for RUN-II. The combined result with systematic error is $\lambda_{3\gamma} = 7.0401 \pm 0.0006(stat.)^{+0.0007}_{0.0009}(sys.)\ \mu s^{-1}$, and total error is 150 ppm.

Results and discussion

Figure 5 shows the summary of the measured decay rate after 1995, three results are obtained with our method and one result is obtained using thin polymer, with which the produced positronium is very slows and the the effect of the thermalized positronium is suppressed. These four results are consistent with each other. and the combined value is $\lambda_{3\gamma} = 7.0401 \pm 0.0007\ \mu s^{-1}$ and shown in the red arrow in the figure. The correlations of the systematic errors are carefully taken into account. The combined result is consistent with the all four results and accuracy is 100 ppm. This result is consistent with the O(α^2) correction[?], and differs from the only up to O(α) by 2.6σ.

HYPER FINE SPLITTING

Current status

The hyper fine splitting (HFS) is the difference of the energy level between the singlet (1^1S_0) state, parapositronium(p-Ps) and the o-Ps. The HFS of the positronium is large because of the following two reasons: (1) The magnetic moment is proportional to the inverse of the mass, and large spin-spin interaction is expected for o-Ps. (2) o-Ps has the same quantum number as photon, and o-Ps has quantum oscillation into virtual photon, o-Ps $\rightarrow \gamma^* \rightarrow$ o-Ps. This frequency of 87GHz contributes only to the o-Ps, and makes HFS larger.

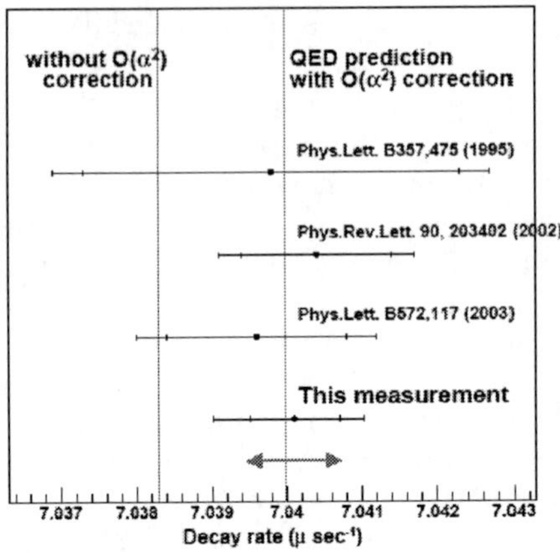

FIGURE 5. Decay rates measured in the last four experiments are listed chronically. The last arrow shows the combined result of the last four measurements. The red and blue lines shows the QED prediction calculated up to O(α^2) and O(α), respectively

The precise measurement of the HFS gives the direct information about the higher order calculation of the QED, especially of the bound state QED. If an unknown light particle (like as axion or millichaged particle) exits, it contributes to the energy level, and makes discrepancy from the QED prediction. Since the quantum oscillation has good sensitivity to the hypothesis particle, whose coupling is super-weak, the precise measurement of HFS is good tool to search for the new physics beyond the Standard Model indirectly.

The precise measurements have been performed in 70's and 80's. These results are consistent with each other and the final precision is 3.6 ppm [12]. No higher order correction($\geq O(\alpha^2)$) has been performed at that time and these results "were" consistent with the 1st order calculation of the QED. New method to calculate the higher order correction on the bound state is established in 2000, and the 2nd and 3rd corrections have been performed[13]. The QED prediction is 203.3917(6) GHz and it differs from the measured value of 203.38910(74) GHz. The discrepancy of 3.5σ is observed. It is still marginal for statistical fluctuation, but there would be possibility of the new physics or the common systematic errors in the previous measurements. (There is also possibility that NRQED approximation is wrong.)

Old experiments and the systematic errors

In the previous all measurements, the HFS transition was not directly measured, since 203GHz was too high frequency to be handled. The static magnetic field makes Zeeman mixing between m_z=0 state of o-Ps and p-Ps, the resultant energy level of m_z=0 state becomes higher than $m_z = \pm 1$. This energy shift is proportional to the HFS energy level. Figure 6 shows the energy levels as function of the static magnetic field. The $m_z = \pm 1$

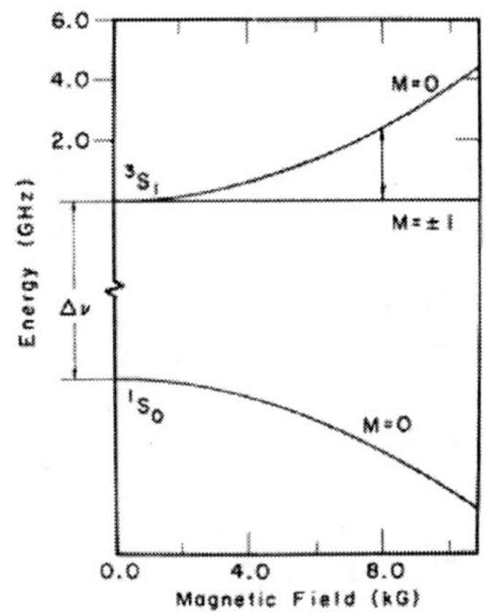

FIGURE 6. The energy level of Ps as function of the static magnetic field. $\Delta\nu$ represents the HFS, and o-Ps state ($^3S_1$) is higher than p-Ps ($^1S_0$). Arrow shows the Zeeman shift between $M_z = 0$ and $M_z = \pm 1$ at B=8kGauss.

state of o-Ps do not couple to the static magnetic field, and do not change the energy level.

Positronium was produced with the β^+ source and gas(N_2, Argon etc) in the RF cavity, in which the high power 2.3GHz microwave was stored. Static magnetic field (about 8kGauss shown with arrow in the figure) was also applied and scanned near the resonance of the Zeeman shift. When the Zeeman energy shift is just on the RF frequency, the transition from $m_z = \pm 1$ to m_z=0 increases, and the m_z=0 state decays into 2γ immediately through p-Ps state. The 2γ decay, which were tagged with back-to-back topology, increases on the resonance. HFS can be determined as center value of the resonance peak by scanning the magnetic field. Since Ps is produced in the gas, the Ps collides with the gas molecule and the electric field of the gas molecule makes the shift of the energy state, which is called as the Stark effect, and about 10ppm shift was observed for 1 atm gas[12]. The HFS in gas were measured by changing the gas pressures. The measured values were extrapolate to zero pressure, and the HFS in the vacuum was obtained. This extrapolation method was the exactly same as in the decay rate measurements[1, 2, 3].

There are two possibilities of systematic errors in this method. (1) First is the effect of the non-thermalized o-Ps, which is the same as in the decay rate measurement. Extrapolation procedure is assumed that Ps is well thermalized and the mean velocity of the Ps is the same for the various pressured gases. We have already shown in the decay rate measurements that this well-thermalized assumption is not satisfied, and that this method makes big systematic errors. The non-thermal Ps would affect also on the HFS measurement. (2) The Ps is widely spread in the RF cavity (size 17cm in diameter) and non-uniformity of the static magnetic field was the main source of the uncertainties of the results. Size of the used magnet were limited and the uniformity of the filed was about 10ppm level, which has been corrected.

FIGURE 7. Superconducting magnet: Bore size is 80 cm and the uniform magnetic field can be applied up to 1T.

Our new methods

We propose the following two new methods in order to solve these systematic uncertainties in the previous measurements.

Conventional Zeeman method using the magnet field

^{22}Na positron source(1MBq) is installed in the thin plastic scintillator(100μm), and the timing information of the positron emission is tagged (positron emits at t=0). There are two benefits to have time information: (1) Many part of positron is just annihilated into 2γ at t=0, and this 2γ annihilation events make S/N seriously worse. The 2γ annihilation background events can be removed dramatically with the requirement of t>10nsec. (2) Thermalization process can be determined with the energy spectrum measured by the germanium semiconducting detectors, as the same as in the measurement of the decay rate. Stark effect (material effect) can be measured changing gas pressure and the non-thermalization effect are measured directly.

Magnetic filed is provided with the superconducting magnet developed for the medical NMR. Figure 7 shows the photograph of the magnet, which has the large bore size (80cm) and strong (about 1T) magnetic field. The magnetic filed is uniform well, we can reduce the uncertainties due to non-uniformity of the magnetic filed. 0.9T of the magnetic filed will be applied with 2.8GHz RF (power 500W) microwave. Cavity with TM110 mode resonance is used to store the RF power.

Figure 8(a) shows the gamma ray detectors and the RF cavity. Six $LaBr_3$ scintillators and two germanium detectors are used to detect gamma ray from the Ps decay. $LaBr_3$ scintillator has the fast signal shape as shown in Fig.8(b). Fast raising time, faster than 10 nsec, provides the good time resolution of about 300psec (FWHM at 511KeV), and

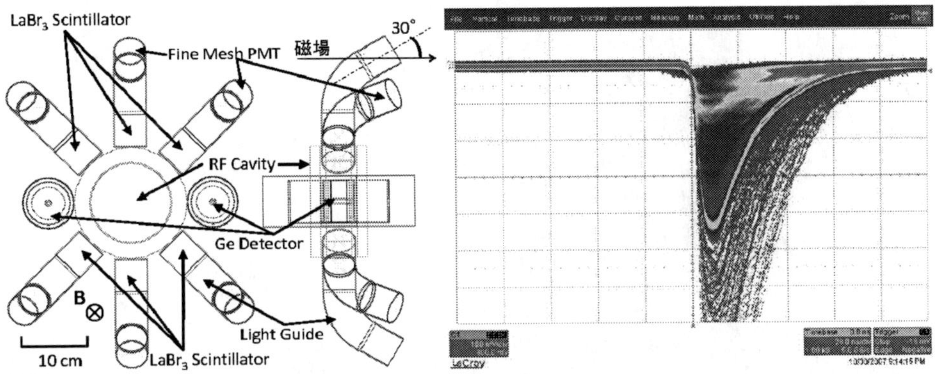

FIGURE 8. (a) The experimental setup of the gamma ray detectors: 6 $LaBr_3$ scintillators and 2 Ge detectors are located around 2.8GHz RF cavity. Since these are located in the magnetic filed, the fine-mesh PMTs are used and the angel between the magnetic field and the PMT axis is set at 30°, at which the magnetic field effect on the PMT gain is about 10. (b) The observed signal shape of the $LaBr_3$ for 667KeV gamma ray emitted from ^{137}Cs.

TABLE 2. The expected numbers of events for 2γ and 3γ decay with the both tagging method. Numbers will be detected on the Zeeman resonance for one day run

	geometrical tagging	Energy tagging
2γ decay	4.3×10^4	3.2×10^6
3γ decay	7×10^3	2×10^6

the fast decay time (26nsec) grantees the pileup effect can be reduced. Energy can be measured with the high resolution (FWHM=3% at 511keV) as shown in Fig.8(b). The energy resolution is better than NaI by factor 2, then 2γ processes can be tagged with the single scintillator using the energy information (Energy tagging). Since the tagging efficiency of the energy tagging is significantly higher than the 'geometrical tagging', in which 2γ decay is tagged with two scintillators located back-to-back. The expected numbers are summarized in table 2. First measurement with this new method will be performed in this Autumn, and a few ppm accuracy will be obtained for a half year.

Direct transition using sub THz RF

(Sub) THz RF is the unexplored filed, since (sub) THz RF has the characters between the optics and the radio wave. Developing the high quality/high power source for the (sub) THz RF is useful and interesting for both science and technology. We can observe the direct transition of the HFS without magnet field, if we have good source of 203GHz RF. It is free from the systematic error due to magnetic filed, and it open the new paradigm of the atomic/particle physics using (sub) THz.

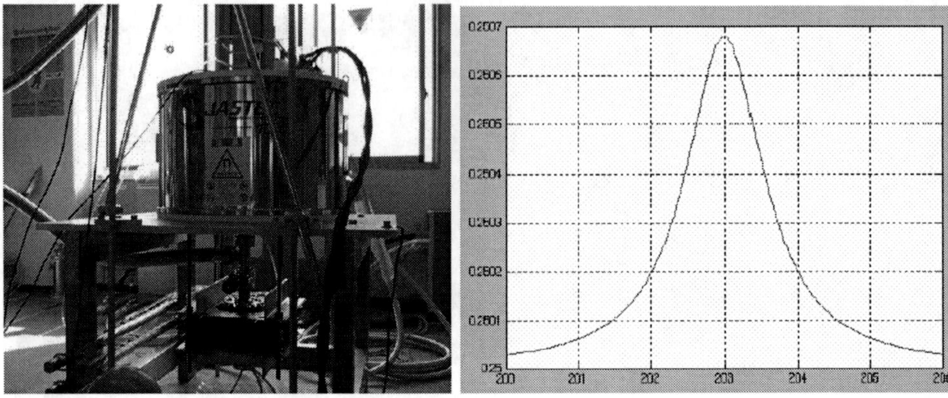

FIGURE 9. (a) Photograph of Gyrotron FU CW-4: High power (500W) and high frequency (400GHz) sub THz wave can be produced. (b) The expected resonance curve with Q=6000 cavity and 100W RF source.

Figure9(a) shows photograph of the gyrotron RF generator[14], in which high power(>100W) sub THz RF can be produced continuously. The accelerated electron emits RF in the strong magnetic filed, and the one/specific frequency is enhanced with the resonance cavity. Arbitrary RF frequency can be picked up with the tuning resonance cavity. Tunable frequency is also big challenge of THz RF source. Improvement stability of the power is in progress and the final design of stability is about 10 ppm. This is big challenge on the THz technology and we can expect various spin off for not only basic science but also technology.

203 ± 1GHz (tunable frequency) CW RF source will be ready in this year. The gamma ray detectors mentioned above are used, and the positron timing is also measured in order to reduce annihilation 2γ and measure the thermalization process. The Fabry-Perot cavity is used for RF storage, since 203GHz has optical characters. Figure 9(b) shows the expected resonance curve with 100W RF power and Q=6000 cavity. The width(FWHM 0.63GHz) is determined with the decay rate of p-Ps, and the center value can be determined with accuracy of about 100 ppm.

CONCLUSION

The decay rate of o-Ps was measured using a direct 2γ correction method in which the thermalization effect of o-Ps is accounted for and integrated into the time spectrum fitting procedure. The new result is $\lambda_{3\gamma} = 7.0401 \pm 0.0006(stat.)^{+0.0007}_{-0.0009}(sys.)\ \mu s^{-1}$, and total error is 150 ppm[9]. It is most accurate result and agrees well with the last three results[6, 7, 8]. The combined result of these four measurements is $7.0401 \pm 0.0007 \mu s^{-1}$, which is consistent well with th NRQED prediction corrected up to $O(\alpha^2)$ term[4], and differ from the result up to $O(\alpha)$ by 2.6σ. It is the first result to validate the $O(\alpha^2)$ correction.

The Hyper Fine Splitting between o-Ps and p-Ps is sensitive to the higher order corrections of the QED calculation, and the discrepancy of 3.5σ is found between the measured values and the 3rd order QED prediction. We point out that there are two possibilities of the common systematic errors in the previous all measurements. (1) Non-thermalized Ps makes error in the extrapolation method as the same as in the decay rate measurements. This was serious source of the orthopositronium lifetime puzzle. (2) Non-uniformity of the magnetic field is also serious source of the systematic error.

We propose new methods to measure HFS precisely without the these uncertainties. Thermalization processes are measured with the $2\gamma/3\gamma$ ratio, and also 2γ annihilation background events can be reduced by requirement of $t > 10$ nsec. $LaBr_3$ scintillators are used for tagging 2γ decay, since its energy resolution is better than the conventional scintillator *i.e.* NaI and CsI, we can tag 2γ decay only with the energy information, not with back-to-back topology. It gains by factor 70 for the tagging efficiency. Large bore superconducting magnet is prepared for new measurement, it provides the uniform/strong magnetic filed.

203GHz high power and stable RF source are developed. It makes possible to transit p-Ps into o-Ps directly. We can measure HFS directly instead of Zeeman shift. It will be the first application of sub THz RF source to the basic science.

REFERENCES

1. C. I. Westbrook, D. W. Gidley, R. S. Conti, and A. Rich, *Phys. Rev. Lett* **58** 1328 (1987).
2. C. I. Westbrook, D. W. Gidley, R. S. Conti, and A. Rich, *Phys. Rev.* **A40** 5489 (1989).
3. J. S. Nico, D. W. Gidley, A. Rich, and P. W. Zitzewitz, *Phys. Rev. Lett.* **65** 1344 (1990).
4. G. S. Adkins, R. N. Fell, and J. Sapirstein, *Phys. Rev. Lett.* **84** 5086 (2000) and *Ann. Phys.* **295** 136 (2002).
5. S. Asai , T. Hyodo, Y. Nagashima, T.B. Chang and S. Orito, *Materials Science Forum,* **619** 175 (1995).
6. S.Asai *New measurement of orthopositronium lifetime*, **Ph. D. thesis**, University of Tokyo (1994); S. Asai, S. Orito, and N. Shinohara, *Phys. Lett.* **B357** 475 (1995).
7. O. Jinnouchi, *Study of bound state QED: precision measurement of the orthopositronium decay rate*, **Ph. D. thesis**, University of Tokyo (2001); O. Jinnouchi, S. Asai, and T. Kobayashi, *Phys. Lett.* **B572** 117 (2003).
8. R.S. Vallery, P.W. Zitzewitz and D.W. Gidley, *Phys. Rev. Lett* **90** 203402 (2002).
9. Y. Kataoka *Test of bound State QED: Higher order correction: Precision measurement of orthopositronium decay rate*, **Ph. D. thesis**, University of Tokyo (2007), `http://tabletop.icepp.s.u-tokyo.ac.jp/oPs-life/main.final.pdf`
10. M.H. Yam, P.O. Egan, W.E. Frieze, and V.M. Hughes, *Phys. Rev.* **A18** 350 (1978).
11. J. A. Rich, *Phys. Rev* **61** 140 (1951).
12. M.W. Ritter, P.O. Egan, V.W.Highes and K.A.Woodle, *Phys. Rev.* **A30** 1331 (1984). A.P. Mills. Jr, *Phys. Rev.* **A27** 262 (1983).
13. G.S. Adkins, R.N. Fell, and P.Mitrikov. *Phys. Rev. Lett.* **79** 3387 (1997). B. kniehl and A.A.Penin, *Phys. Rev. Lett.* **85** 5094 (2000).
14. T. Idehara et al. *Int. J. of Infrared and Millimeter Waves* **28** 433 (2007).

Tests of *CP* and *CPT* symmetry with positronium

T. Namba*, K. Nishihara†, T. Yamazaki†, S. Asai† and T. Kobayashi*

*ICEPP, University of Tokyo, 7-3-1 Hongo, Bunkyo-ku, Tokyo 113–0033, Japan
†Department of Physics, Graduate School of Science, University of Tokyo, 7-3-1 Hongo, Bunkyo-ku, Tokyo 113–0033, Japan

Abstract. The 3γ decay of spin-aligned triplet positronium can be used to test *CP* (*C*=charge conjugation, and *P*=parity operation) invariance in the lepton sector. The angular correlation $(\hat{\mathbf{s}}\cdot\hat{\mathbf{k}_1})(\hat{\mathbf{s}}\cdot\hat{\mathbf{k}_1}\times\hat{\mathbf{k}_2})$ is used for this test, where $\mathbf{s}$ is the positronium spin and $|\mathbf{k}_1|>|\mathbf{k}_2|>|\mathbf{k}_3|$ are the γ momenta.

We designed a new detector to test this symmetry at the 10^{-3} level. This detector consists of a magnet, a β^+ tagging and o-Ps creation system, γ-ray detectors, and a turntable. The detector construction is almost finished and data acquisition will begin in this spring.

With a similar detector, the *CPT* (*T*=time reversal) symmetry can be also tested. In this case, the angular correlation $(\hat{\mathbf{s}}\cdot\hat{\mathbf{k}_1}\times\hat{\mathbf{k}_2})$ is used. This test will be started after the *CP* test is finished.

Keywords: positronium, CP symmetry, CPT symmetry, LYSO scintillator
PACS: 36.10.Dr, 11.30.Er, 29.30.Kv

INTRODUCTION

The matter-antimatter asymmetry in the universe is one of the largest unsolved problems in fundamental physics. *CP* asymmetry may be key to explaining the antimatter abundance. Many experimental searches have been performed over a wide range of energy scales and in many different systems. In the quark sector, the Cabibbo-Kobayashi-Maskawa mixing matrix predicts *CP* asymmetry, and a small *CP* asymmetry was observed mainly in the K and B meson systems. However, this asymmetry ($\sim 10^{-3}$ level) is too small to explain the matter-antimatter asymmetry. On the other hand, in the lepton sector, there are no observations of the *CP* violating phenomena, but the sensitivity of the searches is not very high. Especially in the positronium system (Ps), the *CP* symmetry is confirmed to only 2% [1]. The purpose of our experiment is to test the *CP* symmetry with Ps at the 10^{-3} level. Since positronium consists of two leptons, the electron and the positron, and is described purely by bound state QED, *CP* violation at the 10^{-3} level would mean a discovery of a new physics.

CP TEST WITH POSITRONIUM

The ground state of Ps is composed of three triplet state of total spin $\mathbf{s}=1$ and a $\mathbf{s}=0$ singlet state. The former is so called orthopositronium (o-Ps), and the latter as parapositronium (p-Ps). Since these two systems have different *C* states, they also have different decay modes. p-Ps (*C*=even) decays to an even number γ rays (mainly 2 γ rays) with a 125 ps lifetime, and o-Ps (*C*=odd) decays to an odd number of γ rays (mainly 3 γ

CP1037, *Cold Antimatter Plasmas and Application to Fundamental Physics,* edited by Y. Kanai and Y. Yamazaki

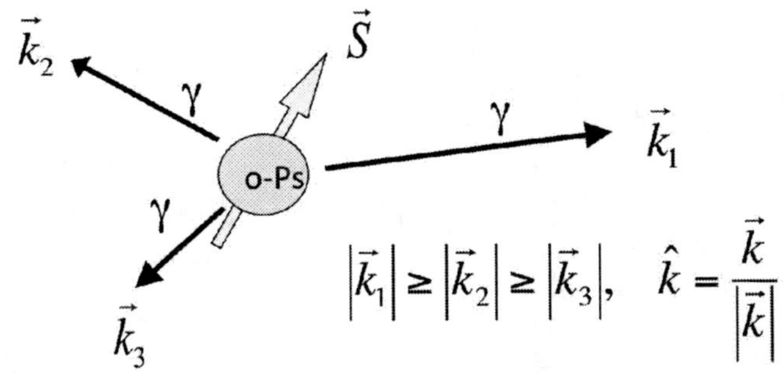

FIGURE 1. 3-γ decay of o-Ps. o-Ps ($\mathbf{s} = 1$) decays to 3 γ's whose momenta are $\mathbf{k}_1$, $\mathbf{k}_2$, and $\mathbf{k}_3$ ($|\mathbf{k}_1| \geq |\mathbf{k}_2| \geq |\mathbf{k}_3|$).

rays) with a 142 ns lifetime.

In our *CP* test, we will use the angular correlations in o-Ps 3-γ decay (Fig. 1) [1, 2, 3],

$$C_{CP}\left(\hat{\mathbf{s}} \cdot \hat{\mathbf{k}_1}\right)\left(\hat{\mathbf{s}} \cdot \hat{\mathbf{k}_1} \times \hat{\mathbf{k}_2}\right), \tag{1}$$

here, C_{CP} is the amplitude of the *CP* asymmetry, $\hat{\mathbf{s}}$ is the direction of the o-Ps spin, and $\hat{\mathbf{k}_1}, \hat{\mathbf{k}_2}, \hat{\mathbf{k}_3}$ are the direction of the emitted 3 γ rays, with

$$|\mathbf{k}_1| \geq |\mathbf{k}_2| \geq |\mathbf{k}_3|. \tag{2}$$

The discrete symmetry properties of Eq. (1) can be seen as follows: Eq. (1) is *P* odd, since there are the three momentum vectors which change sign under the *P* operation, and the two spin axial vectors which do not. Also Eq. (1) is *C* even, since there are 4 particles (o-Ps and 3 γ rays) in this system, and all of them are *C*=1. Therefore, Eq. (1) is odd under the *CP* operation.

DETECTOR DESIGN

To measure the distribution of Eq. (1), we need to observe at least 3 values, the spin of o-Ps (**s**) and the momenta of two of the γ rays ($\mathbf{k}_1$ and $\mathbf{k}_2$). For this purpose, we designed a new detector as shown in Fig. 2. The principal components of the detector are as follows:

- Magnet and return yoke
- o-Ps source and β^+ tagging system
- γ ray detectors
- Turntable

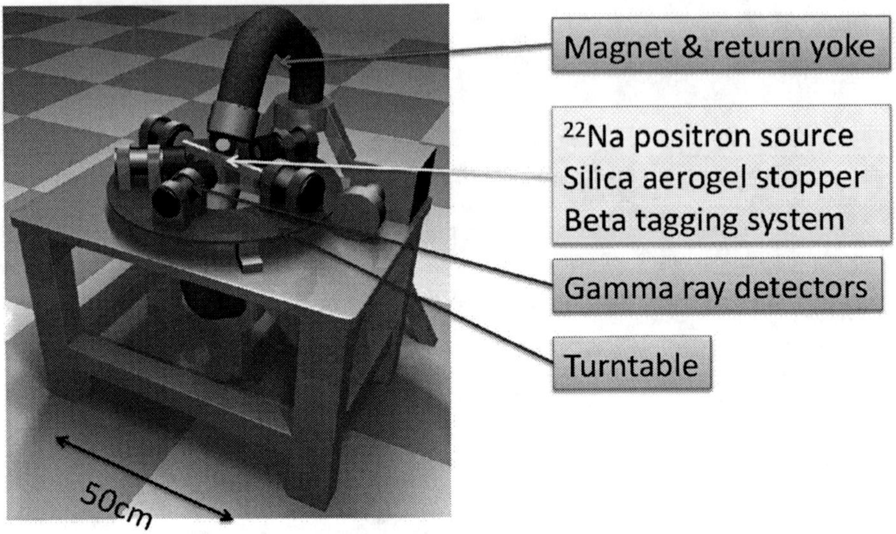

FIGURE 2. Schematic drawing of our detector.

Magnet for spin alignment

For the spin alignment of o-Ps, we will use the magnetic field created by two Neodymium magnets and a return yoke. To be precise, we do not in fact align the spin, but only select the o-Ps with the necessary spin component. When we apply the magnetic field to the o-Ps, the o-Ps's spin along the magnetic field direction (m_z) is quantized into 3 component, $m_z = -1, 0, +1$. As can be seen in Eq. (1), o-Ps's spin s appears twice in the *CP* violating term, so $m_z = \pm 1$ components yield the same value. As such, for our test, all that is required is that the $m_z = \pm 1$ o-Ps be selected. To this end, ortho-para mixing can be used.

When the external magnetic field is applied to the o-Ps, $m_z = 0$ state is no longer an energy eigenstate, and mixes with p-Ps (ortho-para mixing). The ortho-para mixing gives two new states (ψ_+ and ψ_-) of the linear combination of p-Ps and $m_z = 0$ of o-Ps:

$$\psi_+ = \frac{1}{\sqrt{1+Y^2}} \{\psi(\text{o-Ps}, m_z = 0) + Y\psi(\text{p-Ps})\} \tag{3}$$

$$\psi_- = \frac{1}{\sqrt{1+Y^2}} \{\psi(\text{p-Ps}) - Y\psi(\text{o-Ps}, m_z = 0)\}, \tag{4}$$

where the amount of mixing is related to the *B*-field:

$$Y = \frac{X}{1+\sqrt{1+X^2}} \tag{5}$$

$$X = \frac{4\mu_0 B}{\Delta E_{\text{hfs}}} = 0.0276 \times B\,(\text{kG}). \tag{6}$$

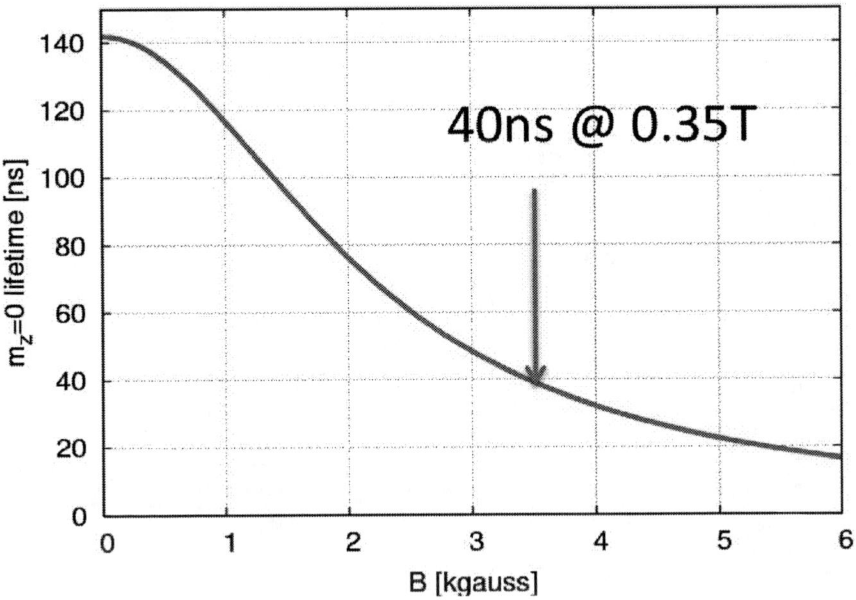

FIGURE 3. *B* dependency of the ψ_+ component lifetime $(1/\lambda_+)$.

Due to this mixing, the decay rates of these two new states differ from the original decay rates. The new decay rates of ψ_+ and ψ_- are

$$\lambda_+ = \frac{1}{1+Y^2}(\lambda_{\text{o-Ps}} + Y^2\lambda_{\text{p-Ps}}) \quad (7)$$

$$\lambda_- = \frac{1}{1+Y^2}(Y^2\lambda_{\text{o-Ps}} + \lambda_{\text{p-Ps}}), \quad (8)$$

where $\lambda_{\text{o-Ps}}(= 7.05\mu\text{s}^{-1})$ is the decay rate of o-Ps, and $\lambda_{\text{p-Ps}}(= 8.0\text{ns}^{-1})$ is that of p-Ps. In Fig. 3, the *B* dependency of ψ_+ component lifetime $(1/\lambda_+)$ is shown. At 0.35T, where the detector operated, the lifetime of ψ_+ is about 40ns.

On the other hand, the $m_z = \pm 1$ components which we need to select are not affected by the magnetic field, and their lifetime is constant (142ns). Therefore, by applying a timing window cut, we can select the $m_z = \pm 1$ components, and can eliminate $m_z = 0$. In Fig. 4, the lifetime spectrum of o-Ps obtained in a test setup is shown. In this setup, the magnetic field is not so strong (0.22T), but the two lifetime components can be seen. In the final detector, the magnetic field will be 0.35T, and the lifetime of ψ_+ will change to 40ns. In that setup, better separation is expected, and the contamination of ψ_+ in the 100–270 ns timing window should be less than 10%.

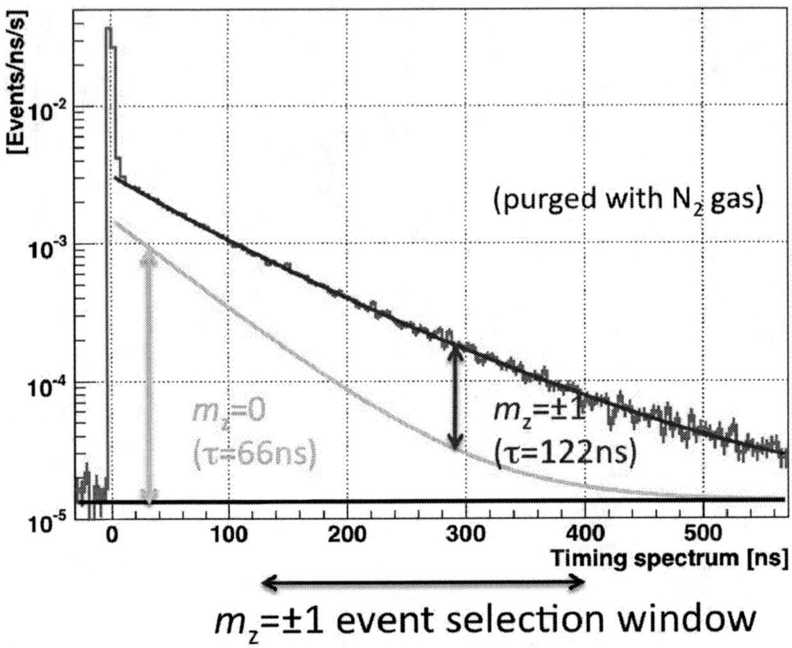

FIGURE 4. o-Ps lifetime spectrum obtained from an N_2 purged system under a 0.22T magnetic field. The peak at $t = 0$ ns is a prompt peak created by e^+ annihilation and ψ_-. After the prompt peak, there are two lifetime components of 122ns (o-Ps in N_2 gas) and 66ns (ψ_+). With a delayed timing window, we can enhance the proportion of $m_z = \pm 1$ events.

o-Ps creation & tagging system

A 1 MBq ^{22}Na source is used as an e^+ source (Fig. 5). Ps is formed in two silica aerogel stoppers located at the both sides of the source. The density of the stopper is 0.1 g/cm^3 and their thickness 0.5 cm. In order to measure the Ps creation timing, two thin plastic scintillators (thickness = 0.1mm) are inserted between the source and the stoppers. Emitted light from the scintillator is guided by an acrylic lightguide to two fine mesh PMTs. β^+ tagging signals are created by the coincidence of these two PMTs.

In Fig. 6, the energy spectra of this tagging system are shown. Comparing these with Geant4 MC results, we found that the light output is about 2.3 keV/p.e., and that the tagging efficiency for Ps creation is almost 100%. Observing the timing difference between the left and the right PMTs, we found that the timing resolution of this tagging system is $\sigma = 0.85$ ns. This value is sufficiently small for the delayed timing window selection.

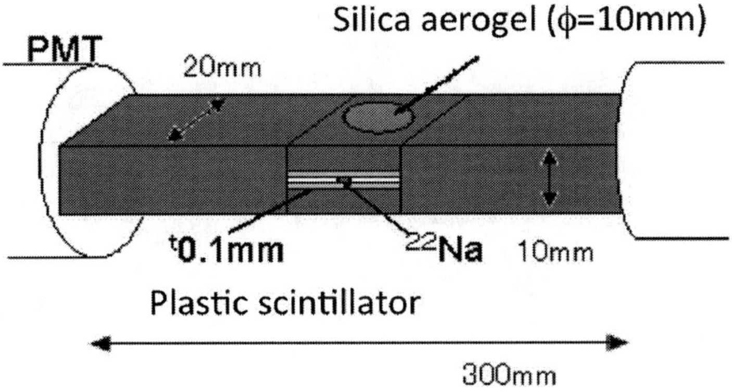

FIGURE 5. Schematic drawing of the o-Ps creation & tagging system. ^{22}Na source is placed at the center of the acrylic lighitguide. The source is surrounded by two thin plastic scintillators and the silica aerogel stoppers.

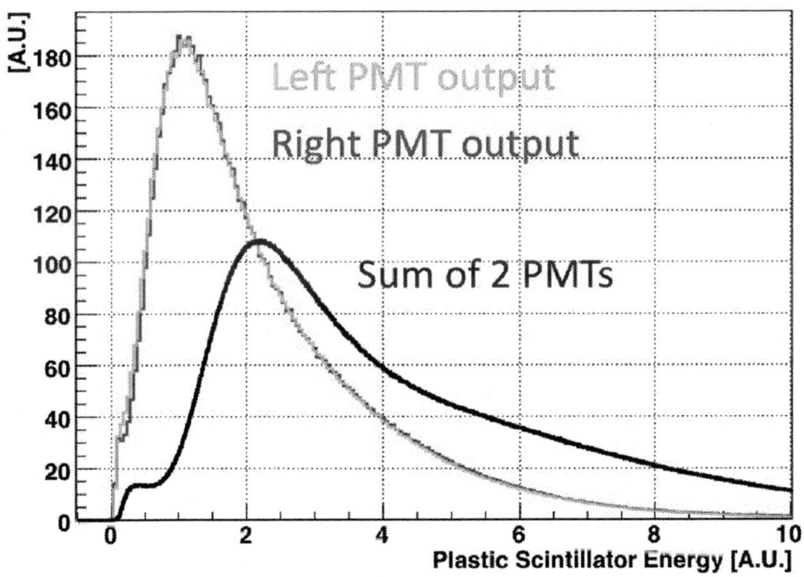

FIGURE 6. Energy spectrum of the β^+ tagging system obtained in a test setup. Light outputs of two PMTs (left and right) are very similar and are enough for the coincidence trigger.

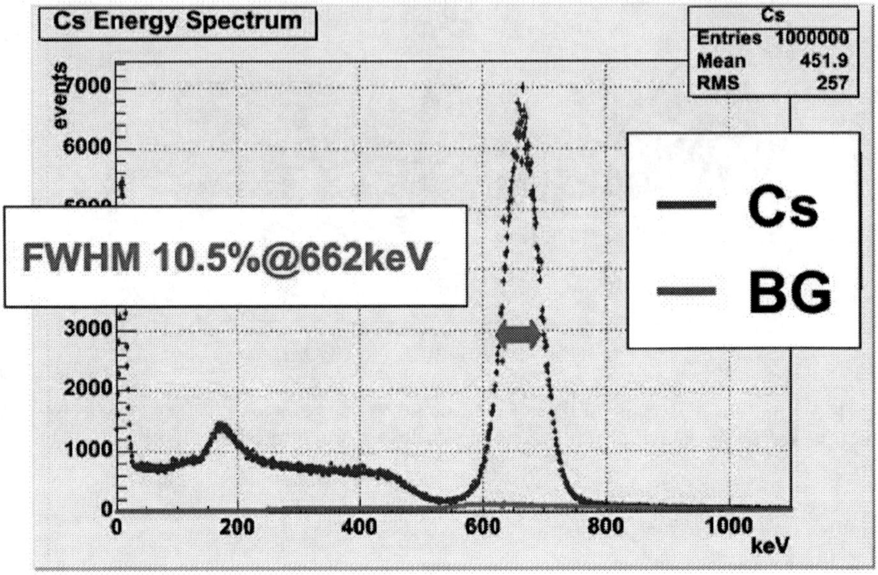

FIGURE 7. ^{137}Cs energy spectrum obtained by an LYSO scintillator.

γ-ray detectors

For our experiments, the γ-ray detectors must satisfy the following requirements.

- Large stopping power for $\sim$ 500 keV γ rays to reduce the compton smearing
- Good energy resolution to separate 3γ decays from 2γ
- Good timing resolution for the delayed timing window selection

Taking account of these requirements, we decided on LYSO ($Lu_{1.8}Y_{0.2}SiO_5$) scintillators. This scintillator contains Lutetium which has atomic number 71, and density 7.1 g/cm^3. Owing to this large atomic number and high density, the attenuation length for a 511 keV γ ray is 1.2cm, which is about 1/3 that of NaI scintillator. LYSO scintillator also has a good photon yield, $\sim$ 32 photons/keV, more than 3.5 times that of BGO scintillator. The timing properties are also good, with its light decay time being 40 ns.

We currently have LYSO scintillators of diameter 3 cm and length 3 cm. Basic studies of this scintillator have already been done. In Fig. 7, a ^{137}Cs spectrum obtained with our LYSO scintillator is shown. Large amounts of 662 keV monochromatic γ rays create a photoelectric peak, and its energy resolution is 10.5% FWHM. In Fig. 8, waveforms of 511 keV signals are shown. You can clearly see the short lifetime (40 ns) of LYSO scintillator, and no long tail component is observed. We also checked the timing resolution for the 511 keV γ rays and found it to be $\sigma = 150$ ps.

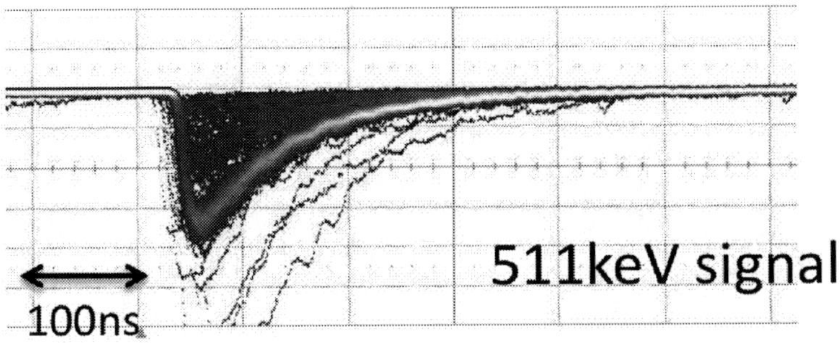

FIGURE 8. 511 keV signal of LYSO scintillator.

One flaw of this scintillator is its RI contaminations. The unstable Lutetium isotope ^{176}Lu contributes considerable background noise. From our measurements, the RI contamination of a (ϕ3cm $\times$ L3cm) crystal is about 5.6 kBq. However in our *CP* test we require triple coincidence for each event (β^+ tagging + 2 γ rays) which reduces this background to a negligible level.

In our final detector, we are planning to use 4 LYSO scintillators. The placement of these 4 scintillators is shown in Fig. 9. With this setup, we have three pairs of detectors which are able to measure decays with a $\mathbf{k}_1$ $\mathbf{k}_2$ opening angle of 150°.

Turntable

To test *CP* symmetry to the 10^{-3} level, we need to observe the distribution of Eq. (1) at a 10^{-4} precision. To achieve this precision, it is critical to minimize small systematic errors such as detector differences and geometrical differences. To prevent such effects, we mount the ^{22}Na source, the β^+ tagging systems, and all γ-ray detectors on a rotating table (Fig. 9). Only the magnet is fixed. Therefore, by rotating the table, we effectively rotate only the magnetic field. Since the decay plane of o-Ps (i.e., the plane of the table) does not move with respect to the magnet, the second term in Eq. (1), $\hat{\mathbf{s}} \cdot (\hat{\mathbf{k}}_1 \times \hat{\mathbf{k}}_2)$ is constant at all table positions for each γ-ray detector pair. On the other hand the first term, $\hat{\mathbf{s}} \cdot \hat{\mathbf{k}}_1$, depends on the angle of $\mathbf{k}_1$ with the magnetic field. Therefore, by observing the count rate dependency on the rotating angle, we can extract the *CP* violating term with the same pairs of the γ-ray detectors.

CURRENT STATUS AND FUTURE PROSPECTS

Currently, almost all detector parts are ready. The β^+-tagging system and the γ-ray detectors are already prepared, and their performance has been tested. As for the turntable, its final design was fixed, and we are now constructing the table. Its motor control system and its position monitor are also ready for operation. The remaining component of our detector is the magnet. Its design is almost finished, and now we are currently tuning

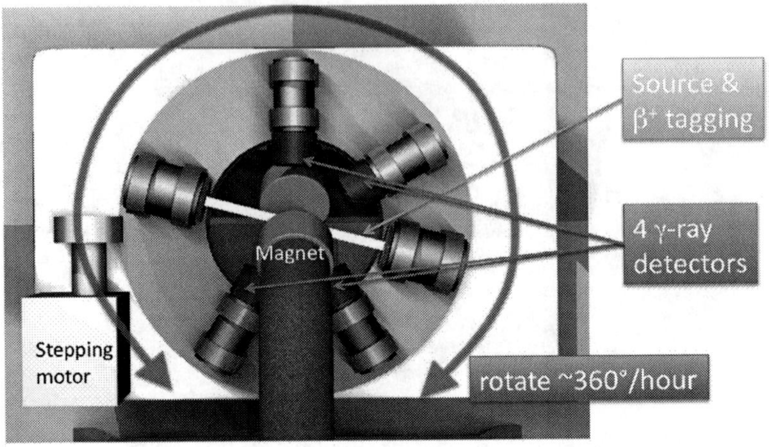

FIGURE 9. Top view of the turntable. The β^+ tagging system and all γ-ray detectors are placed on the rotating table. This table rotates at about 1 rotation/hour, and cancels the systematic errors.

the details of the setup. The magnet construction should be complete by spring, at which time we will be ready to begin data acquisition. From our MC simulation, the expected event rate in the selection window is about 10 Hz, thus we can collect 10^8 events in 100 days. This means that our *CP* test to the 10^{-3} level will be complete by the end of this year.

CPT TEST WITH A SIMILAR SETUP

Similar to the *CP* violating angular correlation, the following angular correlation [2, 3],

$$\hat{\mathbf{s}} \cdot \left(\hat{\mathbf{k}_1} \times \hat{\mathbf{k}_2}\right), \tag{9}$$

violates *CPT* symmetry. Since this correlation contains the o-Ps spin only once, it is necessary to determine the spin direction [4, 5]. To this end a ^{68}Ga β^+ source is ideal since its β^+ is almost polarized ($\langle P \rangle \sim 90\%$) due to its high endpoint energy (1.9 MeV).

To determine $\mathbf{k}_1$ and $\mathbf{k}_2$, the same setup for the *CP* test can be used. LYSO scintillator pairs show good performance for the γ-ray detection, and the turntable will reduce the systematic errors.

After the *CP* test is finished, we will start the *CPT* test with this setup.

CONCLUSION

A new experimental test for *CP* symmetry with spin-aligned o-Ps has been designed, and its detector construction is almost finished. To achieve a 10^{-3} level sensitivity, we use neodymium magnets, a 1 MBq ^{22}Na source, a β^+ tagging system using plastic scintillators, γ-ray detectors with LYSO scintillator, and a turntable. The data acquisition will start from this spring, and *CP* symmetry will be tested to the 10^{-3} level by the end of this year. *CPT* symmetry will also be tested with a similar setup.

ACKNOWLEDGMENTS

The author (T.N.) thank M. Hashimoto for his useful advice. This work has been supported by a Grant-in aid for Scientific Research from the Japanese Ministry of Education, Culture, Sports, Science and Technology.

REFERENCES

1. M. Skalsey and J. Van House, *Phys. Rev. Lett.* **67**, 1993–1996 (1991).
2. M. Skalsey, *Mod. Phys. Lett. A* **7**, 2251–2262 (1992).
3. W. Bernreuther, *et al.*, *Z. phys. C* **41**, 143–158 (1988).
4. B. K. Arbic, *et al.*, *Phys. Rev. A* **37**, 3189–3194 (1988).
5. P. A. Vetter and S. J. Freedman, *Phys. Rev. Lett.* **91**, 263401 (2003).

Fundamental Physics with Cold Positronium

D. B. Cassidy, H. W. K. Tom and A. P. Mills, Jr.

Department of Physics and Astronomy, University of California, Riverside CA 92521, USA

Abstract. The use of positron plasmas to produce intense positron pulses has made it possible to create relatively large instantaneous positronium densities. This in turn provides an opportunity to perform improved precision spectroscopy measurements on positronium that will be limited by systematic rather than statistical effects which will improve present QED positronium tests by orders of magnitude. Further increases in the positronium density will lead to the production of a positronium Bose-Einstein condensate, further increasing the accuracy of such measurements and making it possible to study the phase diagram of a matter-antimatter condensate for the first time.

Keywords: Positron; Positron accumulator; Positronium; Ps_2; Bose-Einstein condensate; QED.
PACS: 36.10.Dr, 34.80.Nz, 82.33.Ln 12.20.Fv.

INTRODUCTION

The ability to accumulate and store positrons in Surko-Penning-Malmberg traps [1] has proved to be of great utility in a number of experimental endeavours, most notably the production of Antihydrogen [2] and measurements of positron scattering from atoms and molecules at very low energies [3]. Similar methods have also made it possible to perform experiments with systems containing more than one positronium (Ps) atom [4].

Studies of Ps-Ps interactions are complicated by two main difficulties. First, Ps has a relatively short lifetime (142 ns or less in the ground state) and so one cannot simply store up Ps atoms. Secondly, positrons are rather scarce and so it is necessary either to create them in some sort of high energy process (e.g., using an electron linac or a nuclear reactor [5]) or capture the output from a radioactive isotope (which must itself be created in a reactor or accelerator). Large scale facilities that may be used to generate Intense positron beams are generally extremely expensive, and as a result the most common way to produce positron beams in the laboratory is to use a beta emitting radioisotope, the standard being ^{22}Na. However, current technology limits the useful activities of such sources to ~ 100 mCi or less, from which slow positron beams of less than 10^7 s^{-1} may be generated. This is obviously too low to study interactions between Ps atoms directly. Since Ps cannot be stored the solution is to collect positrons and then use them all at once to create bursts of Ps atoms, and for this the positron plasmas generated by Surko type traps are well suited.

Up to 10^8 positrons can be stored in our arrangement, and other systems have stored more than 10^9 particles [6]. By rapidly dumping positron plasmas held in a trap we have been able to produce intense, sub-ns, positron bursts with instantaneous currents of more than 10 mA [7]. By implanting these positron bursts into suitable targets, Ps

CP1037, *Cold Antimatter Plasmas and Application to Fundamental Physics,* edited by Y. Kanai and Y. Yamazaki

atoms may be created that have a reasonable chance of interacting with each other, and in this way we have begun to study Ps-Ps interactions.

So far we have observed the formation of molecular positronium (Ps_2) on the internal surfaces of porous silica [8] and on a clean metal surface [9] as well as spin exchange quenching (SEQ) between oppositely polarized ortho-positronium atoms [10]. The main objective of our research, however, is to produce an ensemble of *spin polarised* positronium atoms in a cavity at a density where they might undergo a phase transition and form a Bose-Einstein condensate (BEC) at an experimentally realistic temperature [11].

In this regard SEQ and Ps_2 formation are undesirable, insofar as they represent a loss mechanism that will reduce the final Ps density. Conversely, these processes lead to a critical experimental signal that would be absent if a fully spin polarized beam were used, and at the Ps densities we have been able to achieve so far provide the only evidence we have for Ps-Ps interactions.

Since the incident positron beam used is derived from the beta decay of the isotope ^{22}Na it is naturally partially spin polarized [12], and steps may be taken to increase the net beam polarization at the cost of intensity. The SEQ and/or Ps_2 formation mechanisms may serve as diagnostics in experiments carried out at higher densities in which they are not the main focus. It is therefore necessary to understand the details of these processes, not only because of their intrinsic scientific interest but also because they present a useful tool for the development of procedures to produce a high density spin aligned positronium ensemble, and thence a Ps BEC [13].

In general Bose-Einstein condensation is of interest because it is a macroscopic manifestation of a fundamentally quantum mechanical effect. The essence of the phenomenon is that, below a certain critical temperature T_c, a macroscopic fraction of a collection of identical particles obeying Bose-Einstein statistics (bosons) will "condense" into the single lowest energy state, which will only occur if the density of particles is higher than a certain threshold for a given temperature. The creation of atomic BEC's has been accompanied by an explosion in other technological uses of atom traps, including the atomic fountain clock [14], sensitive gravimeters based on atom interferometry [15], quantum computation schemes [16], and unbreakable quantum cryptography methods [17].

It should be pointed out that, while it is intriguing that the low mass of positronium means that BEC formation may occur at or near room temperature (for achievable densities), it is the novel effects associated with the collective properties of a matter-antimatter system that make the positronium BEC worthy of study. Indeed, the scarcity of antimatter in general, and the concurrent difficulties associated with producing large numbers of positronium atoms far outweigh any advantages conferred by the higher transition temperature, as evidenced by the spectacular success of alkali atom and hydrogen BEC research [18, 19]. However, there are a number of unique aspects to the positronium BEC that are of sufficient interest to justify the efforts needed to create this system:

- First, it is important to note that the total spin 2 ortho-positronium-ortho-positronium scattering length is positive (a_2 = 0.83 Å) [20, 21]. The small positive scattering length makes a collection of spin-polarized o-Ps atoms (all in the m=1 triplet state) a nearly non-interacting ideal gas that may undergo Bose–Einstein condensation [22].

- A positronium atom laser could be created from a positronium BEC that escapes into the vacuum via stimulated tunnelling from a physical cavity through one (or more) narrow channels in the cavity wall. Since the Ps atoms would tunnel without gain or loss of energy, a highly mono-energetic beam would be produced in vacuum (the energy being that of the positronium in the many particle ground state of the cavity).
- A positronium atom laser beam excited (Doppler-free) to a high Rydberg state could be used in a Mach-Zender type interferometer to measure the positronium gravitational redshift, and by inference reveal the gravitational interaction of antimatter [23].
- Since positronium atoms annihilate, it is possible for significant rates of stimulated annihilation to occur in a sufficiently dense positronium BEC in which all the atoms are flipped into the singlet state, providing the basis for an annihilation gamma ray laser [24,25] and a measurement of the Compton wavelength with an uncertainty comparable to its part in 10^{11} linewidth.

Antimatter gravity and gamma ray laser studies are, to say the least, technically demanding experiments, and their realization will almost certainly require some advances in positron generation and accumulation technology. However, underpinning this exciting work are the experiments currently underway, and in particular the most significant advances will be predicated upon the existence of a Ps BEC. The demonstration of such therefore represents a crucial milestone on this experimental path.

As a purely leptonic, matter-antimatter system, positronium is very well suited for the study of quantum electrodynamic (QED) effects in bound systems [26,27,28] via spectroscopy of it's energy levels and measurements of its decay rate [29]. Indeed, theoretical calculations of positronium energy levels and annihilation rates are actually at a higher level of precision than experimental measurements [30]. The main reason for this is that the difficulties associated with producing useful antimatter particles mean that experimental measurements of Ps are generally limited by statistical rather than systematic effects.

Hydrogen and muonium are also systems that offer the possibility of stringent QED tests [31], but each have their own attendant difficulties. In the case of hydrogen, the proton structure limits the applicability of pure QED, and uncertainty in the muon-electron mass ratio means that precision muonium measurements cannot currently be used to test QED to high precision.

Precision measurements of positronium would be made even more accurate using a Ps BEC since the second order Doppler shift would be greatly reduced. For example, the present theoretical value for the 1S-2S interval in positronium is accurate to 0.5 ppb (parts per billion) [32], while the best measurement to date is accurate to 2.6 ppb [33]. Even without a BEC it should be possible to measure this interval to within ~ 50 KHz, or about 0.05 ppb. However, a positronium atom laser would make an ideal source for this measurement, with a possible precision of order < 0.001 ppb. Positronium is believed to be fully described at this level of precision in terms of pure QED theory, and a positronium 1S-2S measurement of this accuracy could be directly related to the fundamental constants, allowing an improved interpretation of more precise standards such as the hydrogen 1S-2S interval. The experimental values of the positronium ground state hyperfine splitting (hfs) (that is, the difference in the ground state energies of the singlet and triplet states) are even less well known, and presently do not agree with the theoretical value (although the discrepancy is only 4 standard deviations and so does not necessarily imply any new physics). The best measurement so far is accurate to 3.6 ppm [34], with another measurement at 7.9 ppm [35] while the most recent theoretical value is known to 0.8 ppm [36]. For some time the experimental values were known to a higher accuracy than the theoretical values, but this is no longer the case; thus, it is timely to re-visit these measurements. Because QED is so fundamental and because it is exactly applicable to these measurements a persistent discrepancy could indicate new physics, which would have profound implications for the standard model.

EXPERIMENTAL TECHNIQUES

The production of positron plasmas from a radioactive source and a buffer gas trap has been discussed at length in the literature [1, 7] and we shall offer only a brief outline here as our arrangement is not very different from the standard Surko trap. A slow mono-energetic positron beam of around 5×10^6 s^{-1} was generated using a ~ 30 mCi ^{22}Na source and a neon moderator [37]. This beam was used to supply a positron trap and accumulator that produced plasmas containing around 5×10^7 positrons in a ~1 mm FWHM spot every minute or so (in a 500 G axial magnetic field). These positron plasmas are long lived, with lifetimes well over 1000 seconds, mostly due to the confinement afforded by a rotating electric field [38, 39] that essentially eliminates cross field transport, a leading loss mechanism [40]. This field is applied using segmented electrodes in the accumulator and as well as aiding confinement also increases the positron density. A schematic of the source, trap and accumulator layout is shown in Fig. 1.

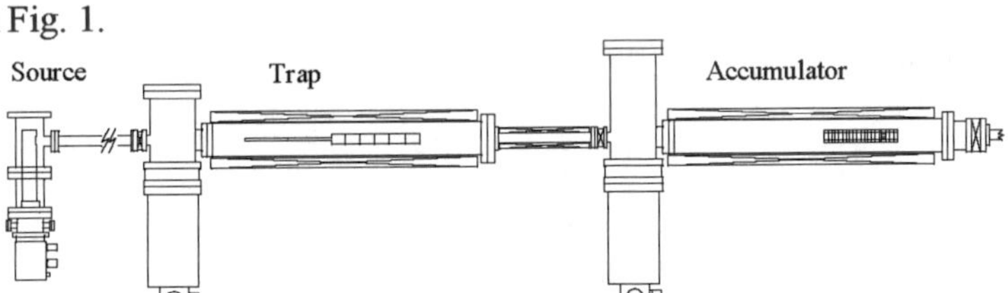

FIGURE 1. A schematic layout of the positron source, accumulator and trap. This system is essentially the same as a standard Surko trap except that the final accumulation region is in a separate vacuum system, which allows us to obtain long lifetimes since the pressure is reduced.

The compression of the positron plasmas in time means that we cannot use the usual detection methods, since the intense sub-ns gamma ray pulse will saturate most detectors. To avoid this we use either a PbF_2 Cherenkov emitter or a $PbWO_4$ scintillator [41]. Both of these have very low light outputs (<1% of that from NaI) and mitigate the problem of saturation. The positron beam density is controlled via the parameters of the "rotating wall" electric field in the accumulator. This allows us to reduce the density by a factor of ~ 5 without changing the magnetic fields, which means that parameters that depend on the field, such as beam transport, detector efficiency and the Zeeman mixing of the positronium m = 0 substates, do not change when the beam density is varied.

The positron plasmas are ejected from the accumulator by a parabolic potential, rapidly applied across the extent of the trapped plasma, which produces a pulse about 30 ns wide (FWHM). A secondary high voltage buncher downstream further compresses this to a sub-ns pulse (typically, 0.8 ± 0.1 ns). A pulsed magnet coil around the target region produces a field of around 1.5 T which compresses the beam spot to ~ 300 microns or less. (NB: these numbers tend to vary depending on the particular experimental requirements, but the maximum final positron beam areal density is always close to $4 \pm 1 \times 10^{10}$ cm^{-2}, which may be due to limitations in the RF signal we can apply to the plasma in the accumulator). The firing sequence for the magnet and bunchers is shown in Fig 2.

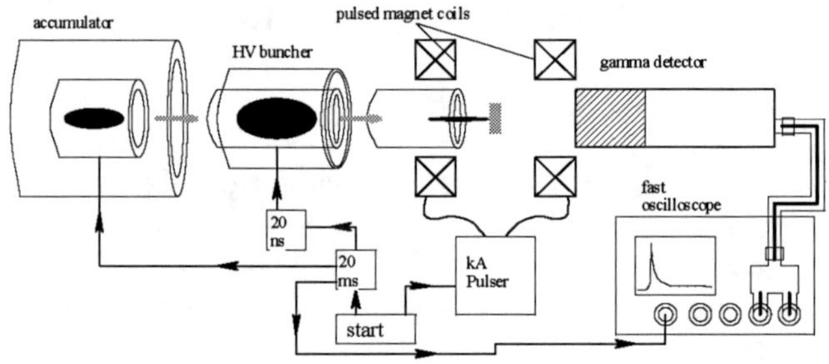

FIGURE 2. The firing sequence for data acquisition using SSPALS. As soon as the positron accumulation phase is complete and all required plasma manipulations have been performed a start signal begins the sequence. Because of its inductance it takes around 20 ms for the magnet to come on, and the HV buncher must be delayed by around 20 ns to account for the transit time of the positron plasma from the accumulator.

The technique of single shot positron annihilation lifetime spectroscopy (SSPALS [42]) involves recording the voltage output from a photomultiplier tube (PMT) directly using a fast oscilloscope following the implantation of an intense positron pulse into a target. The timing resolution of this method is presently ~ 3 ns, mostly due to the response of the PMT. This is more than an order of magnitude worse than may be obtained using conventional lifetime spectroscopy methods [43], but obviously such methods cannot practically be used for measuring what are effectively transitory phenomena. Moreover, the entire system response (including characteristics of the PMT as well as amplifier noise, cable reflections etc) is folded into lifetime spectra obtained in this way, which further complicates later analysis. Despite these limitations, however, changes in the lifetime spectra as a function of the beam density due to Ps-Ps interactions are clearly evident, as is detailed below.

It is not practical to increase the positron beam density by using a stronger magnetic field, and so the next beam upgrade will involve the addition of a remoderation stage [44] before the target. This process involves focusing the beam into a small spot on a remoderator foil (in this case a ~ 100 nm thick Ni foil will be used). The transverse energy given to the beam by the focusing is dissipated, and the phase space of the re-emitted beam is significantly reduced (since the thermalisation process is non-conservative). Even though the total efficiency of the remoderation process is low (around 10-20%) the beam density is greatly increased by this process. In order to do this is it is necessary to extract the beam from the magnetic field, after which electrostatic lenses are used to transport and focus the positrons. We expect that this will increase the beam density by at least two orders of magnitude.

POSITRONIUM-POSITRONIUM INTERACTIONS

When positrons are implanted into certain porous materials (such as silica) positronium may form in the bulk material and diffuse into the internal voids [45]. Ps atoms that diffuse into the voids cannot generally re-enter the bulk material, and so they tend to diffuse through the voids. If these voids are interconnected, and if the incident positron beam density is high enough, it is possible for such Ps atoms to interact with each other, which is the basis for some of our experiments using silica films with both randomly distributed and aligned pores (of around 3-5 nm diameter). In the experiments we have conducted so far we estimate that each Ps atoms visits around 10^4 pores during its lifetime, but that the density corresponds to around 1 Ps atom per 10^5 pores. That is, there is about a 10% chance of Ps atoms interacting with each other. Nevertheless, even this small interaction rate leads to perturbations in the lifetime spectra that are readily apparent, and it has been possible to study Ps-Ps interactions in this way.

Ps atoms in nano-meter sized cavities behave in essentially the same way as they do in vacuum, except that their lifetime may be determined primarily by "pick off" annihilation following interactions with the cavity walls [46] (if the cavity walls contain paramagnetic centers (i.e., unpaired electrons) then this process can drastically reduce the lifetime [47]). In our previous work we observed lifetimes between 20-50 ns, which is long enough for Ps-Ps interactions to occur. When this happens the Ps decay rate may be increased due to various different processes, namely, SEQ and Ps_2 formation. We may describe these processes in the following way:

1. $o-Ps_{m=1}+o-Ps_{m=-1} \rightarrow p-Ps+p-Ps+2E_H$
2. $o-Ps_{m=1}+o-Ps_{m=-1} \rightarrow o-Ps_{m=0}+o-Ps_{m=0}+2E'_H$
3. $X+o-Ps_{m=1}+o-Ps_{m=-1} \rightarrow X+Ps_2+E_b$

Here o-Ps_m refers to the long lived ortho-positronium in the magnetic substate m, p-PS to the short lived para-positronium, E_H is the relevant hyperfine energy difference between the states, X refers to a third body (surface) necessary for Ps_2 formation to conserve energy and momentum and E_b is the Ps_2 binding energy (~ 0.44 eV [48]). In process (1) two triplet atoms are converted to singlets and thus decay rapidly, which increases the decay rate. Similarly, in (2) m = 0 triplets result from the collision, and because of the strong magnetic field these atoms are quenched and decay quickly, also increasing the decay rate. Finally, when molecules are formed (3) the resulting lifetime is 0.25 ns [49], and mean decay rate observed is again increased. Because these are all small changes in the total lifetime spectrum it is not possible to resolve them with our present arrangement, and in effect they all look the same with respect to the timing data. That is to say, using timing data alone we cannot tell the difference between these processes [4].

The difference in the lifetime annihilation spectra caused by Ps-Ps interactions is a non-linear perturbation that may be described by the following differential equation for the density n as a function of time t:

$$\frac{dn}{dt} = -\gamma n(1+\beta n) \tag{1}$$

Here γ is the decay rate, which we assume here is determined only by the pick off annihilation (in reality there may be several components, including one due to the self annihilation, but these may be neglected here) and β represents the non-linear part due to Ps-Ps interactions. If β is equal to zero we have the usual exponential decay (with one component) and this parameter is therefore a measure of the extent to which Ps-Ps interactions occur. Re-write equaton (1) as

$$-dt = \frac{dn}{(\gamma n)(1+\beta n)} = dn\left(\frac{1}{\gamma n} - \frac{\beta}{\gamma(1+\beta n)}\right). \tag{2}$$

Integrating we obtain,

$$-\gamma t + C = \ln(n) - \ln(1+\beta n) \tag{3}$$

$$\frac{n}{1+\beta n} = C_0 \exp(-\gamma t). \tag{4}$$

If the number of atoms at time t = 0 is n_0 we have $C_0 = \frac{n_0}{1+\beta n_0}$. Then,

$$n = \frac{n_0}{(1+\beta n_0)\exp(\gamma t) - \beta n_{0.}} \tag{5}$$

(which of course yields the usual decay rate when β = 0). Figure (3) shows simulated lifetime spectra for various values of β, assuming an arbitrary linear decay rate of 0.02 ns^{-1}. It is clear that as β increases the initial decay rate also increases, as we would expect. Real lifetime spectra are obviously more complicated than this, but the additional components such as the prompt annihilation peak, detector noise, PMT response, other lifetime components etc are constant, so that by taking a difference spectrum between a high density and a low density beam (for which we would expect some approximation to β > 0 and β = 0 respectively) all of these will be subtracted out.

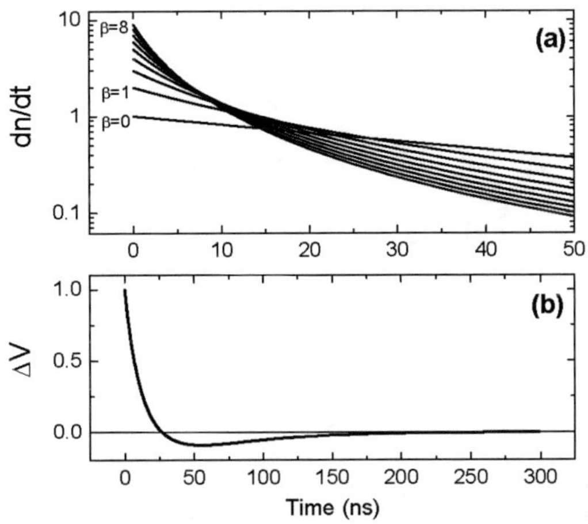

FIGURE 3. (a) A representation of idealized lifetime spectra (area normalized) for different values of the parameter β. (b) Difference spectra obtained by subtracting the interacting case (β =1) from a spectrum in which β =0. If features that do not depend on the beam density are entirely subtracted out then the resulting difference spectra obtained using real data should have a form similar to that shown in (b).

We fit the difference spectra to a function of the form:

$$\Delta V \propto dn(t,\beta)/dt - dn(t,0)/dt = A\gamma\left(\exp(-\gamma t) - \frac{(1+\beta)\exp(\gamma t)}{[(1+\beta)\exp(\gamma t)-\beta]^2}\right) \quad (6)$$

where A is an arbitrary scaling parameter. Fits of this type of function to our data indicate that there is indeed a non-linear, density dependent component to the lifetime spectra and hence also provide further evidence that we are observing Ps-Ps interactions.

In order to distinguish between SEQ and Ps_2 mechanisms, additional data beyond lifetime spectra are required. Such data may be obtained by considering the properties of the sample in which the Ps is formed. In particular, since the formation of Ps_2 requires a third body it is, at least at the present densities obtained, constrained to take place on a surface. Thus, the rate for this process (if it occurs at all) may depend on the population of surface states. It is well known that surface Ps may be thermally desorbed [50] so that, under the right circumstances, the Ps_2 formation rate is a function of sample temperature. No such dependence would be expected with SEQ since this process likely occurs primarily between non-thermal Ps atoms, and in any case would only depend weakly on the temperature. However, SEQ can be affected by the structural properties of the sample since the outgoing states must be able to accommodate the hyperfine triplet-singlet energy difference.

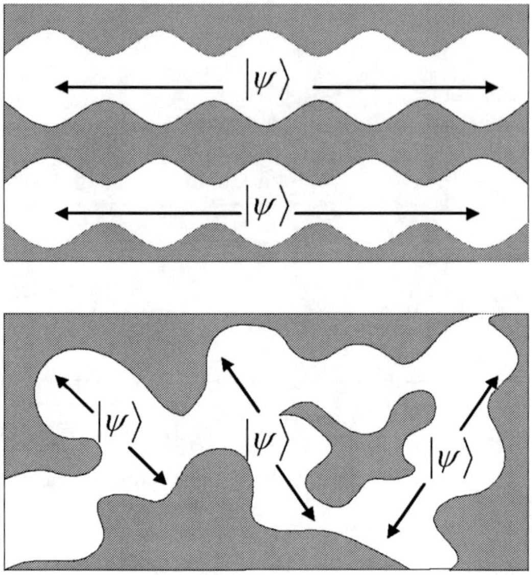

FIGURE 4. Two separate targets with interconnected pores were used. The top panel shows a target in which the pores are aligned into 1-dimensional channels, which means that there is a continuum of accessible eigenstates for Ps atoms contained therein. The bottom panel is a representation of a sample in which the pores are interconnected but randomly orientated; here, at low energies, only discrete eigenstates are accessible.

Figure 4 depicts samples of porous silica films with different pore structures. The top panel shows pores aligned along one dimension, while in the bottom panel the pores are randomly aligned. If the pores are aligned then there will be a continuum of eigenstates accessible to Ps atoms, and thus SEQ will be allowed. For randomly aligned pores this is not the case, and at low energies the accessible eigenstates will be discreet and SEQ will be suppressed. Since the Ps atoms have to diffuse to find each other it is unlikely that SEQ will take place between energetic (that is, non-thermalised) particles, for which this suppression may not occur. At higher densities, however, this will not necessarily be the case.

Although SEQ appears to be suppressed in the randomly aligned sample, we did observe that the amount of long lived Ps present increased with temperature, while the quenching effect Q (which depends on the value of β) decreased. This is shown in figure 5. We interpret these data as being due to the thermal desorption of surface state Ps. When the Ps is desorbed it lives longer in the voids, but is unable to form Ps_2 and, due to the nature of the pore structure, the desorbed Ps is unable to then take part in SEQ interactions. No temperature dependence of the Ps yield or the quenching effect was observed with the aligned sample, indicating that there was no surface state present and therefore no Ps_2 formation therein.

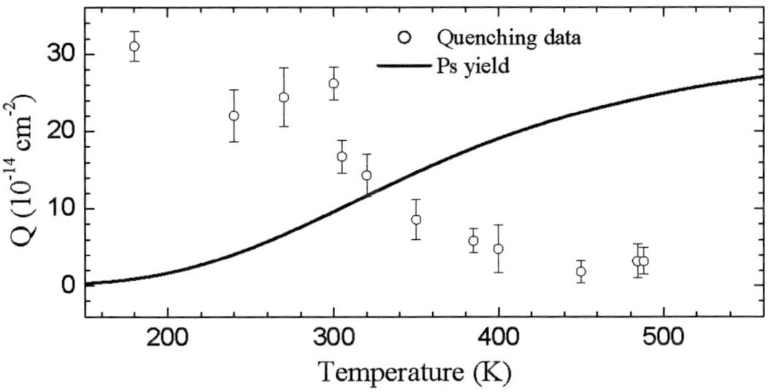

FIGURE 5. The quenching data and Ps yield as a function of temperature. The Ps yield is scaled to match Q which itself depends on the parameter β (see ref [8] for details). The fact that the quenching goes down while the Ps yield goes up with temperature suggests that the quenching process depends directly on the population of the surface state, which indicates the formation of Ps_2 via positronic surface states.

Figure 6 shows lifetime and difference spectra for the two samples described above. As is evident from these data, we have been able to confirm that various Ps-Ps interactions have been observed in more or less the expected manner. While it is interesting to study both the spin exchange quenching process [10] and the formation of Ps_2 [8], the primary objective of our research is to collect an ensemble of spin aligned Ps atoms in a large cavity and form a Bose-Einstein condensate [24]. Even without forming a condensate, however, a dense collection of cold Ps atoms may still be used to perform precision spectroscopy measurements.

In the samples used so far the Ps most likely cools to near room temperatures in ~10-20 ns via collisions with the cavity walls. However, for the larger voids needed to produce a high density Ps cloud the thermalisation rate of Ps is expected to be rather low. Saito and Hyodo have estimated that in a cavity $0.1 \times 0.1 \times 1\ \mu m^3$ (as envisioned for the production of a Ps BEC [11]) the time to cool to the ambient temperature is significantly longer than the lifetime [51]. This suggestion is supported by a number of experiments in which Ps atoms in cold samples (< 50K) do not reach thermal equilibrium, so that below a certain temperature (that depends on the pore size) the decay rate no longer depends on the sample temperature [52, 53].

While the exact thermalisation rate will depend on a number of parameters other than the size of the cavity (such as the internal surface chemistry [54] or the presence of a cooling gas [55]) it seems clear that collisional cooling will be insufficient to produce a Ps BEC unless we are able to achieve densities of around 10^{19} cm^{-3} or much smaller cavities are used. We estimate that with the present experimental arrangement the maximum Ps density obtained is ~ 10^{15} cm^{-3}. In order to increase this we intend to remoderate the beam [44,56] which we expect to increase the density by at least two and possibly as much as three orders of magnitude. However, this methodology is not expected to produce densities higher than this, and so laser cooling of the Ps will be employed.

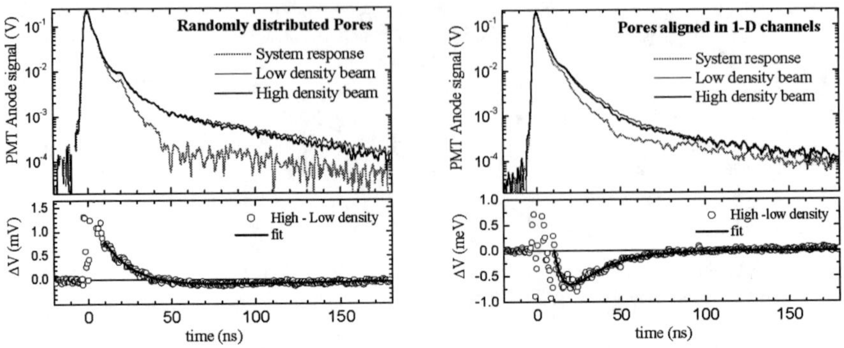

FIGURE 6. Lifetime data obtained using the samples with randomly distributed and aligned pores, as shown in the legends. As explained in the text, the former data are attributed to the formation of Ps_2 and the latter to SEQ. The lower panels show the difference spectra for both cases (*cf* Fig 3 (b)).

LASER COOLING OF POSITRONIUM

Increasing the Ps cooling rate using lasers [24] will significantly relax the Ps density requirements for making a BEC and also allow high precision measurements to be made. Laser cooling of positronium has not yet been demonstrated experimentally, although a number of simulations have indicated that it is feasible [57]. We envisage cooling Ps via the 1S-2P transition using 243 nm light. Ps laser cooling is not intrinsically different from ordinary laser cooling. However, because of its relatively low mass, the Ps 1-photon recoil limit (~0.6 K) is much larger than the Doppler limit (~7.5 mK) and represents the minimum temperature one can realistically expect to obtain. (Compare this with Na, which has a recoil limit of 2.4 μK and a Doppler limit of 240 μK, (using the 3p $^2P_{3/2}$ - 3s $^2S_{1/2}$ transition [58])). Thus, more advanced sub-Doppler cooling techniques (e.g., Sisyphus cooling [59]) cannot be used, although their utility would be limited in any case, due to the short Ps lifetime.

Positronium is emitted into cavities in silica with an energy of ~ 1 eV [60, 61] and, initially, cools rapidly via collisions to ~ 0.2 eV. However, the cooling rate falls off with the temperature, and it may then take many 100's of ns to thermalise [51, 62]. The mean number of absorptions required to laser cool Ps from ~100K to the recoil limit via laser cooling is ~ 30. For a sufficiently powerful laser at > 10 times the saturation intensity (1.3 W/cm^2 per 50 MHz bandwidth), nearly half of the Ps atoms will always be in the 2P state (lifetime>>1μs) and hence the mean lifetime will therefore be almost twice that of the ground state (i.e., 284 ns). Thus, cooling from 100K to 0.6 K will be accompanied by the loss of about half the positronium atoms due to self anihilation.

The spontaneous lifetime of 2P Ps is 3.2 ns but at high optical intensities the fastest the atom can be recycled is 6.4 ns. Thus the minimum time for 30 absorptions is ~200 ns. We will use 300 ns duration, 243 nm, 40 GHz linewidth pulses, made by third-harmonic generation of a ~ 500 ns duration 729 nm pulse from a Q-switched and pulse-stretched alexandrite laser, producing at least 20 mJ/pulse at 243 nm. When focused to a Gaussian diameter of 3 mm, the peak intensity in the 3-d cooling beams will each reach 580 KW/cm^2 over a 40 GHz linewidth, which is 580 times the saturation intensity.

The Doppler width is 380 GHz at 100K, and by detuning the cooling laser to -95 GHz (red-shift) from the line center, we will predominantly decelerate atoms. The laser-cooling spectrum will be further shaped by adding a 24 GHz notch using a high finesse Fabry-Perot interferometer in reflection. This will reduce the probability of absorption for atoms within the ± 2 photon recoil limit and suppress over-cooling. We will further shape the time-profile of the cooling laser pulse using a Pockel's cell pulse slicer. Ending the cooling pulse with high contrast and timing precision is essential in order to do precision optical spectroscopy of the cooled atoms. If the positronium is not confined to a cavity (as it would be for BEC formation) then during the cooling process, a small diameter swarm of triplet Ps atoms will expand from an initial sub mm radius to about 1.7 mm rms radius due to the accumulative effect of about 30 random momentum kicks. The first order Doppler spread of the cooled Ps will be around $\pm 10^{-5}$, small enough for extrapolated measurements of the Ps hyperfine interval to a precision of about 100 kHz (0.5 ppm) and of the 1st-order Doppler free 1S-2S interval to about 30 KHz (3 part in 10^{11}).

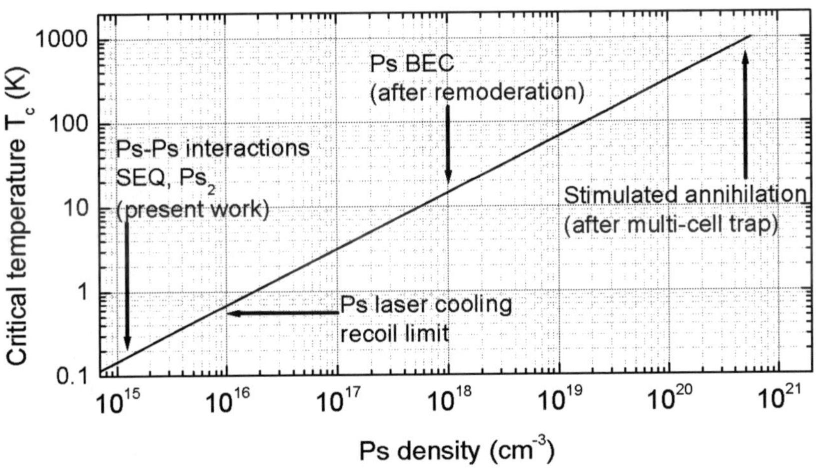

FIGURE 7. The critical temperature for a Ps BEC as a function of density with indications of present and future areas of study.

Figure (7) shows the Ps BEC critical temperature a function of the Ps density. If the Ps is laser cooled to near the recoil limit then only a modest increase of around one order of magnitude in the present Ps density will be required. Thus, with the expected density after remoderation of ~ 10^{18} cm^{-3}, there would be some latitude for variations in the experimental parameters.

PRECISION OPTICAL SPECTROSCOPY

The positron beam in its present state could be utilized for precision optical spectroscopy measurements, and these would be greatly improved as the beam development continues. For the initial experiments the positron beam would be extracted from the magnetic field, in the same way as described above, but without remoderating. Using the ^{22}Na source this would mean extracting a 1 mm diameter beam of ~ 5×10^7 positrons from a 500 G field. This beam could be brought into a field free region and electrostatically focused to a spot with less than 2 mm diameter onto a cooled silica target.

It is known that positronium emitted from powdered SiO_2 has an energy distribution that has a low energy tail, caused by scattering of atoms as they diffuse out into the vacuum [53]. Thus, we may envisage an experiment in which some fraction of the Ps emitted from a silica sample (or some other similar material) will be at temperatures below around 100 K. These atoms would enter a Fabry-Perot cavity tuned for 486 nm in which counter-propagating beams would excite the Doppler-free two-photon 1S-2S transition [63]. Three dimensional laser cooling using the 1S-2P transition as discussed above would be timed to slow the atoms to spend a significant amount of time in the 1S-2S excitation field defined by a 3 mm Gaussian diameter. These excited atoms in the 2S state would then be ionized by a Nd:YAG pulse at 532 nm, and the resulting positrons would be guided to a channel electron multiplier array where they could be detected with high efficiency. If only 1% of the emitted Ps atoms were cold enough (100K) and emitted into the solid angle to be stopped within the 3 mm diameter 1S-2S beam interaction region located 6 mm in front of the SiO_2 sample, and as few as 10% of those were cooled to stay within the 1S-2S beam interaction region, we could excite 5% of the population to the 2S state. If we then wait an additional 142 ns after the cooling pulse ended to assure no remnant of the 2P excitation remained and we photoionize 50% of the atoms in the 2S state, then, with 5×10^7 positrons injected per pulse (and a Ps formation fraction of 0.3) we would expect to detect around 50 positrons per pulse.

Since it takes ~100 seconds to fill the accumulator we could then expect approximately 10^4 counts in a 12 hour run using the procedure outlined above. This would allow a 20 point spectral measurement of the 1.3 MHz natural linewidth (at 200 KHz intervals) to be made to ~ 3% photon-counting uncertainty in a single day, and a 100 point spectral measurement at 40 KHz intervals with 1% uncertainty per point in ~ 50 days. The 1S-2S interval is presently known to 1 MHz accuracy; a 40 KHz per point measurement with curve-fitting could easily locate the line center to within 10 KHz and provide an improvement of two orders of magnitude in the resolution over the current state of the art, and far exceeding the expected accuracy of the currently calculated value.

Using similar techniques it would also be possible to measure the hyperfine splitting (hfs) of the ground state energy levels of positronium [35] in the following way: A 203 GHz, 300 ns duration pulse with ~2X transform-limited bandwidth (<3 MHz linewidth) and 3 mm diameter could be directed into the laser-cooled atoms and double-passed back through the atoms using a mirror. The required 203 GHz pulse could be generated by difference-frequency generation of two 972 nm laser pulses frequency separated by 203 GHz in phase-matched ZnTe, with the two 972 nm pulses formed by amplifying the output of two cw diode lasers each locked to comb lines separated by 203 GHz from a frequency-comb generator. The difference frequency would be controlled via the frequency comb generator, and by an additional ±10 MHz using an acousto-optic modulator on one of the beams. The two 972 nm beams would seed a multi-stage parametric amplifier pumped by the residual ~100 mJ, 400 ns, 364.5 nm SH output of the cooling laser system. (Only the 3rd harmonic is used for cooling). The seed beams would propagate collinearly and overlap spatially to experience identical gain dynamics. The two 400 ns duration 972 nm pulses would be amplified to ~5 mJ pulse energy, and after phase-matched difference-frequency generation, we expect ~12% photon conversion or ~ 0.3 μJ of 203 GHz output.. This optical pulse would be directed to the laser-cooled atoms as in the 1S-2S measurement. Over a 3 mm Gaussian diameter, the intensity at 203 GHz would be 17 W/cm^2 over 3 MHz linewidth.

Under this field Ps atoms in the longer-lived triplet state (142 ns lifetime) would be driven to the shorter lived singlet state (0.125 ns lifetime) and we would detect the absorption of 203 GHz via a change in annihilation rate correlated with the presence of the 203 GHz optical pulse. Because the spectral linewidth will be ~ 3 MHz (shaped by the amplification gain dynamics) we would expect to be able to determine the line center to around 1 MHz accuracy in a short time using measured spectral widths of the 972 nm pulses and simple modeling of the difference-frequency spectrum. This will be relatively accurate because the two 972 nm seed pulses will experience exactly the same amplification dynamics and therefore any amplification-dynamic-induced frequency chirps will be similar for the two pulses and cancel to first order in the difference frequency. Further improvement of the hfs interval to 0.1 MHz accuracy may be possible, but would require higher signal to noise measurements as well as shot-to-shot measurement of the temporal shape of the 972 nm pulse with a fast photodiode and high resolution fast digitizing oscilloscope in order to correct the source lineshape on a shot-to-shot basis.

All of the precision measurements could be improved by using positronium in the BEC state in the form of an atom laser emitted from a cavity. This would be done in essentially the same way as described above, except the cooling lasers would act on positronium confined to a cavity, and the 1S-2S excitation and ionization would occur after the BEC had formed and been emitted into the vacuum. Since all the positronium atoms in the Bose-Einstein condensate in a cavity have the same phase, they will form a plane wave traveling away from the surface of the positronium forming target, essentially forming a positronium atom laser [64]. The atoms will be rapidly emitted from the cavity due to stimulated tunneling through the pores in the cavity wall. The de Broglie wavelength, λ_{dB}, of the emitted positronium atoms will be twice the thickness of the cavity (*i.e.* $\lambda_{dB}=2d$). The diffraction width of the positronium beam will therefore be $\theta\approx 2d/D$. With a cavity diameter D= 4 μm and thickness d= 100 nm, the beam will have a divergence half angle $\theta\approx 25$ mrad. Because of their very low energy and monochromatic velocity distribution, these atoms would be ideally suited for precision two-photon first-order Doppler free measurements of the triplet 1S-2S interval. The second-order Doppler shift would be only 2×10^{-11} for a cavity 100 nm thick. The slow positronium in vacuum could be excited by the standing wave laser field in a high finesse Fabry-Perot cavity and the excited atoms would then be photoionized and the free positrons detected with high efficiency by an electron multiplier. The signal at the peak of the 1S-2S resonance would be about 10^4 counts per pulse, permitting one to achieve a statistical precision of 10^{-3} of the linewidth or 1 part in 10^{12} in 100 positron shots corresponding to about 30 hours of measurements. (NB: for making a Ps BEC the time needed to fill the accumulator would be ~ 1000 seconds). There would thus be sufficient data to study various cavity induced systematic effects in detail. As discussed previously, the laser metrology should enable better than 1 KHz absolute accuracy (1 part in 10^{12}) due to use of the optical frequency reference comb generator [65]. The hfs measurements would also be improved by using a Ps atom laser.

DETECTION OF BOSE-EINSTEIN CONDENSED POSITRONIUM

A collection of spin polarized Ps atoms sufficiently cooled will spontaneously undergo a phase transition and form a BEC, but in many ways the properties of the Bose-Einstein condensed atoms will be similar to those of a non-condensed thermal cloud of atoms slightly above the transition temperature, and the final challenge on the path to produce a Ps BEC is to unambiguously detect its presence.

There will be a negligible difference between the lifetime of a BEC and a cold ensemble of Ps atoms and so annihilation lifetime spectroscopy will be unable to distinguish between them. The most convincing signal of positronium BEC would be a measurement of a sudden narrowing of its parallel momentum distribution to that corresponding to the lowest energy state of the cavity in which it is formed; we intend to observe this via the first order Doppler width of an optical transition. Laser light directed on the Ps in the cavity will have a significant component parallel to the target surface (i.e, along the long axis of the cavity) and thus the Ps Lyman-alpha resonances will be tens of GHz in breadth unless a macroscopic fraction of the atoms are in the cavity ground state. Bose-Einstein condensation will thus be detected from the fact that exactly on the 1S-2P resonance the transmission of the Lyman-alpha light through the cavity will drop significantly if the positronium is in the condensed state.

The Lyman-α probe pulses will be ~ 6 ns in duration and tunable around 235.5 nm, with <30 GHz linewidth. They will be made by sum-frequency generation of the 3rd-harmonic of Nd:YAG (355 nm) and the tunable 701 nm output of a Nd:YAG pumped dye laser in phase-matched BBO. Narrower linewidth pulses (<6 GHz) can be achieved for higher resolution spectroscopy by adding an intracavity etalon in the dye laser. Not only will this make it possible to obtain definitive evidence for the presence of a Ps BEC, it would also allow us to experimentally investigate the Ps phase diagram, which would be a significant advance in experimental antimatter chemistry. Measuring the critical temperature as a function of density is of intrinsic scientific interest, and also has much practical value for later work directed towards initiating and studying stimulated annihilation of a Ps BEC.

CONCLUSIONS AND OUTLOOK

We have started a program of experimentation on many-positron, many-electron systems. So far our work has been limited to observing interactions between pairs of Ps atoms, but the methods used are scalable, and we expect to be able to observe interactions between many thousands of atoms with only minor modifications to the experimental procedure. These techniques, combined with laser cooling of Ps, will make it possible to conduct precision spectroscopic measurements of Ps, and to create a Ps BEC. Using proven technology it seems highly likely that we will be able to create a Ps condensate, and even with the present beam parameters we are confident that, with the right laser systems, we will be able to improve on the best precision 1S-2S measurements by at least an order of magnitude. This area of experimentation is in its infancy, and we have no doubt that there is much exciting work to be done; it is our hope that the modest results we have been able to obtain so far will encourage others to join us in undertaking this work.

ACKNOWLEDGMENTS

DBC thanks the organizing committee of the Workshop on Cold Antimatter Plasmas and Applications to Fundamental Physics (a.k.a Pbar 08) for the invitation to speak and the Institute for Theoretical Atomic Molecular and Optical Physics (ITAMP) at Harvard for financial support. We also thank Dr. Rod Greaves, Dr. Shihu Deng, Mr. Vincent Meligne and Mr. Nick Lopez for discussions and technical assistance. This work was supported in part by the National Science Foundation under grants PHY 0140382, PHY 0537431, PHY 0555701 and DMR 0216927.

REFERENCES

1. C. M. Surko and R. G. Greaves, Phys. Plasmas **11**, 2333 (2004).
2. M. Amoretti *et al,* Nature **456** 419 (2002).
3. S. J. Gilbert, R. G. Greaves, and C. M. Surko, Phys. Rev. Lett. **82**, 5032 (1999).
4. D. B. Cassidy, S. H. Deng, R. G. Greaves, T. Maruo, N. Nishiyama, J. B. Snyder, H. K. Tanaka, and A. P. Mills, Jr., Phys. Rev. Lett. **95**, 195006 (2005).
5. See for example *Positron Beams and their Interactions* edited by edited by P. G. Coleman (World Scientific, Singapore, 2000).
6. L. V. Jørgensen et al, Phys. Rev. Lett. **95**, 025002 (2005).
7. D. B. Cassidy, S. H. M. Deng, R. G. Greaves and A. P. Mills Jr., Rev. Sci. Inst. 77, 073106 (2006).
8. D. B. Cassidy and A. P. Mills, Jr., Nature **449**, 195 (2007).
9. D. B. Cassidy, S. H. M. Deng and A. P. Mills, Jr., Phys. Rev. A **76**, 062511 (2007).
10. D. B. Cassidy and A. P. Mills, Jr., Phys. Rev. Lett. **100**, 013401 (2008).
11. See for example, K. Huang, *Statisitcal Mechanics* (Wiley, New York,1987).
12. L. A. Page and M. Heinberg, *Phys. Rev.* **106**, 1220 (1957).
13. P. M. Platzman and A. P. Mills, Jr., *Phys. Rev. B* **49**, 454-8 (1994).
14. K. Gibble and S. Chu, *Phys. Rev. Lett.* **70**, 1771-1774 (1993).
15. M. Kasevich and S. Chu, *Phys. Rev. Lett.* **67**, 181-184 (1991).
16. See for example D. DiVincenzo, *Fortschr. Phys.* **48**, 771-783 (2000).
17. See for example M. D. Barrett *et al.*, Nature **429**, 737-739 (2004), and refs therein.

18. Anthony J. Leggett, Rev. Mod. Phys. **73**, 307 (2001).
19. See for example *Bose-Einstein Condensation*, edited by A. Griffin, D. W. Snoke and S. Stringari, (Cambridge University Press (1995).
20. I. A. Ivanov, J. Mitroy, and K. Varga, Phys. Rev. Lett. **87**, 063201 (2001).
21. S. K. Adhikari, Phys. Lett. *A* **294**, 308-313 (2002).
22. H. Yabu, Nucl. Instrum. Methods Phys. Res., Sect. B: **221** 144-148 (2004).
23. T. J. Phillips, Hyperfine Interactions. **109** 357 (1997).
24. E. P. Liang and C. D. Dermer, *Optics Commun*, **65**, 419 (1988).
25. A. P. Mills, Jr., D. B. Cassidy and R. G. Greaves, *Materials Science Forum* **445-446**, 424 (2004).
26. A. P. Mills. Jr. and S. Chu, in *Quantum Electrodynamics*, edited by T. Kinoshita (World Scientific, Singapore, 1990) p. 774.
27. S. G. Karshenboim, *Appl. Surf. Sci.* **194**, 307-311 (2002).
28. S. G. Karshenboim, International Journal of Modern Physics *A* **19** 3879 (2004).
29. E. g, A. Rich, Rev. Mod. Phys. **53**, 127 (1981).
30. S. G. Karshenboim, *Physics Reports* **422**, 1-63 (2005).
31. E.g., R. Ley, Appl. Surf. Sci **194** 301 (2002)
32. A. Czarnecki, K. Melnikov, and A. Yelkhovsky, Phys. Rev. A **59**, 4316 (1999).
33. M. S. Fee, A. P. Mills, S. Chu, E. D. Shaw, K. Danzmann, R. J. Chichester, and D. M. Zuckerman, Phys. Rev. Lett. **70**, 1397 (1993).
34. M. W. Ritter, P. O. Egan, V. W. Hughes, and K. A. Woodle, Phys. Rev. A **30**, 1331 (1984)
35. A. P. Mills, Jr., Phys. Rev. A **27**, 262 (1983).
36. Kirill Melnikov and Alexander Yelkhovsky, Phys. Rev. Lett. 86, 1498 (2001)
37. A. P. Mills, Jr. and E. M. Gullikson, Appl. Phys. Lett. **49**, 1121 (1986).
38. F. Anderegg, E. M. Hollmann, and C. F. Driscoll, Phys. Rev. Lett. **81**, 4875 (1998).
39. R. G. Greaves and C. M. Surko, Phys. Rev. Lett. **85**, 1883 (2000).
40. E.g., E. M. Hollmann, F. Anderegg, and C. F. Driscoll, Phys. Rev. Lett. **82**, 4839 (1999).
41. D. B. Cassidy and A. P. Mills, Jr., Nucl. Instrum. And Meth. A. **580** 1338 (2007).
42. D. B. Cassidy, S.H.M. Deng, H.K.M. Tanaka and A.P. Mills Jr., Appl. Phys. Lett. **88** 194105 (2006).
43. See for example, W. Brandt, and A. Dupasquier, (Eds.) *Positron Solid State Physics*. North-Holland Publishing Co., Amsterdam. (1983).
44. A. P. Mills, Jr., Appl. Phys. **23**, 189 (1980).
45. D. W. Gidley, W. E. Frieze, T. L. Dull, A. F. Yee, E. T. Ryan, and H.-M. Ho, Phys. Rev. B **60**, R5157 (1999).
46. D. W. Gidley, K. G. Lynn, M. P. Petkov, J. N. Sun, M. H. Weber, and A. F. Yee, in *New Directions in Antimatter Chemistry and Physics*, edited by C. M. Surko and F. A. Gianturco (Kluwer Academic Publishers, Dordrecht, 2001), p. 151.
47. D. B. Cassidy, K. T. Yokoyama, S. H. M. Deng, D. L. Griscom, H. Miyadera, H. W. K. Tom, C. M. Varma, and A. P. Mills, Jr., Phys. Rev. B **75**, 085415 (2007).
48. S. Bubin and L. Adamowicz, Phys. Rev. A **74**, 052502 (2006).
49. D. M. Schrader, Phys. Rev. Lett. **92**, 043401 (2004).
50. P. Sferlazzo, S. Berko and K. F. Canter, Phys. Rev. B **32**, 6067 (1985).
51. H. Saito and T. Hyodo in *New Directions in Antimatter Chemistry and Physics*, edited by C. M. Surko and F. A. Gianturco, (Kluwer, Dordrecht) (2001).
52. R. F. Kiefl and D. R. Harshman, Phys. Lett. A **98** 447 (1983).
53. A. P. Mills, Jr., E. D. Shaw, R. J. Chichester and D. M. Zuckerman, Phys. Rev. B, **40**, 2045 (1989).
54. C. He *at al*, Phys. Rev B **75** 195404 (2007).
55. R. A. Fox and K. F. Canter, J. Phys. B, **11** L255 (1978).
56. Fujinami & Akahani, *Analytical Sciences* **24**, 73 (2008).
57. T. Kumita et al, Nuclear Inst and Meth B **192** 171 (2002) and references therein.
58. See for example, C. J. Foot, *Atomic Physics* Oxford University Press (2005).
59. J. Dalibard and C. Cohen-Tannoudji, J. Opt. Soc. Am. B **2** 1707 (1985).
60. Y. Nagashima, M. Kakimoto, T. Hyodo, K. Fujiwara, A. Ichimura, T. Chang, J. Deng, T. Akahane, T. Chiba, K. Suzuki, B. T. McKee, and A. T. Stewart, Physical Review A **52** 258 (1995).
61. Y. Nagashima, Y. Morinaka, T. Kurihara, Y. Nagai, T. Hyodo, T. Shidara, and K. Nakahara Phys. Rev. B **58** 12676 (1998).

62. S. Takada, T. Iwata, K. Kawashima, H Saito, Y. Nagashima, and T. Hyodo, Radiat. Phys. Chem. **58** 781 (2000).
63. M. S. Fee, A. P. Mills, Jr., S. Chu, E. D. Shaw, K. Danzmann, R. J. Chichester, and D. M. Zuckerman, Phys. Rev. Lett. **68**, 1397 (1993).
64. B. P. Anderson and M. A. Kasevich *et al.*, Science **282**, 1686 (1998).
65. T. Udem, R. Holzwarth and T. W. Hänsch, Nature **416**, 233 (2002).

New Plasma Tools for Antimatter Science

J. R. Danielson, T. R. Weber and C. M. Surko

Department of Physics, University of California, San Diego La Jolla CA 92093

Abstract. Recent research is described that exploits nonneutral plasma techniques to develop new tools to accumulate, manipulate and store antimatter and to make cold, bright antiparticle beams. Progress is described in three areas: radial compression of single-component plasmas using rotating electric fields in a novel, strong-drive regime; experiments and complementary theoretical modeling of the extraction of antiparticle beams with small transverse spatial extent from single-component plasmas; and work to develop a multicell trap to increase, by orders of magnitude, the capacity for antiparticle storage. Potential applications of these tools and challenges for future research are discussed.

Keywords: antimatter, positrons, cold plasmas, low-emittance beams
PACS: 52.55Dy, 52.27.-h, 52.27Aj, 52.25Fi

I. INTRODUCTION

Much progress has been made recently to create cold antimatter plasmas and to exploit them for a variety of fundamental scientific studies and technological applications [1]. A key focus of this conference has been efforts to create, trap, and study antihydrogen [2-5]. Other examples that require positrons are studies of electron-positron plasmas [6-8], positron interactions with atoms and molecules [9, 10], modeling of astrophysical processes [11, 12], and the characterization of materials [13, 14]. On the horizon, a new generation of more intense positron sources is under development [15-18]. Full utilization of these facilities will require new plasma techniques for a range of applications including study of the positronium molecule, Ps_2 [19, 20], the formation of BEC gasses of positronium atoms [21, 22], and the study of electron-positron plasmas [8, 23].

We describe here recent research to develop new tools to create and manipulate single-component plasmas. Electron plasmas (used for convenience and high data rate) are compressed radially using rotating electric fields [the so-called "rotating wall" (RW) technique]. The work focuses on a regime in which the plasma spins up until the plasma rotation frequency is equal to the applied frequency [24]. This creates a rigidly rotating plasma with a known and constant high density. Studies are also described to extract beams of small transverse spatial extent from the plasma center [25, 26]. The factors determining the radial beam profile and minimum beam width are identified. Finally, work is described to develop a multicell trap capable of storing orders of magnitude more positrons than is currently possible (e.g., particle numbers, $N > 10^{12}$) [27, 28].

CP1037, *Cold Antimatter Plasmas and Application to Fundamental Physics,* edited by Y. Kanai and Y. Yamazaki

II. DESCRIPTION OF THE EXPERIMENTS

Plasmas were confined in a Penning-Malmberg trap shown schematically in Fig. 1 and described in detail in Ref. [29]. It consists of a series of room-temperature, cylindrical electrodes 0.5 m in total length with inner radii, $R_W = 1.27$ cm. A uniform magnetic field, $B = 4.8$ Tesla, is imposed parallel to the electrode axis. In equilibrium, the plasma has a constant density profile with density n_0 (i.e., a "flat top" radial distribution), and undergoes $E \times B$ rotation at a frequency, $f_E = n_0 ce/B$ [18, 19]. The plasma temperature, $T \sim 0.05 - 0.1$ eV, which is uniform across the plasma, is set by the balance between heating sources (due to background drag and/or rotating wall torques) and cyclotron cooling ($\tau_C = 0.16$ s) [18, 19].

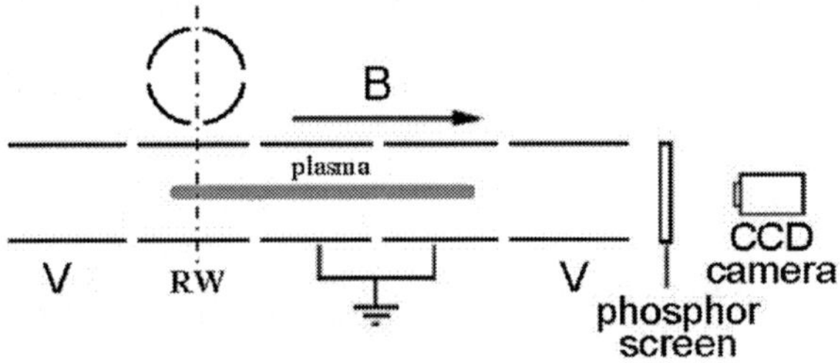

FIGURE 1. Schematic diagram of the Penning-Malmberg trap with segmented rotating wall electrode, phosphor screen and CCD camera.

Radial areal plasma density profiles are measured by releasing the trapped plasma, then accelerating the particles into a phosphor screen where the resulting light is imaged using a CCD camera. The plasma temperature is varied by repeated compression and expansion of the plasma (i.e., changing the plasma length), thereby heating the plasma through Coulomb collisions. The temperature, T, is determined by time-resolved measurement of the number of electrons escaping from the trap when the confining potential, V_C, at one end is slowly lowered [30].

III. ROTATING-WALL COMPRESSION IN THE STRONG DRIVE REGIME

The "rotating wall" (RW) technique, whereby a single-component plasma is compressed radially using a rotating electric field, has proven to be an important tool for antimatter research. This technique was first developed to tailor electron and ion plasmas [31, 32]. More recently, it has become a key tool to manipulate and tailor antimatter plasmas [33, 34]. In the work described here, phased sine waves applied to a sectored electrode are used to generate a rotating electric field with azimuthal mode number, $m_\theta = 1$ [24, 29]. These fields produce a torque on the plasma, thereby compressing the plasma radially in a non-destructive manner. Early experiments achieved good coupling by matching the drive frequency, f_{RW} to Trivelpiece-Gould modes in the plasma [31]. However, this approach has the drawback that compression is only possible at selected frequencies, and careful tuning of the drive frequency to the plasma modes (which change with plasma compression) is required.

We discovered two regimes in which such tuning is unnecessary, first in a positron plasma with buffer-gas cooling when the plasma radius is comparable to the Debye length, λ_D [33], and later in electron plasmas in a high-magnetic-field trap when the drive amplitude is sufficiently large [29]. Shown in Fig. 2 is an example of this later, "strong drive" regime. Above a certain drive amplitude (V_{RW} = 0.7 V p-p in the experiment shown in Fig. 2), the plasma evolves to a high-density steady state in which $f_E \approx f_{RW}$. The protocol for these experiments is such that an initial plasma is created and allowed to come to equilibrium, *then the RW is applied at fixed values of both V_{RW} and f_{RW}.* As illustrated in Fig. 3, the radial density profiles of these plasmas are "flat-top" in shape. Thus the plasma rotates rigidly at frequency f_E and is in an approximate thermal equilibrium state. Experiments at various values of f_{RW} are shown in Fig. 4, illustrating the ability to access a broad range of high-density states in this strong-drive regime.

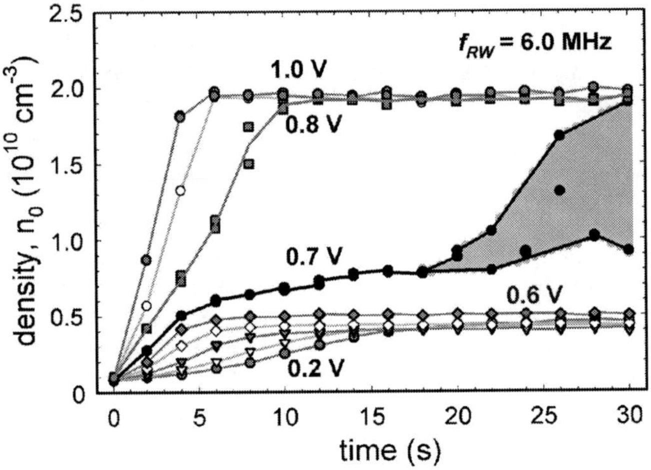

FIGURE 2. Central electron density is shown as a function of time for various amplitudes of applied RW voltage at 6 MHz. Note the bifurcation from a low-density to a high density state as V_{RW} is increased above 0.7 V. Reprinted from Ref. [29].

The central physics question that we addressed recently is what factors determine the ability to access this regime. The key insight came from examining closely behavior such as that illustrated in Fig. 4 (right), where a "step" appears in the data near some characteristic density, in this case $n_0 = 0.4 \times 10^{10}$ cm^{-3}. Close examination indicates that this corresponds to the situation in which a static asymmetry in the laboratory frame drives a Trivelpiece-Gould mode traveling backwards on the rotating plasma. In this case, the mode frequency is zero in the lab frame and referred to as a "zero frequency mode" (ZFM). Coupling of the static asymmetry to this mode causes a drag on the plasma creating an impediment to further compression [29].

In collaboration with T. M. O'Neil, we modeled the drag and drive torques on the plasma to include this ZFM-induce drag, namely [35],

$$\tau = \eta \frac{f_{RW} - f_E}{f_E} V_{RW}^2 - \frac{\beta f_E}{D^2 + f_E^{\ 2}} - \frac{\gamma \delta f_0}{\left(f_E - f_0\right)^2 + \left(\delta f_0\right)^2}, \tag{4}$$

where η, β, γ, and D are constants. The three terms in Eq. (4) represent, respectively, the RW drive, the background drag on the plasma due to trap imperfections, and the drag due to the ZFM. The form of the second term was chosen empirically to model the observed outward transport, and the ZFM is modeled by a Lorentzian of width δf_0, *centered* at frequency f_0.

The predictions of this model for the drive and drag torques are shown in Fig. 5. The plot on the left compares the RW drive torque, for a fixed frequency at various amplitudes, to the sum of the two drag torques from Eq. (4). At the locations where the drive and drag torques match, there is the possibility of a torque-balanced steady state (i.e., an attracting fixed point in the language of nonlinear dynamics). These steady states are shown in the right panel of Fig. 5. Due to the presence of the ZFM torque, there exist two stable steady states (closed circles) – one at high density and one at lower density – and an unstable state (open circles). Which stable state the plasma relaxes to depends on which side of the ZFM the plasma starts. As a consequence, the plasma is predicted to exhibit hysteresis as a function of the RW drive amplitude. Similar hysteresis is also predicted as a function of f_{RW} [35]. The key point is that the ZFM drag has the effect of inhibiting access to the high-density fixed point of the system until a sufficiently large value of V_{RW} is applied. Once in this high-density state, V_{RW} can be reduced significantly without producing much of a decrease

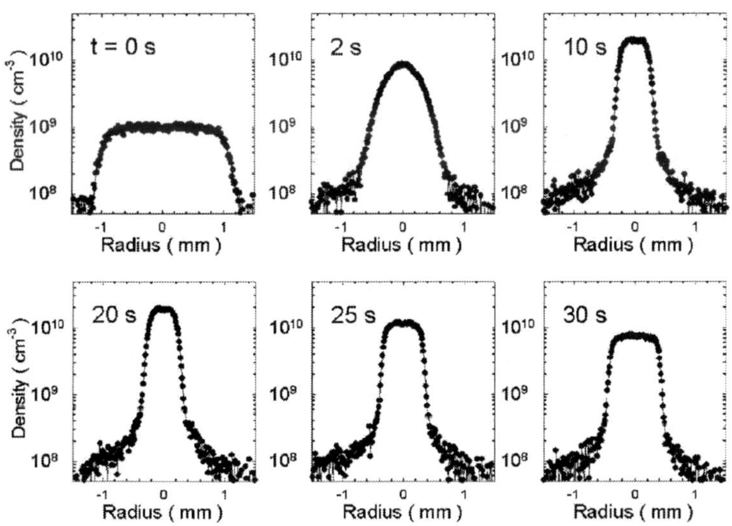

FIGURE 3. Radial profiles obtained by compression at 6 MHz beginning with the plasma at $t = 0$ s. The plasma is in steady-state compression from 10 to 20 s, then allowed to expand with the RW off. All profiles are close to thermal equilibrium with flat-top profiles, except at $t = 2$ s, where the plasma is much hotter, namely $T \sim 3$ eV. This data is reprinted from Ref. [29], where the experiment is described in more detail.

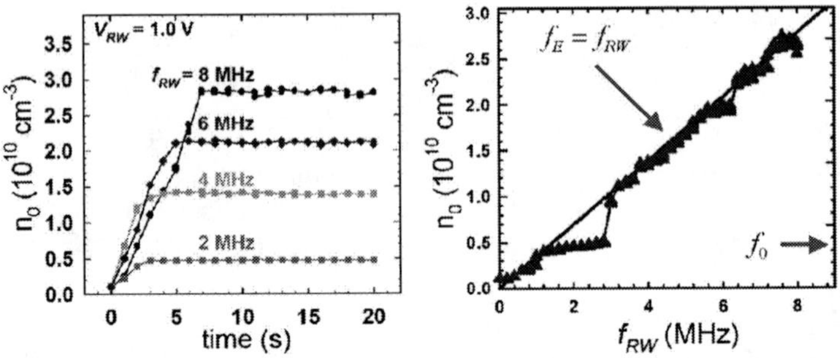

FIGURE 4. (left) central plasma density following application of the RW at various frequencies at V_{RW} = 1.0 V; and (right) steady-state density following the transition to the high-density state as a function of applied RW frequency. The step near n_0 = 0.5 x 10^{10} cm^{-3} is due to a so-called "zero-frequency" mode, which was key to understanding the bifurcation from low- to high-density steady states. Data from Ref. [29].

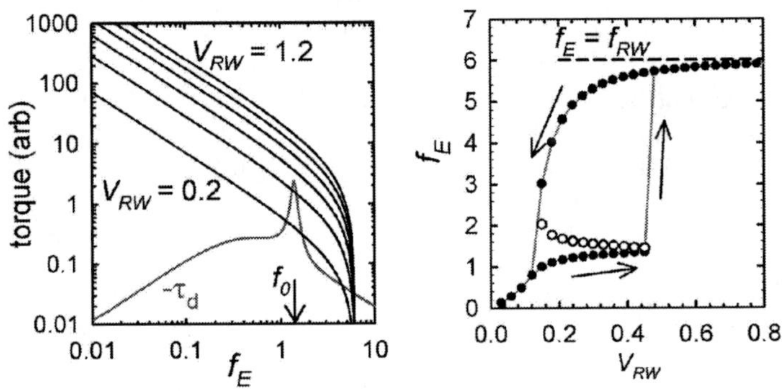

FIGURE 5. (left) the drag and drive torque from Eq. (1), shown at various drive amplitudes, V_{RW}; (right) the resulting torque-balanced steady states, illustrating the hysteretic behavior as a function of V_{RW}. All quantities are in arbitrary units. The predictions of the model are qualitatively in good agreement with the experimental results. This analysis is from Ref. [35]; see this reference for details.

in plasma density (i.e., the system exhibits considerable hysteresis). These predictions are in qualitative agreement with the experimental results, as is a similar hysteresis observed when f_{RW} is varied.

These results provide a physical model of access to the strong-drive regime and the high-density steady states that are achieved. We have made quantitative measurements of the RW torque on the plasma in the high-density state, with the hope that this will motivate a theory of torque in this state. The fact that the initial, final, and intermediate states are all close to thermal equilibrium provides reason to believe that a theory of the compression process might be possible. While admittedly speculative, the plasma evolution might be described, for example, as a series of thermodynamic equilibrium states with the rate-limiting process determined by one variable, such as the momentum relaxation time.

A key practical question now is what limits the compression and the maximum achievable density. Presently we do not know the answer. We are able to work with relative ease up to 8 or 9 MHz and, with some care, we have achieved states up to ~ 18 MHz. However, above 8 MHz, there are imbalances in the four legs of the RW drive circuit likely due to (extrinsic) resonances in the cables and electrodes. This is curable. On the other hand, a generic model of the RW drive predicts that the electric fields due to the RW are screened in the Debye sheath at the plasma edge by particles moving along magnetic field lines. For an electrode of length L, the time to screen such a field (and hence develop a density perturbation upon which the RW electric field acts to produce a torque) is of order $L/2\bar{v}$, where $\bar{v}$ is the thermal particle velocity. As a result, frequencies $f_{RW} \geq \bar{v}/\pi L$ would be expected to be less effective in driving the required torque. Whether this is a practical limitation awaits further study.

IV. EXTRACTION OF BEAMS WITH SMALL TRANSVERSE SPATIAL EXTENT

Specially tailored particle beams have historically found a wealth of applications in science and technology, and this is proving especially true in studies involving antimatter [1]. Applications include the formation of antihydrogen, study of electron-positron plasmas, the formation of Ps_2 and BEC Ps, the use of positrons to characterize materials, and study of the interaction of positrons and Ps atoms with ordinary matter. For applications in which spatial resolution and/or high-quality beam emittance are required, beams with a small energy spread and small transverse spatial extent are desirable. For a single-component plasma in a Penning-Malmberg trap, the space charge potential is largest on the axis of the plasma. Recently, we exploited this fact to create beams of small transverse spatial extent by carefully lowering, in a pulsed manner, one of the confining end-gate potentials, V_C [25, 26].

A schematic diagram of the experimental arrangement is shown in Fig. 6. Of interest is the radial profile of the extracted beam, and in particular, the possible minimum beam diameter and the maximum number of particles that can be extracted in a pulse at this diameter. The potential, V_C, was lowered to carefully predetermined values by applying a 10 μs square-wave pulse with amplitude ΔV. The particles were found to escape in a time ~ 5 μs. Shown in Fig. 7 is an example of the areal plasma

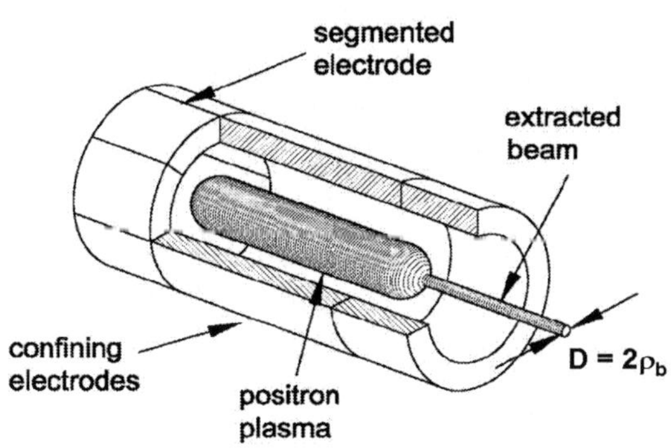

FIGURE 6. Experimental arrangement used to extract beams of small spatial extent from a single-component plasma by slowly lowering an end-gate potential in a Penning-Malmberg trap. The focus of the research described here is determining the minimum beam radius, ρ_b.

density profile before and after beam extraction and examples of radial profiles, $\sigma_b(r)$, for various extracted beams.

The results were analyzed by two methods. For the smallest-amplitude beams, we developed a simple analytic expression predicting Gaussian areal density profiles with a radius to 1/e of $\rho_b = 2\lambda_D$. For larger beams the expression for the width becomes [26]

$$\rho_b = 2\lambda_D\left[1+\xi\right]^{1/2}, \tag{2}$$

where

$$\xi = \frac{N_b}{N_0}\left(\frac{r_p}{2\lambda_D}\right)^2 = \frac{e\Delta\phi}{T} \tag{3}$$

is the critical parameter determining the growth of ρ_b with the amplitude of the extracted beam, with N_b/N_0 the fraction of the plasma extracted per pulse, and r_p is the plasma radius. As indicated in Eq. (3), $\xi = e\Delta\phi/T$, which is the ratio of the difference in space-charge potential across the beam, $\Delta\phi$, due to the N_b particles to the plasma temperature.

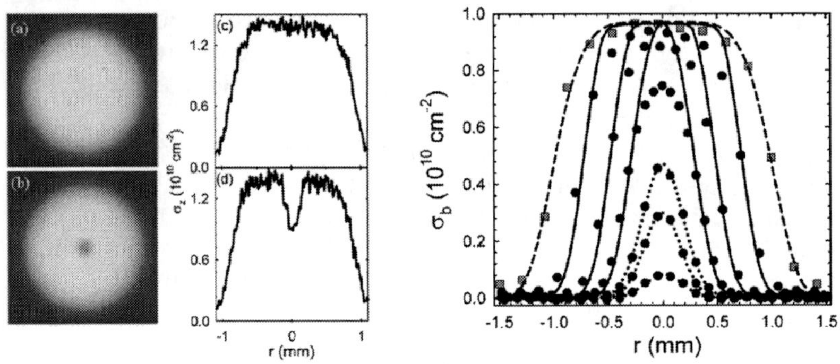

FIGURE 7. (left) CCD camera images of the areal density profiles (a) before; and (b) 10 μs after beam extraction, and in (c) and (d) the corresponding radial profiles, $\sigma_z(r)$; (right) profiles, $\sigma_b(r)$, of extracted beams for $\xi \approx 0.1, 0.3, 0.5, 1.0, 1.9, 2.8$. Shown for the three smallest beams are Gaussian fits (···), while the three largest beams are fit (–) to numerical solutions. The initial plasma profile, $\sigma_z(r)$, is also shown (■). Reprinted from Ref. [26].

When the factor $\xi \geq 0.5$, the profiles depart from the Gaussian predictions. In this case, the profiles were calculated numerically with the assumption that particles exit the plasma before scattering or radial diffusion occurs. As shown in Fig. 7, the profiles of small-amplitude beams fit well with the Gaussian predictions, and the profiles of larger amplitude beams are in reasonable agreement with the numerical calculations. As shown in Fig. 8, the predicted *beam widths* from Eqs. (2) and (3) are in good agreement with the data over the wide range of beam amplitudes studied.

A key result of these studies is that the beam diameter is limited to be $\geq 4\lambda_D$. Another important result is the identification of ξ as the parameter determining beam widths. Equations (2) and (3) set a limit on the fraction of beam particles that can be extracted at the minimum diameter of $4\lambda_D$. Physically, this is the condition that the areal density within the pulse must be smaller than the areal density of the parent plasma, namely $(N_b/2\lambda_D)^2 < N_0/(r_p)^2$. Equations (2) and (3) also quantify the beam widths for larger beam pulses.

In practical applications, one would like to convert a trapped plasma into a train of approximately identical pulses. Shown in Fig. 9 is an example where over 50% of a trapped plasma was converted to a train of 20 nearly identical pulses. This was

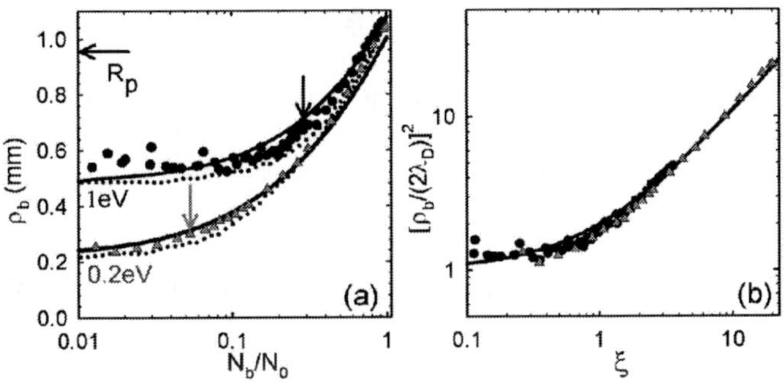

FIGURE 8. (a) Beam-width parameter, ρ_b, plotted vs. N_b/N_0 for T = 1.0 eV (●), and 0.2 eV (▲). Predictions (–) from Eqs. (2) and (3) with no fitted parameters, and (···) a numerical calculation from [26]. Arrows correspond to beams with ξ = 1. (b) Data from (a) plotted as $(\rho_b/2\lambda_D)^2$, vs. ξ, demonstrating the scaling predicted by Eqs. (2) and (3). Reprinted from Ref. [26].

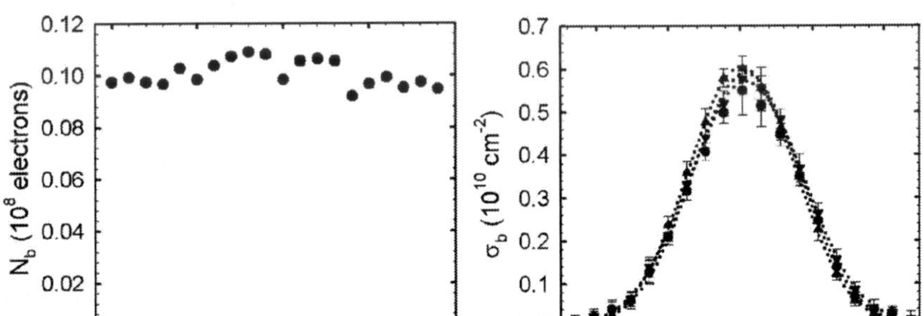

FIGURE 9. (left) Amplitudes, N_b, for 20 pulses extracted consecutively with $\langle N_b \rangle = 0.1 \pm 0.005 \times 10^8$. $\xi \approx 0.2$. (right) areal density profiles for the 1st (●), 10th (▲), and 20th (▼) extracted beams. $N_b \approx 0.1 \times 10^8$ for all beams. The beam width remains approximately constant, λ_D is maintained invariant during the extractions; Gaussian fits to the profiles are also shown. Reprinted from Ref. [26].

accomplished by maintaining the central plasma density constant throughout the extraction process using the RW. After the extraction, the "holes" left by the extracted pulses propagate coherently to the plasma edge in a time < 1 ms. This returns the plasma to a rigid rotor, thermal equilibrium state that would, in principle, allow pulse extractions at a kilohertz rate. Whether this is possible, given the fact that the plasma must be compressed with the RW to maintain constant n *and* λ_D, is an important topic for future research.

There are a number of potential applications of this technique. For example, there are situations in which an *electrostatic* as opposed to a magnetically guided beam is desirable. While the latter is most compatible with accumulation and storage in a Penning-Malmberg trap, the former admits to the use of electrostatic focusing and exploitation of ballistic trajectories in three dimensions. In Ref. [26], we provide an example of extracting a cold beam ($T \sim 10$ meV) from a plasma in a 5 tesla field. Assuming it is guided to a region where $B \sim 5$ gauss (i.e., 5×10^{-4} T) and then extracted from the field, this produces an electrostatic beam with a diameter of 1 mm and an energy spread of 10 meV (essentially all in the parallel direction). Such a beam would be a considerable advance for positron-atomic physics scattering experiments, and could potentially be used to form microbeams for materials analysis. For example, typical electrostatic beams used for atomic physics studies have energy spreads ≥ 200 meV and transverse extents ≥ 5 mm. Given the constraint on ξ (i.e., $\xi < 1$), the maximum number of particles per pulse for the 10 meV beam described here is $N_b \sim 4 \times 10^3$.

V. PROGRESS TOWARDS A MULTICELL TRAP

For a number of applications, one would like very large quantities of positrons – amounts that exceed the capabilities of current accumulation and storage techniques. In particular, typical cylindrical plasmas in Penning-Malmberg traps are space-charge limited at $\sim 10^9 \tilde{V}_C$ particles per centimeter of plasma length, where $\tilde{V}_C$ is the confining potential in kV [27]. Examples of potential applications requiring such large numbers of positrons include the study of electron-positron plasmas and the creation of a Bose-condensed gas of positronium atoms. Currently, new intense sources of positrons are planned or coming online at several laboratories throughout the world, and the ability to accumulate, cool and manipulate large amounts of positron plasma would facilitate efficient multiplexing of end-user experiments at these facilities.

We have begun development of a multicell Penning-Malmberg trap that circumvents this space-charge limit by arranging multiple Penning-Malmberg-trapped plasmas ("cells") in the same vacuum system and magnetic field, with these plasmas shielded from each other by copper electrodes [27, 28]. A schematic diagram of this trap is shown in Fig. 10. Positrons would first be accumulated using a buffer-gas trap, then transferred to a high-field trap (e.g., $B \sim 5 - 10$ T), where they would be cooled by cyclotron radiation in UHV. We propose to build a 100 cell trap with $\tilde{V}_C \sim 1 - 2$ kV which would accommodate $N > 10^{12}$ positrons in ~ 100 cells. A convenient arrangement is 5 banks of 19 cells in the direction perpendicular to $B \sim 5$ T.

Recently, we have made considerable progress in establishing the technical basis for such a trap [26]. The confinement of $N > 10^{10}$ electrons in a plasma 10 cm in length with $\tilde{V}_C = 1$ kV was demonstrated. The ability to operate two in-line cells simultaneously was also established. In addition, the multicell trap requires that a method be developed to move plasma efficiently across the magnetic field to fill off-axis cells with plasmas from the buffer-gas trap. This was accomplished using autoresonant excitation of the diocotron mode, wherein a plasma at the center of a cylindrical electrode is driven off axis by a frequency-swept signal applied to a segmented electrode. The plasma phase-locks to the drive signal, and so it can be deposited at a precise radial and azimuthal location (i.e., required to fill off-axis cells in a multicell trap). Data demonstrating this capability are shown in Fig. 11. Positioning accuracy is ~ ± 0.1 mm, and it was demonstrated that plasmas can be

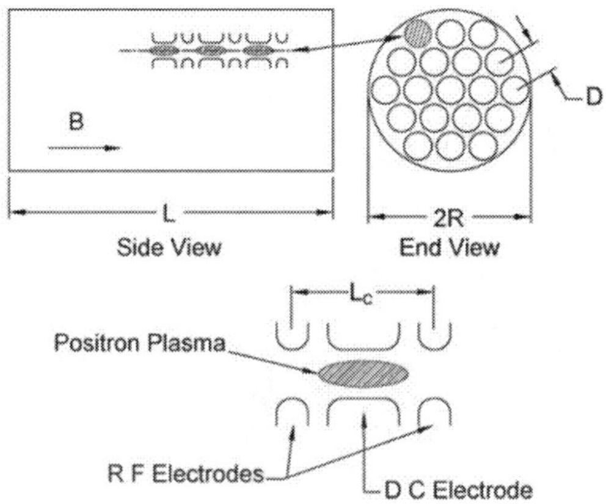

FIGURE 10. Conceptual design of a 95-cell trap showing the arrangement of cells parallel and perpendicular to B. The device consists of 19, hexagonally close-packed cells perpendicular to B and 5 in-line cells in the field direction.

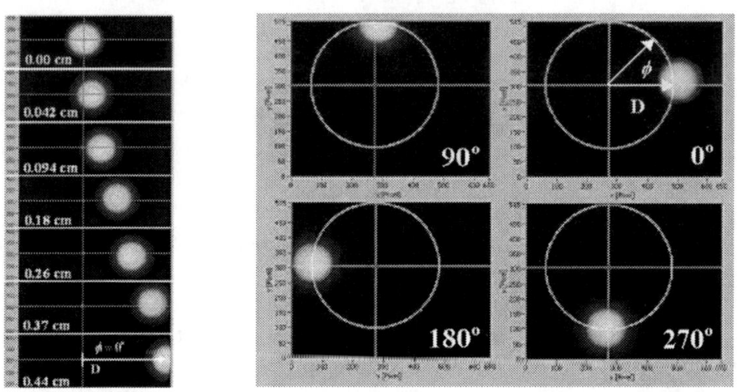

FIGURE 11. CCD camera images of displaced plasmas dumped on the phosphor screen using autoresonant, diocotron-mode excitation. (a) radial displacements at fixed azimuthal position; (b) azimuthal displacements at fixed radial position. From Ref. [28].

translated radially out to 80% of the electrode inner diameter. A schematic diagram showing the master cell for acceptance of plasmas from the buffer-gas trap and translation off axis and two banks of 19 storage cells is illustrated in Fig. 12.

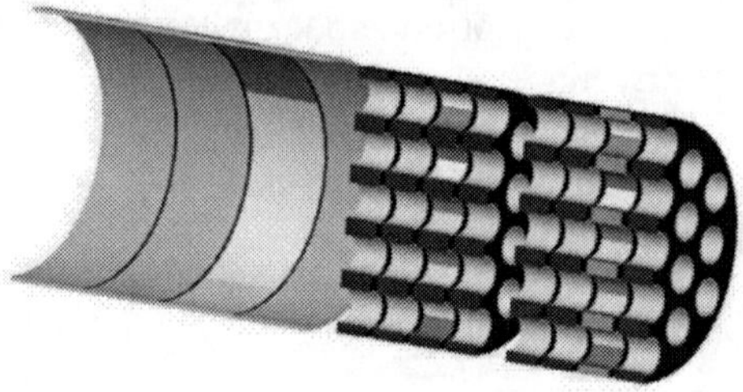

FIGURE 12. Cutaway view of the electrode structure design for a multicell trap: Large diameter electrodes form the master plasma-manipulation cell to receive plasmas from a buffer-gas trap and to displace them for deposition into off-axis cells. Also shown are two banks of 19 cells (38 cells total). In a 95 cell trap, there would be three more 19-cell banks at the right. From Ref. [28].

One remaining issue to be addressed is establishing that plasmas can be well confined (i.e., using RW compression to mitigate outward transport) in cells a significant distance off-axis and in regions of only modest magnetic field homogeneity (e.g., near the ends of the solenoid that provides the axial magnetic field). Experiments of this type are planned. We also plan experiments to establish the minimum electrode diameters that will be required for good confinement. This latter consideration will determine how compact such a device can be. Ultimately, one would like to have portable antimatter traps, but this will likely require advances in magnet technology to avoid the necessity of cryogenic cooling of the high-field magnet.

VI. CONCLUDING REMARKS

In this paper, we have summarized recent work on the development of nonneutral plasma-physics tools for antimatter studies. We have also highlighted important issues for future research. In the area of RW compression of plasmas in the strong-drive regime, it is now clear that this technique is robust and quite generally applicable to single component plasmas. It is particularly convenient for compression of positron-mass particles, where cooling can be accomplished under UHV conditions by cyclotron radiation. One outstanding issue is the density limit for this technique and, at present, this is an open question.

In the area of beam formation by pulsed extraction from the center of a single-component plasma, many of the questions have been answered, but other challenges remain. They include establishing beam extraction from cryogenic plasmas and demonstrating the ability to extract such beams from the magnetic field for further tailoring or use with electrostatic manipulation techniques. Finally, a number of the technical requirements for a practical, multicell positron trap have been established. One might envision construction of such a trap for $\sim 10^{12}$ particles in the next few years.

ACKNOWLEDGMENTS

We wish to acknowledge the collaboration of T. M. O'Neil and M. W. Anderson in developing a theoretical model of RW compression in the strong-drive regime and the expert technical assistance of E. A. Jerzewski. This work was supported by the U. S. National Science Foundation, grant PHY 07-13958.

REFERENCES

1. C. M. Surko and R. G. Greaves, *Phys. Plasmas* **11**, 2333-2348 (2004).
2. M. Amoretti, C. Amsler, G. Bonomi, A. Bouchta, P. Bowe, C. Carraro, C. L. Cesar, M. Charlton, M. Collier, M. Doser, V. Filippini, K. Fine, A. Fontana, M. Fujiwara, R. Funakoshi, P. Genova, J. Hangst, R. Hayano, M. Holzscheiter, L. Jorgensen, V. Lagomarsino, R. Landua, D. Lindelof, E. L. Rizzini, M. Macri, N. Madsen, G. Munuzio, M. Marchesotti, P. Montagna, H. Pruys, C. Regenfus, P. Riedler, J. Rochet, A. Rotondi, G. Rouleau, G. Testera, A. Variola, T. Watson, and D. VanderWerf, *Nature* **419**, 456-459 (2002).
3. G. Gabrielse, N. Bowden, P. Oxley, A. Speck, C. Storry, J. Tan, M. Wessels, D. Grzonka, W. Oelert, G. Schepers, T. Sefzick, J. Walz, H. Pittner, T. Hansch, and E. Hessels, *Phys. Rev. Lett.* **89**, 213401 1-4 (2002).
4. G. Gabrielse, N. Bowden, P. Oxley, A. Speck, C. Storry, J. Tan, M. Wessels, D. Grzonka, W. Oelert, G. Schepers, T. Sefzick, J. Walz, H. Pittner, T. Hansch, and E. Hessels, *Phys. Rev. Lett.* **89**, 233401 1-5 (2002).
5. J. Ellis, *Nature* **424**, 631-634 (2003).
6. R. G. Greaves and C. M. Surko, *Phys. Plasmas* **4**, 1528-1543 (1997).
7. S. J. Gilbert, D. H. E. Dubin, R. G. Greaves, and C. M. Surko, *Phys. Plasmas* **8**, 4982-4994 (2001).
8. T. S. Pederson, A. H. Boozer, W. Dorland, J. P. Kremer, and R. Schmitt, *J. Phys. B: At. Mol. Opt.* **36**, 1029-1039 (2003).
9. M. Charlton and J. W. Humberston, *Positron Physics* (Cambridge Univ. Press, Cambridge, U. K., 2001).
10. C. M. Surko, *New Directions in Antimatter Chemistry and Physics,* C. M. Surko and F. A. Gianturco, eds. (Kluwer Scientific Publishers, the Netherlands, 2001).
11. K. Iwata, R. G. Greaves, and C. M. Surko, *Can. J. Phys.* **51**, 407-410 (1996).
12. N. Guessoum, P. Jean, and W. Gillard, *Appl. Surf. Sci.* **252**, 3352 - 3361 (2006).
13. J. N. Sun, D. W. Gidley, Y. F. Hu, W. E. Frieze, and S. Yang, *Mat. Res. Soc. Symp. Proceedings* **726**, Q10.5 (2002).
14. B. Barbiellini, M. Weber, K. Lynn, P. Sterne, and A. Denison, *Mat. Res. Soc. Proceedings* **704** (2002).
15. C. Hugenschmidt, K. Schreckenbach, M. Stadlbauer, and B. Straßer, *Appl. Surf. Sci.* **252**, 3098-3105 (2006).
16. R. Krause-Rehberg, S. Sachert, G. Brauer, A. Rogov, and K. Noack, *Appl. Surf. Sci.* **252**, 3106-3110 (2006).
17. H. M. Chen, Y. C. Jean, C. D. Jonah, S. Chemerisov, A. F. Wagner, D. M. Schrader, and A. W. Hunt, *Appl. Surf. Sci.* **252**, 3159-3165 (2006).
18. A. G. Hathaway, M. Skalsey, W. E. Frieze, R. S. Vallery, and D. W. Gidley, *Nucl. Instrum. and Meth. in Phys. Res. A* **579**, 538-541 (2007).
19. D. B. Cassidy, S. H. M. Deng, R. G. Greaves, and A. P. Mills, Jr., *Rev. Sci. Instrum.* **77**, 073106-08 (2006).
20. D. B. Cassidy and A. P. Mills, Jr., *Nature* **449**, 195-197 (2007).
21. P. M. Platzman and A. P. Mills, Jr., *Phys. Rev. B* **49**, 454-458 (1994).
22. A. P. Mills, Jr., *Nucl. Instrum. Methods B* **192**, 107-116 (2002).
23. T. S. Pedersen, A. H. Boozer, J. P. Kremer, and R. G. Lefrancois, *Fusion Sci. and Tech.* **46**, 200-208 (2003).
24. J. R. Danielson and C. M. Surko, *Phys. Rev. Lett.* **95**, 035001-035004 (2005).
25. J. R. Danielson, T. R. Weber, and C. M. Surko, *Appl. Phys. Lett.* **90**, 081501 1-3 (2007).
26. T. R. Weber, J. R. Danielson, and C. M. Surko, *Phys. Plasmas* **15**, 012106 1-10 (2008).
27. C. M. Surko and R. G. Greaves, *Rad. Chem. and Phys.* **68**, 419 (2003).
28. J. R. Danielson, T. R. Weber, and C. M. Surko, *Phys. Plasmas* **13**, 123502 1-10 (2006).
29. J. R. Danielson and C. M. Surko, *Phys. Plasmas* **13**, 055706 1-10 (2006).
30. D. L. Eggleston, C. F. Driscoll, B. R. Beck, A. W. Hyatt, and J. H. Malmberg, *Phys. Fluids B* **4**, 3432-3439 (1992).
31. E. M. Hollmann, F. Anderegg, and C. F. Driscoll, *Phys. Plasmas* **7**, 2776-2789 (2000).
32. X. P. Huang, J. J. Bollinger, T. B. Mitchell, and W. M. Itano, *Phys. Rev. Lett.* **80**, 73-76 (1998).
33. R. G. Greaves and C. M. Surko, *Phys. Rev. Lett.* **85**, 1883-1886 (2000).
34. L. V. Jorgensen, M. Amoretti, G. Bonomi, P. D. Bowe, C. Canali, C. Carraro, C. L. Cesar, M. Charlton, and M. Doser, *Phys. Rev. Lett.* **95**, 025002 1-4 (2005).
35. J. R. Danielson, C. M. Surko, and T. M. O'Neil, *Phys. Rev. Lett.* **99**, 135005 1-4 (2007).

Antiproton compression and radial measurements

G.B. Andresen*, W. Bertsche†, P.D. Bowe*, C.C. Bray**, E. Butler†, C.L. Cesar‡, S. Chapman**, M. Charlton†, J. Fajans**, M.C. Fujiwara§, R. Funakoshi¶, D.R. Gill§, J.S. Hangst*, W.N. Hardy||, R.S. Hayano¶, M.E. Hayden††, A.J. Humphries†, R. Hydomako‡‡, M.J. Jenkins†, L.V. Jørgensen†, L. Kurchaninov§, R. Lambo‡, N. Madsen†, P. Nolan§§, K. Olchanski§, A. Olin§, R.D. Page§§, A. Povilus**, P. Pusa§§, F. Robicheaux¶¶, E. Sarid***, S. Seif El Nasr||, D.M. Silveira‡, J.W. Storey§, R.I. Thompson‡‡, D.P. van der Werf†, J.S. Wurtele** and Y. Yamazaki†††

**Department of Physics and Astronomy, Aarhus University, DK-8000 Aarhus C, Denmark*
†Department of Physics, Swansea University, Swansea SA2 8PP, United Kingdom
***Department of Physics, University of California at Berkeley, Berkeley, CA 94720-7300, USA*
‡Instituto de Física, Universidade Federal do Rio de Janeiro, Rio de Janeiro 21941-972, Brazil
§TRIUMF, 4004 Wesbrook Mall, Vancouver BC, Canada V6T 2A3
¶Department of Physics, University of Tokyo, Tokyo 113-0033, Japan
||Department of Physics and Astronomy, University of British Columbia, Vancouver BC, Canada V6T 1Z4
††Department of Physics, Simon Fraser University, Burnaby BC, Canada V5A 1S6
‡‡Department of Physics and Astronomy, University of Calgary, Calgary AB, Canada T2N 1N4
§§Department of Physics, University of Liverpool, Liverpool L69 7ZE, United Kingdom
¶¶Department of Physics, Auburn University, Auburn, AL 36849-5311, USA
****Department of Physics, NRCN-Nuclear Research Center Negev, Beer Sheva, IL-84190, Israel*
†††Atomic Physics Laboratory, RIKEN, Saitama 351-0198, Japan

Abstract. Control of the radial profile of trapped antiproton clouds is critical to trapping antihydrogen. We report detailed measurements of the radial manipulation of antiproton clouds, including areal density compressions by factors as large as ten, achieved by manipulating spatially overlapped electron plasmas. We show detailed measurements of the near-axis antiproton radial profile, and its relation to that of the electron plasma. We also measure the outer radial profile by ejecting antiprotons to the trap wall using an octupole magnet.

Keywords: Antihydrogen, Antiprotons, Rotating Wall, Radial Profile
PACS: 36.10.Dk, 34.80.Lx, 52.20.Hv

INTRODUCTION

The current generation of antihydrogen experiments [1, 2] aims to trap $\overline{\mathrm{H}}$ atoms. Neutral $\overline{\mathrm{H}}$ atoms have a small permanent magnetic moment, and can be trapped in a magnetic minimum. Traps based on this effect are called Minimum-B traps [3]. To trap both charged and neutral species simultaneously, Minimum-B and Penning-Malmberg traps must be co-located. The compatibility of Minimum-B and Penning-Malmberg traps remains controversial [4, 1, 2], but it is clear that the two are most compatible if the

CP1037, *Cold Antimatter Plasmas and Application to Fundamental Physics,* edited by Y. Kanai and Y. Yamazaki

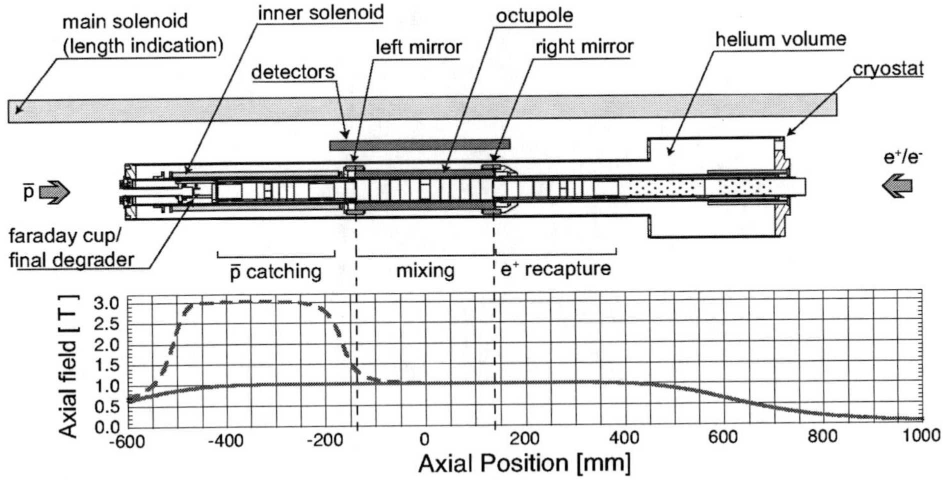

FIGURE 1. Schematic diagram of the ALPHA apparatus. A moveable probe on the right alternately inserts an electron gun and a MCP/Phosphor assembly. The graph below the schematic plots the axial magnetic field in the trap; the dashed line includes the field of the inner solenoid, while the solid line is the field without the inner solenoid.

$\bar{p}$'s and e^+'s are held close to the trap axis where the perturbations from the Minimum-B fields are smallest [5, 6, 7]. This minimizes fast (due to ballistic processes) and slow (due to diffusion) losses from the trap. Minimizing the slow losses also reduces the temperature of the $\bar{p}$'s, as slow charged particle expansion converts self-electrostatic energy into kinetic energy. Keeping the $\bar{p}$'s close to the axis also helps make the $\bar{H}$ that is synthesized more trapable. It reduces the islanding [6] which occurs as $\bar{p}$'s bounce back and forth, and can disrupt the collisional cooling of the $\bar{p}$'s on electron plasmas, thereby keeping the $\bar{H}$'s warm. It slows the $\mathbf{E} \times \mathbf{B}$ drifts of the $\bar{p}$'s which increase the kinetic energy of the $\bar{H}$'s that form from the $\bar{p}$'s, and can make the $\bar{H}$'s impossible to confine in the very shallow Minimum-B traps. Finally, it assures that any $\bar{H}$ that is synthesized is created very near the trap's minimum magnetic field, thereby retaining the maximum possible trap depth.

Successful $\bar{p}$ compression has been briefly reported elsewhere [8, 9, 10]; here we present carefully controlled and quantitative characterization of the process [11], as well as accurate measurements of the near-axis radial distribution of the $\bar{p}$'s. We have demonstrated areal density increases by as much as a factor of ten, and have produced $\bar{p}$ clouds with radii as small as 0.29 ± 0.06 mm. Clouds of this size are far from the loss limits [5] of our combined trap [1], and promise to be much easier to confine.

A schematic drawing of our apparatus is shown in Fig 1. The experiment synthesizes $\bar{H}$ using the well-known procedures pioneered by the ATHENA [12] and ATRAP [13] collaborations. Of special note here is the electron plasma used to cool the antiprotons [14] in the "$\bar{p}$ catching" section of the trap, and the octupole in the "mixing" (Minimum-B) section of the trap.

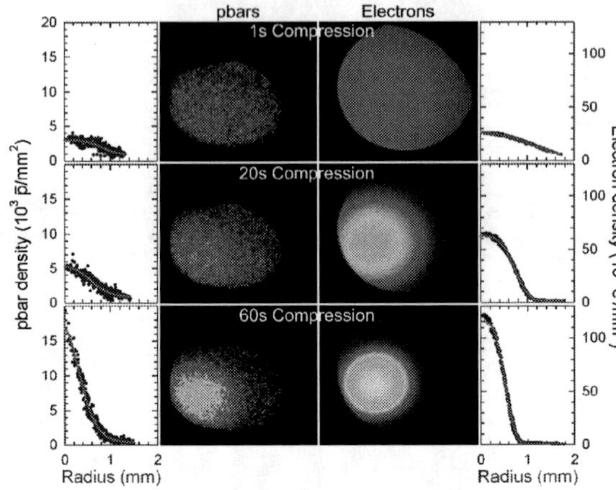

FIGURE 2. $\bar{p}$ and e^- images showing the effects of compression, and the resulting radial profiles. The red lines are Gaussian-like (i.e. $\exp(-|r/r_0|^k)$, where $k \approx 2$) fits to the radial profiles. The sharp boundaries are caused by apertures.

IMAGING

We image (see Fig. 2) the e^-'s and $\bar{p}$'s by allowing them to escape along the magnetic field lines onto a microchannel plate (MCP)/phosphor screen assembly. We capture the resultant images with a CCD camera [15]. Some typical images are shown in Fig. 2. As the MCP is in a B-field of 0.024T, which is much lower than the trap field of 3T, the cross-sectional area covered by the particles expands by a factor of 125 as they follow the field lines to the MCP. The details of the expansion process, as well as the brightness calibration of the image, will be described in a forthcoming paper.

ROTATING WALL

The radial size of the electron plasma used to cool the incoming $\bar{p}$'s can be controlled by applying a rotating electrostatic potential to the plasma [16, 17]. By varying the size of this plasma we find, as expected, that the size of the resulting $\bar{p}$ cloud scales with the size of the e^- plasma (see Fig. 3). Moreover the fraction of the trapped $\bar{p}$'s that are cooled (the cooling efficiency) also increases with the size of the e^- plasma [9]. This result is consistent with the model that $\bar{p}$'s do not make large radial excursions, and cool only if they traverse regions of high e^- density.

COMPRESSION

As shown in Fig. 3, we have to use a fairly large radius e^- plasma to cool the majority of the initially captured $\bar{p}$'s. Such a large radius plasma is fatter than we would like for injecting into our octupole magnet. Consequently, we would prefer to start with a large

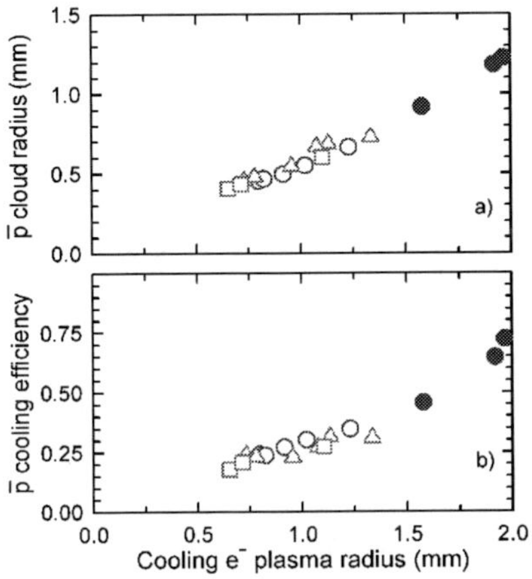

FIGURE 3. a) The cooled $\bar{p}$ cloud radius as a function of the radius of the e$^-$ plasma. The various symbols correspond to trials with differing total electron numbers (100-165M). b) The cooling efficiency (see text) as a function of the e$^-$ plasma radius.

radius e$^-$ plasma to maximize cooling, and compress the $\bar{p}$'s after cooling to optimize injection. We can accomplish this by compressing a mixed e$^-$–$\bar{p}$ plasma with a rotating electrostatic potential. Images of typical results are shown in Fig. 2. We observe (Figs. 4 and 5) that, so long as we compress slowly, the radius of the $\bar{p}$'s tracks that of the e$^-$'s. We have observed compressions as large as a factor of ten; the data series in Fig. 4 shows compression by a factor of five. We do not get compression for a pure $\bar{p}$ plasma. This is likely because the rotating wall technique is presumed to require a cooling mechanism; no such mechanism exists for a pure $\bar{p}$ plasma. We note, however, that the ASACUSA group does obtain compression of some of their $\bar{p}$'s in pure $\bar{p}$ plasmas [8, 9, 10]. Their results are hard to reconcile with ours.

OUTER RADIAL PROFILE MEASUREMENTS

Our MCP diagnostic only images $\bar{p}$'s that are near the trap axis. While the $\bar{p}$ clouds that are best for synthesizing and trapping $\bar{H}$ lie in the imaged region, it is very useful, for diagnostic purposes, to be able to determine the profile of $\bar{p}$ clouds that extend well outside the imaged region. We can find the profile of such clouds using our octupole magnet [18]. To understand how this diagnostic works, it is helpful to visualize the field lines generated by the solenoid and octupole coils. The field lines originating from a circular locus of points in the plane transverse to the solenoid axis form a four-fluted cylindrical surface; the flutes at each end are rotated by 45° with respect to each other (Fig. 6). Figure 7 shows an image of one quadrant of the field lines, generated by passing e$^-$'s through the octupole and onto our MCP/Phosphor screen [11].

Antiprotons confined by the electrostatic well within the octupole bounce back and forth while following the magnetic field lines. Antiprotons on field lines that extend to

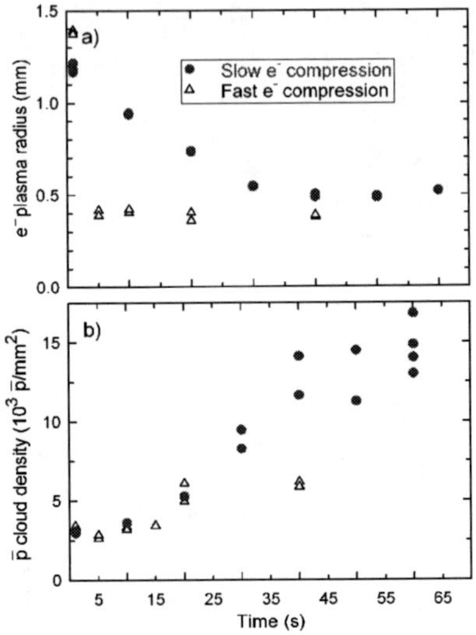

FIGURE 4. The a) e^- radius and b) $\bar{p}$ density as a function of time, for fast and slow compression. Note that the $\bar{p}$ density does not track the e^- compression if the latter is fast.

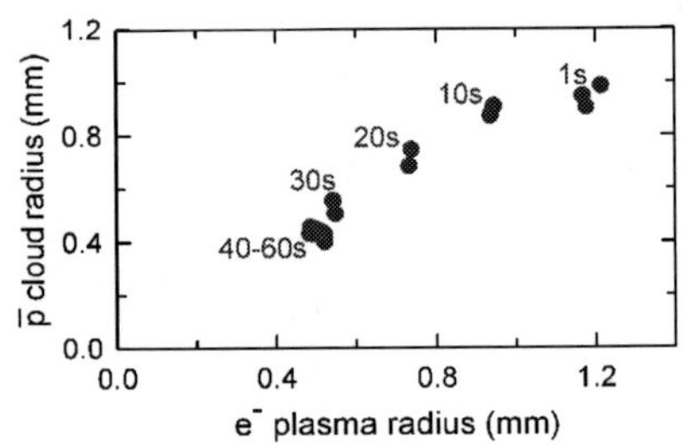

FIGURE 5. The $\bar{p}$ cloud radius as a function of the e^- plasma radius for the slow compression data shown in Fig. 4.

the physical trap wall at R_w before reaching one of the electrostatic walls will follow them there and annihilate. For a given end-to-end bounce length L, field lines lying outside of a critical radius r_c at the trap center will hit the wall, while those lying inside the critical radius will not. The normalized critical radius is [5, 19]:

$$\frac{r_c}{R_w} = \frac{1}{\sqrt{1 + \frac{B_w}{B_z}\frac{L}{R_w}}}, \tag{1}$$

where B_w is the octupole strength at the wall, and B_z is the solenoid field strength. The longer the trap, and the stronger the octupole field, the smaller the critical radius. The normalized critical radius is never smaller than about 7 mm for our magnets and wells

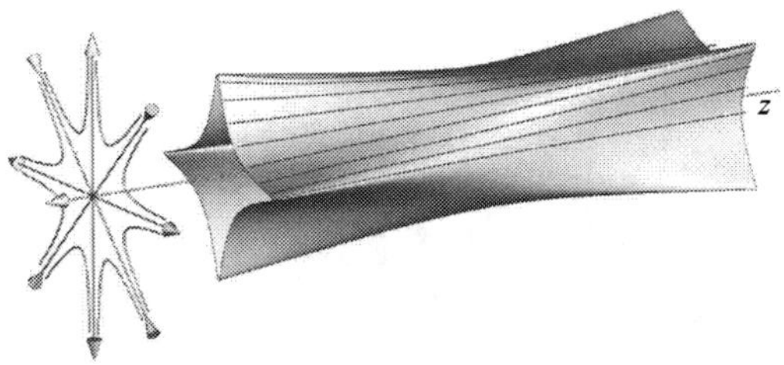

FIGURE 6. Magnetic field generated by the octupole and solenoid coils. The vectors on the left represent the directions of the axially-invariant field from these coils. The surface is created by following the field lines from a radially centered circular locus of points; the lines shown within the surface are field lines.

because the octupole field, which scales as r^3/R_w^3, is very weak near the trap axis relative to its strength at the wall.

The ballistic loss of particles on trap walls in the presence of a multipole field was first identified with electrons in a quadrupole magnet [4]. This process is easier to study with $\bar{\text{p}}$'s than with e^-'s, however, because individual $\bar{\text{p}}$ annihilations can be detected and localized on the trap wall with a position sensitive detector. The detector [20] comprises three layers of silicon cylindrically arrayed around the trap axis just outside of the octupole magnet (see Fig. 1). It is not yet fully deployed, but, using a partial system consisting of 10% of the full system, we observe (Fig. 8) that $\bar{\text{p}}$'s hit the wall at the ends of the electrostatic well. We expect to observe this type of loss pattern, as it is at the ends of the trap that the accessible field lines extend furthest outward; we note, however, that annihilations tend to occur at the ends of the electrostatic well even in the absence of an octupole field [21].

For the experiments reported here, annihilations were detected by scintillators coupled to Avalanche Photo Diodes (APDs). As with the silicon detector, the scintillators are cylindrically arrayed around the trap axis just outside of the octupole magnet. To measure the size of a $\bar{\text{p}}$ cloud, we first transfer it into an electrostatic well in the octupole field region; the octupole field is turned off during the transfer. We then measure the $\bar{\text{p}}$'s

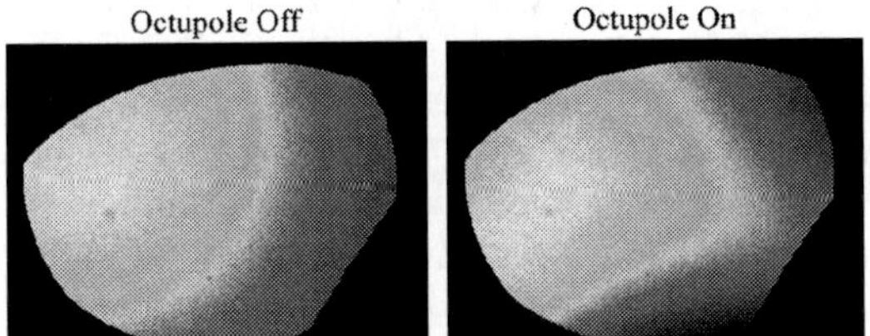

FIGURE 7. Field lines imaged by passing a circular e^- plasma through the octupole with the octupole off and on. Apertures [11] form the image boundaries and limit us to viewing only one quadrant of the octupole field map. The distortion evident in the right-hand image corresponds to one of the flutes at the end of the magnetic surface shown in Fig. 6.

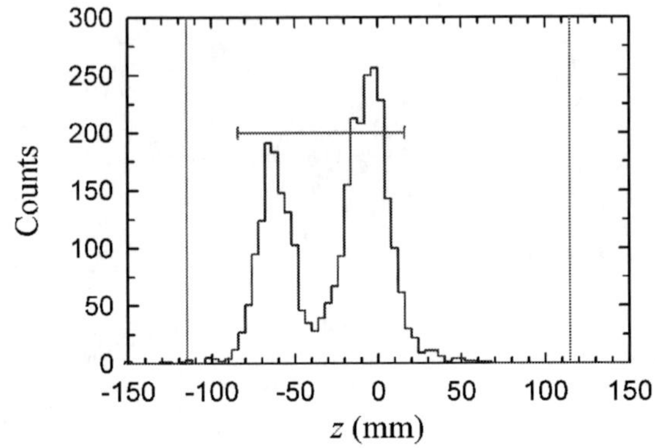

FIGURE 8. Axial positions where the $\bar{p}$'s hit the trap wall under the influence of the octupole. The horizontal bar indicates the the axial extent and position of the electrostatic well confining the $\bar{p}$'s. The loss is greatest near the ends of the confining electrostatic well. The positions are determined by a position-sensitive particle detector which monitors the $\bar{p}$ annihilation products; the vertical bars indicate the axial extent of the detector.

kinetic energy by monitoring the rate at which the $\bar{p}$'s escape as we slowly lower one endwall of the electrostatic well [22]. Typically we find that the energy is between 1 and 15 eV; the energy depends on the details of the transfer process and the electrostatic well potentials. This measurement is destructive, but since the energy is largely set by the electrostatics, not by the $\bar{p}$ radial profile, it is sufficient to measure this energy once for a series of profile measurements. From this energy, we determine the bounce length L of the $\bar{p}$'s in the electrostatic well. The uncertainty (and spread) of the $\bar{p}$ energy sets the uncertainty in the orbit lengths. Finally, for each $\bar{p}$ cloud that we want to analyze, we slowly ramp up the octupole field B_w while monitoring the losses. From the time history of the losses, we can invert Eq. (1) to reconstruct the radial distribution of $\bar{p}$'s.

We have verified that, for identically prepared $\bar{p}$ clouds, the reconstructed profile is independent of the electrostatic well shape and length, and the octupole ramp time [18]. The data for the latter test are shown in Fig. 9.

We have used our new diagnostic to characterize our $\bar{p}$ manipulation sequences, and to study interesting physics issues. An example is shown in Fig. 10, where we show radial profiles for a mixed e^+-$\bar{p}$ plasma. As the density of the e^+ plasma is increased, $\bar{p}$'s appear to be transported outward. The interpretation of these results is complicated by cooling of the $\bar{p}$'s on the e^+'s. Cooling will cause some charge to appear at falsely low radii, and this very likely causes us to underestimate the outward movement of the $\bar{p}$'s.

A possible explanation of the outward movement shown in Fig. 10 is that it is the result of the formation of highly excited $\overline{\text{H}}$ that is either 1) ionized at the radial edge of the e^+ plasma by its self consistent electric field, where the field is strongest, or 2) ionized by the vacuum electrostatic well fields. Note that the $\bar{p}$'s from $\overline{\text{H}}$ that was ionized within the e^+ plasma radius would have the opportunity to recombine into $\overline{\text{H}}$ again, while those at larger radii would orbit unperturbed. With time, the $\bar{p}$'s remaining in the e^+ plasma would be swept out to larger radii. Unpublished simulations of realistic

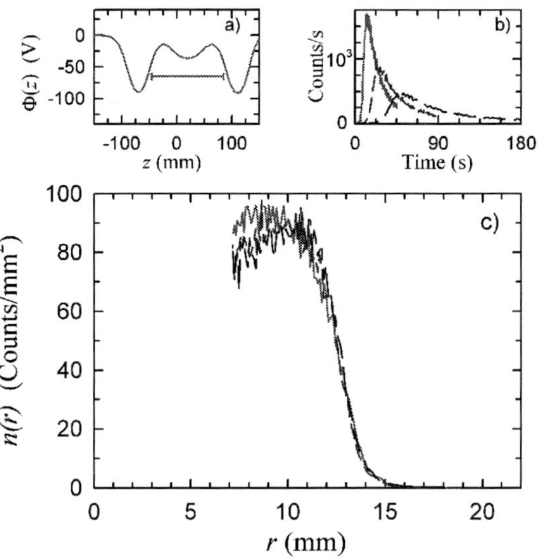

FIGURE 9. Comparison of the radial profiles with octupole ramps of 45 (green solid), 90 (red short-dash), and 180 s (blue long-dash). Panel a) shows the electrostatic well potential $\Phi(z)$; the horizontal bars indicate the 130 ± 5 mm axial extent and position of the $\overline{\mathrm{p}}$ orbits before the application of the octupole field. Panel b) shows the time history of the $\overline{\mathrm{p}}$ annihilations as the octupole field is ramped up. Panel c) shows the resulting radial profiles. The maximum B_w at the end of the ramp was 1.54 T, and $B_z = 1.03$ T.

antihydrogen formation/field ionization cycles, using the code described in [23], found similar transport. We do not yet have any other direct experimental evidence that this cycling occurs.

CONCLUSIONS

In this paper, we report detailed measurements of trapped $\overline{\mathrm{p}}$ radial compression [11]. We can compress the $\overline{\mathrm{p}}$ density by a factor of ten and decrease their radii to 0.29 mm. These clouds are 10-20 times smaller in radius than the clouds reported by ATHENA [21] and ATRAP [24]. Control of the radial profile of the $\overline{\mathrm{p}}$'s is critical to their survival in a Minimum-B trap. We have also studied the effect of the e^- plasma radius on the cooling of hot $\overline{\mathrm{p}}$'s. Finally, we have developed [11, 18] diagnostics that give the detailed radial profile of $\overline{\mathrm{p}}$'s near the trap axis, and from about 7 mm to the trap wall.

ACKNOWLEDGMENTS

This work was supported by CNPq, FINEP (Brazil), ISF (Israel), MEXT (Japan), FNU (Denmark), NSERC, NRC/TRIUMF (Canada), DOE (USA), EPSRC and the Leverhulme Trust (UK) and HELEN/ALFA-EC.

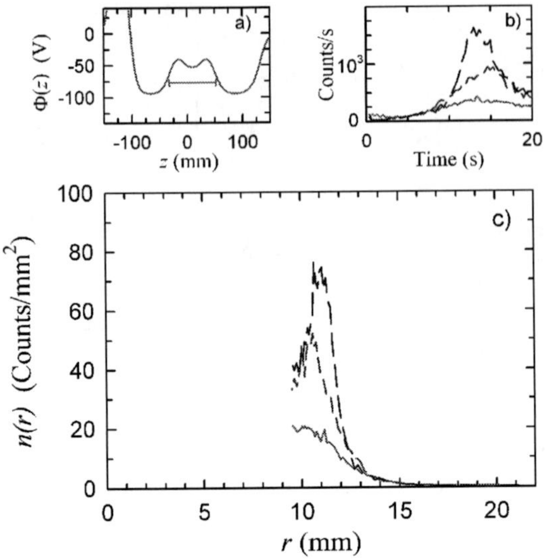

FIGURE 10. Comparison of the $\overline{\mathrm{p}}$ radial profile with different density positron plasmas. The green, solid curve shows the profile with no e^+, and the red short-dash and blue long-dash curves show the profile with 13 million and 25 million e^+ respectively. The well length was 85 ± 5 mm, the maximum field was 1.20 T, and the ramp time was 20 s. The graph descriptions and all other parameters are the same as in Fig. 9.

REFERENCES

1. G. Andresen, W. Bertsche, A. Boston, P. D. Bowe, C. L. Cesar, S. Chapman, M. Charlton, M. Chartier, A. Deutsch, J. Fajans, M. C. Fujiwara, R. Funakoshi, D. R. Gill, K.Gomberoff, J. S. Hangst, R. S. Hayano, R. Hydomako, M. J. Jenkins, L. V. Jørgensen, L. Kurchaninov, N. Madsen, P. Nolan, K. Olchanski, A. Olin, A. Povilus, F. Robicheaux, E. Sarid, D. M. Silveira, J. W. Storey, H. H. Telle, R. I. Thompson, D. P. van der Werf, J. S. Wurtele, and Y. Yamazaki, *Phys. Rev. Lett.* **98**, 023402 (2007).
2. G. Gabrielse, P. Larochelle, D. L. Sage, B. Levitt, W. S. Kolthammer, I. Kuljanishvili, R. McConnell, J. Wrubel, F. M. Esser, H. Glückler, D. Grzonka, G. Hansen, S. Martin, W. Oelert, J. Schillings, M. Schmitt, T. Sefzick, H. Soltner, Z. Zhang, D. Comeau, M. C. George, E. A. Hessels, C. H. Storry, M. Weel, A. Speck, F. Nillius, J. Walz, and T. W. Hänsch, *Phys. Rev. Lett.* **98**, 113002 (2007).
3. D. E. Pritchard, *Phys. Rev. Lett.* **51**, 1336 (1983).
4. J. Fajans, W. Bertsche, K. Burke, S. F. Chapman, and D. P. van der Werf, *Phys. Rev. Lett.* **95**, 155001 (2005).
5. J. Fajans, W. Bertsche, K. Burke, A. Deutsch, S. F. Chapman, K. Gomberoff, D. P. van der Werf, and J. S. Wurtele, "Simple loss scaling laws for quadrupoles and higher-order multipoles used in antihydrogen traps," in *Non-Neutral Plasma Physics VI: Workshop on Non-Neutral Plasmas*, edited by M. Drewsen, U. Uggerhøj, and H. Knudsen, AIP, Melville, N.Y., 2006, vol. 862, p. 176.
6. J. Zhang, C. L. Taylor, and F. Robicheaux, *J. Phys. B* **40**, 1019 (2007).
7. J. Fajans, and A. Schmidt, *Nucl. Instr. Meth. Phys. Res. A* **521**, 318 (2004).
8. Y. Yamazaki, *Hyperfine Interactions* **138**, 141 (2001).
9. N. Kuroda, H. A. Torii, M. Shibata, Y. Nagata, D. Barna, D. Horváth, M. Hori, J. Eades, A. Mohri, K. Komaki, and Y. Yamazaki, "Control of plasmas for production of ultraslow antiproton beams," in *Physics with Ultra Slow Antiproton Beams*, edited by Y. Yamazaki, and M. Wada, American Institute

of Physics, Wako, Japan, 2005, vol. CP793, p. 307.
10. H. A. Torii, N. Kuroda, M. Shibata, Y. Nagata, D. Barna, M. Hori, A. Mohri, K. Komaki, and Y. Yamazaki, "Extraction of ultra-slow antiproton beams and their physics application," in *Low Energy Antiproton Physics-LEAP '05*, edited by D. Grzonka, R. Czyzykiewicz, W. Oelert, T. Rozek, and P. Winter, Amer. Inst. of Physics, 2005, vol. CP796, p. 413.
11. G. Andresen, W. Bertsche, P. D. Bowe, C. C. Bray, E. Butler, C. L. Cesar, S. Chapman, M. Charlton, J. Fajans, M. C. Fujiwara, R. Funakoshi, D. R. Gill, J. S. Hangst, W. N. Hardy, R. S. Hayano, M. E. Hayden, R. Hydomako, M. J. Jenkins, L. V. Jørgensen, L. Kurchaninov, R. Lambo, N. Madsen, P. Nolan, K. Olchanski, A. Olin, A. Povilus, P. Pusa, F. Robicheaux, E. Sarid, S. S. E. Nasr, D. M. Silveira, J. W. Storey, R. I. Thompson, D. P. van der Werf, J. S. Wurtele, and Y. Yamazaki, "Compression of antiproton clouds for antihydrogen trapping" (2008), submitted to Phys. Rev. Lett.
12. M. Amoretti, C. Amsler, G. Bonomi, A. Bouchta, P. Bowe, C. Carraro, C. Cesar, M. Charlton, M. J. T. Collier, M. Doser, V. Filippini, K. S. Fine, A. Fontana, M. C. Fujiwara, R. Funakoshi, P. Genova, J. S. Hangst, R. S. Hayano, M. H. Holzscheiter, L. V. Jørgensen, V. Lagomarsino, R. Landua, D. Lindelöf, E. Lodi-Rizzini, M. Macrì, N. Madsen, G. Manuzio, M. Marchesotti, P. Montagna, H. Pruys, C. Regenfus, P. Riedler, J. Rochet, A. Rotondi, G. Rouleau, G. Testera, A. Variola, T. L. Watson, and D. P. van der Werf, *Nature* **419**, 456 (2002).
13. G. Gabrielse, N. S. Bowden, P. Oxley, A. Speck, C. H. Storry, J. N. Tan, M. Wessels, D. Grzonka, W. Oelert, G. Schepers, T. Sefzick, J. Walz, H. Pittner, T. W. Hänsch, and E. A. Hessels, *Phys. Rev. Lett.* **89**, 213401 (2002).
14. G. Gabrielse, X. Fei, L. A. Orozco, R. L. Tjoelker, J. Haas, H. Kalinowsky, T. A. Trainor, and W. Kells, *Phys. Rev. Lett.* **63**, 1360 (1989).
15. A. J. Peurrung, and J. Fajans, *Rev. Sci. Instrum.* **64**, 52 (1993).
16. X.-P. Huang, F. Anderegg, E. M. Hollmann, C. F. Driscoll, and T. M. O'Neil, *Phys. Rev. Lett.* **78**, 875 (1997).
17. R. G. Greaves, and C. M. Surko, *Phys. Rev. Lett.* **85**, 1883 (2000).
18. G. Andresen, W. Bertsche, P. D. Bowe, C. C. Bray, E. Butler, C. L. Cesar, S. Chapman, M. Charlton, J. Fajans, M. C. Fujiwara, R. Funakoshi, D. R. Gill, J. S. Hangst, W. N. Hardy, R. S. Hayano, M. E. Hayden, A. J. Humphries, R. Hydomako, M. J. Jenkins, L. V. Jørgensen, L. Kurchaninov, R. Lambo, N. Madsen, P. Nolan, K. Olchanski, A. Olin, A. Povilus, P. Pusa, F. Robicheaux, E. Sarid, S. S. E. Nasr, D. M. Silveira, J. W. Storey, R. I. Thompson, D. P. van der Werf, J. S. Wurtele, and Y. Yamazaki, "A novel antiproton radial diagnostic based on octupole induced ballistic loss" (2008), submitted to Phys. Plasmas.
19. J. Fajans, N. Madsen, and F. Robicheaux, "Critical loss radius in a Penning trap subject to multipole fields" (2008), submitted to Phys. Plasmas.
20. M. Fujiwara, "Detecting antihydrogen: the challenges and the applications," in *Physics with ultra slow antiproton beams*, edited by Y. Yamazaki, and M. Wada, AIP, Wako, Japan, 2005, vol. 793, p. 111.
21. M. C. Fujiwara, M. Amoretti, G. Bonomi, A. Bouchta, P. D. Bowe, C. Carraro, C. L. Cesar, M. Charlton, M. Doser, V. Filippini, A. Fontana, R. Funakoshi, P. Genova, J. S. Hangst, R. S. Hayano, L. V. Jørgensen, V. Lagomarsino, R. Landua, E. Lodi-Rizzini, M. Marchesotti, M. Macri, N. Madsen, G. Manuzio, P. Montagna, P. Riedler, A. Rotondi, G. Rouleau, G. Testera, A. Variola, D. P. van der Werf, and Y. Yamazaki, *Phys. Rev. Lett.* **92**, 65005 (2004).
22. M. Amoretti, C. Amsler, G. Bonomi, A. Bouchta, P. D. Bowe, C. Canali, C. Carraro, C. L. Cesar, M. Charlton, M. Doser, V. Filippini, A. Fontana, M. C. Fujiwara, R. Funakoshi, P. Genova, J. S. Hangst, R. S. Hayano, L. V. Jørgensen, A. Kellerbauer, V. Lagomarsino, R. Landua, D. Lindelof, E. L. Rizzini, M. Macrí, N. Madsen, G. Manuzio, P. Montagna, H. Pruys, C. Regenfus, A. Rotondi, G. Testera, A. Variola, L. Venturelli, D. P. van der Werf, and Y. Yamazaki, *Phys. Lett. B* **590**, 133 (2004).
23. F. Robicheaux, *Phys. Rev. A* **70**, 022510 (2004).
24. P. Oxley, N. S. Bowden, R. Parrott, A. Speck, C. H. Storry, J. N. Tan, M. Wessels, G. Gabrielse, D. Grzonka, W. Oelert, G. Schepers, T. Sefzick, J. Walz, H. Pittner, T. W. Hänsch, and E. A. Hessels, *Phys. Lett. B* **595**, 60 (2004).

Electrons Confined with an Axially Symmetric Magnetic Mirror Field

H. Higaki, K. Ito, K. Kira and H. Okamoto

Graduate School of Advanced Sciences of Matter, Hiroshima University
1-3-1 Kagamiyama, Higashi-Hiroshima, Hiroshima 739-8530 Japan

Abstract. Low energy non-neutral electron plasmas were confined with an axially symmetric magnetic mirror field and an electrostatic potential to investigate the basic confinement properties of a simple magnetic mirror trap. As expected the confinement time became longer as a function of the mirror ratio. The axial electrostatic oscillations of a confined electron plasma were also observed. Obtained results suggested an improved scheme to accumulate low energy charged particles with the use of a magnetic mirror field, which would enable the investigation of electron-positron plasmas.

Keywords: magnetic mirror trap, non-neutral plasma, electron-positron plasma
PACS: 52.55.Jd, 52.27.Jt, 52.27.Ep

INTRODUCTION

A magnetic mirror configuration can be observed in many magnetized plasmas, such as large devices for studying the thermonuclear fusion, electron cyclotron resonance ion sources, the earth's magnet sphere, etc. Also, a special kind of magnetic mirror field has been studied for the production and confinement of anti-hydrogen atoms [1,2]. It is clear that a large amount of low energy charged particles confined in a trap have opened new research fields in physics. For examples, low energy non-neutral plasmas have been confined with a uniform magnetic field, a toroidal magnetic field and a linear Paul trap for basic plasma physics research [3]. Argon ions confined in a linear Paul trap have been used to study nonlinear resonances of a space charge dominated charged particle beam [4,5]. Positrons accumulated in a Penning-Malmberg trap provided intense pulsed low energy positron beams [6].

Here, properties of non-neutral electron plasmas confined with a magnetic mirror field were investigated experimentally. Since the confinement time of a non-neutral electron plasma in a simple magnetic mirror was found to be much longer than its plasma oscillation periods, it will be possible to investigate basic properties of electron-positron plasmas confined with a magnetic mirror trap.

In the following, a magnetic mirror confinement of charged particles is explained at first. Some properties of non-neutral electron plasmas confined in a magnetic mirror are reported with experimental results. Then, an improved scheme for accumulating low energy charged particles with a magnetic mirror trap is described. Also, the possibility of investigating electron-positron plasmas with the use of a magnetic mirror configuration is discussed.

CP1037, *Cold Antimatter Plasmas and Application to Fundamental Physics,* edited by Y. Kanai and Y. Yamazaki

PRINCIPLES OF MAGNETIC MIRROR CONFINEMENT

Principles of a single charged particle confinement in a magnetic mirror with the mirror ratio $R = B_1/B_0$ (> 1) can be found in a textbook on plasma physics [7]. Here, B_0 and B_1 denote the minimum and maximum magnetic flux density in a magnetic mirror field, respectively. The magnetic moment μ of a charged particle (charge:e, mass:m) in a local magnetic flux density B is given by

$$\mu = IS = \frac{e\omega_c}{2\pi}\pi\rho^2 = \frac{mv_\perp^2}{2B} \tag{1}$$

with the cyclotron frequency ω_c, Larmor radius ρ, and the perpendicular velocity of the particle $v_\perp$. This magnetic moment is the adiabatic invariant under the condition that the magnetic field changes slowly, which is expressed by

$$\omega_c << \frac{L}{v_{//}}. \tag{2}$$

Here, L denotes the scale length of the magnetic field gradient. The conservation of energy and magnetic moment in the magnetic mirror field give the following equations.

$$\begin{aligned} \varepsilon &= m\frac{v_{//0}^2 + v_{\perp 0}^2}{2} = m\frac{v_0^2}{2} = m\frac{v_{//1}^2 + v_{\perp 1}^2}{2} \\ \mu &= \frac{mv_{\perp 0}^2}{2B_0} = \frac{mv_{\perp 1}^2}{2B_1} \end{aligned} \tag{3}$$

The axial confinement is accomplished if $v_{//1}^2 = 0$. Thus, the loss cone angle θ_0 is determined with the mirror ratio by the following equation.

$$\frac{1}{R} = \frac{B_0}{B_1} = \frac{v_{\perp 0}^2}{v_{\perp 1}^2} = \frac{v_{\perp 0}^2}{v_0^2} \equiv sin^2\theta_0 \tag{4}$$

It means that a particle is confined if the pitch angle θ at the minimum magnetic field B_0 is larger than θ_0 and is lost if $\theta < \theta_0$. In FIG.1 (b), the gray region in a local velocity space is called a loss cone. For example, $R = 5$ gives $\theta_0 \approx 27°$.

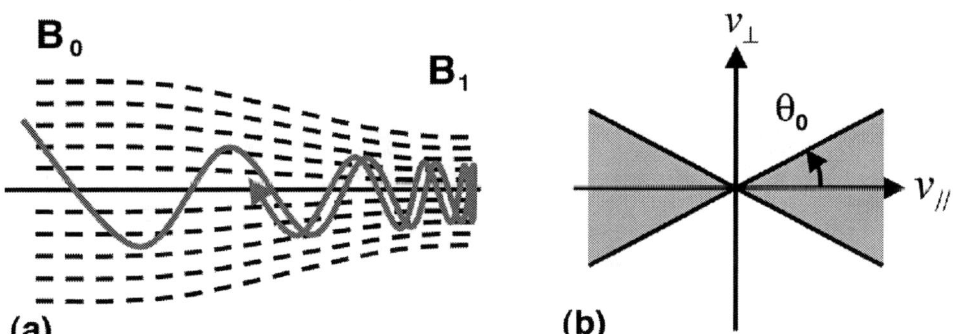

FIGURE 1. (a) A schematic of a single charged particle confinement in a magnetic mirror. (b) A velocity space at the minimum magnetic field B_0. A particle with a pitch angle $\theta < \theta_0$ is not confined with the mirror ratio $R = 1/\sin^2\theta_0$.

When there is an electrostatic potential $\phi(z)$ due to a self field of trapped charged particles and/or externally applied electric field, the loss cone is modified. The conservation of energy and magnetic moment are given by the following equations.

$$\varepsilon = m\frac{v_{//}^2 + v_\perp^2}{2} + e\,\phi(z) = const.$$
$$\mu = \frac{mv_\perp^2}{2B(z)} = const. \tag{5}$$

Then, a condition for the axial confinement inside the magnetic mirror field is given by the following equation.

$$\begin{aligned}\frac{mv_{//1}^2}{2} &= \varepsilon - \mu B_1 - e\phi_1 \\ &= m\frac{v_0^2}{2} - \frac{B_1}{B_0}\frac{mv_{\perp 0}^2}{2} + e(\phi_0 - \phi_1) \\ &= \frac{mv_{//0}^2}{2} - (R-1)\frac{mv_{\perp 0}^2}{2} + e\phi_c < 0\end{aligned} \tag{6}$$

Therefore, the loss cone boundary is modified with the presence of the confinement potential $\phi_c = \phi_0 - \phi_1$ as follows.

$$sin^2\theta = \frac{1}{R}(1 + \frac{2e\,\phi_c}{mv_0^2}) \tag{7}$$

As an example, schematics of the boundaries for protons and electrons are shown in FIG.4 for $\phi_c > 0$. In a simple magnetic mirror field, the ion confinement time $\tau_i \sim 1/\nu_{ii}$ and the electron confinement time $\tau_e \sim 1/\nu_{ee}$ are denoted by the ion-ion collision frequency ν_{ii} and electron-electron collision frequency ν_{ee}. Since ν_{ii} is much less than ν_{ee} in general, the electrons are scattered into the loss cone more rapidly than ions. Thus, the plasma becomes ion rich and ϕ_c becomes positive. This positive ϕ_c tends to confine electrons and expel ions as shown in FIG.2. As a result, the loss rates of ions and electrons balance at a certain point. This feature is called ambipolar diffusion.

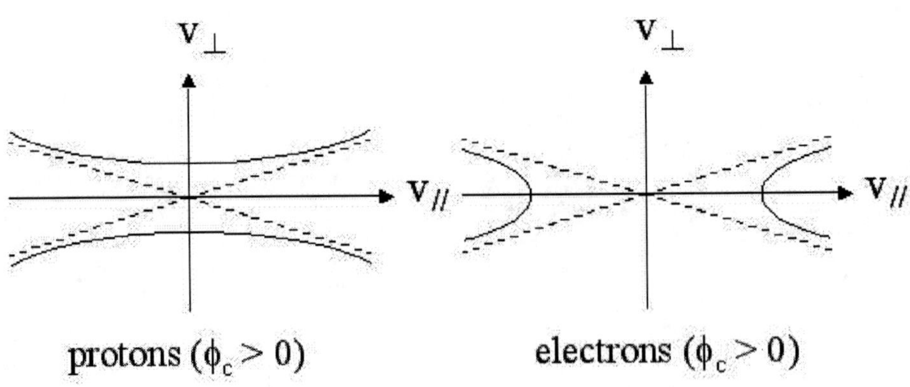

FIGURE 2. The schematics of loss cone distributions modified with the self filed potential $\phi_c > 0$, which are usually observed in a mirror confined plasma for thermonuclear fusion.

In this case, the confinement time is expressed by [8]

$$\tau_M = 0.78\,\tau_i \ln\left[\frac{R}{1+e\,\phi_c/T_i}\right] \quad with \quad \tau_i = \left(\frac{\pi}{2}\right)^{3/2} \frac{m^{1/2}(k_B T_i)^{3/2}}{ne^4 \ln\Lambda}, \tag{8}$$

with the Coulomb logarithm $\ln\Lambda$.

NON-NEUTRAL ELECTRON PLASMAS CONFINED WITH A MAGNETIC MIRROR FIELD

Non-neutral electron plasmas were confined with a magnetic mirror field [9]. The field configuration to confine a large number of electrons was shown in FIG.3. A magnetic field was produced by two series of magnetic coils, which are composed of 14 normal conductor solenoids. For example, keeping a current of coils on the low field side to 10 A, a current of the other coils at high field side could be increased from 10 to 80 A. This results in the magnetic mirror field with $R \sim 5$, as shown FIG.3, where a measured magnetic field strength on the axis of symmetry is plotted. For applying the electrostatic potential, so called multi-ring trap was installed along the axis of symmetry. It contains 45 ring electrodes with the inner diameter of 7 cm and they are placed from –7 to 64 cm. With these electrodes, this trap is very flexible in applying electrostatic potentials. Electrons are injected from electron gun installed at high field side. A Faraday cup with the diameter of 8cm is placed at low field side to measure the number of electrons.

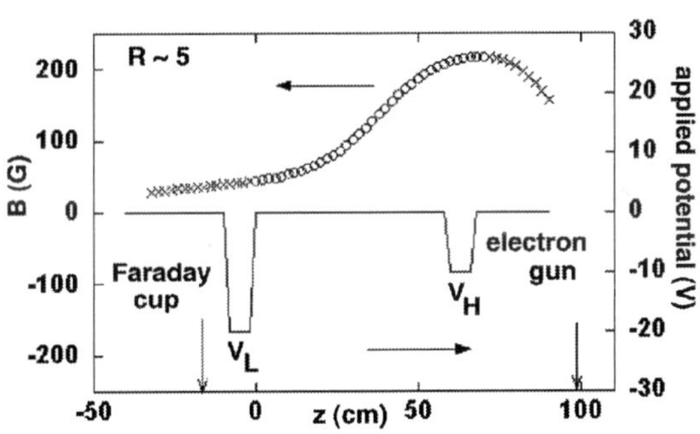

FIGURE 3. The measured magnetic flux density on the axis of symmetry is plotted along with the applied electrostatic potentials.

When a large number of electrons are to be accumulated, the following procedure was employed. First, electrons are injected and confined axially with electrostatic potentials on both ends. The number of electrons becomes 10^9 at the maximum. Then, the potential V_H at high field side is grounded to start a confinement with a magnetic mirror. Holding electrons for a certain period of time, the potential V_L in front of the Faraday cup is grounded to measure the number of electrons N_e. In Fig.4 (a), N_e is plotted as a function of time for various R. The confinement time τ_c is determined with an exponential fitting and τ_c is plotted as a function of R for three different field strength in Fig.4 (b). It is seen that the confinement time can be longer than 50 ms for $R \sim 5$.

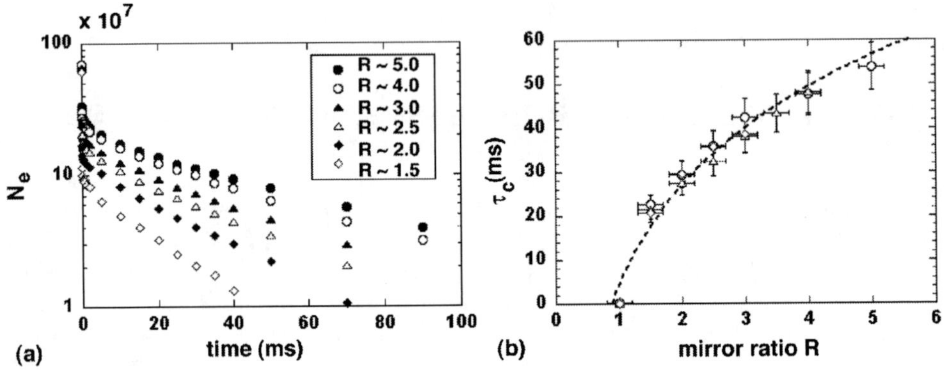

FIGURE 4. (a) The number of electrons as a function of time for various R. (b) The confinement time becomes longer as a function of R.

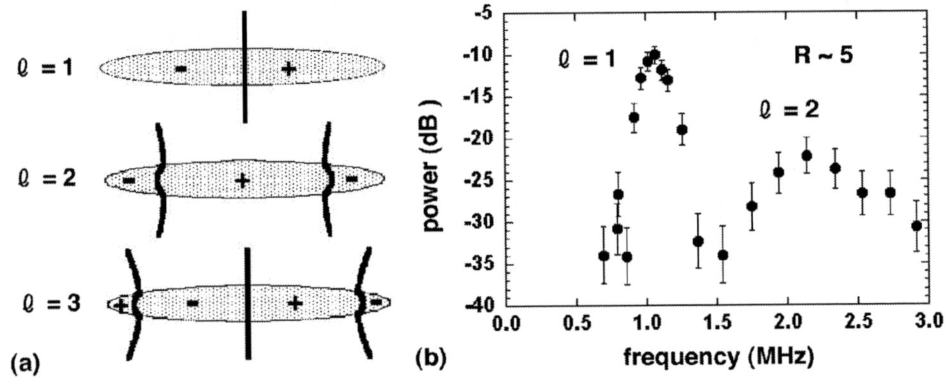

FIGURE 5. (a) The schematics of axial electrostatic oscillations in a non-neutral plasma confined in a Penning trap. Solid curves denote nodes of oscillations. (b) Measured spectrum of electrostatic oscillations in a mirror confined non-neutral plasma.

Electrostatic oscillations of a non-neutral plasma in a trap are important, because they are used as nondestructive measurement of confined plasmas. The axial electrostatic oscillations of a non-neutral plasma in Malmberg traps and Penning traps are standing waves with open ends, as schematically shown in FIG.5 (a). In general, their frequencies are functions of the plasma density and temperature. This is also true in case of a mirror confined plasma. Here, peak power of the excited oscillations is plotted as a function of the excitation frequency. When the excitation frequency was changed, the observed peak frequency also shifted and the spectrum shown in FIG.5 (b) was obtained. Two peaks correspond to axial electrostatic oscillations with the mode number 1 and 2. One important point here is that the oscillation period is shorter than 1 μs. This is much shorter than the confinement time of 50 ms. Another point is that magnetic mirror fields can confine both positive and negative charged particles. Therefore, it is expected that an electron-positron plasma can be confined with a magnetic mirror field and electron-positron plasma oscillations can be observed, even if its confinement time is short. Thus, if possible, it is favorable to accumulate positrons directly in a magnetic mirror trap in a high vacuum.

ACCUMULATING LOW ENERGY CHARGED PARTICLES WITH A MAGNETIC MIRROR FIELD

The accumulation of low energy positrons with a magnetic mirror was tried more than 25 years ago [10]. A magnetic mirror trap was used to increase the number of positrons in a bunch of positron beam to enhance the generation of positronium. Authors mentioned the trapping efficiency of ~ 50 % and $\tau_c \sim 100$ μs. This confinement time is much shorter than 50 ms observed in FIG.4 (b). It seems the details of the accumulation method has not been reported, so far. It may be possible to accumulate more positrons with some improvements.

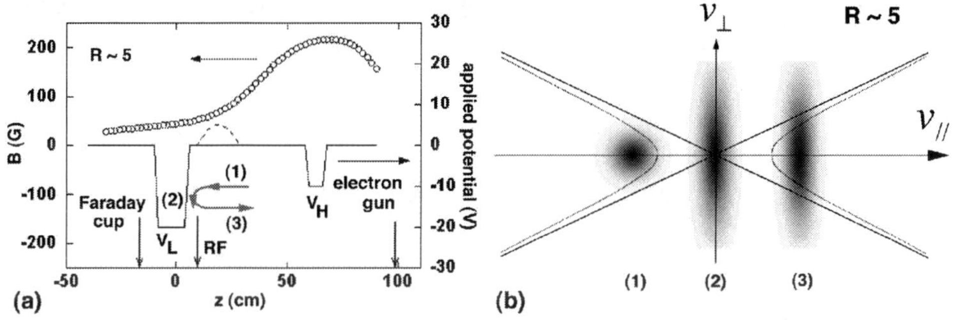

FIGURE 6. (a) When injected electrons are reflected by V_L, cyclotron motions are excited by RF field. (b) A distribution in the velocity space during the excursion. The darker tone means the higher density.

To confirm the basic properties of the accumulation scheme, a low energy electron beam with a current of 2 ~ 40 pA was used as a test. A typical experimental procedure is as follow. The electrostatic potentials V_L and V_H are applied to ring electrodes at first as shown in Fig.6 (a). Then, electrons are injected from the electron gun placed outside of the magnetic mirror field (FIG.6 (b)(1)). Electrons which can overcome the potential V_H flow into the magnetic mirror trap from the high field side. While injecting electrons, a radio frequency (RF) perturbation is applied to an azimuthally segmented electrode to excite a cyclotron motion of electrons. It should be noted that the RF field should be applied at lower field side because the loss cone angle becomes larger at higher field side [11], which makes the initial magnetic mirror confinement less effective. All the incoming electrons are reflected back at the potential V_L. Some electrons gain the transverse energy during the excursion (FIG.6 (b)(2)). Since the magnetic moment is an adiabatic invariant in the present experimental condition, a part of electrons are trapped by the magnetic mirror field (FIG.6 (b)(3)). Then, a part of mirror trapped electrons lose their energy through collisions with background neutral particles and can be trapped electrostatically by V_H. Others are scattered into the loss cone of the magnetic mirror confinement, before they lose enough energy. After the electron injection, the potential V_L is grounded to measure N_e.

In fact, it was found experimentally that the RF field made τ_c shorter. In case of a simple magnetic mirror with $V_H = 0$, the original τ_c~50 ms without RF field became ~ 3 ms. This confinement time was improved by applying the potential V_H. Although the improved τ_c~ 50 ms was more than two orders of magnitude longer than ~100 μs, it is not enough for accumulating useful amount of positrons. This was further improved by separating the accumulation region from the region where RF filed was applied, as shown by the dotted line in FIG.6 (a). The resultant confinement time became longer than a few hundreds seconds. Details of the experimental results will be reported elsewhere [12].

SIMULTANEOUS CONFINEMENT OF ELECTRONS AND POSITRONS IN A MAGNETIC MIRROR FIELD

Here, a simultaneous confinement of electrons and positrons with magnetic mirror fields is considered. A basic procedure for the simultaneous confinement is as follows. At first the low energy electrons and positrons are confined by a nested Penning trap inside magnetic mirror fields as shown in FIG.7 (a). Then, both particles are expanded adiabatically along the magnetic field by reducing the confinement potential. The adiabatic expansion results in anisotropic velocity distribution [13]. The perpendicular energy becomes larger than parallel energy, which enhances the particle confinement in the magnetic mirror field at the beginning.

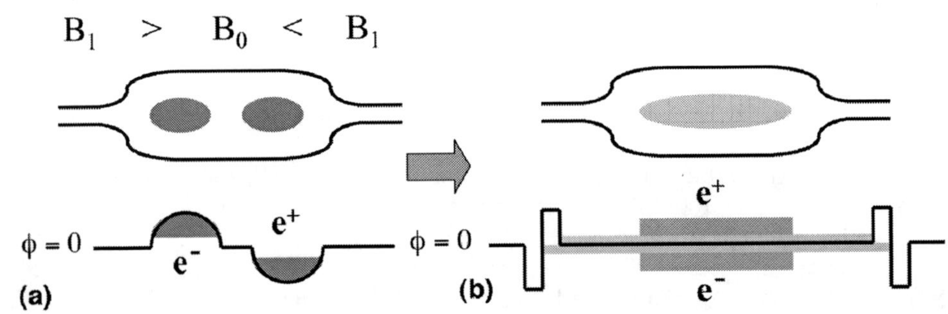

FIGURE 7. Schematics of a possible procedure for a simultaneous confinement of electrons and positrons.

When an equal amount of electrons and positrons are confined in a magnetic mirror trap, a potential due to self field can be ignored. Thus, a reduction of trapped particles observed at the beginning of the magnetic mirror confinement in FIG.4 (a) will be avoided. Then, the confinement time will be in the same order with that of a simple magnetic mirror confinement (Eq.8 with ϕ_c=0). Under the condition that the collisions between charged particles and background neutral particles are ignorable, the confinement time is determined by the collision frequencies between electrons and positrons. Assuming the possible plasma density of $10^5 \sim 10^6$ cm^{-3} and temperature of 0.1 ~ 1.0 eV, the collision frequency is in the order of $1 \sim 10^2$ s^{-1}, as shown in FIG.8. It is thought that electrostatic oscillations of an electron-positron plasma can be investigated with the confinement time of 0.01sec.

The confinement time will be further improved by applying both positive and negative potentials outside of magnetic mirrors, as shown in FIG.7 (b). These potentials reflect back escaping electrons and positrons, which are about 10 % of trapped particles for $R \sim 5$. The confinement time of an electron-positron plasma can be much longer than that of a simple magnetic mirror confinement.

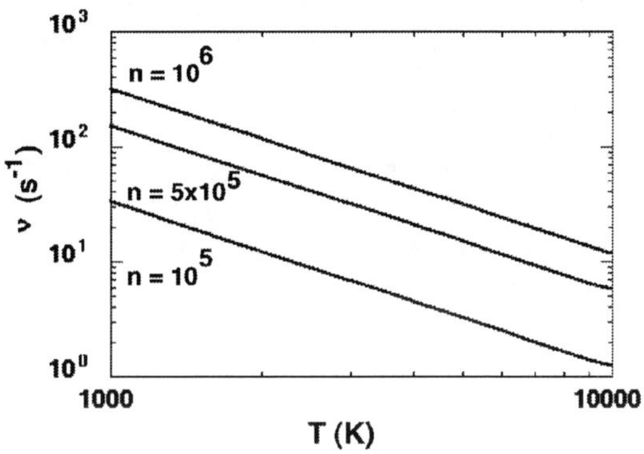

FIGURE 8. Calculated collision frequencies between electrons and positrons for possible plasma parameters.

SUMMARY

Low energy non-neutral electron plasmas were confined with an axially symmetric magnetic mirror field and an electrostatic potential. It was observed experimentally that the confinement time was longer than 50 ms for $R \sim 5$. The electrostatic oscillations with frequencies of a few MHz were also observed. It was confirmed that low energy electrons were accumulated with a magnetic mirror field and electrostatic potentials. Obtained results suggested a possible scheme to investigate electron-positron plasmas with a magnetic mirror field.

ACKNOWLEDGMENTS

The authors acknowledge Prof. Y. Kiwamoto at Kyoto University for his support of our experimental setup. One of the authors (H. Higaki) is also grateful for Dr. A. Mohri who originally designed the trap used in the present experiments.

REFERENCES

1. J. Fajan, W. Bertsche, K. Burke, S. F. Chapman, and D. P. van der Werf, *Phys. Rev. Lett.* **95**, 085002 (2005).
2. G. Andresen *et al.*, *Phys. Rev. Lett.* **98**, 023402 (2007).
3. *NON-NEUTRAL PLASMA PHYSICS VI* edited by M. Drewsen, U. Uggerhoj and H. Knudsen (American Institute of Physics, New York, 2006) **862**.
4. R. Takai, K. Nakayama, W. Saiki, K. Ito, and H. Okamoto, J. Phys. Soc. Jpn. **76**, 014802 (2007).
5. H. Higaki, K. Ito, R. Takai, K. Nakayama, W. Saiki, K. Izawa, and H. Okamoto, Hyperfine Interactions **174**, 77 (2007).
6. D. B. Cassidy, S. H. M. Deng, R. G. Greaves and A. P. Mills, Jr., *Rev. Sci. Instrum.* 77, 073106 (2006).
7. F. F. Chen, *Introduction to Plasma Physics and Controlled Fusion*, New York: Plenum, 1984.
8. D. V. Sivukhim, "Coulomb collisions in a fully ionized plasma," in *Reviews of Plasma Physics,* **4** edited by M. A. Leontovich, New York: Consultants Bureau, 1966, pp. 93-241.
9. H. Higaki, K. Ito, W. Saiki, Y. Omori, and H. Okamoto, *Phys. Rev. E* **75**, 066401 (2007).
10. S. Chu and A. P. Mills, Jr., *Phys. Rev. Lett.* **48**, 1333 (1982).
11. J. Fajans, *Phys. Plasmas* **10** 1209 (2003).
12. H. Higaki, K. Ito, K. Kira, and H. Okamoto, submitted to *Appl. Phys. Express.*
13. A. W. Hyatt, C. F. Driscoll, and J. H. Malmberg, *Phys. Rev. Lett.* **59** 2975 (1987).

Charged-Particle Impact Ionization of Atoms

Klaus Bartschat and Xiaoxu Guan

Department of Physics and Astronomy, Drake University, Des Moines, IA 50311, USA
e-mail: klaus.bartschat@drake.edu; xiaoxu.guan@drake.edu

Abstract. We have developed a hybrid method to treat charged-particle impact ionization of complex atoms and ions. The essential idea is to describe the interaction between a fast projectile and the target perturbatively, up to second order, while the initial bound state and the ejected-electron–residual-ion interaction can be handled via a convergent R-matrix with pseudo-states (close-coupling) expansion. Example results for ionization of the heavy noble gases (Ne–Xe) by positron and electron impact are presented. The general scheme for a distorted-wave treatment of ionization by heavy-particle impact is described.

Keywords: electron, positron, antiproton impact ionization, distorted-wave Born approximation, close-coupling, R-matrix
PACS: 34.80.Uv,34.80.Dp,25.43.+t

INTRODUCTION

In recent years, enormous progress has been made regarding the theoretical and computational treatment of impact ionization processes by charged particles, with particular emphasis on electron impact. Non-perturbative approaches, such as exterior complex scaling (ECS) [1], time-dependent close-coupling (TDCC) [2, 3], convergent close-coupling (CCC) [4], or R-matrix with pseudo-states (RMPS) [5-7], are now able to predict the total (TCS), single-differential (SDCS), double-differential (DDCS) and triple-differential (TDCS) cross section very accurately for the atomic hydrogen target. Similar success has been achieved, particularly by the CCC method, for ionization of helium — provided the residual ion is left in the $He^+(1s)$ ground state, i.e., one of the two target electrons is essentially a spectator [8, 9].

The theoretical situation, however, does not look as good for more complex targets, such as the heavy noble gases Ne–Xe, or more complicated processes, such as simultaneous ionization-excitation or double-ionization of helium by charged-particle impact. Especially for the heavy noble gases, non-perturbative methods based on the distorted-wave approach have been used for many years. Recently, much work has been done by Madison and co-workers [10-13], who are now able to account for the proper asymptotic three-body Coulomb boundary condition in the final state [14]. Unfortunately, however, these methods often suffer from the approximations made in the treatment of the target structure and the ejected-electron–residual-ion interaction, as will be further discussed below.

In order to address these problems, we have developed a hybrid method to treat electron-impact ionization of atoms and ions. The essential idea is to describe the interaction between a fast projectile and the target perturbatively, up to second order, while the initial bound state and the ejected-electron–residual-ion interaction are han-

CP1037, *Cold Antimatter Plasmas and Application to Fundamental Physics*, edited by Y. Kanai and Y. Yamazaki

dled via a convergent R-matrix (close-coupling) expansion [15, 16]. A code for calculating total and single-differential (with respect to energy loss) cross sections, using a first-order distorted-wave approach for the projectile, was published in 1993 [17]. Within the approximations made in the model (no exchange for electron impact, no positronium formation for positron impact), the use of this code for positron impact is straightforward [18]. As long as a plane-wave description is used for the projectile, evaluating the Bethe integral [19] instead of the partial-wave sum over the distorted waves also yields results for heavy-particle impact, after properly accounting for the reduced masses and the size of the projectile charge.

The essential ingredients of our model will be described in the next section. This is followed by some example results and an outlook towards further developments, especially towards a perturbative treatment for heavy projectiles. More details can be found in the references provided.

THEORETICAL FORMULATION

The basic theoretical and computational methods for a fist-order distorted-wave treatment of the process have been outlined before [15-17], as have the extensions to account for second-order effects in the projectile–target interaction [20] and the use of a plane-wave description for high incident energies [21]. Instead of repeating the details here, we will focus on a general discussion of the ingredients in a first-order theory and specify some details of the current model for electron-impact ionization of heavy noble gases.

In our model, the first-order amplitude is given by [15]

$$f_{L_0M_0S_0M_{S_0},\mu_0\to L_fM_fS_fM_{S_f},\mu_1\mu_2}(\boldsymbol{k}_0,\boldsymbol{k}_1,\boldsymbol{k}_2) = -\frac{1}{(2\pi)^{5/2}}\langle \varphi^{(-)}_{\boldsymbol{k}_1\mu_1}(x)\,\Psi^{\boldsymbol{k}_2\mu_2(-)}_{L_fM_fS_fM_{S_f}}(X)\,|V(x,X)|\,\Psi_{L_0M_0S_0M_{S_0}}(X)\,\varphi^{(+)}_{\boldsymbol{k}_0\mu_0}(x)\rangle. \quad (1)$$

Here

$$X = \{\boldsymbol{r}_1,\sigma_1;\boldsymbol{r}_2,\sigma_2;\ldots;\boldsymbol{r}_{N+1},\sigma_{N+1}\} \quad (2)$$

denotes a set of electronic spatial and spin coordinates in the $(N+1)$-electron atom, while $x = \{\boldsymbol{r},\sigma\}$ represents the coordinates of the colliding projectile. The Coulomb potential

$$V(x,X) = \sum_{n=1}^{N+1}\frac{-Q}{|\boldsymbol{r}-\boldsymbol{r}_n|} + \frac{ZQ}{r} \quad (3)$$

describes the interaction between the projectile with charge Q and the atomic electrons as well as the nucleus of charge Z. The integration in the matrix element is performed over the (x,X) coordinates of all $N+2$ particles (the target electrons plus the projectile).

We use a non-relativistic model and also neglect exchange between the "fast" projectile and the target electrons. In a plane-wave treatment, which is usually a reasonable approximation for the He target, the integral over the coordinates of the fast electron can be performed analytically through the Bethe integral [19, 21]. In a distorted-wave treatment for the projectile, which is critical for a target such as the heavier noble gases at the relatively low incident energies we will consider, we use a partial-wave description for $\varphi^{(+)}_{\boldsymbol{k}_0\mu_0}(x)$ and $\varphi^{(-)}_{\boldsymbol{k}_1\mu_1}(x)$. This actually simplifies the angular integration over the projectile coordinates in the second-order term. As described by Reid *et al.* [20], however, we still need to make approximations to evaluate the second-order term. These include the use of only the pole term in a principal-value integral, effectively dropping the real part of the Green's function for propagating through the intermediate states, the choice of an average excitation energy for the intermediate state, and limiting the evaluation of integrals to within the R-matrix sphere.

Another important aspect is the treatment of the ejected-electron–residual-ion interaction, i.e., the function $\Psi^{\boldsymbol{k}_2\mu_2(-)}_{L_fM_fS_fM_{S_f}}$ in equation (1). In principle, coupling a large number of discrete and pseudo-states will ultimately lead to a converged result for the electron–residual-ion collision problem, as well as the initial bound state. As mentioned above, this is the basic idea behind the CCC [4] and RMPS [5-7] methods.

Typical simplifications in the treatment of this part of the problem include (i) limiting the number of coupled states, (ii) using a distorted wave for the ejected electron, with exact, approximate, or no exchange accounted for, or even just an unperturbed Coulomb wave. With increasing projectile and ejected-electron energies, some of these approximations are expected to become less critical.

Returning to the target description, we note that the initial bound state for complex targets is often represented by single-configuration Hartree-Fock (HF), multi-configuration Hartree-Fock (MCHF), or frozen-core multi-configuration Hartree-Fock (FC-MCHF) descriptions. A variant of the latter method is achieved in the R-matrix approach by running the electron–ion collision problem with modified boundary conditions to obtain the bound states of the system. To obtain the results for neon and argon shown below, we used the multi-configuration ionic target descriptions given by Burke and Taylor [22] for the corresponding photoionization problems. For Kr and Xe, similar descriptions were generated [18]. In addition to the 2-state models, in which the $(ns^2np^5)^2\mathrm{P^o}$ and $(nsnp^6)^2\mathrm{S}$ of the final ionic states are closely coupled, we also performed 1-state calculations for direct ionization to the $(ns^2np^5)^2\mathrm{P^o}$ states. Furthermore, we employed single-configuration (SC) rather than configuration-interaction (CI) descriptions of both the initial bound state and the final ionic states. This allows us to assess the sensitivity of the final results on the target description.

RESULTS: IONIZATION OF HEAVY NOBLE GASES BY POSITRON IMPACT

Figure 1 presents our combined cross sections for direct ionization of Ar($3p$) plus Ar($3s$) by positron impact. We also show the results obtained in the cases when only the $(3s^23p^5)^2P^o$ state is included in the close-coupling expansion and when single-configuration descriptions of the initial bound state and the final ionic state are used. We then compare with the corresponding results for electron impact and, finally, we show what happens when the distortion potential for the projectile is set to zero, i.e., the distorted wave is replaced by a plane wave (pw). As can be seen from the figure, the sensitivity of the theoretical results is limited. This lack of sensitivity likely explains

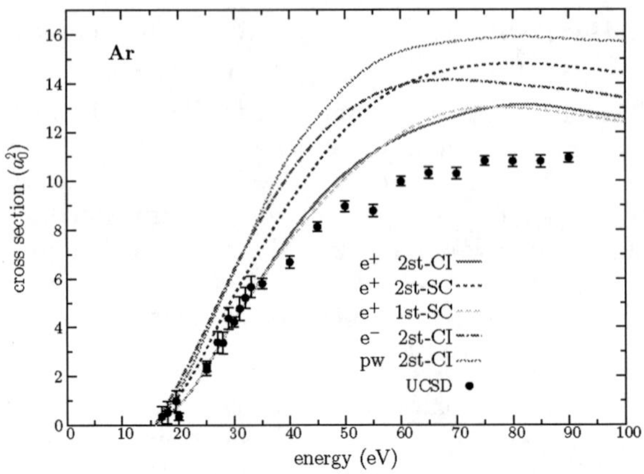

FIGURE 1. Direct ionization cross sections for Ar($3p$) and Ar($3s$) combined by positron and electron impact (from [18]). The experimental data are those of Marler *et al.* [25] obtained at the University of California at San Diego (UCSD).

why the present predictions, as well as those of purely perturbative models [23, 24] (not shown), are all in reasonable agreement with the experimental data of Marler *et al.* [25]. We refer to the latter work for a more detailed comparison with other theoretical and experimental datasets.

Figures 2–4 present the corresponding results for the Ne, Kr, and Xe targets. As a general trend, using multi-configuration interaction descriptions of the initial bound state and the final ionic target states lowers the theoretical results, presumably since the ionization potential in the structure model increases. However, this better computational

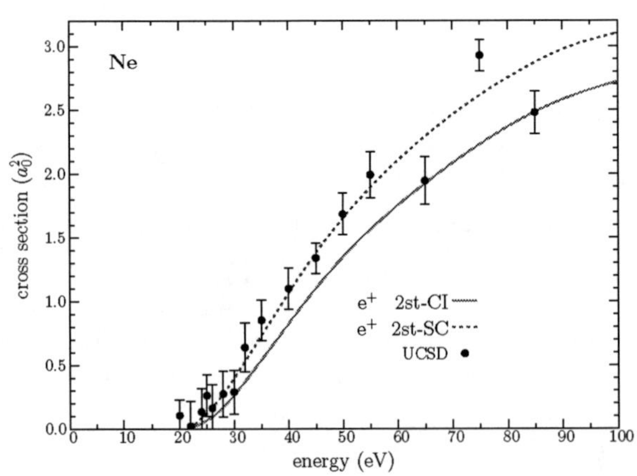

FIGURE 2. Direct ionization cross section for Ne($2p$) and Ne($2s$) combined by positron impact, as obtained in the DW-2st-CI and DW-2st-SC methods. The experimental data are those of Marler *et al.* [25].

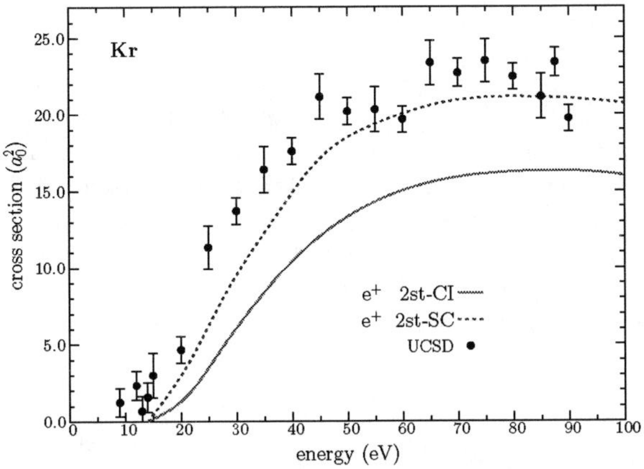

FIGURE 3. Direct ionization cross sections for Kr(4p) and Kr(4s) combined by positron impact, as obtained in the DW-2st-CI and DW-2st-SC methods. The experimental data are those of Marler *et al.* [25].

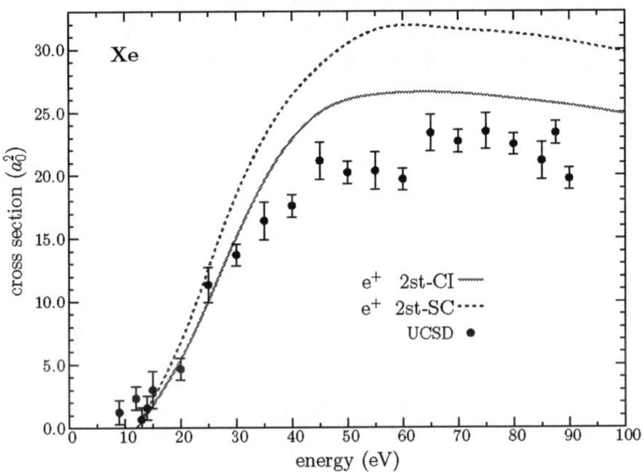

FIGURE 4. Direct ionization cross sections for Xe(5p) and Xe(5s) combined by positron impact, as obtained in the DW-2st-CI and DW-2st-SC methods. The experimental data are those of Marler *et al.* [25].

model does not necessarily yield better agreement with the experiment data. For Ne and Kr, the predictions from the single-configuration models are in nice agreement with the data of Marler *et al.* [25], and hence improving the target structure actually worsens the agreement between theory and experiment. For the Xe target, similar to the Ar case, the decrease in the predicted direct ionization cross section by the CI model brings the results into closer agreement with experiment, but the theoretical results are still systematically too large.

Heavy-Particle Impact Beyond the Plane-Wave Born Approximation.

In a recent paper, Foster *et al.* [26] suggested the need to develop a distorted-wave theory even for the description of a heavy incident projectile. The basis for this somewhat surprising proposal is the need to explain several recent experimental results for charged-particle impact excitation, in which the ejected electron is observed in a direction out of the plane defined by the incident and outgoing projectile momenta.

Let us briefly summarize the requirements to perform such a calculation. As an illustration, we use recent work on electron-impact ionization of Mg($3s^3$) [26-28]. In the DWBA framework, the dynamical aspects of the process are determined by the radial integrals

$$M^{\lambda}_{\ell_0\ell_1\ell_2} \equiv \int_0^\infty dr_1 \int_0^\infty dr_2 F_{\ell_1}(k_1,r_1) F_{\ell_2}(k_2,r_2) \frac{r_<^\lambda}{r_>^{\lambda+1}} P_{3s}(r_2) F_{\ell_0}(k_0,r_1), \tag{4}$$

where $r_<$ $(r_>)$ = min(max)$\{r_1,r_2\}$ and the F_{ℓ_i} are the radial parts of the distorted waves for the projectile. The above (real) matrix elements are then combined with the proper spherical harmonics, phase factors, and angular-momentum coupling coefficients to calculate the scattering amplitudes and the fully-differential cross section (FDCS), which is measured by detecting the outgoing projectile and the ejected electron in coincidence.

Details of the DWBA approach can be found in many publications; see, for example, Chen and Madison [29] and references therein. As mentioned previously, however, it is important to remember, and hence to investigate, the *additional* approximations, *beyond* a first-order description of the projectile interaction with the target electrons, that are often made to describe the process. Recall that these include the description of the various bound states, the ejected-electron–residual-ion interaction, and the treatment of the projectile as a plane or distorted wave.

Another important aspect of any numerical calculation is the amount of computational effort that is needed to obtain accurate results. If many partial-wave contributions have to be summed up to obtain convergence, one might ask how many of those partial waves actually need to be calculated in a distorted-wave or an even better approximation. If that number is comparatively small and some regularities are seen in the numerically calculated matrix elements, it might be possible to use extrapolation techniques or to combine approximations of different complexity to obtain partial-wave converged results in an efficient manner.

Figure 5 exhibits first-order DWBA results for selected matrix elements $M^{\lambda}_{\ell_0\ell_1\ell_2}$ for ionization of Mg($3s^2$) by electron impact [28]. The incident projectile energy is 1,500 eV, and the two final-state electrons have energies of 1,430 eV and 62 eV, respectively. The final ionic state is $Mg^+(3s)$. Because of selection rules, the orbital angular momentum of the ejected electron ℓ_2 in this case is equal to the multipole component λ in the expansion of the Coulomb interaction between the projectile and the target that causes the transition.

Looking at Fig. 5, we see indeed that the matrix elements calculated in the distorted-wave or plane-wave description of the projectile become indistinguishable on the graph for $\ell_0, \ell_1 \gtrsim 20$. This is not unexpected, since the centrifugal barrier corresponds to

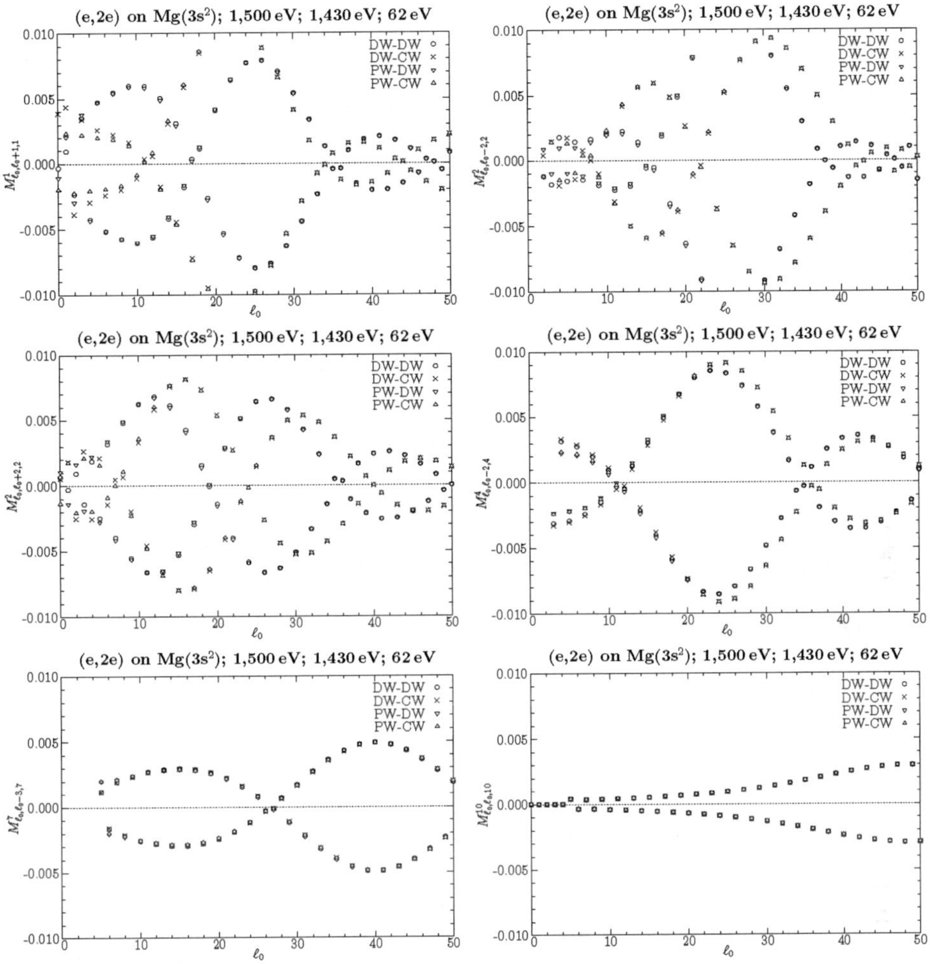

FIGURE 5. Selected matrix elements for ionization of Mg by electron impact at 1,500 eV, with the two outgoing electrons having energies of 1,430 eV and 62 eV, respectively [28]. The abbreviations in the legend specify the description of the projectile by a distorted wave (DW) or a plane wave (PW) and of the ejected electron by a distorted wave or a Coulomb wave (CW). The results are plotted as a function of ℓ_0 for fixed values of $\ell_0 - \ell_1$ and $\ell_2 = \lambda$.

approximately a distance of $b_p = \ell/k \approx 2$ for $\ell = 20$ and a wavenumber of $k = 10$ for the projectile. This result confirms the classical impact-parameter argument of Foster [26] (see their Fig. 4). Furthermore, for $\ell_2 = 7$ and $\ell_2 = 10$ (bottom part of Fig. 5), there is apparently no significant difference in the results obtained with a pure rather than a distorted Coulomb wave. Once again, this is no surprise, since the centrifugal barrier for an ejected electron energy of 62 eV and $\ell = 5$ would already be $b_e \approx 2$.

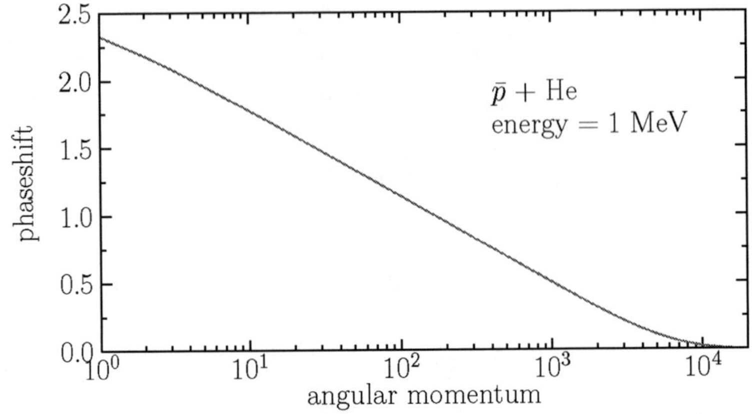

FIGURE 6. Phaseshifts for the scattering of antiprotons from He($1s^2$) at an incident energy of 1 MeV.

Although the computational effort to calculate individual partial waves and the corresponding matrix elements in either the distorted-wave or the plane-wave approximation for the projectile are comparable, the overall computational efficiency can be increased by using the above findings in connection with the "Born subtraction technique". As is well known, the exact first-order plane-wave amplitude for the projectile can be calculated by making use of the "Bethe integral" [19] and then replace a number of partial waves with low orbital angular momentum by distorted waves. Indeed, Bartschat *et al.* [28] demonstrated how the above method could be used to speed up the calculation of the (e,2e) process on Mg.

We believe that a similar idea can be used for heavy-particle impact as well. However, there are several obstacles to overcome. These include: (i) A large number of partial waves may have non-vanishing phaseshifts and hence need to be treated accurately before the Born-subtraction method can be applied. (ii) Some of the integrals over the coordinates of the projectile are highly oscillatory and require special measures for their evaluation (see below). (iii) The number of integrals to be calculated and, ultimately, to be combined in the partial-wave sum is expected to be very large. Note however, that the problem is well adaptable to modern parallel computer architecture, since each integral can be calculated on a single processor.

As an example, Fig. 6 presents the phaseshifts for the scattering of antiprotons on He($1s^2$) at an incident energy of 1 MeV. This is still a regime of relatively small perturbation $|Z/v| \approx 0.16$, i.e., a distorted-wave treatment seems promising. As can be seen, the phaseshifts are significantly nonzero (magnitudes bigger than 0.05) for partial-wave angular momenta $\ell \lesssim 10{,}000$. This clearly indicates the difficulty of performing such calculations.

Due to the highly oscillatory behavior of the distorted waves at high energies, it is essentially impossible to use straightforward conventional methods to obtain the radial integrals (4) involving these waves accurately. To avoid this problem, we follow a

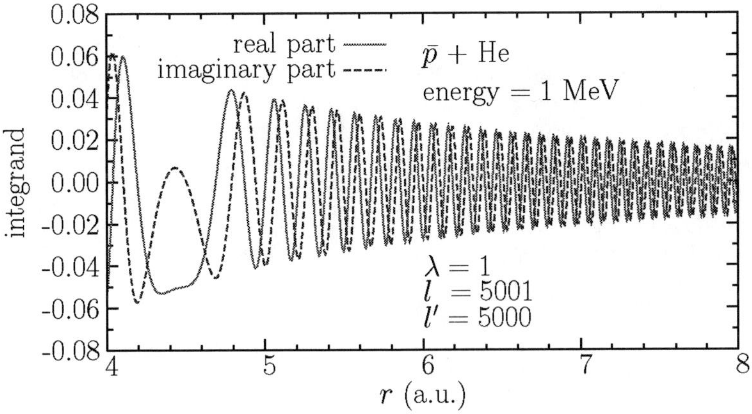

FIGURE 7. Example of an integrand in the DWBA formulation along part of the real axis.

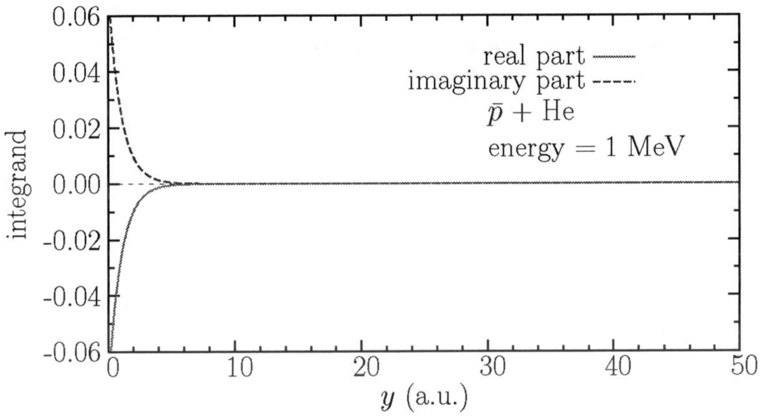

FIGURE 8. The integrand corresponding to fig. 7 along the imaginary axis.

suggestion by Sil *et al.* [30]. The essential idea is to first convert the problematic integral

$$C^{\lambda}_{\ell_0\ell_1} \equiv \int_a^{\infty} dr_1 F_{\ell_1}(k_1,r_1)\frac{1}{r_1^{\lambda+1}}F_{\ell_0}(k_0,r_1), \tag{5}$$

over the projectile coordinate r_1 to the complex plane. Here a is the radius, at which the projectile interaction with the target has become negligible (essentially the R-matrix box). Then the integral is evaluated along the imaginary rather than the real axis, since the integrand falls off exponentially instead of being highly oscillatory. Figures 7 and 8 show part of a typical integrand along the real axis and the complete integrand along the imaginary axis. Not surprisingly, the integral can now be evaluated with standard integration schemes using just a few terms.

We are currently in the process of developing a computer code along the above lines. The principal goal will be to calculate fully-differential cross sections, in and out of the reaction plane, to address questions raised by experiments such as those reported by Schulz *et al.* [31]. As a first step, however, we plan to check the program in the region of "small perturbations" for impact ionization of He and the heavier noble gases by anti-proton impact. Since the charge-exchange reaction channel is not open in this case, the process is actually easier to compute than proton impact. By comparing the results with those from fully non-perturbative treatments such as the recent work of Foster *et al.* [32], we will assess how low in energy the method is still appropriate, i.e., the problem is still tractable by a perturbative approach.

SUMMARY AND OUTLOOK

We have presented a hybrid approach to treat charged-particle impact ionization of complex targets. Channel-coupling and exchange effects in the interaction of the ejected electron and the residual ion can, in principle, be treated to convergence, while the projectile–target interaction is limited to first-order or at most second-order processes. Highly sophisticated multi-configuration expansions can readily be used for the initial bound and the final ionic states, which is an important aspect when treating more complex targets than atomic hydrogen and helium. The basic ideas of how to extend the method to heavy-particle impact were presented, with some preliminary results for antiproton collisions with helium. Further work in this direction is currently in progress.

ACKNOWLEDGMENTS

This work was supported by the United States National Science Foundation under grant No. PHY-0222470.

REFERENCES

1. T.N. Rescigno, M. Baertschy, W.A. Isaacs, and C.W. McCurdy, Science **286** (1999) 2474.
2. M.S. Pindzola and F. Robicheaux, Phys. Rev. A **54** (1996) 2142.
3. J.P. Colgan, M.S. Pindzola, F. Robicheaux, D.C. Griffin, and M. Baertschy, Phys. Rev. A **65** (2002) 042721.
4. I. Bray, D.V. Fursa, A. Kheifets, and A.T. Stelbovics, J. Phys. B **35** (2002) R117.
5. K. Bartschat, E.T. Hudson, M.P. Scott, P.G. Burke, and V.M. Burke, J. Phys. B **29** (1996) 115.
6. K. Bartschat and I. Bray, Phys. Rev. A **54** (1996) R1002.
7. K. Bartschat, Comp. Phys. Commun. **114** (1998) 168.
8. I. Bray, Phys. Rev. Lett. **89** (2002) 273201.
9. A.T. Stelbovics,I. Bray, D.V. Fursa, and K. Bartschat, Phys. Rev. A **71** (2005) 052716.
10. D.A. Biava, H.P. Saha, E. Engel, R.M. Dreizler, R.P. McEachran, M.A. Haynes, B. Lohmann, C.T. Whelan, and D.H. Madison, J. Phys. B **35** (2002) 293.
11. D.A. Biava, K. Bartschat, H.P. Saha, and D.H. Madison, J. Phys. B **35** (2002) 5121.
12. A. Prideaux and D.H. Madison, Phys. Rev. A'**67** (2003) 052710.
13. A. Prideaux, D.H. Madison, and K. Bartschat, Phys. Rev. A **72** (2005) 032702.
14. M. Brauner, J.S. Briggs, and H. Klar, J. Phys. B **22** (1989) 2265.
15. K. Bartschat and P.G. Burke, J. Phys. B **20** (1987) 3191.

16. K. Bartschat and P.G. Burke, J. Phys. B **21** (1988) 2969.
17. K. Bartschat, Comp. Phys. Commun. **75** (1993) 219.
18. K. Bartschat, Phys. Rev. A **71** (2005) 032718.
19. S. Geltman, *Topics in Atomic Collision Theory*, (Krieger, Florida, 1997).
20. R.H.G. Reid, K. Bartschat, and A. Raeker, J. Phys. B **31** (1998) 563; corrigendum: J. Phys. B **33** (2000) 5261.
21. Y. Fang and K. Bartschat, J. Phys. B **34** (2001) L19.
22. P.G. Burke and K.T. Taylor, J. Phys. B **8** (1975) 2620.
23. R.I. Campeanu, R.P. McEachran, and A.D. Stauffer, Can. J. of Phys. **79** (2001) 1231.
24. R.I. Campeanu, R.P. McEachran, and A.D. Stauffer, Can. J. of Phys. **81** (2001) 919.
25. J.P. Marler, J.P. Sullivan, and C.M. Surko, Phys. Rev. A **71** (2005) 022701.
26. M. Foster, J.L. Peacher, M. Schulz, D.H. Madison, Z. Chen, and H.R.J. Walters, Phys. Rev. Lett. **97** (2006) 093202.
27. R.W. van Boeyen, N. Watanabe, J.W. Cooper, J.P. Doering, J.H. Moores, and M.A. Coplan, Phys. Rev. A **73** (2006) 032703.
28. K. Bartschat, X. Guan, and D. Weflen, J. Phys. B **40** (2007) 3231.
29. Z. Chen and D.H. Madison, J. Phys. B **38** (2005) 4195.
30. N.C. Sil, M.A. Crees, and M.J. Seaton, J. Phys. B **17** (1984) 1.
31. M. Schulz, R. Moshammer, D. Fischer, H. Kollmus, D.H. Madison, S. Jones, and J. Ullrich, Nature (London) **422** (2003) 48.
32. M. Foster, J. Colgan, and M.S. Pindzola, Phys. Rev. Lett. **100** (2008) 033201.

Testing *CPT* Invariance with Antiprotonic Helium Atoms [1]

Dezső Horváth

KFKI Research Institute for Particle and Nuclear Physics, H–1525 Budapest, Hungary and Institute of Nuclear Research (ATOMKI), Debrecen, Hungary

Abstract. The structure of matter is related to symmetries at every level of study. CPT symmetry is one of the most important laws of field theory: it states the invariance of physical properties when one simultaneously changes the signs of the charge and of the spatial and time coordinates of free elementary particles. Although in general opinion CPT symmetry is not violated in Nature, there are theoretical attempts to develop CPT-violating models. The Antiproton Decelerator at CERN has been built to test CPT invariance. The ASACUSA experiment compares the properties of particles and antiparticles by studying the antiprotonic helium atom via laser spectroscopy and measuring the mass, charge and magnetic moment of the antiproton as compared to those of the proton.

Keywords: antimatter, laser spectroscopy, antiproton charge, antiproton mass, magnetic moment
PACS: 14.20.Dh, 36.10.-k, 32.70.Jz

INTRODUCTION: SYMMETRIES IN PARTICLE PHYSICS

Symmetries in particle physics are even more important than in chemistry or solid state physics. Just like in any theory of matter, the inner structure of the composite particles are described by symmetries, but in particle physics everything is deduced from the symmetries (or invariance properties) of the physical phenomena or from their violation: the conservation laws, the interactions and even the masses of the particles (see [1] for all general references).

The conservation laws are related to symmetries: the Noether theorem states that a global symmetry leads to a conserving quantity. The conservation of momentum and energy are deduced from the translational invariance of space-time: the physical laws do not depend upon where we place the zero point of our coordinate system or time measurement; and the fact that we are free to rotate the coordinate axes at any angle is the origin of angular momentum conservation.

Spin is one of the most important properties of the particles: those having half-integer spins are the fermions whereas the integer-spin particles are bosons. The different symmetries of the fermions and bosons lead to dramatic differences in their behaviour, e.g. the numbers of fermions are conserved whereas the numbers of bosons are not. The basic building blocks of the visible matter of the Universe, the quarks and leptons are fermions and all known interactions are mediated by bosons.

All fermions have *antiparticles*, anti-fermions which have identical properties but

---

[1] Invited talk presented at PBAR'08, Workshop on Cold Antimatter Plasmas and Application to Fundamental Physics, Naha, Okinawa, Japan, 20-22 February 2008

CP1037, *Cold Antimatter Plasmas and Application to Fundamental Physics,* edited by Y. Kanai and Y. Yamazaki

with opposite charges. The different abundance of particles and antiparticles in our Universe is one of the mysteries of astrophysics: apparently there is no antimatter in the Universe in significant quantities, see, e.g., [2]. If there were antimatter galaxies they would radiate antiparticles and we would see zones of strong radiation at their borders with matter galaxies, but the astronomers do not see such a phenomenon anywhere.

An extremely interesting property of *antiparticles* is that they can be treated mathematically as if they were *particles* of the same mass and of oppositely signed charge of the same absolute value *going backward in space and time.* This is the consequence of one of the most important symmetries of Nature: *CPT* invariance [3]. *CPT* reflection means the following simultaneous operations:

- charge conjugation (i.e. changing particles into antiparticles), $C\psi(r,t) = \overline{\psi}(r,t)$;
- parity change (i.e. mirror reflection), $P\psi(r,t) = \psi(-r,t)$, and
- time reversal, $T\psi(r,t) = \psi(r,-t)$.

The principle of *CPT* invariance states that this does not change the physical properties of the system. This means that, e.g., the annihilation of a positron with an electron can be described as if an electron came to the point of collision, irradiated two or three photons and then went out backwards in space-time. Although the weak interaction, unlike the other interactions, violates both *P* and *CP* conservation, *CPT* invariance is assumed to be absolute. In the author's opinion, the strongest argument for the validity of *CPT* invariance is the fact that using Feynman diagrams gives theoretical results in quantum electrodynamics which agree with experiment with an incredible precision.

TESTING *CPT* INVARIANCE

This principle requires, e.g., that particles and antiparticles have the same mass and have additive quantum numbers (like charge) of the same absolute value but opposite sign. Thus a straightforward *CPT* test is measuring the mass and charge of particles and antiparticles (the best candidates being the proton and the antiproton as the heaviest stable particles).

All such laws have to be and are checked experimentally. *CPT* invariance is so deeply embedded in field theory that many theorists claim it is impossible to test within the framework of present-day physics. Indeed, in order to develop *CPT*-violating models one has to reject such fundamental axioms as Lorentz invariance or the locality of interactions (i.e. causality) or unitarity [4, 5, 6].

As far as we know, the Standard Model is valid up to the Planck scale, $\sim 10^{19}$ GeV. Above this energy scale we expect to have new physical laws which may allow for Lorentz and *CPT* violation as well [4]. Quantum gravity [5, 6] could cause fluctuations leading to Lorentz violation, or loss of information in black holes which would mean unitarity violation. Also, a quantitative expression of Lorentz and *CPT* invariance needs a Lorentz and *CPT* violating theory [4]. On the other hand, testing *CPT* invariance at low energy should be able to limit possible high energy violation. This makes experimental *CPT* tests physically valuable in spite of the fact that most of us do not expect its violation.

CPT invariance is so far fully supported by the available experimental evidence and it is absolutely fundamental in field theory. Nevertheless, there are many experiments trying to test it. The simplest way is to compare the mass or charge of particles and antiparticles. The most precise such measurement is that of the relative mass difference of the neutral K meson and its antiparticle: it is less than 10^{-18} [3].

In spite of its huge success in describing all experimental data the Standard Model of elementary particles has several theoretical difficulties. Gravity does not fit into its system of gauge interactions. We do not know why there are exactly three fermion families and what interaction causes the mixing of neutrino types. The prevalence of matter against antimatter in the Universe and the left-right asymmetry of weak interactions are also unexplained and the Universe seems to be dominated by the mysterious dark matter and the even more mysterious dark energy. There are extensions of the Standard Model trying to explain these effects, but so far we have no experimental evidence supporting any of them in spite of great efforts of the particle physics community.

Phenomenologically, the oscillations observed for solar and atmospheric neutrinos can be incorporated using two neutrino mass differences and a mixing matrix with three angles and a phase [3]. The *Liquid Scintillator Neutrino Detector (LSND)* experiment at Los Alamos found a strange excess of $\overline{\nu}_e$ in a $\overline{\nu}_\mu$ beam [8] which cannot be explained within the Standard Model with three neutrinos, but can be interpreted in terms of Lorentz violation [9]. Recently, the Kostelecký group developed a Lorentz-violating extension of the Standard Model called *Tandem model* [10] which could explain all of the observed neutrino oscillations using massive neutrinos in a Hamiltonian with Lorentz-violating terms. Another recent attempt [11] deduced the neutrino mass differences from unitarity-violating quantum-gravity induced decoherence. However, in 2007 MiniBooNE experiment, designed to test the LSND anomaly, announced [12] that their study excludes the LSND result assuming the same mechanisms for neutrino and antineutrino oscillation.

Moreover, Lorentz violation should involve a dependence of the result of an experiment on its spatial direction. The BaBar collaboration made a systematic sidereal study of $B^0 - \overline{B}^0$ oscillations and excluded any Lorentz violation [7] with a significance of 2.8σ.

CERN has constructed its *Antiproton Decelerator (AD)* facility [13] in 1999 (Fig. 1) in order to test the *CPT* invariance by comparing the properties of proton and antiproton and those of hydrogen and antihydrogen. The AD was constructed mainly using outside funds and started to operate at the end of 1999. By the end of 2000 it was brought to specifications. 25 GeV/c protons from the Proton Synchrotron are shot in an iridium target where they produce particle–antiparticle pairs. Antiprotons are collected at 3.5 GeV/c momentum and slowed down in the AD ring in three steps to 100 MeV/c using stochastic and electron cooling.

The aim of the present work is to briefly summarize some of the results of the ASACUSA [14] experiment; the interested reader can find a comprehensive review in [15]. The other two AD-experiments testing the CPT invariance are ALPHA [16] and ATRAP [17], they are preparing spectroscopic studies of antihydrogen, the bound state of an antiproton with a positron [18].

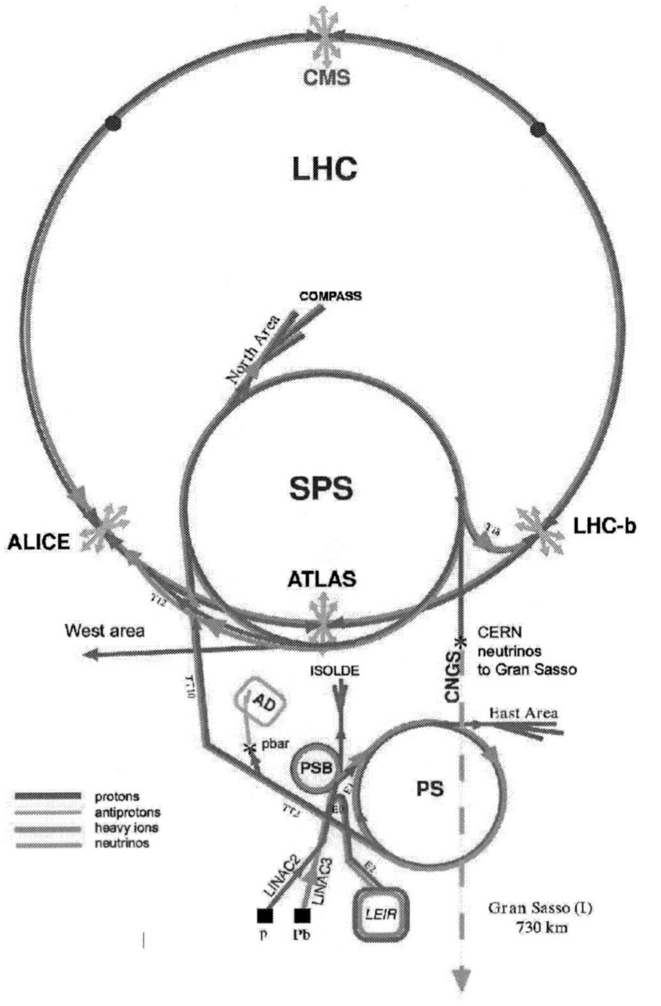

FIGURE 1. The accelerator complex of CERN. The LINAC2 linear accelerator and the PSB booster feed protons into the PS proton synchrotron, which accelerates them to 25 GeV/c and passes them to the experiments in the East Area or to the SPS super proton synchrotron for further acceleration and once every 100 seconds into an iridium target to produce antiprotons. The antiprotons are collected at 3.5 GeV/c by the AD where they are decelerated in three steps to 100 MeV/c. The PS also accelerates heavy ions for the SPS North Area experiments and until 2000 it did accelerate electrons and positrons for the LEP Large Electron Positron collider which was dismounted to be replaced by the LHC Large Hadron Collider.

ANTIPROTONIC HELIUM ATOMS

An exotic atom is formed when a fast negative particle — muon, pion, kaon or antiproton — penetrates matter: it first slows down in atomic collisions (mostly via ionization), then

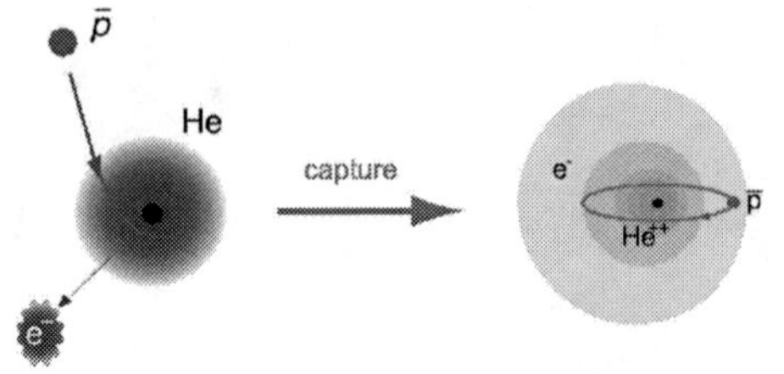

FIGURE 2. Formation of the metastable antiprotonic helium atom, a three-body bound state.

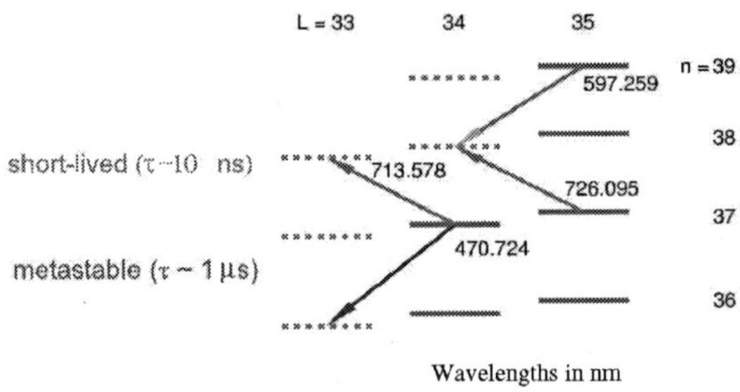

FIGURE 3. Principle of laser spectroscopy: forcing fast annihilation by inducing transitions between long-lived and short-lived antiprotonic states.

gets captured in an atomic orbit *replacing* the last knocked–out electron. The capture cross–section is related to the overlap between the wave functions of the particle and the atomic electron so the heavy particle will initially populate atomic states with radii close to that of the electrons. Thus an antiproton captured, e.g., in a helium atom will initially populate the $\bar{p}He^+$ states with principal quantum numbers $n_0 = \sqrt{M/m} \approx 38$ where $M \approx 0.8m_p$ and $m \approx m_e$ are the reduced masses of the $\bar{p}He^{++}$ and e^-He^{++} systems[19]. A high n, naturally, involves orbital quantum numbers in the region $0 \leq \ell \leq n-1$; and although experiments found deviations of the initial populations from a purely statistical $2\ell+1$ distribution, the states with higher ℓ will be populated with higher probability.

The freshly formed, highly excited exotic atom has two basic ways to step down. Between high-n states, where the energy spacing is low, the Auger mechanism dominates whereas lower lying levels will preferably decay via radiation. Approaching the ground state a strongly interacting hadron like the antiproton gets absorbed by the nucleus from higher nS levels and it hardly reaches ground state in heavier atoms. Both in condensed media and in gases at higher pressures (about standard conditions) slowing down, atomic capture, de-excitation and nuclear absorption proceeds quite fast: theoretical calculations and experimental measurements agree upon total lifetimes below or around 1 ps (10^{-12} s).

The only exception is helium: while 97% of the antiprotons stopped in a dense helium target annihilate with the usual short lifetimes, 3% live as long as several microseconds, sufficiently long to do laser spectroscopy. They form a $\bar{p}He^+$ 3–body system (Fig. 2)

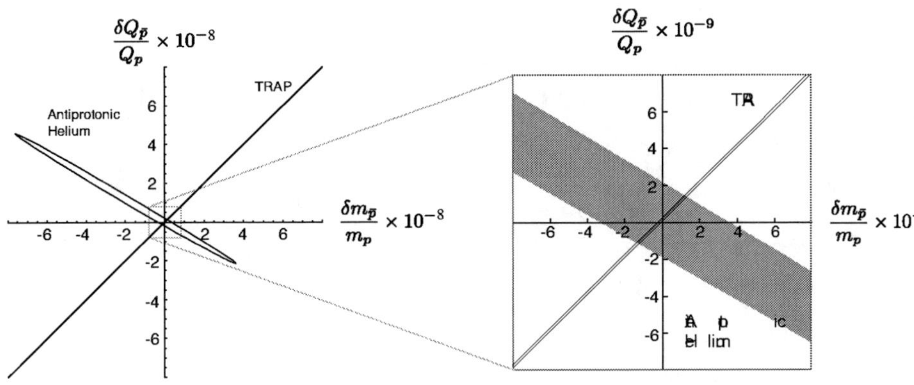

FIGURE 4. Limits on the difference of mass and charge between proton and antiproton. The charge/mass (Q/M) ratio was measured by the TRAP group [23] whereas $M \cdot Q^2$ by ASACUSA [24, 27, 28, 29]. With the improvement of the experimental technique the allowed region was step-by-step reduced: the present limit is 2 ppb (2×10^{-9}) [29].

where the antiproton orbit is protected against collisions by the electron, and the antiprotonic states of the same n but different ℓ lose the energy degeneracy and so cannot undergo Stark transitions. The model and its experimental proof are described in reviews [15, 20, 21, 22].

The principle of the spectroscopy method is simple. We stop a bunch of antiprotons in helium, wait until the promptly annihilating states disappear and then stimulate the transition from a long–living state to a short–lived one with a tunable laser system. At the resonant frequency corresponding to the transition energy the laser shot will be followed by immediate annihilation (Fig. 3).

The Mass and Charge of the Antiproton

The TRAP group measured the charge/mass ratio of the proton and the antiproton at LEAR [23]. They kept a single antiproton and a single H^- ion in the same Penning–trap simultaneously at different orbits and measured their cyclotron frequencies. After having made corrections for the H^- — p deviation they limited the relative difference to $9 \cdot 10^{-11}$. Achieving this precision took 10 years' work.

The aim of the ASACUSA experiment is to supply additional data for facilitating the separation of charge and mass information. This is done via studying antiprotonic transition energies in the $\overline{\mathrm{p}}\mathrm{He}^+$ system as those are proportional to $m(\overline{\mathrm{p}}) \cdot q(\overline{\mathrm{p}})^2$:

$$E_n \approx -\frac{m_{\mathrm{red}} c^2 (Z\alpha)^2}{2n^2}$$

The precision of determining the transition frequencies is limited among others by the laser bandwidth, the density shift of the lines and the Doppler effect. Our first

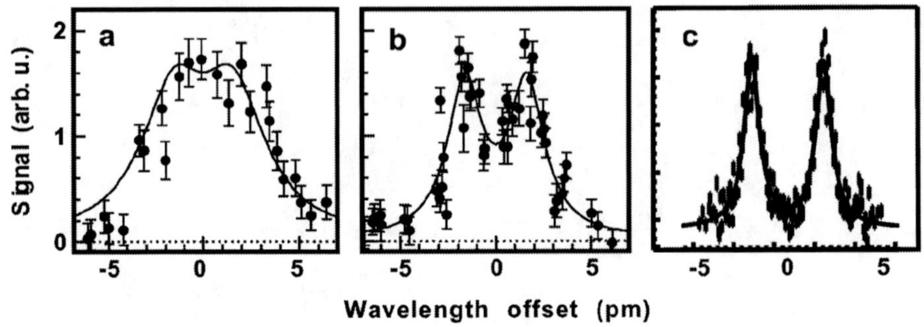

FIGURE 5. The development of the energy resolution of $\overline{\text{p}}$He spectroscopy. The $(n,\ell) = (37,35) \rightarrow (38,34)$ split transition at 726.1 nm as measured in (a) 2000; (b) 2002 and (c) 2004.

measurement was performed at LEAR for different helium densities and we extrapolated to zero [24] in order to make the comparison with the theoretical calculations for isolated atoms [25, 26]. The way to obtain a limit on the antiprotonic mass and charge is illustrated in Fig. 4: the intersection of the regions allowed by the two measurements constitutes the limit. The result was $[m(\text{p}) - m(\overline{\text{p}})]/m(\text{p}) < 5 \cdot 10^{-7}$ and similarly $[q(\text{p}) - q(\overline{\text{p}})]/q(\text{p}) < 5 \cdot 10^{-7}$.

In 2000, the first year of the AD we lowered the deteriorating effects of the laser bandwidth and gained almost an order of magnitude in precision [27], but the collisional effects were still significant. In 2002 we have installed a radio-frequency quadrupole post-decelerator (RFQD) which decelerated the AD beam from 5.6 MeV to 100 keV with a $\sim$ 30% efficiency. The RFQD made it possible to substantially reduce the density effect by using a low-pressure ($<$ 1 mbar), cryogenic ($T = 6 - 10$ K) target (Fig. 5). Our latest published limit of possible *CPT*-violation on the antiproton charge and mass is 2 ppb ($< 2 \times 10^{-9}$) [29]. The new laser system and the newest results are described in Naoya Ono's contribution to this conference.

Further improvement is expected from two-photon spectroscopy which helps to reduce the Doppler broadening. In a counterpropagating geometry two laser beams of the same frequency can completely eliminate the first-order Doppler effect. However, this is not feasible to use in our case as it should involve unattainably high laser powers [15]. In the practical case shown in Fig. 6 the frequencies of the two laser beams are adjusted so that their sum, $\nu_1 + \nu_2$ is tuned to the two-photon transition $(n,\ell) = (36,34) \rightarrow (34,32)$ involving an angular momentum change of $\Delta\ell = 2$, and the virtual intermediate state involved in the two-photon transition is tuned close (within a few GHz) to a real state $(n,\ell) = (35,33)$. The nearness of the virtual state to a real one should lead to an enhancement of the two-photon transition probability by several orders of magnitude compared to the case $\nu_1 = \nu_2$ and laser energy densities of $\sim$ 1 mJ/cm^2 would be sufficient to efficiently drive the two-photon transition. The observed width $\Delta\nu_{2\gamma}$ of the resonance line will decrease by a factor 20 compared to the thermal Doppler width $\Delta\nu_D$ according to the equation,

$$\Delta\nu_{2\gamma} = \left| \frac{\nu_1 - \nu_2}{\nu_1 + \nu_2} \right| \Delta\nu_D. \tag{1}$$

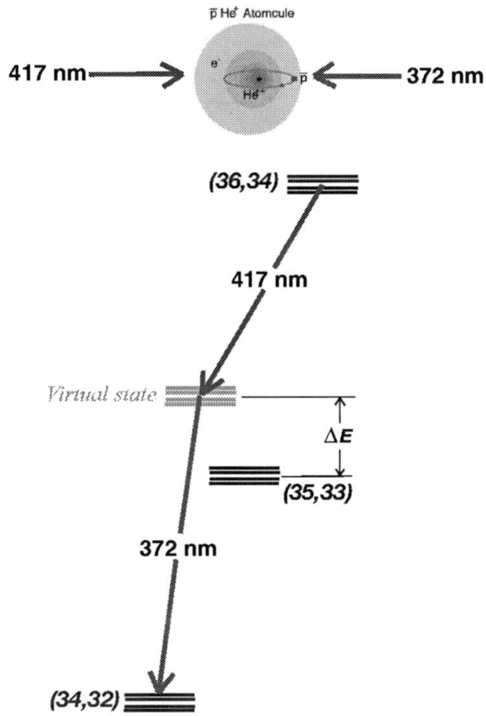

FIGURE 6. Portion of the energy level diagram of $\overline{p}^4\text{He}^+$, indicating the principle of sub-Doppler two-photon spectroscopy. The ΔE energy difference should be as small as possible, but larger than the involved laser frequency uncertainties.

Level Splitting and Magnetic Moment

As seen in Fig. 5 the $(n = 37, \ell = 35) \rightarrow (38, 34)$ line is split due to interaction between the antiproton magnetic moment and the electron spin. This resonance was used by the ASACUSA Collaboration to measure the magnetic moment of the antiproton: of course, as that is in a highly excited state, the measured momentum is mostly orbital.

The level scheme is presented in Fig. 7. Both levels involved in the transition are split, scanning the laser frequency will show the difference between the transition frequencies f_- and f_+. We have measured the splitting ν_{HF} directly by emptying one of the split levels with a suitably tuned laser pulse, irradiating the system with a variable microwave pulse and again with a laser pulse of the same frequency as before. When the microwave was correctly tuned to the resonance, the second resonance had a reduced amplitude. The result is presented in Fig. 7 [30]. It agrees within 6×10^{-5} with the recent theoretical calculations performed with the properties of the proton assumed. We also checked the possible effect of collisions on the result by studying its density dependence and found it was negligible.

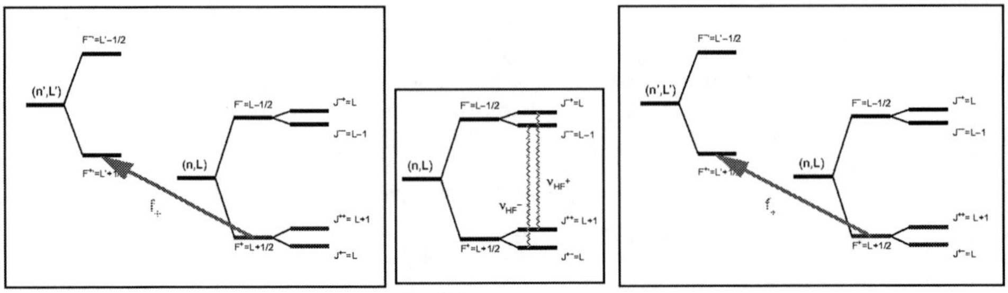

Step 1: depopulation of F⁺ doublet with f_+ laser pulse

Step 2: equalization of populations of F⁺ and F⁻ by microwave

Step 3: probing of population of F⁺ doublet with 2nd f_+ laser pulse

FIGURE 7. The scheme of a split transition line in the $\overline{p}$-$^4He^+$ atom as studied using the laser–microwave–laser resonance method in order to measure the magnetic moment of the antiproton [30] in the atomic bound state.

The experimental details and newest results are reported in Thomas Pask's contribution.

CONCLUSION

Our latest published limit of *CPT* violation is 2×10^{-9} (2 ppb). However, I must point out that these experiments can only confirm the validity of *CPT*, nobody would believe us if we reported discovering a violation as there are theoretical calculations are involved in the analysis. On the other hand, if we assume the validity of *CPT*, our data can be used to increase the precision of the proton mass: the official CODATA uncertainty for those measurements decreased from 0.46 ppb to 0.43 ppb as a result of including our data.

ACKNOWLEDGEMENTS

The present work was supported by Monbukagakusho (Grant No. 15002005), by the Hungarian National Research Foundation (Contracts OTKA T046095, K72172 and NK67974) and the FP6 project MKTD-CT-2004-509252.

REFERENCES

1. Halzen, F. and Martin, A.D., *Quarks and Leptons: An Introductory Course in Modern Particle Physics*, Wiley, New York, 1984.
2. Cohen, A.G., De Rujula, A., Glashow, S.L., *A Matter - Antimatter Universe?*, Astrophys. J., **495**, 539 (1998).
3. Particle Data Group, W.-M. Yao *et al.*, *Review of Particle Physics*, J. Phys. G **33**, 1 (2006). (URL: http://pdg.lbl.gov).

4. V. A. Kostelecký, *Gravity, Lorentz Violation, and the Standard Model*, Phys.Rev.D **69**, 105009 (2004).
5. N. E. Mavromatos, *CPT Violation and Decoherence in Quantum Gravity*, Lecture Notes in Physics, **669**, 245 (2005).
6. F. R. Klinkhamer, Ch. Rupp, *Space-Time Foam, CPT Anomaly, and Photon Propagation*, Phys.Rev.D **70**, 045020 (2004).
7. The BaBar Collaboration, B. Aubert et al., *Search for CPT and Lorentz Violation in B0-B0 Oscillations with Dilepton Events*, Preprint arXiv:0711.2713v1, 27 Nov 2007.
8. A. Aguilar *et al.* [LSND Collaboration], *Evidence for neutrino oscillations from the observation of anti-nu/e appearance in a anti-nu/mu beam,* Phys. Rev. D **64**, 112007 (2001).
9. L. B. Auerbach *et al.* [LSND Collaboration], *Tests of Lorentz violation in anti-nu/mu –> anti-nu/e oscillations,* Phys. Rev. D **72**, 076004 (2005).
10. T. Katori, A. Kostelecky and R. Tayloe, *Global three-parameter model for neutrino oscillations using Lorentz violation,* Phys. Rev. D **74**, 105009 (2006).
11. G. Barenboim, N. E. Mavromatos, S. Sarkar and A. Waldron-Lauda, *Quantum decoherence and neutrino data*, Nucl. Phys. B **758**, 90 (2006).
12. A. A. Aguilar-Arevalo et al. [The MiniBooNE Collaboration], *A Search for electron neutrino appearance at the* $\Delta m \sim 1\text{eV}^2$ *scale*, Phys. Rev. Lett. **98**, 231801 (2007).
13. The Antiproton Decelerator (AD) at CERN, `http://psdoc.web.cern.ch/PSdoc/acc/ad/index.html`
14. ASACUSA (*Atomic Spectroscopy And Collisions Using Slow Antiprotons*) Collaboration [2], `http://www.cern.ch/ASACUSA`.
15. R.S. Hayano, M. Hori, D. Horváth, E. Widmann, *Antiprotonic helium and CPT invariance*, Reports on Progress in Physics **70**, 1995-2065 (2007).
16. ALPHA (*Antihydrogen Laser PHysics Apparatus*) Collaboration, `http://alpha.web.cern.ch/alpha`.
17. ATRAP (*Antimatter TRAP*) Collaboration, `http://hussle.harvard.edu/ atrap/`.
18. M. Charlton, J. Eades, D. Horváth, R. J. Hughes, C. Zimmermann, *Antihydrogen Physics*, Physics Reports **241**, 65-117 (1994).
19. K. Tőkési, B. Juhász, J. Burgdörfer, *Exotic Rydberg atom formation in low-energy antiproton-helium collisions*, J. Phys. B: At. Mol. Opt. Phys. **38**, 401-407 (2005).
20. D. Horváth, *Metastable antiprotonic helium atoms*, Radiochimica Acta **77**, 75-86 (1997).
21. J. Eades, J.F. Hartmann, *Forty years of antiprotons*, Rev. Mod. Phys. **71**, 373-420 (1999).
22. T. Yamazaki, N. Morita, R.S. Hayano, E. Widmann, J. Eades, *Antiprotonic Helium*, Physics Reports, **366**, 183-329 (2002).
23. G. Gabrielse, A. Khabbaz, D.S. Hall, C. Heimann, H. Kalinowsky, W. Jhe, *Precision mass spectroscopy of the antiproton and proton using simultaneously trapped particles*, Phys. Rev. Lett. **82**, 3198-3201 (1999).
24. H. A. Torii, R. S. Hayano, M. Hori, T. Ishikawa, N. Morita, M. Kumakura, I. Sugai, T. Yamazaki, B. Ketzer, F. J. Hartmann, T. von Egidy, R. Pohl, C. Maierl, D. Horváth, J. Eades, and E. Widmann, *Laser measurements of the density shifts of resonance lines in antiprotonic helium atoms and stringent constraint on the antiproton charge and mass*, Phys. Rev. **A 59**, 223-229 (1999).
25. V.I. Korobov, *Variational calculation of energy levels in* $\bar{p}\text{He}^+$ *molecular systems*, Phys. Rev. **A 54**, R1749-R1752 (1996).
26. V.I. Korobov, D.D. Bakalov, *Energies and relativistic corrections for the metastable states of antiprotonic helium atoms* Phys. Rev. Lett. **79**, 3379-3382 (1997).
27. M. Hori, J. Eades, R. S. Hayano, T. Ishikawa, J. Sakaguchi, E. Widmann, H. Yamaguchi, H. A. Torii, B. Juhász, D. Horváth, T. Yamazaki, *Sub–ppm laser spectroscopy of antiprotonic helium and a CPT–violation limit on the antiprotonic charge and mass*, Phys. Rev. Lett., **87**, 093401 (2001).
28. M. Hori, J. Eades, R. S. Hayano, T. Ishikawa, W. Pirkl, E. Widmann, H. Yamaguchi, H. A. Torii, B. Juhász, D. Horváth, T. Yamazaki, *Direct measurement of transition frequencies in isolated pbar-He+ atoms, and new CPT violation limits on the antiproton charge and mass, Phys. Rev. Lett.* **91**, 123401

[2] Asakusa is one of the oldest districts of Tokyo; the name was proposed by our non–Japanese collaborators to honour the dominant Japanese contribution to the experiment

(2003).
29. M. Hori, A. Dax, J. Eades, K. Gomikawa, R. S. Hayano, N. Ono, W. Pirkl, E. Widmann, H. A. Torii, B. Juhász, D. Barna, D. Horváth, *Determination of the antiproton-to-electron mass ratio by precision laser spectroscopy of* $\overline{p}$*He*, *Phys. Rev. Lett.* **96**, 243401 (2006).
30. E. Widmann, R.S. Hayano, T. Ishikawa, J. Sakaguchi, H. Yamaguchi, J. Eades, M. Hori, H.A. Torii, B. Juhász, D. Horváth, T. Yamazaki, *Hyperfine structure of antiprotonic helium revealed by a laser-microwave-laser resonance method*, Phys. Rev. Lett., **89**, 243402 (2002).
31. K. Y. Franzen, N. Kuroda, H. A Torii, M. Hori, Z. Wang, H. Higaki, S. Yoneda, B. Juhász, D. Horváth, A. Mohri, K. Komaki, Y. Yamazaki, *Transport beam line for ultraslow monoenergetic antiprotons*, Rev. Sci. Instrum. **74**, 3305 (2003).
32. N. Kuroda, H. A. Torii, K. Yoshiki Franzen, Z. Wang, S. Yoneda, M. Inoue, M. Hori, B. Juhász, D. Horváth, H. Higaki, A. Mohri, J. Eades, K. Komaki, Y. Yamazaki, *Confinement of a large number of antiprotons and production of an ultraslow antiproton beam*, Phys. Rev. Lett. **94**, 023401 (2005).
33. E. Widmann, R. S. Hayano, M. Hori and T. Yamazaki, *Measurement of the hyperfine structure of antihydrogen,* Nucl. Instrum. Meth. B **214**, 31-34 (2004).
34. R. Bluhm, V. A. Kostelecky and N. Russell, *CPT and Lorentz tests in hydrogen and antihydrogen,* Phys. Rev. Lett. **82**, 2254-2257 (1999)
35. CODATA *Internationally recommended values of the fundamental physical constants* 2006. http://physics.nist.gov/cuu/Constants/index.html.

Determination of the antiproton-to-electron mass ratio by high precision laser spectroscopy of $\bar{p}He^+$

Naoya Ono*,†, Masaki Hori†,*, Ryugo S. Hayano* and the ASACUSA collaboration**

*Department of Physics, University of Tokyo, 7-3-1 Hongo, Bunkyo-ku, Tokyo 113-0033, Japan
†Max-Planck-Institut für Quantenoptik, Hans-Kopfermann-Strasse 1, 85748 Garching, Germany
**CERN, CH-1211 Geneva 23, Switzerland

Abstract. The ASACUSA collaboration has carried out laser spectroscopy of an antiprotonic helium atom ($\bar{p}He^+$: a Coulomb three-body system consisted of an antiproton, electron, and a helium nucleus) at CERN's Antiproton Decelerator facility and measured transition frequencies of $\bar{p}He^+$. By comparing the measured frequencies with three-body QED calculations, the antiproton-to-electron mass ratio $m_{\bar{p}}/m_e$ was determined. We recently developed a new laser system consisting of single-mode continuous-wave lasers, a femtosecond optical frequency comb, a pulse-amplified dye laser, and a chirp compensation, and achieved a precision of $\sim 1 \times 10^{-8}$ on the $\bar{p}He^+$ transition frequencies, which was about one order of magnitude improvement. The comparison of the measured frequencies and the three-body QED calculations yielded the antiproton-to-electron mass ratio of $m_{\bar{p}}/m_e = 1836.152\,674(5)$. Combined with a constraint on the proton and antiproton charge-to-mass ratio of the TRAP group, a limit on any *CPT*-violating difference between the antiproton and proton charges and masses was also set as $|Q_p + Q_{\bar{p}}|/Q_p \sim |m_p - m_{\bar{p}}|/m_p < 2 \times 10^{-9}$ at a confidence level of 90 %.

Keywords: CPT symmetry, antiproton, antiprotonic helium
PACS: 36.10.-k, 06.20.Dk, 14.20.Dh, 32.70.Jz

INTRODUCTION

The *CPT* symmetry is one of the most basic properties of the standard model of the particle physics. According to this symmetry, physical laws are invariant against the combined operations of charge conjugation, parity, and time reversal. Among the consequences of this is that the values of the charge, mass, lifetime of a particle are equal to those of its antiparticle. All experiments are presently consistent with the *CPT* symmetry.

Recently, various theoretical works about the extensions of the standard model have been developed[1, 2] and experimental tests have become important. ATRAP[3, 4, 5], ALPHA[6], AEGIS[7], and previously ATHENA[8] collaborations are attempting to carry out precision tests of the *CPT* symmetry by studying antihydrogen at CERN's Antiproton Decelerator facility (AD). Meanwhile, the ASACUSA collaboration studies antiprotonic helium atoms ($\bar{p}He^+$), which are Coulomb three-body systems consisting of an antiproton, electron, and a helium nucleus[9, 10]. By measuring the transition frequencies of $\bar{p}He^+$ and comparing them with three-body QED calculations[11, 12, 13, 14], a limit was set on any *CPT*-violating difference between the antiproton and proton

CP1037, *Cold Antimatter Plasmas and Application to Fundamental Physics,* edited by Y. Kanai and Y. Yamazaki

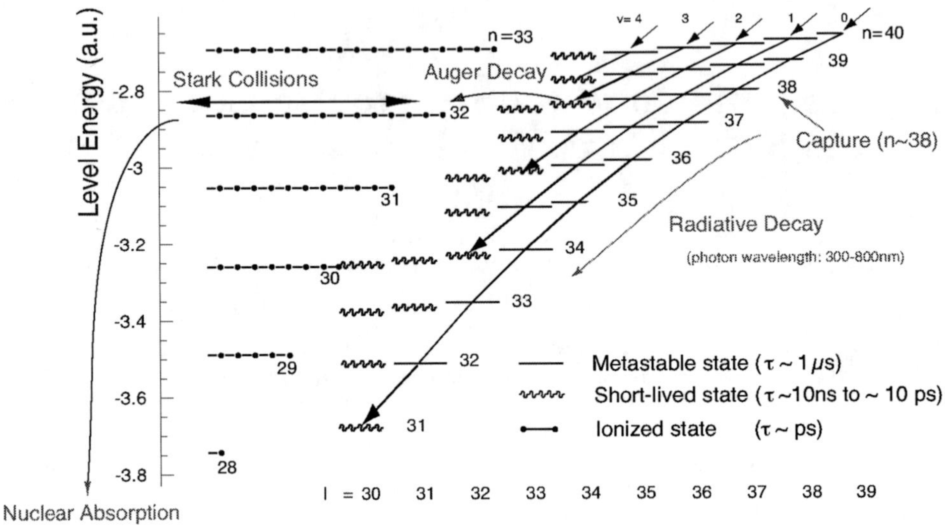

FIGURE 1. A portion of the energy level diagram of $\bar{p}^4He^+$. The principal (n) and the angular momentum (l) quantum numbers are indicated. Metastable, short-lived, and l-degenerate ionic states are indicated by solid lines, wavy lines, and dotted lines. A laser beam induces a transition from a metastable state to a short-lived state. When a $\bar{p}He^+$ reaches to a short-lived state, an Auger electron is emitted and the antiproton is absorbed in a helium nucleus[10].

charges and masses. If we assume the validity of *CPT* invariance at the present experimental precision, the results can be interpreted as a determination of the antiproton-to-electron mass ratio ($m_{\bar{p}}/m_e$).

This article reviews our latest experiment [15], in which we used an optical frequency comb[16, 17, 18] and a continuous-wave pulse amplified dye laser and achieved about one order of magnitude better precision compared to the previous experiment [19].

FORMATION OF ANTIPROTONIC HELIUM ATOMS AND LASER SPECTROSCOPY

Some 3 % of antiprotons injected into a helium target forms metastable $\bar{p}He^+$ with a mean lifetime of 3 μs in the following way. The antiprotons reduce their kinetic energy by exciting and ionizing helium atoms. After the energy of an antiproton falls roughly below the ionization energy of a helium atom, $\bar{p}He^+$ is formed by replacing one of the ground-state electrons.

The metastable $\bar{p}He^+$ is formed when the antiproton is initially captured in the state with the large principal quantum number, $n \sim \sqrt{M^*/m_e} \sim 38$, where M^* is the reduced mass of the $\bar{p}$-He system, and the large angular momentum quantum number $l \sim n-1$[9, 20, 21]. This is due to two reasons; (i) The remaining electron of $\bar{p}He^+$ removes the degeneracy of the $\bar{p}He^+$ states. The electron also screens $\bar{p}He^+$ from the surrounding

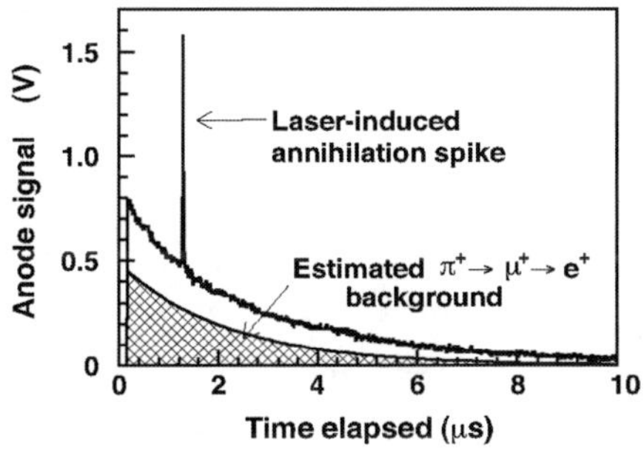

FIGURE 2. Delayed annihilation time spectrum of the $\bar{p}He^+$ transition $(n,l) = (39,35) \rightarrow (38,34)$ at wavelength $\lambda = 597.3$ nm. The hatched area represents the estimated contribution of $\pi^+ \rightarrow \mu^+ \rightarrow e^+$ decay produced by the pions stopped in materials surrounding the helium target[22].

helium atoms by the Pauli principle. Therefore, the collisional Stark effect is suppressed. (ii) Because of the large ionization energy of the remaining electron (~ 25 eV) compared to the level spacing of the n-different states of $\bar{p}He^+$ (~ 2 eV), transitions from a state with large l to an ionic state require a large angular momentum change Δl up to ~ 6. Qualitatively, the Auger lifetime τ_A is theoretically expected to relate to Δl as $\tau_A \sim 10^{-17+3\Delta l}$[9]. Considering the typical lifetime of these states against radiative transitions $\tau = 1 \sim 2\ \mu$s, the Auger transitions from states indicated by solid-line in Fig. 1, which require $\Delta l \geq 4$ ($\tau_A > 10\mu$s), are suppressed.

The main cascade proceeds by preserving the vibrational quantum number $v \equiv n - l - 1$[22]. When the antiproton reaches an Auger-dominated short-lived state indicated by wavy lines in Fig. 1, the remaining electron is ejected and the system proceeds to an ionic state, $\bar{p}He^{2+}$[23, 24, 25]. The ionic state is subject to collisional Stark mixing which causes transitions to states of small l at high n. Because the wavefunction of the helium nucleus now overlaps with that of the antiproton, the antiproton is absorbed by the helium nucleus via strong interaction, and annihilates.

Laser spectroscopy is performed by inducing a transition between a metastable state and a short-lived state, for example, between $(n,l) =$ (39, 35) and (38, 34) of $\bar{p}^4He^+$ with a transition wavelength of 597.3 nm (Fig. 1). In case the lower state is a short-lived one, e.g. $(n,l) =$ (38, 34), the $\bar{p}He^+$ decays immediately via a fast Auger transition. Thus the $\bar{p}He^+$ atom annihilates following the laser irradiation. By detecting the charged pions emerging from these annihilations, a delayed annihilation time spectrum, i.e. the distribution of the number of annihilations as a function of the time elapsed since $\bar{p}He^+$ formation, can be obtained (Fig. 2). The peak intensity relative to the laser frequency was measured to obtain a resonance profile from which the transition frequency could be determined.

EXPERIMENTAL SETUP

Figure 3 shows the layout of the experiment. Antiprotons of momentum $p = 100$ MeV/c extracted from the AD were decelerated by a radiofrequency quadrupole de-

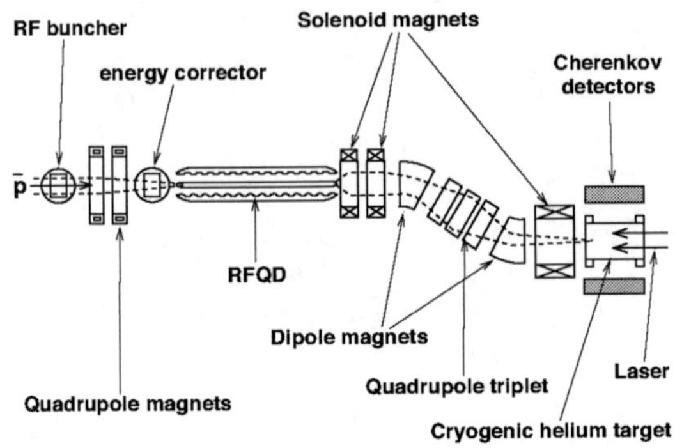

FIGURE 3. A layout of the experimental setup. 5-MeV antiprotons from the AD were decelerated to 100 keV by using the RFQD. The decelerated antiprotons were injected into a helium target through the momentum analyzer and formed $\bar{p}He^+$. A laser pulse from the opposite side irradiated the $\bar{p}He^+$ and induced the annihilation. Charged pions produced by the annihilations of antiprotons were detected by Cherenkov counters[19].

celerator (RFQD). The RFQD was developed by the ASACUSA collaboration[19] and CERN[26]. The RFQD consisted of a set of four rod electrodes with a length of 3.4 m, which were excited by an RF field of frequency $f = 202.5$ MHz. First, an RF buncher shaped the 100-ns-long antiproton pulse into a train of 20 micropulses, each with a pulse length of 0.3 ns and pulse-to-pulse separation of 5 ns. The RFQD could be biased with a dc potential to vary the output energy from 10 to 120 keV. An energy-correcting RF cavity at the input of the RFQD was used to compensate for the changes in the energy of the incident beam resulting from this biasing, and variations in the energy of the antiprotons extracted from the AD. The deceleration efficiency was about 30 %, whereas the remaining 70 % emerged with little or no deceleration. The decelerated antiprotons were diverted by an achromatic momentum analyzer and focused into a 10-mm-diameter entrance of a cryogenic helium target. The diversion made it possible to avoid large background which could be produced by the undecelerated antiprotons.

The antiprotons entered the ^{4}He or ^{3}He target through a 0.8-μm thick polyethylene foil. A target chamber with a diameter of 150 mm and a length of 350 mm was isolated by a vacuum and cooled by liquid helium to about 10 K. Impurity gases were frozen at this temperature so that the quenching of the $\bar{p}He^+$ by collisions with these impurities were avoided[27]. The target density was kept at about 10^{18} cm^{-3} and this avoided collisional shifts in the measured transition frequencies[19]. The $\bar{p}He^+$ was irradiated by a laser through a fused silica window from the direction opposite to that of the antiproton beam. Charged pions emerging from the annihilations were detected using Cherenkov detectors[28].

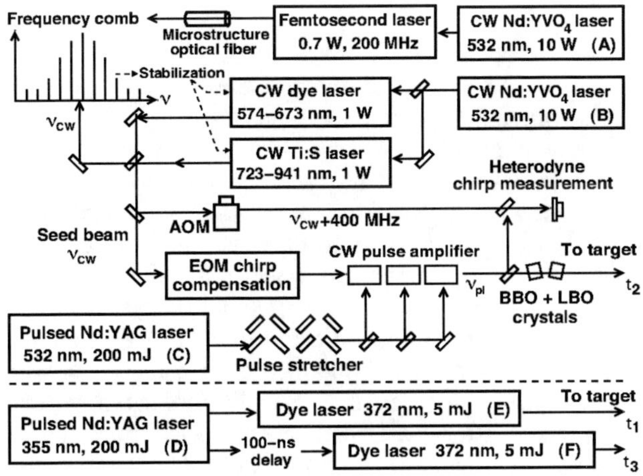

FIGURE 4. Top: A layout of the stabilization of the cw pulse-amplified dye laser to femtosecond frequency comb. Bottom: additional dye lasers used to measure the 417.8 nm transition[15].

HIGH PRECISION SPECTROSCOPY USING AN OPTICAL FREQUENCY COMB AND A CW PULSE-AMPLIFIED DYE LASER

In the previous experiment [19, 29], we used commercial pulsed lasers because only pulsed lasers could provide megawatt-scale intensities which were needed to induce the $\bar{p}He^+$ transitions. These lasers, however, had large linewidths (at least ~ 600 MHz in the best case[19, 29]) and the output beam consisted of several modes whose intensities fluctuated shot by shot[30]. Furthermore, the wavelength meter used in these experiments was calibrated over the wide range of $\bar{p}He^+$ transition wavelengths $\lambda = 264.7$-726.1 nm using molecular and atomic calibration lines[30], but the precisions of these were around 10^{-6}.

In order to achieve higher precision on the $\bar{p}He^+$ transition frequency, we developed a new laser system consisting of single-mode continuous-wave (cw) lasers, a femtosecond optical frequency comb, a pulse-amplified dye laser, and a chirp compensation as shown in Fig. 4.

First, we used a single-mode cw ring Ti:Sapphire or dye laser with linewidth $\Gamma \sim 1$ or 4 MHz and power $P \sim 1$ W to produce a seed beam. The frequency of the cw lasers were measured and stabilized using a femtosecond optical frequency comb[16, 17, 18]. We here briefly recall the principle of the frequency comb. A mode-locked femtosecond laser essentially consists of a cavity with a length L, in which a single pulse is circulating. The laser produces a train of pulses with a repetition rate $\nu_{\rm rep} = c/2L$. The phase of the carrier wave of the laser pulse is shifted from that of the envelope of the pulse train due to the difference between the group velocity of the envelope and the phase velocity of the carrier wave. In the frequency domain, the pulse train can be seen as a comb of laser modes separated by the repetition rate $\nu_{\rm rep}$. The frequency of the nth mode is expressed as

$$\nu_n = \nu_{\mathrm{off}} + n\nu_{\mathrm{rep}}. \tag{1}$$

The cw laser can be locked to one of the modes of the comb with the frequency difference ν_{diff}. By stabilizing ν_{off}, ν_{rep}, and ν_{diff} against a 10-MHz quartz oscillator, which was referenced to an atomic clock signal from the global positioning system, the frequency of the cw laser was measured and locked to an absolute precision $< 4 \times 10^{-10}$.

Second, the seed beam was amplified in three dye cells pumped by a pulsed Nd:YAG laser with pulse length ~ 3 ns. In order to decrease the Fourier-limited linewidth of the amplified laser pulse, the pump beam was stretched to ~ 20 ns by splitting it into seven beams and merging them after incremental delays. As a result, laser pulses with an energy E =5–20 mJ and a linewidth $\Gamma \sim 60$ MHz were produced. The shorter wavelengths $\lambda = 264.7$–470.7 nm were obtained by frequency doubling or tripling the fundamental light in beta-barium borate (BBO) or lithium triborate (LBO) crystals.

Although the frequency of the cw laser beams was stabilized precisely, that of the amplified laser pulses could change time-dependently. This phenomenon is called 'frequency chirp' and is caused by nanosecond-scale changes in the refractive index n_c of the dye. The change in n_c shifts the phase ϕ_c of the amplified laser pulse, whereas the time derivative of ϕ_c is equal to the chirp following the equation,

$$\Delta\nu_c(t) = -\frac{1}{2\pi}\frac{d\phi_c}{dt}. \tag{2}$$

The time evolution of the chirp $\Delta\nu_c(t)$ in the laser pulse was derived in the following way[31, 32, 33, 34]. The cw seed laser beam was split and its frequency was shifted by 400 MHz using an acousto-optic modulator (AOM). The frequency-shifted cw laser beam was merged with the pulse-amplified laser beam and a heterodyne beat signal was measured using a photodiode. The chirp was derived by applying Fourier analysis to the beat signal. In addition, the chirp was actively compensated using an electro-optic modulator (EOM) which shifted the frequency of the seed laser before the amplification.

METASTABLE-TO-METASTABLE TRANSITION

Because of the 10-ns lifetimes of the short-lived daughter states, the natural width $\Gamma \sim 20$ MHz of these transitions would limit the experimental precision on the transition frequency. We introduced a different technique to measure a metastable-to-metastable transition, (n, l) = (36, 34) $\rightarrow$ (35, 33) of $\bar{p}^4He^+$ with the natural width $\Gamma_n \sim 100$ kHz[15].

Three successive laser pulses were used to measure this transition. A laser pulse tuned at the resonance wavelength $\lambda = 372.6$-nm of the lower transition first depleted the state (35, 33). A second laser pulse with $\lambda = 417.8$ nm then mixed the population of the states (36, 34) and (35, 33). Another irradiation of 372.6 nm laser pulse was used to measure the population of the (35, 33) state after the mixing.

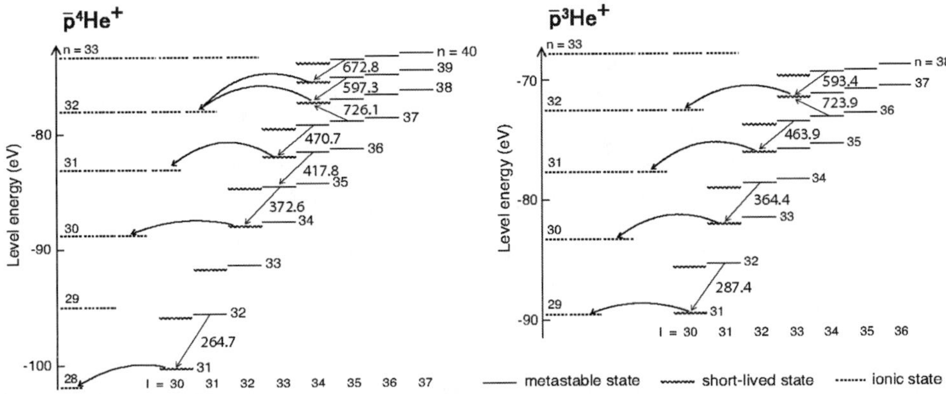

FIGURE 5. Twelve transitions measured with the new laser system. The transition wavelengths are indicated in nanometers.

DETERMINATION OF THE ANTIPROTON-TO-ELECTRON MASS RATIO

We measured twelve $\bar{p}He^+$ transitions (Fig. 5). Typical resonance profiles are shown in Fig. 6. Each state of (n, l) is split into doublet states by the coupling of the electron spin and the antiproton orbital angular momentum, and these are further split by the coupling of the antiproton spin. The ^3He spin is also taken into account in the $\bar{p}^3He^+$ case. As a result, each state has four (in $\bar{p}^4He^+$) or eight (in $\bar{p}^3He^+$) hyperfine substates[14, 35]. Arrows in Fig. 6 indicate the calculated positions of the most intense non-spin-flip transitions[35].

We simulated these resonance profiles using the optical Bloch equation. The following data were used in the calculation. The frequency of the laser was expressed as,

$$\nu_{\text{laser}}(t) = \nu_{\text{off}} + n\nu_{\text{rep}} + \nu_{\text{diff}} + \Delta\nu_{\text{c}}(t), \tag{3}$$

including the time-dependent residual chirp $\Delta\nu_c(t)$. The chirp was reduced to < 10 MHz over most of the pulse using the active compensation of the EOM. The Rabi frequency $\Omega_m(t)$ for each magnetic substate was calculated using the dipole moment of the $\bar{p}He^+$ transitions and the time profile of the laser power measured with a photodiode. All possible transitions including those that flipped the spin of the constituent particles, which were typically a factor 50 weaker than non-spin-flip ones, were calculated simultaneously. The thermal Doppler broadening of the resonance lines was also taken into account using the Maxwell-Boltzmann distribution of $\bar{p}He^+$ thermalized to $T = 10$ K. The obtained lineshape was then fitted to the resonance profile (solid line in Fig. 6) to determine ν_{exp}. Finally, frequency shifts due to collisions with surrounding helium atoms were corrected using our previous results[29].

Figure 7 shows the results on transition frequencies ν_{exp} of $\bar{p}^4He^+$ and $\bar{p}^3He^+$ determined in the way described above. The one standard deviation error $\sigma_{\text{exp}} = 4 - 15$ MHz shown here was the quadratic sum of the statistical one (3 – 13 MHz) and sys-

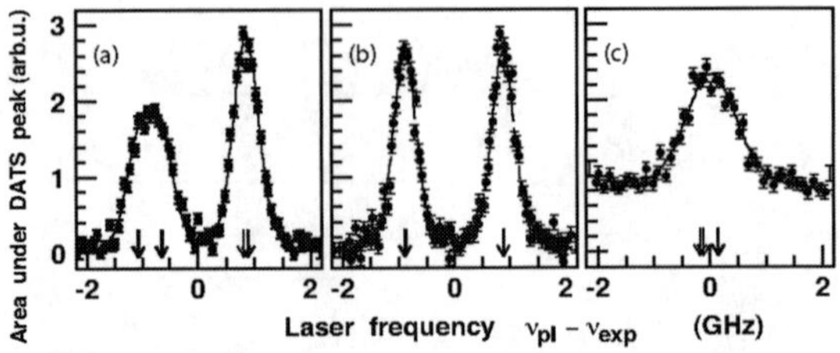

FIGURE 6. The resonance profiles of the transition (a) (36, 34) → (37, 33) in $\bar{p}^3He^+$, (b) (37, 35) → (38, 34) and (c) the metastable-to-metastable transition (36, 34) → (35, 33) in $\bar{p}^4He^+$. The arrows indicate the calculated positions of the most intense non spin-flip transitions[15].

tematic ones arising from the chirp $(2-4$ MHz), collisional shifts $(0.1-2$ MHz), and the harmonic generation $(1-2$ MHz). Two sets of theoretical calculations ν_{th} using the 2002 CODATA recommended values for the fundamental constants[36] are compared in the same figure. Uncertainties on ν_{th} caused by the finite size effects of the antiproton and the helium nucleus were negligibly small for these vibrational transitions. The values from [14] scatter within 7-8 standard deviations of ν_{exp}. Those of [12] agree within $< 1\times 10^{-8}$ with four highest-precision measurements in $\bar{p}^4He^+$ and $(36,34)\rightarrow(37,33)$ in $\bar{p}^3He^+$. Four values of [12] in $\bar{p}^3He^+$ were lower than ν_{exp} by ~ 2 standard deviations.

The comparison was interpreted from two different perspectives. In the following analysis, only the latest values of [12] were used because they are claimed to have a precision $\sigma_{th} =$ 1-2 MHz, while the authors of older values [14] claim $\sigma_{th} =$ 5-100 MHz, which are much larger than the experimental errors.

If we assume that the *CPT* invariance is not violated at the present experimental precision, the comparison of ν_{exp} and ν_{th} yields $m_{\bar{p}}/m_e$ using other constants known with better precision such as Rydberg constant or alpha particle-to-proton mass ratio. From this standpoint, $Q_p + Q_{\bar{p}} = 0$ was set first. The transition frequencies were calculated as a function of $m_{\bar{p}}/m_e$, $\nu_{th}(m_{\bar{p}}/m_e)$. Performing the χ^2-minimization with

$$\chi^2 = \sum \frac{(\nu_{th}(m_{\bar{p}}/m_e) - \nu_{exp})^2}{\sigma^2_{exp} + \sigma^2_{th}}, \tag{4}$$

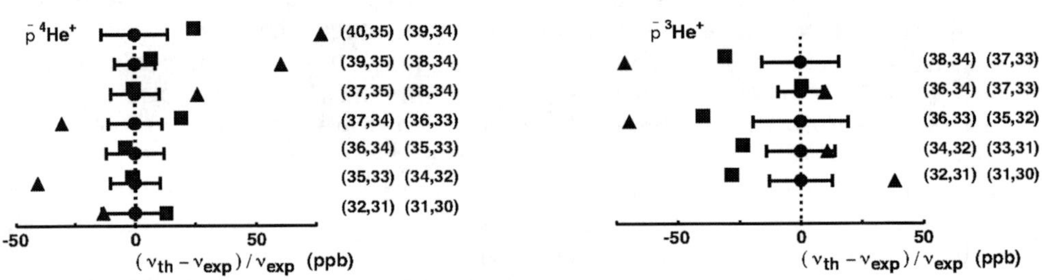

FIGURE 7. Experimental results (filled circles with one standard deviation errors) compared to the theoretical calculations, [12] (filled squares) and [14] (filled triangles)[15].

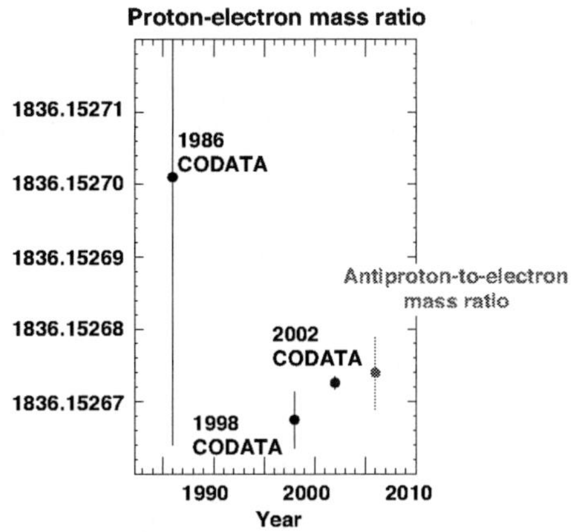

FIGURE 8. The antiproton-to-electron mass ratio obtained in this experiment compared with the recommended values for the proton-to-electron mass ratio of CODATA[10].

the best-fit value was obtained as

$$m_{\bar{p}}/m_e = 1836.152\,674(5). \tag{5}$$

The error is the quadratic sum of statistical errors (~ 4), systematic errors (~ 3, mainly coming from the frequency chirp of the laser and collisional effects), and an uncertainty of the theoretical calculations (~ 2).

The limit on the *CPT* symmetry was also set from the results[37]. In this case, we let both antiproton mass and charge vary. In the simple Bohr model, transition frequencies are proportional to $m_{\bar{p}}Q_{\bar{p}}^2$ so that the correlation between the fractional mass $\delta m_{\bar{p}}/m_{p}$ and charge $\delta Q_{\bar{p}}/Q_{p}$ is expected to be

$$\delta m_{\bar{p}}/m_{p} = -2\delta Q_{\bar{p}}/Q_{p}. \tag{6}$$

Naively, transition frequencies of $\bar{p}He^+$ are expected to show a similar relation. The χ^2-minimization yielded a *CPT* limit which correlates along the $\delta m_{\bar{p}}/m_{p} \sim -2\delta Q_{\bar{p}}/Q_{p}$. Using a constraint on the proton and antiproton charge-to-mass ratio[38], which is the most precise CPT test on the antiproton, a limit on the equality of antiproton and proton masses and charges was obtained as,

$$\frac{|Q_{p}+Q_{\bar{p}}|}{Q_{p}} \sim \frac{|M_{p}-M_{\bar{p}}|}{M_{p}} < 2\times 10^{-9}, \tag{7}$$

at a confidence level of 90 %.

ACKNOWLEDGMENTS

We are indebted to the CERN AB division for providing a stable antiproton beam. We would like to thank D. Bakalov, Y. Kino, and V. I. Korobov for theoretical discussions, and K. S. E. Eikema, M. Fischer, R. Holzwarth, and K. Jungmann for invaluable expertise and advice concerning the lasers. This work was supported by the Grant-in-Aid for Creative Basic Research (No. 15002005) of Monbukagakusho of Japan, the Hungarian Scientific Research Fund (OTKA T046095), and the EURYI award of the European Science Foundation and the Deutsche Forschungsgemeinschaft.

REFERENCES

1. D. Colladay, and V. A. Kostelecký, *Phys. Rev.D* **55**, 6760–6774 (1997).
2. V. A. Kostelecký, *Phys. Rev. D* **69**, 105009 (2004).
3. G. Gabrielse, N. S. Bowden, P. Oxley, A. Speck, C. H. Storry, J. N. Tan, M. Wessels, D. Grzonka, W. Oelert, G. Schepers, T. Sefzick, J. Walz, H. Pittner, T. W. Hänsch, and E. A. Hessels, *Phys. Rev. Lett.* **89**, 213401 (2002).
4. C. H. Storry, A. Speck, D. L. Sage, N. Guise, G. Gabrielse, D. Grzonka, W. Oelert, G. Schepers, T. Sefzick, H. Pittner, M. Herrmann, J. Walz, T. W. Hänsch, D. Comeau, and E. A. Hessels, *Phys. Rev. Lett.* **93**, 263401 (2004).
5. G. Gabrielse, P. Larochelle, D. L. Sage, B. Levitt, W. S. Kolthammer, I. Kuljanishvili, R. McConnell, J. Wrubel, F. M. Esser, H. Glückler, D. Grzonka, G. Hansen, S. Martin, W. Oelert, J. Schillings, M. Schmitt, T. Sefzick, H. Soltner, Z. Zhang, D. Comeau, M. C. George, E. A. Hessels, C. H. Storry, M. Weel, A. Speck, F. Nillius, J. Walz, and T. W. Hänsch, *Phys. Rev. Lett.* **98**, 113002 (2007).
6. G. Andresen, W. Bertsche, A. Boston, P. D. Bowe, C. L. Cesar, S. Chapman, M. Charlton, M. Chartier, A. Deutsch, J. Fajans, M. C. Fujiwara, R. Funakoshi, D. R. Gill, K. Gomberoff, J. S. Hangst, R. S. Hayano, R. Hydomako, M. J. Jenkins, L. Jorgensen, L. Kurchaninov, N. Madsen, P. Nolan, K. Olchanski, A. Olin, A. Povilus, F. Robicheaux, E. Sarid, D. M. Silveira, J. Storey, H. H. Telle, R. I. Thompson, D. P. van der Werf, J. S. Wurtele, and Y. Yamazaki, *Phys. Rev. Lett.* **98**, 023402 (2007).
7. AEGIS collaboration, Proposal for the AEGIS experiment at the CERN antiproton decelerator (Antimatter Experiment: Gravity, Interferometry, Spectroscopy), Proposal CERN-SPSC-2007-017 (2007).
8. M. Amoretti, C. Amsler, G. Bonomi, A. Bouchta, P. Bowe, C. Carraro, C. L. Cesar, M. Charlton, M. J. T. Collier, M. Doser, V. Filippini, K. S. Fine, A. Fontana, M. C. Fujiwara, R. Funakoshi, P. Genova, J. S. Hangst, R. S. Hayano, M. H. Holzscheiter, L. V. Jørgensen, V. Lagomarsino, R. Landua, D. Lindelöf, E. Lodi Rizzini, M. Macrì, N. Madsen, G. Manuzio, M. Marchesotti, P. Montagna, H. Pruys, C. Regenfus, P. Riedler, J. Rochet, A. Rotondi, G. Rouleau, G. Testera, A. Variola, T. L. Watson, and D. P. van der Werf, *Nature* **419**, 456–459 (2002).
9. T. Yamazaki, N. Morita, R. S. Hayano, E. Widmann, and J. Eades, *Phys. Rep.* **366**, 183–329 (2002).
10. R. S. Hayano, M. Hori, D. Horváth, and E. Widmann, *Rep. Prog. Phys.* **70**, 1995 (2007).
11. V. I. Korobov, *Phys. Rev. Lett.* **67**, 62501 (2003), erratum Phys. Rev. A 68, 019902.
12. V. I. Korobov, "Precise spectroscopy of antiprotonic helium and sensitivity of transitions to the antiproton mass," in *Proceedings of the International Conference on Exotic Atoms and Related Topics (EXA'05), Vienna, 2005*, edited by A. Hirtl, J. Marton, E. Widmann, and J. Zmeskal, Austrian Academy of Sciences Press, Vienna, 2005, pp. 391–400.
13. Y. Kino, N. Yamanaka, N. Kamimura, P. Froelich, and H. Kudo, *Hyperfine Interactions* **138**, 179–182 (2001).
14. Y. Kino, H. Kudo, and M. Kamimura, *Modern Physics Letters A* **18**, 388–397 (2003).
15. M. Hori, A. Dax, J. Eades, K. Gomikawa, R. S. Hayano, N. Ono, W. Pirkl, E. Widmann, H. A. Torii, B. Juhász, D. Barna, and D. Horváth, *Phys. Rev. Lett.* **96**, 243401 (2006).
16. R. Holzwarth, T. Udem, T. W. Hänsch, J. C. Knight, W. J. Wadsworth, and P. S. J. Russell, *Phys. Rev. Lett.* **85**, 2264–2267 (2000).
17. D. J. Jones, S. A. Diddams, J. K. Ranka, A. Stentz, R. S. Windeler, J. L. Hall, and S. T. Cundiff, *Science* **288**, 635 (2000).
18. T. Udem, R. Holzwarth, and T. W. Hänsch, *Nature* **416**, 233–237 (2002).
19. M. Hori, J. Eades, R. S. Hayano, T. Ishikawa, W. Pirkl, E. Widmann, H. Yamaguchi, H. A. Torii, B. Juhász, D. Horváth, and T. Yamazaki, *Phys. Rev. Lett.* **91**, 123401 (2003).

20. K. Sakimoto, *Phys. Rev. A* **76**, 042513 (2007).
21. M. Hori, J. Eades, E. Widmann, H. Yamaguchi, J. Sakaguchi, T. Ishikawa, R. S. Hayano, H. A. Torii, B. Juhász, D. Horváth, and T. Yamazaki, *Phys. Rev. Lett.* **89**, 093401 (2002).
22. M. Hori, J. Eades, E. Widmann, T. Yamazaki, R. S. Hayano, T. Ishikawa, H. A. Torii, T. von Egidy, F. J. Hartmann, B. Ketzer, C. Maierl, R. Pohl, M. Kumakura, N. Morita, D. Horváth, and I. Sugai, *Phys. Rev. A* **70**, 012504 (2004).
23. H. Yamaguchi, T. Ishikawa, J. Sakaguchi, E. Widmann, J. Eades, R. S. Hayano, M. Hori, H. A. Torii, B. Juhász, D. Horváth, and T. Yamazaki, *Phys. Rev. A* **66**, 022504 (2002).
24. H. Yamaguchi, R. S. Hayano, T. Ishikawa, J. Sakaguchi, E. Widmann, J. Eades, M. Hori, H. A. Torii, B. Juhász, D. Horváth, and T. Yamazaki, *Phys. Rev. A* **70**, 012501 (2004).
25. M. Hori, J. Eades, R. S. Hayano, W. Pirkl, E. Widmann, H. Yamaguchi, H. A. Torii, B. Juhász, D. Horváth, K. Suzuki, and T. Yamazaki, *Phys. Rev. Lett.* **94**, 063401 (2005).
26. A. M. Lombardi, W. Pirkl, and Y. Bylinsky, "First operating experience with the CERN decelerating RFQ forantiprotons," in *Proceedings of the 2001 Particle Physics Accelerator confenrece*, edited by P. Lucasa, and S. Webber, IEEE, Piscataway, NJ, 2001, pp. 585–587.
27. B. Juhász, D. Barna, J. Eades, H. Fuhrmann, R. S. Hayano, M. Hori, D. Horváth, H. A. Torii, E. Widmann, H. Yamaguchi, T. Yamazaki, and J. Zmeskal, *Chemical Physics Letters* **427**, 246 (2006).
28. M. Hori, K. Yamashita, R. S. Hayano, and T. Yamazaki, *Nucl. Instrum. Methods Phys. Res. A* **496**, 102 (2003).
29. M. Hori, J. Eades, E. Widmann, H. Yamaguchi, J. Sakaguchi, T. Ishikawa, R. S. Hayano, H. A. Torii, B. Juhász, D. Horváth, and T. Yamazaki, *Phys. Rev. Lett.* **87**, 093401 (2001).
30. M. Hori, R. S. Hayano, E. Widmann, and H. Torii, *Optics Letters* **28**, 2479–2481 (2003).
31. K. S. E. Eikema, W. Ubachs, W. Vassen, and W. Hogervorst, *Phys. Rev. A* **55**, 1866–1884 (1997).
32. M. S. Fee, K. Danzmann, and S. Chu, *Phys. Rev. A* **45**, 4911–4924 (1992).
33. I. Reinhard, M. Gabrysch, B. Fischer von Weikersthal, K. Jungmann, and G. zu Putlitz, *Appl. Phys. B* **63**, 467–472 (1996).
34. V. Meyer, S. N. Bagayev, P. E. G. Baird, P. Bakule, M. G. Boshier, A. Breitrück, S. L. Cornish, S. Dychkov, G. H. Eaton, A. Grossmann, D. Hübl, V. W. Hughes, K. Jungmann, I. C. Lane, Y.-W. Liu, D. Lucas, Y. Matyugin, J. Merkel, G. zu Putlitz, I. Reinhard, P. G. H. Sandars, R. Santra, P. V. Schmidt, C. A. Scott, W. T. Toner, M. Towrie, and K. Träger, *Phys. Rev. Lett.* **84**, 1136–1139 (2000).
35. V. I. Korobov, *Phys. Rev. A* **73**, 022509 (2006).
36. P. J. Mohr, and B. N. Taylor, *Rev. Mod. Phys.* **77**, 1–107 (2005).
37. R. Hughes, and B. I. Deutch, *Phys. Rev. Lett.* **69**, 578–580 (1992).
38. G. Gabrielse, A. Khabbaz, D. S. Hall, C. Heimann, H. Kalinowsky, and W. Jhe, *Phys. Rev. Lett.* **82**, 3198–3201 (1999).

Collisional Effects on the Antiprotonic Helium Hyperfine Structure Measurement

T. Pask*, D. Barna†, A. Dax†, R.S. Hayano†, M. Hori[1,†,**], D. Horváth[‡,§], B. Juhász*, C. Malbrunot[2,*], J. Marton*, N. Ono†, K. Suzuki*, J. Zmeskal* and E. Widmann*

*Stefan Meyer Institute for Subatomic Physics, Austrian Academy of Sciences, Boltzmanngasse 3, A-1090 Vienna, Austria.
†Department of Physics, University of Tokyo, 7-3-1 Hongo, Bunkyo-ku, Tokyo 113-0033, Japan.
**CERN, CH-1211 Geneva 23, Switzerland.
‡KFKI Research Institute for Particle and Nuclear Physics, H-1525 Budapest, PO Box 49, Hungary.
§Institute of Nuclear Research of the Hungarian Academy of Sciences, H-4001 Debrecen, PO Box 51, Hungary.

Abstract.
We report preliminary results from a systematic study of the experimental conditions for a measurement of the hyperfine (HF) structure of antiprotonic helium $(n,l) = (37,35)$. Improvements to the laser system have increased the peak-to-total signal and allowed the delay of the second laser to be extended. We have completed the first experimental measurements of the relaxation collision rate between HF states, by which antiprotonic helium atoms undergo inelastic spin exchange collisions. We show that a microwave induced $\pi-$pulse between HF states is achieved and that a higher signal-to-noise ratio than our previous results is possible. The study was carried out at the Antiproton Decelerator (AD) at CERN.

Keywords: Antiprotonic helium, Microwave spectroscopy, Hyperfine structure, CPT
PACS: 36.10.-k, 32.10.Fn, 33.40.+f

INTRODUCTION

It was first discovered at KEK, in 1991 [1], and later measured at the Low Energy Antiproton Ring (LEAR) [2, 3] and the Antiproton Decelerator (AD) at CERN [4], that a small fraction ($\sim 3\%$) of antiprotons ($\overline{\mathrm{p}}$), stopped in a helium target, survive to form antiprotonic helium ($\overline{\mathrm{p}}\mathrm{He}^+$): an exotic metastable atom consisting of a helium nucleus, an electron and an antiproton ($\mathrm{He}^{++} + \mathrm{e}^- + \overline{\mathrm{p}}$) [5–7]. A *hyperfine structure (HFS)* [3] arises from the coupling of the electron and the antiproton magnetic moments. This paper presents the initial results of a current systematic study where the method of measuring the HFS of $\overline{\mathrm{p}}\mathrm{He}^+$ has been improved.

When an antiproton approaches a helium atom, at an energy of the order of the helium ionization energy, it can, simultaneously, eject one of the two ground state electrons

---

[1] Present address: Max-Planck-Institut für Quantenoptik, Hans-Kopfermann-Strasse 1, D-85748 Garching, Germany.
[2] Present address: TRIUMF, 4004 Wesbrook Mall, Vancouver, BC, V6T 2A3, Canada.

CP1037, *Cold Antimatter Plasmas and Application to Fundamental Physics,* edited by Y. Kanai and Y. Yamazaki

and become captured. While the remaining electron occupies the ground (1s) state, the antiproton is most likely occupy the $n = n_0 \equiv \sqrt{M^*/m_e} = 38$ state (where M^* is the reduced mass of the antiproton-helium nucleus system and m_e is the electron mass). In the region of principle quantum number $n = 32$ - 40 and vibrational quantum number $v = 0$ - 3 (where $v = n - l - 1$), the antiproton undergoes a cascade where it radiatively decays, with a lifetime of the order of microseconds, to an adjacent state [3].

As the neutral system retains one electron it is protected from external atoms by the Pauli exclusion principle. The presence of the electron also removes the l degeneracy for the same n therefore protecting it against Stark mixing. The Auger decay of the remaining electron is suppressed by the large ionization energy ($\sim$ 25 eV) compared to the $n \rightarrow n-1$ level spacing of a few eV. At the end of the cascade, when Auger decay is no longer suppressed, the second electron is ejected in less than 10 ns, degeneracy occurs Stark mixing becomes dominant and the electron annihilates in the nucleus.

There are two types of electric dipole transitions: favored [Δ v $= 0$; $(n,l) \rightarrow (n-1, l-1)$] and unfavored [$\Delta$ $v = 2$; $(n,l) \rightarrow (n+1, l-1)$] [3]. The dipole moment of the unfavored transitions is an order of magnitude smaller than that of the favored. In 2001 only the unfavored ones had a hyperfine line splitting larger than the laser bandwidth. Improvements to the laser system have reduced the bandwidth but the resolution remains limited by Doppler broadening. From among the candidate transitions, the $(37, 35) \rightarrow (38, 34)$ unfavored transition was chosen for study because the parent state is relatively highly populated.

In 1996, at LEAR, a precise laser spectroscopy scan at 726.10 nm was made, resolving, for the first time, the two hyperfine peaks of the $(37,35) \rightarrow (38,34)$ transition [8]. However, to accurately determine the HF splitting it is necessary to induce a transition between the HFS substates. While it is possible to induce a transition between (n,l) states with laser light, a magnetic M1 resonance is required to move between the HFS substates.

In 2001, a laser-microwave-laser measurement was performed, resolving the hyperfine structure of the (37, 35) state to a precision of 300 kHz [9]. By comparing the experimentally measured transition frequency with three body QED calculations, the antiproton spin magnetic moment can be extracted. Because this is small compared to the other contributions, an improvement in the HFS measurement is not directly proportional to the same improvement in the spin magnetic moment.

THE HYPERFINE STRUCTURE OF ANTIPROTONIC HELIUM

The HFS of $\bar{p}He^+$ arises from the interactions of the magnetic moments of its constituents and has been calculated by Korobov and Bakalov [19]. To the order α^4, the total spin-dependent interactions can be expressed by the sum of two-body interactions composed of three terms: the spin orbit interaction $U_{ij}^{(so)}$, the spin-spin contact interaction $U_{ij}^{(ssc)}$ and the spin-spin tensor interaction $U_{ij}^{(sst)}$ [19].

$$U = \sum_{i \neq j} U_{ij}^{(so)} + U_{ij}^{(ssc)} + U_{ij}^{(sst)}, \tag{1}$$

$$U_{ij}^{(so)} = -\alpha^2 \left[\frac{z_i(z_j - 2\mu_j)}{2m_j^2} \frac{\mathbf{r}_{ij} \times \mathbf{p}_j \cdot \mathbf{s}_j}{r_{ij}^3} + \frac{z_i \mu_j}{m_i m_j} \frac{\mathbf{r}_{ij} \times \mathbf{p_i} \cdot \mathbf{s}_j}{r_{ij}^3} \right], \tag{2}$$

$$U_{ij}^{(ssc)} = -\alpha^2 \frac{8\pi}{3} \frac{\mu_i \mu_j}{m_i m_j} (\mathbf{s}_i \cdot \mathbf{s}_j) \delta(\mathbf{r}_{ij}), \tag{3}$$

$$U_{ij}^{(sst)} = -\alpha^2 \frac{\mu_i \mu_j}{m_i m_j} \left[3 \frac{(\mathbf{r}_{ij} \cdot \mathbf{s}_i)(\mathbf{r}_{ij} \cdot \mathbf{s}_j)}{r_{ij}^5} - \frac{(\mathbf{s}_i \cdot \mathbf{s}_j)}{r_{ij}^3} \right]. \tag{4}$$

In the above equations $\mathbf{r}_i$, $\mathbf{p}_i$, $\mathbf{s}_i$, m_i, z_i and μ_i denote, respectively, the position vector, momentum, spin, mass, electric charge and magnetic moment of the antiproton ($\overline{\mathrm{p}}$), helium nucleus (He^{++})and the electron (e^-).

Bakalov and Korobov constructed an effective Hamiltonian,

$$H^{\text{eff}} = E_1(\vec{L} \cdot \vec{S}_e) + E_2(\vec{L} \cdot \vec{S}_{\overline{\mathrm{p}}}) + E_3(\vec{S}_e \cdot \vec{S}_{\overline{\mathrm{p}}}),$$

$$+E_4\{2L(L+1)(\vec{S}_e \cdot \vec{S}_{\overline{\mathrm{p}}}) - 6[(\vec{L} \cdot \vec{S}_e)(\vec{L} \cdot \vec{S}_{\overline{\mathrm{p}}})]\}. \tag{5}$$

Due to the large orbital angular moment of the antiproton ($L \sim 35$), the dominant HF splitting arises from the interaction of angular momentum $\vec{L}$ with the electron angular spin $\vec{S}_{\mathrm{e}}$ (term 1: $\overline{\mathrm{p}}$ - e^- spin-orbit splitting). The contributions of the next three terms cause a superhyperfine splitting (SHFS) of each level of the HF doublet. These contributions are more or less equal. The contact spin-spin (term 3) and the tensor spin-spin (term 4) terms act to cancel eachother out leaving the coupling of the angular momentum and antiproton spin $\vec{S}_{\overline{\mathrm{p}}}$ (term 2: $\overline{\mathrm{p}}$ spin-orbit splitting) as the dominant contribution to the SHFS. The resulting quadruplet structure for $\overline{\mathrm{p}}^4\mathrm{He}^+$ is shown in Fig. 1a where ν_{HF} = 10 - 15 GHz is two orders of magnitude larger than the SHF splitting (ν_{SHF} = 150 - 300 MHz) which is too small to be resolved by single photon spectroscopy.

In the case of $\overline{\mathrm{p}}^3\mathrm{He}^+$, because of the additional coupling to the $^3\mathrm{He}$ nuclear spin $\vec{S}_{\mathrm{h}}$, each level is split in to an octet, shown in Fig. 1b. Although the helion magnetic moment is smaller than that of the antiproton, its overlap with the electron cloud is larger so the helion spin contributes to the quadruplet superhyperfine splitting of $\overline{\mathrm{p}}^3\mathrm{He}^+$ while the antiproton spin results in the octet super-superhyperfine structure SSHF.

The HF doublet is described by the quantum number $\vec{F} = \vec{L} + \vec{S}_e$ with components

$$F_+ = L + \frac{1}{2}, \tag{6a}$$

$$F_- = L - \frac{1}{2}, \tag{6b}$$

and the $\overline{\mathrm{p}}^4\mathrm{He}^+$ quadruplet by $\vec{J} = \vec{F} + \vec{S}_{\overline{\mathrm{p}}}$ with components

$$J_{-+} = F_- + \frac{1}{2}, \tag{7a}$$

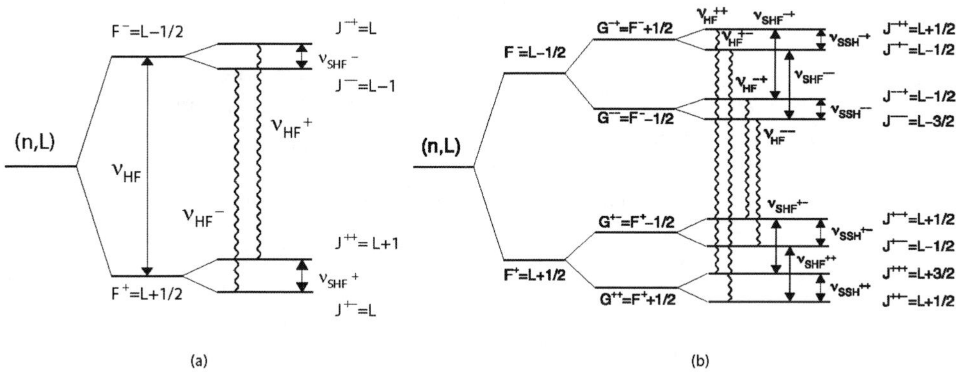

FIGURE 1. Hyperfine splitting of a state (n,l) of a) $\overline{\mathrm{p}}^4\mathrm{He}^+$ and of b) $\overline{\mathrm{p}}^3\mathrm{He}^+$. The wavy lines denote allowed M1 transitions that can be induced by microwave radiation.

$$J_{--} = F_- - \frac{1}{2}, \tag{7b}$$

$$J_{++} = F_+ + \frac{1}{2}, \tag{7c}$$

$$J_{+-} = F_+ - \frac{1}{2}, \tag{7d}$$

while the $\overline{\mathrm{p}}^3\mathrm{He}^+$ quadruplet is described by $\vec{G} = \vec{F} + \vec{S}_\mathrm{h} = \vec{L} + \vec{S}_\mathrm{e} + \vec{S}_\mathrm{h}$ and the octet by $\vec{J} = \vec{G} + \vec{S}_{\overline{\mathrm{p}}} = \vec{L} + \vec{S}_\mathrm{e} + \vec{S}_\mathrm{h} + \vec{S}_{\overline{\mathrm{p}}}$

Transitions between SHF states and SSHF are in the microwave frequency spectrum. It requires three orders of magnitude more power to induce both the electron and antiproton spin flip than a single electron spin flip so the former is referred to as unfavored while the latter is favored. (Due to power limitations only the favored transitions were measured). In Fig. 1 the favored M1 transitions that can be induced by an oscillating magnetic field are shown.

The favored and unfavored transitions for $\overline{\mathrm{p}}^4\mathrm{He}^+$ are given below:
Favored (electron spin flip transition):

$$\nu_{\mathrm{HF}}^{+} : J_{++} = F_+ + \frac{1}{2} \leftrightarrow J_{-+} = F_- + \frac{1}{2} \tag{8a}$$

$$\nu_{\mathrm{HF}}^{-} : J_{+-} = F_+ - \frac{1}{2} \leftrightarrow J_{--} = F_- - \frac{1}{2} \tag{8b}$$

Unfavored (antiproton and electron spin flip transition):

$$\nu_{\mathrm{HF}}^{\mathrm{unfav}} : J_{-+} = F_+ - S_{\overline{\mathrm{p}}} \leftrightarrow J_{+-} = F_- + S_{\overline{\mathrm{p}}} \tag{8c}$$

NUMERICAL MODELS

The experimental method is a three step laser-microwave-laser process. A population asymmetry is induced through laser depopulation of one of the doublet substates. A population transfer between the substates is induced with a microwave resonant pulse. Then a second laser is used to measure the transferred population. The microwave frequency is scanned to determine the frequency of the HF transitions.

Both the laser and microwave transitions were numerically simulated by evolving the Bloch equations. A depopulation efficiency of 80% for the first laser and 70% for the second was calculated for an unatentuated beam. The microwave resonance induced population transfer between the HF substates was determined by evolving Eq. 9a derived from the optical Bloch equations.

$$\frac{\mathrm{d}}{\mathrm{dt}}\rho = \mathbf{M}\rho, \tag{9a}$$

$$\rho = \begin{pmatrix} \rho_{-+} \\ \rho_{++} \\ \rho_{x+} \\ \rho_{y+} \\ \rho_{--} \\ \rho_{+-} \\ \rho_{x-} \\ \rho_{y-} \\ \rho_{u38} \\ \rho_{u39} \end{pmatrix}, \tag{9b}$$

$$\mathbf{M} = \begin{pmatrix} -\lambda_c & \lambda_i & 0 & \frac{1}{2}\Omega_m^+ & \lambda_i & 0 & 0 & 0 & \gamma_{u38} & 0 \\ \lambda_i & -\lambda_c & 0 & \frac{1}{2}\Omega_m^+ & 0 & \lambda_c & 0 & 0 & \gamma_{u38} & 0 \\ 0 & 0 & -\lambda_T & -\Delta\omega_+ & 0 & 0 & 0 & 0 & 0 & 0 \\ -\Omega_m^+ & \Omega_m^+ & \Delta\omega_+ & -\lambda_T & 0 & 0 & 0 & 0 & 0 & 0 \\ \lambda_i & 0 & 0 & 0 & -\lambda_c & \lambda_i & 0 & \frac{1}{2}\Omega_m^- & \gamma_{u38} & 0 \\ 0 & \lambda_i & 0 & 0 & \lambda_i & -\lambda_c & 0 & \frac{1}{2}\Omega_m^- & \gamma_{u38} & 0 \\ 0 & 0 & 0 & 0 & 0 & 0 & -\lambda_T & -\Delta\omega_- & 0 & 0 \\ 0 & 0 & 0 & 0 & -\Omega_m^- & \Omega_m^- & \Delta\omega_- & -\lambda_T & 0 & 0 \\ 0 & 0 & 0 & 0 & 0 & 0 & 0 & 0 & -\gamma_{u38} & \gamma_{u39} \\ 0 & 0 & 0 & 0 & 0 & 0 & 0 & 0 & 0 & -\gamma_{u39} \end{pmatrix}, \tag{9c}$$

where $\rho_{-+}, \rho_{++}, \rho{--}$, and $\rho{+-}$ represent the different time dependent populations of the four SHF states of the (n,l) = (37, 35) state. In a field free region $\Omega_m = 0$, these populations simply decay radiatively γ_r at a rate of 0.715 MHz to the (36, 34) state and 8.5 kHz to the (37, 34) state [3]. When there is a population asymmetry and the an external magnetic field is present transfer between the states can be induced. The complex dependency of the transitions is represented by $\rho_{x\pm}$ and $\rho_{y\pm}$ for the real and imaginary parts respectively. Broadening effects are dependent on both the radiative decay rate γ_r and the elastic collisional frequency λ_e: $\lambda_T = \gamma_r + \lambda_e$.

The four SHF states and are refilled as the upper states ρ_{u38} (38,36) and ρ_{u39} (39,37) decay at a rate of γ_{u38} = 0.655 MHz and γ_{u39} = 0.588 MHz into the lower as part of the cascade [3]. The initial populations of these states have been experimentally measured [17, 18]. Through inelastic relaxation collisions λ_i the atoms return to an equilibrium. In Eq.(9a) $\lambda_c = 2\lambda_i + \gamma_r$ and collisions which result in the spin flip of more than one particle are ignored [15].

$\Delta\omega = \pi(\nu_{\rm M} - \nu_{\rm HF\pm})$ and represents the offset between the microwave $\nu_{\rm M}$ and transition $\nu_{\rm HF}$ frequencies. The Rabi frequency is dependent upon the magnetic field strength B and the atom's magnetic dipole moment μ_m

$$\Omega_m = \frac{\mu_m B(x,y,t)}{\hbar}, \tag{10}$$

$$\mu_m = \langle n',L',F',J',m|\mu_M|n,L,F,J,m\rangle, \tag{11}$$

which can be solved using Wigner's 3-j and Racah's 6-j coefficients

$$\begin{aligned}\langle n,L,F^-,J',m|\mu|n,L,F^+,J,m\rangle &= (-1)^{J'+m}\begin{pmatrix} J & 1 & J' \\ m & 0 & -m \end{pmatrix}_{3-j} \\ &\times\sqrt{(2J+1)(2J'+1)}\left|\begin{Bmatrix} F^- & J' & \frac{1}{2} \\ J & F^+ & 1 \end{Bmatrix}_{6-j}\right| \\ &\times\sqrt{(2F^+ +1)(2F^- +1)}\left|\begin{Bmatrix} \frac{1}{2} & F^- & L \\ F^+ & \frac{1}{2} & 1 \end{Bmatrix}_{6-j}\right| g_e\mu_b\langle\frac{1}{2}|s_e|\frac{1}{2}\rangle ,\end{aligned} \tag{12}$$

where n, L, F, J, m are the corresponding quantum numbers.

Equation (9a) was set to a range of 60 equally spaced different detuning frequencies to simulate the microwave scan. The population positional distribution and magnetic field variance were modelled with a Monte Carlo positional simulation. The magnetic field distribution at the target is dependant on the pill box cavity and varies spatially with respect to the radial r and angular cylindrical ϕ co-ordinates [11]. Apart from edge effects the z cylindrical component is constant. The radial B_r and angular B_ϕ components of the magnetic field is given by

$$B_r(r,\phi) = B_0\frac{J_1(kr)}{kr}sin(\phi), \tag{13a}$$

$$B_\phi(r,\phi) = B_0 J_1'(kr)cos(\phi), \tag{13b}$$

where k is the wave number and J_1 is the Bessel function of the first kind. The population distribution of the $\bar{\rm p}{\rm He}^+$ is Gaussian.

These simulations provided an insight into the size of the signal-to-noise ratio expected from the experiment. They determined the optimal laser diameter to be 5 mm and that a laser delay of T = 350 ns was the best compromise between line width and signal-to-noise ratio, although T = 500 ns was shown to be achievable.

When the experimentally determined relaxation frequency had been measured that parameter was set in the model to achieve a simulated result which had the same width

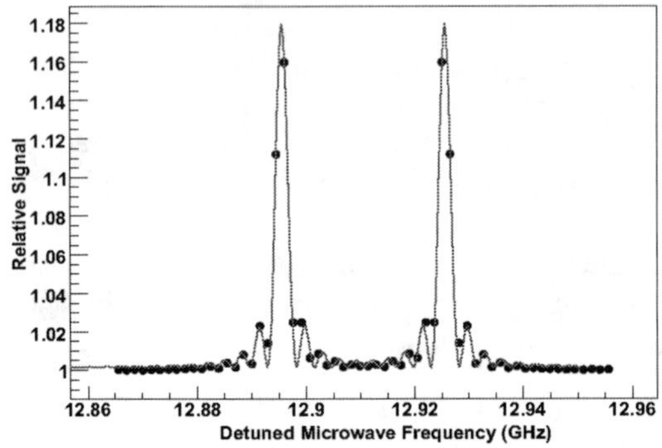

FIGURE 2. Simulated microwave scan with microwave exposure time of 350 ns and power of 5 W. The elastic collision rate $\lambda_e = 3.5$ MHz and the inelastic collision rate $\lambda_i = 1.4$ MHz. The points are fitted with Eq.(14).

as the expected results from a microwave scan, shown in Fig. 2. By this method an estimate of the elastic collision broadening of ~ 3.5 MHz was possible.

The line shape for a two level system under the effects of an external oscillating field for time T is given by Eq. (14). The optimum case is when the system undergoes a π–pulse, $|b|T = \pi/2$. Parameter $X(\omega)$ is the probability of transferring an atom to the other HF state and b is related to the half-width of the distribution:

$$X(\omega) = \frac{|2b|^2}{|2b|^2 + (\omega_0 - \omega)^2} \sin^2\left\{\frac{1}{2}\left[|2b|^2 + (\omega_0 - \omega)^2\right]^{\frac{1}{2}} T\right\}, \tag{14}$$

where T is the time spent in the oscillating field and b is the Rabi frequency.

COLLISION INDUCED RELAXATION RATE

To achieve the highest degree of precision it is necessary to make a high statistic measurement under optimum conditions. The line width is limited by the Fourier transform of the microwave pulse which is in turn determined by the laser delay time T. Ideally, to make the line width as narrow as possible, the maximum laser delay should be chosen. However, if T is increased indefinitely, long before all the metastable atoms annihilate, the asymmetry, induced by the first laser, is lost due to relaxation due to inelastic collisions. Once an asymmetry has been created, the particles begin to return to equilibrium through spin exchanging inelastic collisions.

Cross sections of a single spin flip are each of the same order of magnitude. They are four orders of magnitude less than elastic collision cross sections. The cross sections of double spin flip transitions are three orders of magnitude less than the cross sections of single spin flip transitions [15]. The elastic collisions occur at a frequency calculated to be $\lambda_c/2\pi = 2.66 - 5.92$ MHz and result in a broadening of the line, while the relaxation

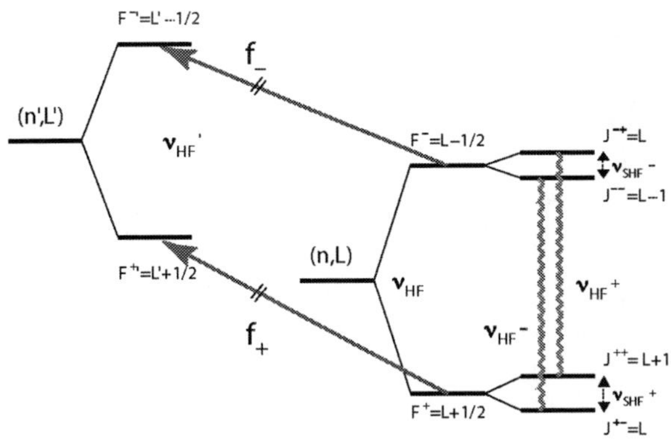

FIGURE 3. Schematic view of the splitting of $\bar{p}He^+$ for the unfavored electric dipole transitions. The state drawn on the right is the radiative decay dominated parent (n, L), and the left state is the Auger decay dominated daughter (n', L'). The laser transitions, from the parent to daughter doublets, are indicated by straight lines and the microwave transitions, between the quadruplets of the parent, by wavy ones. For this experiment $(n, L) = (37, 35)$ and $(n', L') = (38, 34)$.

time lies theoretically between 160.5 ns and 325 ns. The theoretical results stated are for a target density of 250 mbar and a temperature of 6 K [15].

In 2001 a maximum of only 150 ns laser splitting was technically achievable but the theory suggests that after a longer time the asymmetry would have already returned to equilibrium. To determine the optimum laser delay and whether the short relaxation time would indeed restrict the laser time delay, it was experimentally measured.

The laser spectroscopy method takes advantage of the charged pions produced when the antiproton annihilates in the helium nucleus. An exponentially decaying background signal is ever present as the antiprotons, in various levels of the cascade, decay and eventually annihilate in the nucleus. A measurement of the population of a particular state can be made by transferring it to a fast-decaying one. The population of such a state annihilates quickly, producing a sharp signal against the background.

The general setup is described in Ref. [11]. Two narrow-band lasers were used, with a time delay T of the second laser. Both were tuned to the f^+ transition between the radiative decay dominated parent state (37, 35) and the Auger decay dominated daughter state (38, 34) shown in Fig. 3. This produced the first sharp annihilation peak against the exponentially decaying background (Fig. 4). The ratio of the peak area to this background (peak-to-total) indicated the size of the population transferred from the parent to the daughter state.

The second laser was pulsed after a time delay T following the first. An equal number of measurements were taken when (1) both lasers were fired (Fig. 4a) and (2) only the second was fired (Fig. 4b). When both lasers were fired, the peak-to-total signal produced by the second laser (ptt_2^{++}) was smaller than that of the peak-to-total (ptt_2^{0+}) when only the second laser was fired. The difference between ptt_2^{0+} and ptt_2^{++} provided information about the refilling and laser depopulation efficiency. As T was increased the difference between the two signals decreased due to relaxation collisions refilling

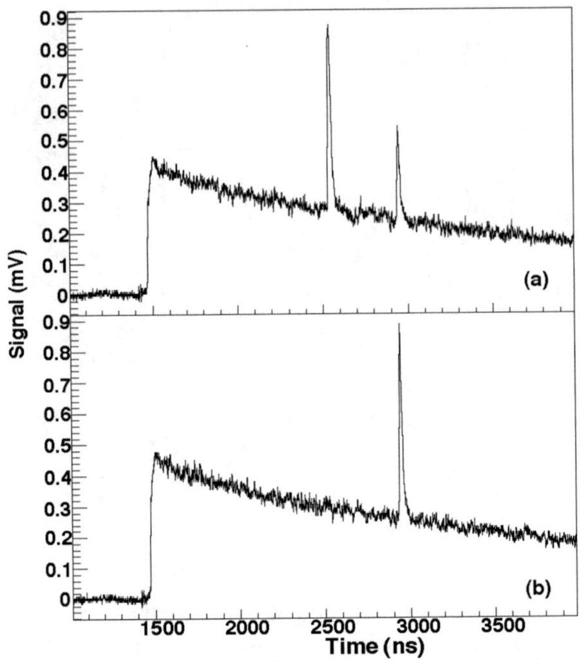

FIGURE 4. ADATS signal for the two different situations (a) where both laser induced annihilation peaks were recorded (b) where only the second annihilation peak was recorded.

from upper states and radiative decay. The signal of the first laser (ptt_1^{++}) was used for normalization.

These measurements were carried out at 3 different densities (500 mbar, 250 mbar and 150 mbar) and four different laser energy fluences (34, 17, 10, 5 mJ/cm^2 for the first laser and 30, 15, 9, 4.5 mJ/cm^2 for the second). The ratio ptt_2^{++}/ptt_1^{++} was plotted vs laser delay and fitted with Eq. (16). An example of such a plot is shown in Fig. 5. The fit was derived from the integral of the following two equations[16]:

$$\frac{\mathbf{d}\rho_+}{\mathbf{dt}} = g_+ - (\lambda_{+-} + \gamma_r)\rho_+ + \lambda_{-+}\rho_-, \tag{15a}$$

$$\frac{\mathbf{d}\rho_-}{\mathbf{dt}} = g_- - (\lambda_{-+} + \gamma_r)\rho_- + \lambda_{+-}\rho_+, \tag{15b}$$

where $\rho_\pm$ is the population density of the HF states, $g_\pm$ is the refilling from the above states, which can be assumed to be constant for $T \leq 1000$ ns for the (37, 35) state [17]. λ_{+-} and λ_{-+} are the relaxation rates from ρ_+ to ρ_- and ρ_- to ρ_+ respectively and γ_r =0.7149 MHz is the radiative decay rate, see Fig. 6.

If $\lambda_{+-} = \lambda_{-+} \equiv \lambda_i$, $g_+ = g_- \equiv g$ and the relative populations before the first laser pulse are $\rho_{(\pm)} = \frac{1}{2}$ then $ptt_1^{++}(t) = I_0\frac{1}{2}\varepsilon_1$ and $ptt_2^{++}(t) = I_0\rho_+(t)\varepsilon_2$, where I_0 is a

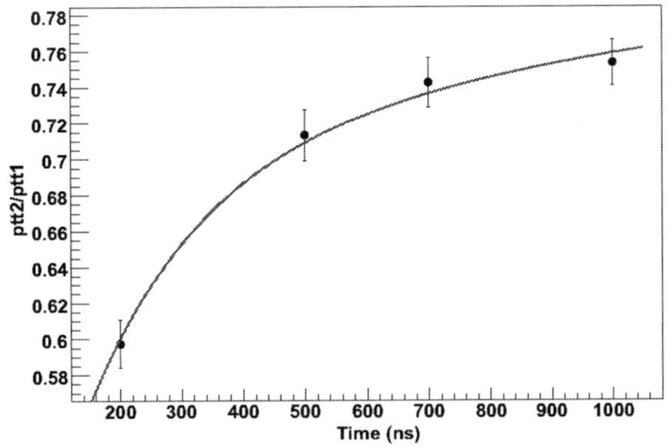

FIGURE 5. An example of one of several Laser Depopulation Efficiency plots used to determine the relaxation collision rate, fitted with Eq. (16). The target pressure was 250 mbar and the laser energy density was 9 mJ/cm^2.

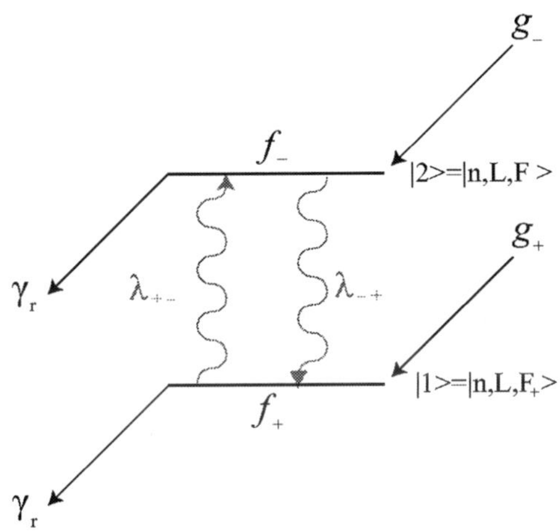

FIGURE 6. Energy level diagram of part of the cascade showing the refilling $g_\pm(t)$ from above states, decay to lower states γ_r and the relaxation collision rate $\lambda_{\pm\rightarrow\mp}$.

normalization constant, $\varepsilon_{1\ \&\ 2}$ is the laser depopulation efficiency for the first and second laser. Substituting these values into Eq. (15a & 15b) and intergrating gives:

$$\frac{\mathrm{ptt}_2^{++}(\mathrm{t})}{\mathrm{ptt}_1^{++}(\mathrm{t})} = \frac{\varepsilon_2}{\varepsilon_1}[e^{-\gamma_r t}(1-\frac{\varepsilon_1}{2}-\frac{\varepsilon_1}{2}e^{-\lambda_i t}-\frac{g}{\gamma_r})+\frac{g}{\gamma_r}]. \tag{16}$$

Table 1 shows the results for several densities. The error is high, because of the large fluctuations of the $\overline{\mathrm{p}}$ beam and limited measurement points, but clearly shows the relaxation rate to be much lower than suggested by theory.

TABLE 1. Relaxation Collision data, where p is the target pressure, ρ is the target density, λ_i is the relaxation collision rate and τ_i is the relaxation time $(1/\lambda_i)$.

p (mbar)	ρ (10^{20} cm^{-3})	λ_i (MHz)	τ_i (ns)
150	1.726 ± 0.040	1.50 ± 1.30	670 ± 130
250	2.912 ± 0.060	1.72 ± 0.36	580 ± 150
500	6.177 ± 0.13	4.60 ± 1.20	215 ± 120

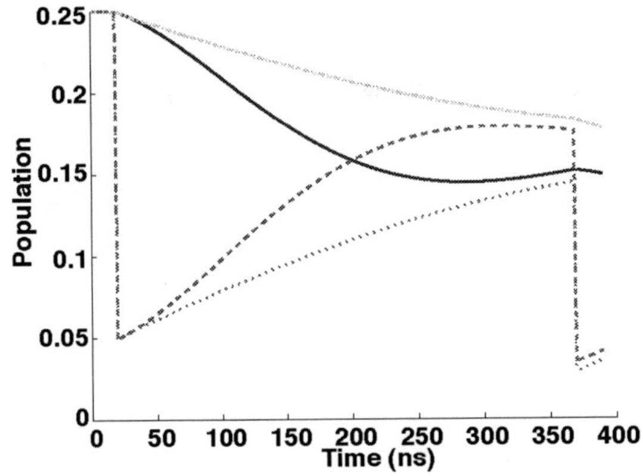

FIGURE 7. Total population evolution of the laser-microwave-laser transition for the ν_{HF+} resonant case, where the laser delay time $T = 350$ ns. The microwave starts after the first laser at time $t = 20$ ns. The second laser shoots at $t = 370$ ns. The solid line represents the J_{-+}, the dashed the J_{++}, the dash-dotted line the J_{--}, and the dotted line the J_{+-} state. Note that damping and refilling effects prevent the population from being perfectly inverted.

CREATING A π-PULSE

To maximize the population transfer it is necessary to induce a π−pulse (half a Rabi oscillation) that inverses the populations of the two quadruplets. Figure 7 shows the simulated time evolution of population density transfer between the four quantum states where conditions have been optimized to induce a π−pulse with the 350 ns microwave pulse.

Both lasers remained fixed to the f^+ transition but now a microwave pulse was used to drive the magnetic transitions between the quadruplets. The microwave frequency was fixed to one of either two frequencies: the ν^+_{HF} transition or a single off-resonant frequency. The microwave power was ramped over a range of powers.

At very low power P ($P = 0.01$ W), or off-resonance, the microwave has no effect on the atoms. However when the microwave is on-resonance, and in the region of P = 1-30 W, Rabi oscillations between the quadruplets are induced and a population transfer occurs. An equal number of measurements were carried out with microwave powers ranging from $P = 0.01$ W to 30 W and one off-resonant frequency. The microwave power

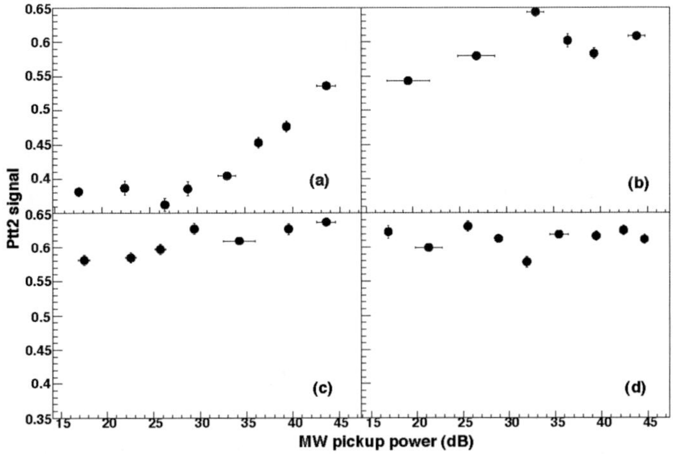

FIGURE 8. Microwave Power Dependence results for (a) T = 150 ns, (b) T = 500 ns, (c) T = 700 ns and (d) T = 1000 ns at 250 mbar.

was limited to $P = 30$ W as, although the TWTA could produce an amplified signal up to $P = 2$ kW, sparking would occur at higher powers.

The scans were repeated over four different laser delays (T = 150ns, 500 ns, 700 ns, 1000 ns) with a laser energy fluence of 30 mJ/cm^2. The four separate power dependence plots are shown in Fig. 8 where each data point was measured eight times. A measurement for T = 350 ns was made under different conditions and is therefore not shown in Fig. 8. In Figs. 8b and 8c it can be observed that a π-pulse is achieved where the first maximum occurs. This is neither visible in Fig. 8a because there are no points after the maximum to ensure the maximum has been reached nor in Fig. 8d where the collisions have removed the asymmetry so much that the signal is too weak.

From these graphs it was possible to determine the optimum power for each T where a π−pulse occurred and maximum population transfer was induced. Taking the ratio between peak-to-total for the π-pulse (ptt_2^{π}) and the off-resonant ($\mathrm{ptt}_2^{\mathrm{off}}$) for each laser delay time, allowed the information to be compressed into one time dependant graph (Fig. 9) where the point at T = 350 ns has been added from microwave scans under the same conditions.

A numerical model was used to calculate each point using the measured rate of inelastic collisions λ_i = 1.4 MHz for a pressure of 250 mbar and an elastic collision rate λ_e = 3.5 MHz, determined from the numerical model shown in Fig. 2. Calculated this way the theoretical points match the experimental results for T = 350 ns, 500 ns, 700 ns, and 1000 ns (Fig. 9). However for T = 150 ns it reveals that it is theoretically possible to achieve a much higher signal than experimentally observed. From Fig. 8 a signal of 0.65 occurs at the maximum population transfer for all situations apart from T = 150 ns therefore it is assumed that $P > 30$ W would have revealed a higher population transfer.

Because of the narrow width observed during simulations, the high signal to noise ratio and the reduced sensitivity to laser overlap, the 350 ns delay between laser 1 and 2

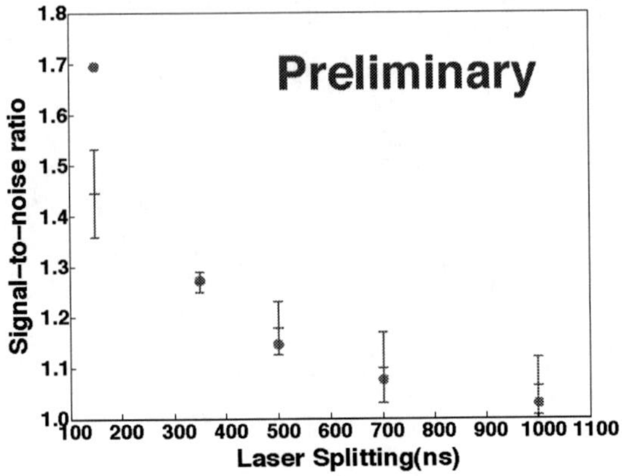

FIGURE 9. A graph showing the optimum signal-to-noise ratio for each time delay T between both lasers. The experimental results are shown as a vertical cross and the simulated points as a circle.

was chosen for the microwave scans.

CONCLUSIONS

A systematic study of the HF splitting of the (37, 35) state of $\bar{p}He^+$ has been used to determine the optimal parameters for microwave scans. Improvements to the laser system have increased the peak-to-total signal and allowed the delay of the second laser to be extended. The first experimental measurements of the relaxation frequency between HF states show that the collisions occur at a rate approximately a factor of two slower than the lower limit predicted by the theory [15]. However the statistics are low and more data points are required to determine an accurate value for these measurements.

A comparison between the measured transition frequencies and three body QED calculation can be used to determine the magnetic moment of the antiproton. Bakalov and Widmann [22] calculated that the precision required to reach the limit of the accuracy of the theory of the (37, 35) state is 33 kHz. A measurement to this accuracy would, if it agreed with theory, provide the most precise determination of the antiproton proton magnetic moment ratio to date. 0.1% accuracy can be achieved with a density and power dependence microwave spectroscopy measurement like the numerical simulation shown in Fig. 2.

ACKNOWLEDGMENTS

We are grateful to three undergraduate students: Andrea Puhm, Martin Leitgab and Gilbert Hangel who worked on this project. We thank the AD operators for providing the antiproton beam. This work was supported by Monbukagakusho (grant no. 15002005), by the Hungarian National Research Foundation (NK67974 and K72172) and by the Austrian Federal Ministry of Science and Research.

REFERENCES

1. M. Iwasaki, S.N. Nakamura, K. Shigaki, Y. Shimizu, H. Tamura, T. Ishikawa, R.S. Hayano, E. Takada, E. Widmann, H. Outa, M. Aoki, P. Kitching, T. Yamazaki, *Phys. Rev. Lett.* **67** 1246 (1991).
2. T. Yamazaki, E. Widmann, R.S. Hayano, M. Iwasaki, S.N. Nakamura, K. Shigaki, F. J. Hartmann, H. Daniel, T. von Egidy, P. Hofmann, Y.S. Kim, and J. Eades, *Nature (London)* **361**, 238 (1993).
3. T. Yamazaki, N. Morita, R.S. Hayano, E. Widmann, J. Eades, *Physics Reports* **366** 183-392 (2002).
4. R.S. Hayano, M. Hori, D. Horváth, and E. Widmann, *Rep. Prog. Phys.* **70** 1?71 (2007).
5. T. Yamazaki and K. Ohtsuki, *Phys. Rev. A* **45**, 7782 (1992).
6. G. T. Condo, *Phys. Rev. Lett.* **9**, 65 (1996).
7. J. E. Russell, *Phys. Rev. Lett.* **23**, 63 (1969); *Phys. Rev.* **188**, 187 (1969); *Phys. Rev. A* **1**, 721 (1970); *Phys. Rev. A* **1**, 735 (1970); *Phys. Rev. A* **1**, 742 (1970); *J. Math. Phys.* **12**, 1906 (1971); *Phys. Rev. A* **6**, 2488 (1972).
8. E. Widmann, J. Eades, T. Yamazaki, H.A. Torii, R.S. Hayano, M. Hori, T. Ishikawa, M. Kumakura, N. Morita, I. Sagai, F.J. Hartman, T. von Egidy, B. Ketzer, C. Maierl, R. Pohl, D. Horváth, *Phys. Lett. B* **404**, 15 (1997).
9. E. Widmann, J. Eades, T. Ishikawa, J. Sakaguchi, T. Tasaki, H. Yamaguchi, R.S. Hayano, M. Hori, H.A. Torii, B. Juhász, D. Horváth, T. Yamazaki, *Phys. Rev. Lett.* **89** 243402 (2002).
10. V. Korobov an D. Bakalov, *J. Phys. B* **34**, L519 (2001).
11. J. Sakaguchi, J. Eades, R. S. Hayano, M. Hori, D. Horváth, T. Ishikawa, B. Juhász, H. A. Torii, E. Widmann, H. Yamaguchi and T. Yamazaki *Nucl. Intr. Meth. A* **533** 598 (2004).
12. M. Hori, *Rev. Sci. Inst.* **76** 113303 (2005).
13. M. Hori, K. Yamashita, R.S. Hayano and T. Yamazaki, *Nucl. Inst. Meth. A* 496 102 (2003).
14. M. Hori, A. Dax, J. Eades, K. Gomikawa, R. S. Hayano, N. Ono, W. Pirkl, E. Widmann, H. A. Torii, B. Juhász, D. Barna,6,2 and D. Horváth, M. Hori *Phys. Rev. Lett.* **96**, 243401 (2006).
15. G.Ya. Korenman, S.N. Yudin, *J. Phys. B* **39**, 1473(2006).
16. Chloé Malbrunot Diploma Thesis, Technische Universität Wien, 2007, unpublished.
17. M. Hori, J. Eades, R.S. Hayano, T. Ishikawa, J. Sakaguchi, T. Tasaki, E. Widmann, H. Yamaguchi, H.A. Torii, B. Juhász, D. Horváth, T. Yamazaki, *Phys. Rev. Lett.* **89** 093401 (2002).
18. M. Hori, J. Eades, E. Widmann, T. Yamazaki, R.S. Hayano, T. Ishikawa, H.A. Torii, T. von Egidy, F.J. Hartmann, B. Ketzer, C. Maierl, R. Pohl, M. Kumakura, N. Morita, D. Horváth, I. Sugai, *Phys. Rev. A* **70**, 012504 (2004).
19. D. Bakalov and V. Korobov, *Phys. Rev. A* **57**, 1662 (1998).
20. N.Yamanaka, Y.Kino, H.Kudo, M.Kamimura, *Phys. Rev. A* **63**, 012518 (2001).
21. Y.Kino, N.Yamanak, MKamimura, H.Kudo, in Proceedings of the 3rd European Conference on Atomic Physics at Accellerators (APAC2001), Aarhus, Denmark, (2001).
22. D. Bakalov and E. Widmann, *Phys. Rev. A* **76** 012512 (2007).

Conclusions from recent pionic–atom experiments

D. Gotta*, F. Amaro†, D. F. Anagnostopoulos**, S. Biri‡, D. S. Covita†, H. Gorke§, A. Gruber¶, M. Hennebach*, A. Hirtl¶, T. Ishiwatari¶, P. Indelicato‖, Th. Jensen‖, E.–O. Le Bigot‖, J. Marton¶, M. Nekipelov*, J. M. F. dos Santos†, S. Schlesser‖, Ph. Schmid¶, L. M. Simons††,†, Th. Strauch*, M. Trassinelli‖, J. F. C. A. Veloso† and J. Zmeskal¶

**Institut für Kernphysik, Forschungszentrum Jülich, D–52425 Jülich*
†Dept. of Physics, Coimbra University, P–3000 Coimbra, Portugal
***Dept. of Material Science and Engineering, University of Ioannina, Ioannina, GR–45110, Greece*
‡Institut of Nuclear Research, Hungarian Academy of Sciences, H–4001 Debrecen, Hungary
§Zentrallabor für Elektronik, Forschungszentrum Jülich, D–52425 Jülich
¶Stefan Meyer Institut, Austrian Academy of Sciences, A–1090 Vienna
‖Lab. Kastler Brossel, UPMC–Paris 6, ENS, CNRS; 4 place Jussieu, F–75005 Paris, France
††Paul Scherrer Institut, Villigen PSI, CH–5232 Villigen, Switzerland

Abstract. The most recent pionic–hydrogen experiment marks the completion of a whole series of measurements, the main goal of which was to provide conclusive data on pion–nucleon interaction at threshold for comparison with calculations from Chiral perturbation theory. The precision achieved for hadronic shift and broadening of 0.2% and 2%, respectively, became possible by comprehensive studies of cascade effects in hydrogen and other light exotic atoms including results from the last years of LEAR operation. In order to obtain optimum conditions for the Bragg crystal spectrometer, the cyclotron trap II has been used to provide a suitable X–ray source. To characterize the bent crystal spectrometer, the cyclotron trap has been modified to operate as an electron–cyclotron resonance source, which produces with high intensity narrow X-ray transitions in the few keV range originating from highly charged ions.

Keywords: Exotic atoms, pion–nucleon intercation, X–ray spectroscopy, crystal spectrometer
PACS: 36.10.-k, 36.10.Gv, 13.75.Gx, 32.30.Rj, 07.85.Nc

INTRODUCTION

Calculations within the modern low–energy approach of QCD – Chiral Perturbation Theory (χPT) – have reached an accuracy of a few per cent. This in turn requires data at a similar level of precision. χPT is based on the fact that the (almost) massless quarks allow a quantitative description by effective degrees of freedom circumventing a treatment based on the elementary constituents quarks and gluons [1]. A fundamental treatment in terms of color QCD is expected to be feasible in future by lattice gauge calculations, but nowadays effective field theories like χPT must be used in the non perturbative regime. Nonetheless, the underlying chiral symmetry arising from the vanishing mass of the (current) quarks u, d, and s determines the interaction of the effective fields like pions and kaons. They are identified to be the Goldstone bosons originating from the so called spontaneous breaking of chiral symmetry. Basic low–energy theorems are formulated in

CP1037, *Cold Antimatter Plasmas and Application to Fundamental Physics,* edited by Y. Kanai and Y. Yamazaki

the chiral limit, i. e., vanishing quark masses. Finite masses as well as electromagnetic interaction are included as perturbation quantifying the breaking of the symmetry. In this scheme, mass and interaction of the "Goldstone bosons" are related to the degree of symmetry breaking [2, 3, 4, 5, 6].

At threshold the interaction of the pion–nucleon system is given by two real numbers, the scattering lengths according to the two isospin combinations $I = 1/2$ and 3/2 occuring for the πN system. The isospin even and odd combinations read

$$a^+ = \frac{1}{3}(a_{1/2} + 2a_{3/2}) \tag{1}$$

$$a^- = \frac{1}{3}(a_{1/2} - a_{3/2}). \tag{2}$$

The leading order result of χPT for a^+ and a^- is obtained from current algebra, which revealed already as an important consequence of the underlying chiral symmetry a simple relation between the isospin contributions [7, 8, 9]. E. g., it causes the isoscalar combination a^+ to vanish exactly:

$$a^+ = 0 \tag{3}$$

$$a^- = a_{1/2} = -2a_{3/2} = -0.079/m_\pi \tag{4}$$

Deviations from these values represent the higher orders in χPT [10, 11, 12, 13, 14, 15, 16, 17, 18].

The scattering lengths $a^\pm$ are related to elementary πN reactions accessible by experiment:

$$a^\pm = \frac{1}{2}(a_{\pi^- p\to\pi^- p} \pm a_{\pi^+ p\to\pi^+ p}). \tag{5}$$

In the limit of isospin symmetry, the intercations with charged pions yield directly the charge exchange process

$$a_{\pi^- p\to\pi^- p} - a_{\pi^+ p\to\pi^+ p} = -\sqrt{2}a_{\pi^- p\to\pi^0 n}. \tag{6}$$

The scattering lengths $a_{\pi^- p\to\pi^- p}$ and $a_{\pi^- p\to\pi^0 n}$ are related to the ground state strong–interaction shift ε and broadening Γ of the pionic–hydrogen atom and are measured by means of energy and line width of characteristic X–radiation from the πH atom (Fig. 1). Shift and width may be expressed by Deser–type formulae[1] [19, 20, 21]:

$$\frac{\varepsilon_{1s}}{B_{1s}} = -\frac{4}{r_B}a_{\pi^- p\to\pi^- p}(1+\delta_\varepsilon) \tag{7}$$

$$\frac{\Gamma_{1s}}{B_{1s}} = 8\frac{q_0}{r_B}(1+\frac{1}{P})[a_{\pi^- p\to\pi^0 n}(1+\delta_\Gamma)]^2. \tag{8}$$

---

[1] $q_0 = 0.1421 fm^{-1}$ is the centre–of–mass momentum of the π^0 in the charge–exchange reaction $\pi^- p \to \pi^0 n$ and $P = 1.546 \pm 0.009$[22] the branching ratio of charge exchange and radiative capture (Panofsky ratio).

Obviously $\varepsilon_{1s} \propto a^+ + a^-$ and, in the limit of isospin invariance, $\Gamma_{1s} \propto (a^-)^2$ holds in leading order. The contribution of charge exchange to the total width is obtained after correction for radiative capture $\pi^- p \to \gamma n$ using the well known Panofsky ratio P [22]. Electromagnetic and higher order corrections encoded by the parameters δ_ε and δ_Γ have been continuously improved especially within the framework of χPT [13, 15, 23, 24, 25, 26, 27].

More important quantities are closely related to the πN scattering lengths. Using the Goldberger–Miyazawa–Oehme (GMO) sum rule, the isovector scattering length a^- yields the πN coupling constant $f^2_{\pi N}$ [28, 29, 30]. The coupling constant $f^2_{\pi N}$ itself may be used to determine the induced pseudovector coupling g_P, a quantity accessible otherwise by muon capture experiments $\mu^- p \to n\nu_\mu$ [31, 32, 33]. The electric dipole amplitude E_{0+} for π^- photo production $\gamma n \to \pi^- p$ is also directly connected to a^- [9, 34]. The isoscalar contribution a^+ is related to the πN σ–term which is an important contribution to the nucleon mass [35, 36].

χPT, being an effective field theory, contains at any order of expansion coupling constants, so called low–energy constants (LECs), which represent the short range part of the interaction. The LECs have to be determined from experiment and not all of them are known to sufficient accuracy. In the case of the πH ground state shift, the precision for $a^+ + a^-$ is limited to $\pm 3\%$ by one particular LEC f_1 [25], whereas experiment is one order of magnitude more precise (see below and [37, 38]). The LEC f_1 does not appear in the expression relating Γ_{1s} to a^- and, hence, the theoretical uncertainty is quoted to be only $\pm 0.2\%$ [26].

Combining πH ground state shift, broadening and πD ground state shift, the scattering lengths a^+ and a^- are overdetermined, because in leading order $\varepsilon_{1s}(\pi\mathrm{D}) \propto a^+$ [39, 40, 41, 42, 43, 44]. Precise πH and πD data together even yield a constraint on f_1 [45].

Because of the almost cancellation of the $\pi^- p$ and $\pi^- n$ interaction, the πD system shows an outstanding sensitivity to isospin breaking contributions for reactions involving charged pions [45, 46]. The hadronic broadening in πD is related by detailed balance to s–wave pion production in the reaction $pp \to d\pi^+$ [47]. Consequently, to improve on the results of previous measurements [48, 49], a high–precision measurement of the πD system was part of the experimental program [50, 51].

ATOMIC CASCADE

Pionic hydrogen

After capture into high–lying atomic states, the de–excitation cascade of pionic hydrogen is governed by collisional processes. Due to a finite target density numerous elastic and inelastic collisions occur leading among others to Stark mixing and Coulomb de–excitation. The upper part of the atomic cascade is determined by external Auger effect and Coulomb de–excitation. Between low–lying states, X–ray transitions dominate (Fig. 1).

Due to the fact that exotic hydrogen is electrically neutral, Stark mixing plays an outstanding role, because the atom approaches easily the nuclei of the target molecules experiencing there a strong Coulomb field. In the presence of strong interaction, the mixing of different angular momentum states ℓ for given main quantum number n leads to nuclear reactions whenever an s–wave contribution is admixed and, consequently, the X–ray cascade is strongly depleted with increasing density [52]. Typical line yields for pionic hydrogen are a few per cent only at target densities equivalent to pressures of a few bar [53]. More details on capture, cascade and strong–interaction effects may be found in [54, 55].

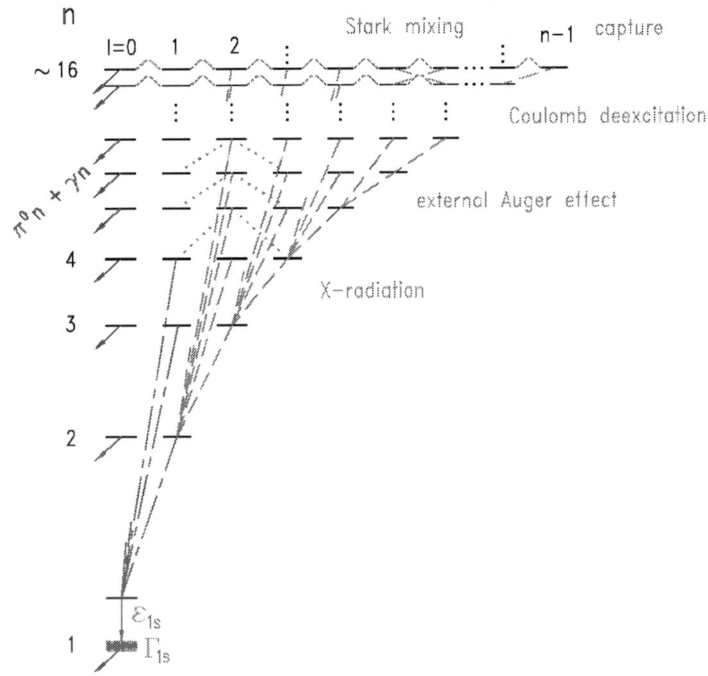

FIGURE 1. Atomic cascade in pionic hydrogen. In this experiment, the transitions πH$(2p-1s)$, πH$(3p-1s)$, and πH$(4p-1s)$ with the energies of 2.437, 2.885, and 3.043 keV have been studied. The hadronic shift ε_{1s} is given by the difference $E_{exp}-E_{QED}$ of the measured and calculated purely electromagnetic energy for the X–ray transition. Hadronic shift and broadening Γ_{1s} of the ground state are about +7 and 1 eV [37], respectively. In the case of πH, the positive shift reveals the attractive π^-p interaction. For πD, the sum of the π^-p and π^-n interaction leads to a small negative value for the shift (≈ -2.5 eV), where the broadening is 1.2 eV, i. e., comparable to the πH case [48, 49].

Coulomb de–excitation is particularly important in the case of the line width measurement. In distributing the energy gain of a cascade step as kinetic energy to the collision partners (generally the πp system and an H atom from an H_2 molecule), the pionic hydrogen atom is accelerated. The acceleration was observed directly for the first time from a broadening in neutron time–of–flight spectra stemming from the two–body reaction $\pi^-p \rightarrow \pi^0 n$ [59]. Alike a Doppler broadening of the X–ray line must be expected from such a moving source and has to be quantified preciseley for a proper extraction of the hadronic width. Whereas charge exchange occurs from ns states, mainly with

$n = 3-5$ of the π^-p system, in the case of X–ray emission np levels are the initial states and preceding cascade steps may be completely different. Therefore, a dedicated study for the X–ray case is necessary.

Acceleration caused by Coulomb de–excitation is not included in the standard cascade model (SCM), where the kinetic energy is a state independent parameter fitted to the measured X-ray line yields [56]. In the extended cascade model (ESCM), the development of the kinetic energy through the de–excitation steps has been included [57, 58]. The kinetic energy distribution for the πH atom being in an atomic state (n,ℓ) is obtained by following the competition of acceleration by Coulomb de–excitation and deceleration in the foregoing cascade steps. In order to test the predictions of the ESCM, Coulomb de–excitation was studied by measuring the $3p-1s$ transition in muonic hydrogen, where Doppler broadening is better visible because of the absence of strong interaction (see below).

Another process to be considered is molecular formation, possibly affecting the extraction of the hadronic shift. As known from muon–catalyzed fusion, complexes like $(\mu H)_{nl}+H_2 \rightarrow [(pp\mu)_{nvj}\,p]2e^-$ are formed during the collision [60]. Similarly, molecular formation should occur with pions. The quantum numbers v and j denote vibrational and total angular momentum of the 3–body molecular state. Though the 3–body system $(pp\pi)_{nvj}$ is assumed to de–excite mainly by Auger emission it cannot be excluded beforehand that a fraction of the πH atoms bound into such molecules decays radiatively to the ground state. Small line shifts – here to lower energies – cannot be resolved and, hence, falsify the extracted hadronic shift [61, 62, 63]. It is assumed that the formation rate depends on the number of collisions, i. e., on density, and consequently a density dependent shift of the line center indicates radiative de–excitation of molecules. Therefore, in the experiment a possible energy dependence of the X–ray energy was thoroughly studied.

Light pionic atoms

Exotic–atom X–ray transitions are usable as calibration standards [64]. In low and medium Z atoms the electron shell is completely removed by Auger emission when refilling of electrons from neighbouring atoms is suppressed [65, 66]. Hence, in the medium part of the cascade the pionic atom is a hydrogen–like system, and because the small overlap of the pion and nucleus wave functions, strong–interaction effects are negligibly small. Level energies are then calculable from QED to a precision of a few meV [67]. This was exploited for the transition combination $\pi H(3p-1s)$ and $\pi^{16}O(6h-5g)$.

Radiative widths in exotic atoms are much smaller than in electronic systems which suggests to use the πO transition also for the determination of the apparative response. However, for molecules at first the binding electrons are quickly removed, while the heavy and highly charged ions are still at a distance of about the molecular bond length. Finally, the pion is attached to one of the ions which, due to Coulomb repulsion, is accelerated. The subsequent X–ray transition is then Doppler broadenend. Coulomb explosion was observed directly by comparing the line widths of X–ray transitions in

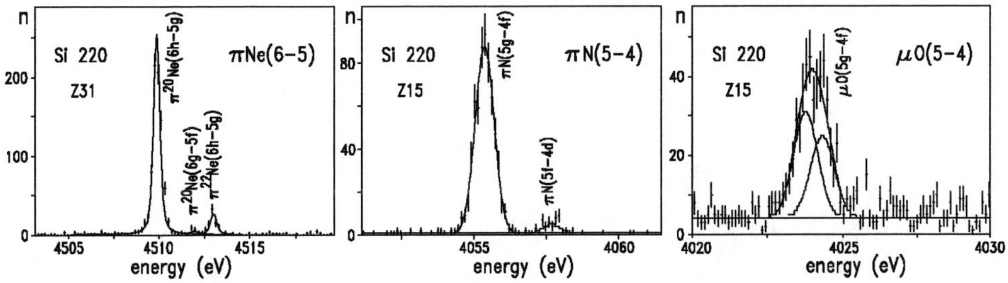

FIGURE 2. Comparison of the line widths of πNe, πN, and μO transitions. The additional broadening for the atoms formed with molecules is due to a Doppler broadening caused by Coulomb explosion (from [68]).

pionic neon to pionic nitrogen and muonic oxygen [68] (Fig. 2).

Consequently, a precision determination of the crystal resolution would be restricted to noble gases, however, at first transition energies do not fit and, secondly, count rates are very limited. Therefore, a new approach using highly charged ions had to be used (see below), because intense γ rays from radioactive sources are also not available in the few keV range (see below and [64]).

EXPERIMENT

X–ray energies of the ground–state transitions in muonic and pionic hydrogen and deuterium are 2–3 keV, where hadronic shift and width are of the order of eV. Therefore, a reflection–type spectrometer must be used equipped with quartz or silicon Bragg crystals, which are the only manageable materials reaching a resolution of about 10^{-4} in this energy range.

The determination of the hadronic effects at the per cent level requires both high statistics and ultimate energy resolution, which is usually mutually exclusive. Even with a high flux facility as available at the Paul Scherrer Institut (PSI), exotic–atom X–ray rates are at the lower edge for ultimate–resolution spectroscopy. Hence, a compromise has to be found for setting up a high–resolution device and an acceptable count rate. The best approach is a Johann–type spectrometer using spherically bent crystals (Fig. 3. Because the overall efficiency of such a spectrometer is still very low, the cyclotron trap is essential to provide a bright X–ray source. A combination of cyclotron trap and Johann spectrometer was used first at LEAR in the measurement of the Lα transitions (1.7 and 2.4 keV) in antiprotonic hydrogen and deuterium [69].

X-ray source

PSI provides low–energy pions of high intensity which are produced by means of the 590 MeV proton beam (up to 1.8 mA). The short πE5 beam line, set up for 112 MeV/c momentum, leads up to a few $10^8\,\pi^-/s$ to the experimental area, where the pions

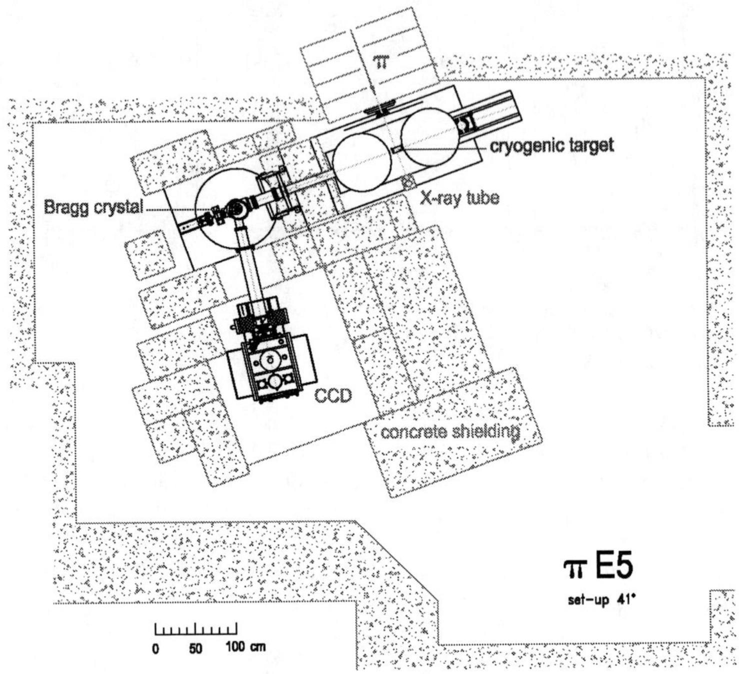

FIGURE 3. Setup in the πE5 area at PSI of cyclotron trap, crystal spectrometer and X–ray detector (CCD) for the πH$(3p-1s)$ measurement. The massive concrete shielding is essential to achieve a good peak–to–background ratio.

enter the gap between two superconducting coils (cyclotron trap II). The cyclotron trap provides a magnetic field with focussing properties which, with the help of degraders, guides the beam towards its center [70] (Fig. 4).

To study the cascade effects, a wide range must be covered for the hydrogen density. This was achieved by using a cryogenic target. At a gas pressure of 1 bar, about 0.5% of the incoming pions are stopped in a thin–walled hydrogen filled cell. The stop efficiency increases linearly with density up to 30 bar pressure equivalent, counterbalancing the decreasing X–ray line yields. The cell diameter of about 5 cm represents the extension of the X–ray source. The large gap of the cyclotron trap II allows also an efficient formation of muonic atoms. The muons originate from the decay of slow pions inside the magnet's chamber. The stop efficiency for muons is about a factor of 20 lower than for pions, which is partly compensated by the higher μH line yields [71].

Due to pion absorption a high neutron induced background occurs. With a massive concrete shielding, a significant improvement of the background level was achieved in the X–ray spectra compared to the previous experiment [37].

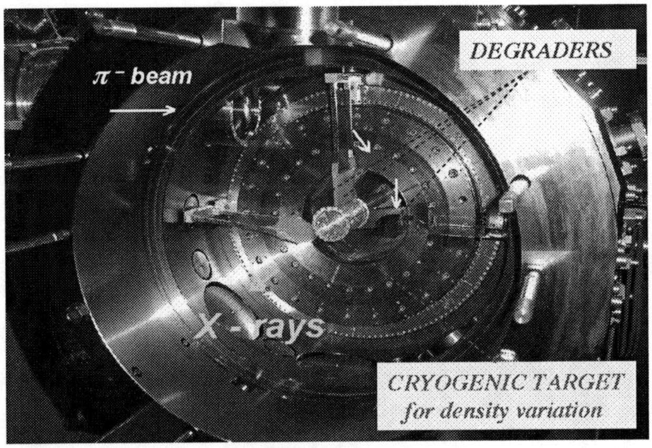

FIGURE 4. Interior of the cyclotron trap with the second coil and its iron yoke removed. Degraders reduce the momentum that after about 2 turns a fraction of the beam stops in gas of the target cell. The wall of the cell is made of 50 μm thick KAPTON foil. The exit window towards the crystal spectrometer has a thickness of 7.5 μm.

Crystal spectrometer

A bent crystal setup (Johann–type spectrometer) allows the simultaneous measurement of an energy interval, the width of which is given by the extension of the X–ray source in the direction of dispersion. A crystal diameter of about 6–10 cm and a curvature of about 3 m constitutes a compromise between count rate and abberations due to the finite size of the crystal. The use of spherically bent crystals leads in addition to a partial vertical focussing enhancing further the count rate. The total height of the reflection is about 100 mm is then mostly covered by the detector's vertical extension of 75 mm. The efficiency of the spectrometer, given by geometry and reflectivity of the crystal, is of the order 10^{-6}. Count rates of 10–50 per hour were achieved for the various pionic–hydrogen transitions.

In the direction of dispersion, the detector width has to cover about the width of the target window. E. g., for the πH$(3p-1s)$ transition at 2.885 keV and using a Si 111 crystal yielding a Bragg angle of 43.2°, the width of the detector array of 50 mm corresponds to 45 eV. This allows the simultaneous measurement of the π^{16}O$(6h-5f)$ transition of 2.880 keV used for energy calibration (Fig. 8). At higher target densities, when oxygen freezes out, hydrogen and oxygen were measured alternately. Because of the detector size, none of the spectrometer parts had to be moved thus preserving the energy calibration.

X–ray detector

At facilities producing hadron induced background, charge–coupled devices (CCDs) are most suitable X–ray detectors because they combine a build–in two–dimensional

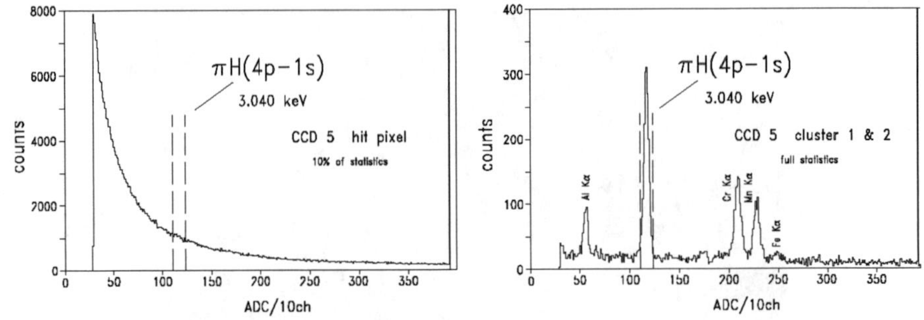

FIGURE 5. ADC (charge) spectra from the πH($4p-1s$) experiment: raw data (left) and data after eliminating clusters except single and adjacent two–pixel hits (right). Position spectra are then generated only from the events of the cleaned data lying inside the ADC cut corresponding to the energy of the πH($4p-1s$) X–rays.

position sensitivity with the good energy resolution owing to semiconductor devices of 200 eV or better at a few keV. Where X–rays produce short–ranged photo electrons depositing charge in one or two pixels only, background events stemming from high energy photons following neutron capture cause larger patterns. As the pixel structure allows a reconstruction of the hit pattern, background is reduced by at least an order of magnitude (Fig. 5). In the recent pionic–atom experiment, a 2×3 array was used built up of 6 CCDs with 600×600 pixels of 40 μm size and cooled to $-100°$C [72]. The CCD's cryostat volume was separated from the spectrometer vacuum by a 5 μm thick MYLAR window.

ECRIT

The response function of a bent crystal spectrometer is given by the convolution of the geometrical imaging and the intrinsic properties of the crystal material. The total reflection is the sum over all source points contributing to the reflection. The intrinsic properties are given by the so called rocking curve, which is reliably calculable for plane crystals by diffraction theory [73]. The geometry of the spherically bent crystal setup is modeled by Monte Carlo ray–tracing calculations. Rocking curve, assuming an ideal material, and ray tracing yields the "ideal" response of the setup. Due to the non trivial superposition of geometry and crystal properties an experimental verification of the response is mandatory.

In order to quantify precisely deviations from the ideal case, systematic high–statistics studies of the crystal response funcion were performed. For the first time narrow X–ray lines from highly ionised atoms, produced in a dedicated device, have been used. A permanent hexapole magnet was inserted in between the coils of the cyclotron trap to set up an electron–cyclotron resonance ion trap (ECRIT) [74]. Atoms of a dilute gas were heated and ionised by means a 6.4 GHz HF emitter coupled to the inner chamber of the ECRIT. In such a device, usually operated as a source for highly charged atoms, the

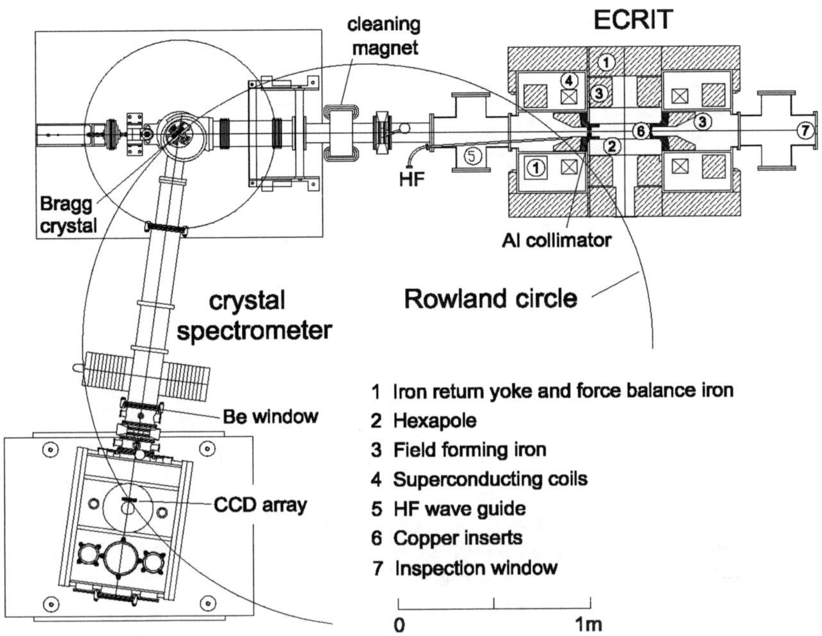

FIGURE 6. Setup of the PSI ECRIT and the Bragg crystal spectrometer (from [76]).

ions are rather slow as measured by optical methods and, hence, no noticable Doppler broadening occurs [75]. For the large crystal bending radii used in this experiment, it turned out, that quartz and silicon behave as expected like ideal crystals and, altogether, a realistic picture is obtained from diffraction theory and ray tracing. A small additional contribution of 10–20%, described by a Gaussian, takes into account the deviation from the ideal resolution.

For the pionic–hydrogen experiment, the long–living M1 transitions in helium–like sulphur, chlorine, and argon atoms were measured (Fig. 7), the energies of which coincide almost with the energies of the πH$(2p-1s)$, πH$(3p-1s)$, and πH$(4p-1s)$ lines. With the ECRIT, an M1 intensity of up to 30000 counts per hour was achieved allowing systematic studies of focal conditions and apertures for our complete set of Bragg crystals. Energy resolutions of 400–500 meV (FWHM) were measured [76].

RESULTS

Muonic hydrogen

The line shape of the μH$(3p-1s)$ reveals directly the influence of Coulomb de–excitation by a significant broadening when compared to the response extracted from the ECRIT measurement (Fig. 8–left). In a numerical analysis three different contributions are identified: a high and medium energy component from the $(4-3)$ and $(5-4)$

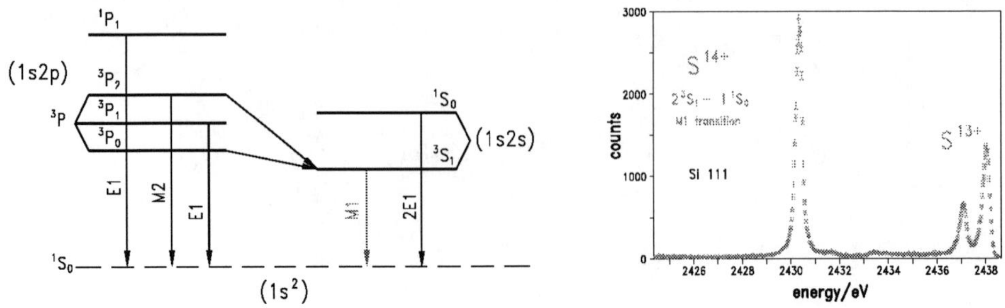

FIGURE 7. Left: level scheme of helium–like medium Z ions. Right: X–ray spectrum around the narrow M1 transition in sulphur used to calibrate the response function of the crystal spectrometer for the πH$(2p-1s)$ transition.

Coulomb transitions and a strong low–energy part representing kinetic energies below 2 eV. The low–energy part originates from the moderation of fast μH atoms by elastic and inelastic collisions. For the μH$(3p-1s)$ transition it amounts to about 60% [78].

ESCM calculations are not able to reproduce such a high fraction of slow μH atoms, which suggests that processes during the upper part of the cascade are missing or not treated correctly. A further complication is that calculations hardly take into account the molecular structure in the μH+H_2 collisions. Therefore, present cascade calculations are not able to constrain the relative strength of the Coulomb de–excitation steps for fitting the πH X–ray line shapes. Extending the calculations to molecules and increasing the initial level to start the cascade modeling are planned [79].

Molecular–formation calculations on the radiative de–excitation of molecular levels suggest, that satellites may appear several eV apart at the low–energy side of μH X–ray transitions, which stem from Auger stabilised molecular complexes [63]. However, no satellites are identified at the few per cent level at the low–energy side of the μH$(3p-1s)$ transition. This is in line with the non observation of such satellites in πH.

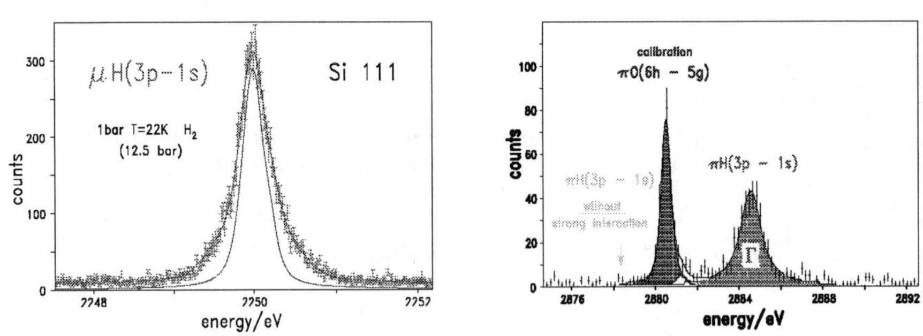

FIGURE 8. Left: muonic hydrogen – the narrow solid line corresponds to the line shape as derived from the resolution function obtained from the ECRIT measurements (from [38]). Right: pionic hydrogen – the pionic oxygen line is used for the energy calibration.

Another result is that the μH 1s hyperfine states $^3S_1$ and $^1S_0$ are statistically populated ($P(^3S_1):P(^1S_0)=3:1$). Fitting a doublet to the line shape and fixing the hyperfine splitting to the calculated value of 183 meV [80] yields a ratio of 2.9 ± 0.3. The result confirms for the first time experimentally the assumption used in muon–capture experiments [33].

Density dependence of the transition energy

The πH$(3p-1s)$ transition energy was measured in the a density range corresponding to hydrogen pressures from 3.5 bar to liquid (being equivalent to 700 bar). The energy was always deduced from the angular difference to the nearby π^{16}O$(6h-5g)$ line (Fig. 8–right). No density dependence was found within the experimental accuracy. It is concluded that no radiative de–excitation occurs from molecular states. The strong–interaction shift is calculated as the weigthed average of the results for all densities [77] to be

$$\varepsilon_{1s}^{\pi H} = 7120 \pm 8\,(stat.) \pm 6\,(sys.)\ \text{meV} \quad (preliminary). \tag{9}$$

The systematic uncertaintainty is dominated by the the line shape model used to fit the π^{16}O$(6h-5g)$ line and the parallel transitions π^{16}O$(6g-5f)$ and π^{16}O$(6g-5d)$. Only a contribution as small as 1 meV arises from the calculation of the electromagnetic transition energy [67].

Line shape and hadronic width

With decreasing main quantum number of the X–ray transition's initial state an increasing line width is expected because additional Coulomb de–excitation steps with increasing energy gain can contribute. This picture is confirmed by the total line width increasing from the πH$(4p-1s)$ over the πH$(3p-1s)$ to the πH$(2p-1s)$ transition (after subtraction of the experimental contribution). The total width of the πH$(4p-1s)$ line constitutes already an upper limit for the hadronic broadening of $\Gamma < 850$ meV [77].

Based on the experience from the analysis of the μH$(3p-1s)$ line shape, the Lorentzian contribution to the three πH lines measured is extracted by fitting the relative strengths of the Doppler contributions contained in these transitions. A preliminary analysis yields as weighted average (statistical error only)

$$\Gamma_{1s}^{\pi H} - 823 \pm 19\text{meV}. \tag{10}$$

Noteworthy, that a systematic analysis of the numerical procedures revealed a strong correlation of background level and Lorentzian width leading to a systematic discrepancy. Detailed Monte–Carlo studies were performed to quantify the correction for the determination of the hadronic width. A first result is available for the πH$(4p-1s)$ line yielding $\Gamma_{1s} = 765 \pm 56$ meV, which includes a correction due to the correlation of -5% [81]. A similar analysis for the πH$(3p-1s)$ and πH$(2p-1s)$ transitions is in progress.

Pionic deuterium

In the last data taking period, the pionic deuterium $(3p-1s)$ transition was measured. Again, the pressure dependence was studied, because molecular formation calculations predict an enhanced radiative de–excitation compared to hydrogen [63]. Data were taken for 3, 10, and 30 bar equivalent density, but also in this case no energy dependence of the line energy nor satellite structures could be found. The accumulated statistics will allow to improve the accuracy for the hadronic shift and broadening from 2% to about 0.5% and from 12% to about 4%, respectively [82].

For the $\pi D(3p-1s)$ transition, no suitable calibration line is available from a hydrogen–like pionic atom like the combination $\pi H/\pi O$. Therefore, the Ga Kα_2 fluorescence line was used, which is reflected in 3rd order under the same Bragg angle as the $\pi D(3p-1s)$ line in 1st order (Fig. 9). Important was to confirm, that the tabulated value for the Ga fluorescence X–ray energy [83] was obtained with the same gallium compound (GaAs) as used in this experiment in order to avoid a possible difference of the chemical shift.

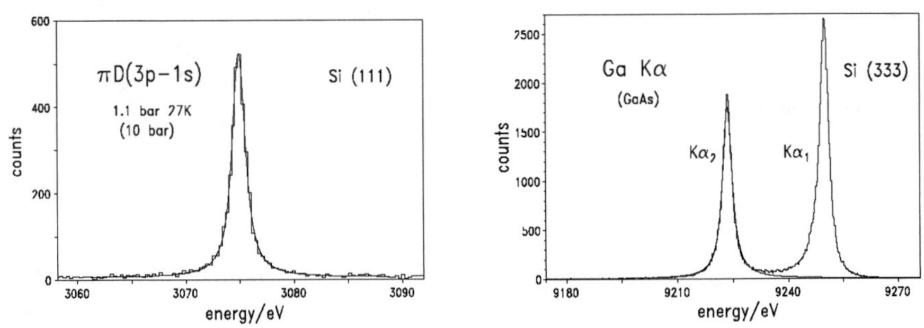

FIGURE 9. Pionic deuterium and Ga Kα_2 calibration line.

SUMMARY

A variety of techniques has been developed in the course of approaching an accurate determination of the strong–interaction effects in pionic hydrogen and deuterium. For pion beams, the cyclotron trap is essential to achieve stop densities to form a sufficiently bright X–ray source suitable for ultimate–resolution spectroscopy. By using X–rays from highly charged ions, produced in a dedicated electron–cyclotron resonance source, the complex imaging properties of a Johann–type crystal spectrometer were systematically studied.

Various cascade effects have been identified and quantified in order to extract reliably hadronic shift and broadening. Coulomb explosion of molecules and Coulomb de–excitation in hydrogen have been directly observed revealing among others a still incomplete modeling of the atomic cascade.

Regarding antiprotonic atoms, the newly developed techniques will allow to achieve accuracies for strong–interaction effects comparable to the ones in pionic atoms. Cascade studies using high–resolution devices were not performed during the LEAR era. E. g., processes like Coulomb explosion are expected to be directly observable in antiprotonic atoms, in particular because Auger emission is even more efficient [84, 85]. A summary and perspectives for such kind of measurements may be found elsewhere [55, 86, 87].

In particular, the forthcoming FLAIR facility [88] is designed for antiproton fluxes comparable to the ones at LEAR being essential to resume crystal spectrometer experiments hardly possible at the AD [89], which is the only presently operating low–energy antiproton facility. In the case that ultra–low energy beam lines will be realised at FLAIR, bright X–ray sources may be achieved without using a cyclotron trap by direct injection of about 100 keV beams into a gas cell or by atomic beams crossing antiprotons stored in a trap.

ACKNOWLEDGMENTS

It is a pleasure to thank the organizers for hospitality and support during a pleasant stay (D. G.).

REFERENCES

1. S. Weinberg, Physica A **96**, 327 (1979).
2. M. Gell-Mann, R. Oakes, and B. Renner, Phys. Rev. **122**, 2195 (1968).
3. J. Gasser and H. Leutwyler, Ann. Phys. **158**, 142 (1984).
4. J. Gasser and H. Leutwyler, Nucl. Phys. B **250**, 465 (1985).
5. V. Bernard, N. Kaiser, Ulf-G. Meissner, Int. J .Mod. Phys. E **4**, 193 (1995).
6. G. Ecker, Prog. Part. Nucl. Phys. **35**, 1 (1995).
7. S. Weinberg, Phys. Rev. Lett. **17**, 616 (1966).
8. Y. Tomozawa, Nuovo Cim. A, 707 (1966).
9. T. E. O. Ericson and W. Weise, *Pions and Nuclei* (Clarendon, Oxford 1988).
10. V. Bernard, N. Kaiser, Ulf–G. Meißner, Phys. Lett. B **309**, 421 (1993).
11. V. Bernard, N. Kaiser, Ulf–G. Meißner, Phys. Rev. C **52**, 2185 (1995).
12. V. Bernard, N. Kaiser, Ulf–G. Meißner, Nucl. Phys. A **615**, 483 (1997).
13. N. Fettes, U.–G. Meißner, S. Steininger, Nucl. Phys. A **640**, 199 (1998).
14. M. Mojžiš, Eur. Phys. J. C **2**, 181 (1998).
15. N. Fettes, U.–G. Meißner, Nucl. Phys. A **676**, 311 (2000).
16. N. Fettes, U.–G. Meißner, Nucl. Phys. A **693**, 693 (2001).
17. A. W. Thomas, W. Weise, *The Structure of the Nucleon* (WILEY–VCH, Berlin 2001).
18. S. Scherer, *Introduction to Chiral Perturbation Theory, Advances in Nuclear Physics* **27**, ed. by J. W. Negele, E. W. Vogt (Springer, 2003).
19. S. Deser, L. Goldberger, K. Kaufmann, and W. Thirring, Phys. Rev. **96**, 774 (1954).
20. T. L. Trueman, Nucl. Phys. **26**, 57 (1961).
21. G. Rasche and W. S. Woolcock, Nucl. Phys. A **381**, 405 (1982).
22. J. Spuller et al., Phys. Lett. B **67**, 479 (1977).
23. V. E. Lyubovitskij, A. Rusetsky, Phys. Lett. B **494**, 9 (2000).
24. V. E. Lyubovitskij, Th. Gutsche, A. Faessler, R. Vinh Mau, Phys. Lett. B **520**, 204 (2001).
25. J. Gasser *et al.*, Eur. Phys. J. C **26**, 13 (2002).

26. P. Zemp, In *Proc. of Chiral Dynamics 2003*, p. 128, Bonn, Germany, September 8–13 (2003), arXiv:hep–ph/0311212v1 and 'Pionic Hydrogen in QCD+QED: Decay Width at NNLO', thesis univ. of Bern (2004).
27. T. E. O. Ericson, B. Loiseau, A. W. Thomas, Phys. Lett. B **595**, 76 (2004).
28. M. L. Goldberger, H. Miyazawa, and R. Oehme, Phys. Rev. **99**, 986 (1955).
29. T. E. O. Ericson, B. Loiseau, A. W. Thomas, Phys. Rev. C **66**, 014005 (2002).
30. V. V. Abaev, P. Metsä, and M. E. Sainio, Eur. Phys. J. A **32**, 321 (2007).
31. V. Bernard, N. Kaiser, Ulf–G. Meißner, Phys. Rev. D **50**, 6899 (1994).
32. N. Kaiser, Phys. Rev. C **67**, 027002 (2003).
33. V. A. Andreev et al., Phys. Rev. Lett. **99**, 032002 (2007).
34. V. Bernard, N. Kaiser, Ulf–G. Meißner, Phys. Lett. B **383**, 116 (1996).
35. J. Gasser, H. Leutwyler, M. P. Locher, M. E. Sainio, Phys. Lett. B **213**, 85 (1988).
36. M. E. Sainio, *Pion–nucleon σ-term – a review, Proc. of the 9th Symp. on Meson–Nucleon Physics and the Structure of the Nucleon (MENU'01), πN newsletter* **16**, 138 (2002), and references therein.
37. H.–Ch. Schröder et al., Eur. Phys. J. C **21**, 433 (2001).
38. D. Gotta et al., *Pionic Hydrogen*, Lect. Notes Phys. **745**, 165 (2008).
39. S. R. Beane, V. Bernard, T.–S. Lee, U.–G. Meißner, Phys. Rev. C **57**, 424 (1998).
40. A. Deloff, Phys. Rev. C **64**, 065205 (2001).
41. S. R. Beane, V. Bernard, E. Epelbaum, U.–G. Meißner, D. R. Phillips, Nucl. Phys. A **720**, 399 (2003).
42. B. Burasoy, H. W. Grießhammer, Int. J. Mod. Phys. E **12**, 65 (2002).
43. M. Döring, E. Oset, M. J. Vicente Vacas, Phys. Rev. C **70**, 045203 (2004).
44. U. G. Meißner, U. Raha, A. Rusetski, Eur. Phys. J. C **41**, 213 (2005).
45. U.–G. Meißner, U. Raha, A. Rusetsky, Phys. Lett. B **639**, 478 (2006).
46. U.–G. Meißner, S. Steininger, Phys. Lett. B **419**, 403 (1998).
47. V. Lensky et al., Eur. Phys. J. A **27**, 37 (2006).
48. D. Chatellard et al., Nucl. Phys. A **625**, 855 (1997).
49. P. Hauser et al., Phys. Rev. C **58**, R1869 (1998).
50. PSI experiment R-98.01, `http://www.fz-juelich.de/ikp/exotic-atoms`.
51. PSI experiment R-06.03.
52. T. B. Day, G. A. Snow, J. Sucher, Phys. Rev. Lett. **3**, 61 (1959); Phys. Rev. **118**, 864 (1960).
53. A. J. Rusi El Hassani et al., Z. Phys. A **351**, 113 (1995).
54. *Proc. of the Fifth Course of the International School of Physics of Exotic Atoms*, ed. L. M. Simons, D. Horvâth, G. Torelli, Erice, Italy, May 14–20, 1989, (Plenum Press, New York 1990), and references therein.
55. D. Gotta, Prog. Part. Nucl. Phys. **52**, 133 (2004).
56. E. Borie and M. Leon, Phys. Rev. A **21**, 1460 (1980).
57. T. S. Jensen and V. E. Markushin, Eur. Phys. J. D **19**, 165 (2002); Eur. Phys. J. D **21**, 261 (2002); Eur. Phys. J. D **21**, 271 (2002).
58. T. S. Jensen, Eur. Phys. J. D **31**, 11 (2004)
59. A. Badertscher et al., Europhys. Lett. **54**, 313 (2001).
60. D. Taqqu, *Muon Catalyzed Fusion*, AIP Conference Proceedings**181**, 217 (1989).
61. S. Jonsell, J. Wallenius, P. Froelich, Phys. Rev. A **59**, 3440 (1999).
62. E. Lindroth, J. Wallenius, S. Jonsell, Phys. Rev. A **68**, 032502 (2003).
63. S. Kilic, J.–P. Karr, L. Hilico, Phys. Rev. A **70**, 052506 (2004).
64. D. F. Anagnostopoulos, D. Gotta, P. Indelicato, L. M. Simons, Phys. Rev. Lett. **91**, 240801 (2003).
65. R. Bacher et al., Phys. Rev. Lett. **54**, 2087 (1985).
66. K. Kirch et al., Phys. Rev. A **59**, 3375 (1999).
67. P. Indelicato, priv. comm.
68. T. Siems et al., Phys. Rev. Lett. **84**, 4573 (2000).
69. D. Gotta et al., Nucl. Phys. A **660**, 283 (1999).
70. L. M. Simons, Physica Scripta **T22**, 90 (1988); Hyperfine Int. **81**, 253 (1993).
71. B. Lauss et al., Phys. Rev. Lett. **80**, 3041 (1998).
72. N. Nelms et al., Nucl. Instr. Meth. A **484**, 419 (2002).
73. M. Sanchez del Rio and F. Cerrina, Rev. Sci. Instrum. **63**, 936 (1992).
74. S. Biri, L. M. Simons, D. Hitz, Rev. Sci. Instrum. **71**, 1116 (2000).
75. C. Bernard, PhD thesis, Université Claude Bernard, Lyon, 1996.

76. D. F. Anagnostopoulos et al., Nucl. Instr. Meth. A **545**, 217 (2005).
77. M. Hennebach, *Precision Measurement of Ground State Transitions in Pionic Hydrogen*, PhD thesis, Universität zu Köln, Cologne, 2003.
78. D. S. Covita et al., in preparation.
79. T. S. Jensen, V. Popov, V. Pomerant'sev, priv. comm.
80. R. N. Faustov, A. P. Martynenko, J. Exp. Theor. Phys. **98**, 39 (2004).
81. A. Hirtl, *Determination of the Strong Interaction Ground State Width in Pionic Hydrogen*, PhD thesis, Technische Universität Wien, Vienna 2008.
82. T. Strauch et al., *Proc. of 11th Int. Conference on Meson–Nucleon Physics and the Structure of the Nucleon (MENU 2007)*, to be published.
83. R. Deslattes et al., Rev. Mod. Phys., **75**, 35 (2003).
84. R. Bacher et al., Phys. Rev. A **38**, 4395 (1988).
85. D. Gotta et al., Eur. Phys. J. D **47**, 11 (2008).
86. D. Gotta, AIP conf. proc. **793**, 169 (2005).
87. H. Gorke et al., AIP conf. proc. **793**, 341 (2005).
88. *Facility for Antiproton and Ion Research*, `http://www.oeaw.ac.at/smi/flair`.
89. *The Antiproton Decelerator (AD) at CERN*, `http://psdoc.web.cern.ch/PSdoc/acc/ad`.

Atoms and plasmas in a high-magnetic-field trap

G. Raithel, B. Knuffman, M. H. Shah, C. Hempel, E. Paradis, R. Mhaskar, X. Zhang*, J.-H. Choi†, A. P. Povilus** and J. R. Guest‡

*FOCUS Center, Department of Physics, University of Michigan, Ann Arbor, Michigan 48109, e-mail: graithel@umich.edu
†MIT-Harvard Center for Ultracold Atoms, Department of Physics, Massachusetts Institute of Technology, Cambridge, MA 02139.
**Department of Physics, University of California, Berkeley, CA 94720
‡Center for Nanoscale Materials, Argonne National Laboratory, Argonne, IL 60439

Abstract. We investigate cold rubidium plasmas in a particle trap that has the unique capability to simultaneously laser-cool and trap neutral atoms as well as to confine plasmas in magnetic fields of about three Tesla. The atom trap is a high-field Ioffe-Pritchard laser trap, while the plasma trap is a Ioffe-Penning trap that traps electrons and ions in separate wells. The observed plasma dynamics is characterized by a breathing-mode oscillation of the positive (ionic) plasma component, which feeds back on the behavior of the negative (electron) component of the plasma. At higher densities, the observed oscillations become nonlinear. The electron component has been found to undergo rapid cooling. We further report on the recombination of magnetized plasmas into Rydberg atoms in transient traps and quasi-steady-state traps. In transient traps, large numbers of recombined Rydberg atoms in high-lying states are observed. In quasi-steady-state traps, the measured numbers of recombined atoms are lower and the binding energies higher.

Keywords: Magnetized plasma, Rydberg atoms, Optical cooling of atoms
PACS: 52.25.Xz, 52.27.Aj, 32.80.Pj

1 INTRODUCTION

Strong magnetic (B) fields drastically change the physics of charged-particle systems, leading to diverse single-particle and collective spectroscopic phenomena in atomic, plasma, and condensed matter physics. When a B-field is applied to a free electron, states with a constant energy separation of $\hbar\omega_c$, known as Landau states, become solutions of the Schrödinger equation [1] (ω_c denotes the cyclotron frequency). In strongly magnetized condensed-matter systems, the states of the quasi-free electron gas condense into Landau bands, leading to fundamental alterations in the transport properties of the whole electron gas [2]. In atomic systems, spectral effects of high B-fields on highly excited Rydberg atoms [3, 4]), such as quasi-Landau resonances [5], are well known and have been studied extensively (see [6, 7] and references therein).

In a strong, homogeneous magnetic field $\mathbf{B} = B\hat{\mathbf{z}}$, the single-atom Hamiltonian of a highly excited Rydberg atom,

$$H = \frac{p^2}{2} + \frac{m^2 - 1/4}{2\rho^2} + \frac{B}{2}(m + m_s) + \frac{B^2\rho^2}{8} - \frac{1}{\sqrt{\rho^2 + z^2}} \quad , \tag{1}$$

is dominated by the diamagnetic term, $B^2\rho^2/8$ [atomic units]. There, ρ is the coordinate transverse to $\mathbf{B}$, m is the conserved magnetic quantum number, and m_s is the magnetic

CP1037, *Cold Antimatter Plasmas and Application to Fundamental Physics,* edited by Y. Kanai and Y. Yamazaki

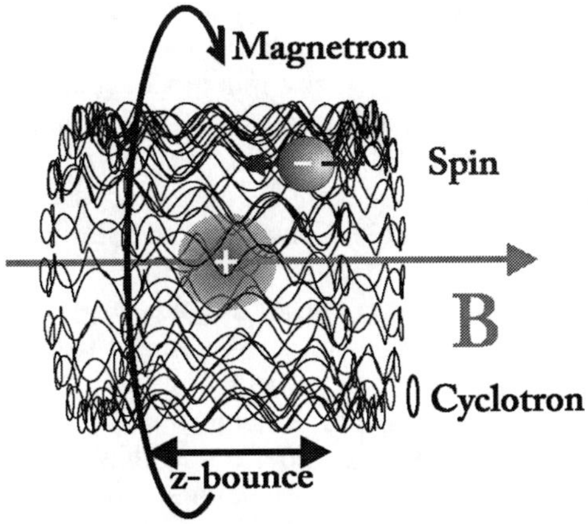

FIGURE 1. Semi-classical picture of a high-angular-momentum Rydberg atom in a strong magnetic field. The motion adiabatically separates into cyclotron, spin, z-bounce and magnetron motion (in descending order of frequency). For detailed information on the quantum mechanics of these states, see [12, 13] and references therein.

quantum number of the electron spin. At fields and energies where the Coulomb interaction presents only a weak perturbation, exotic diamagnetic atomic states are found. The largest fraction of phase space of these atoms corresponds to large negative values of m, where the dynamics is regular and characterized by drift-orbit solutions, also known as guiding-center solutions [8, 9, 10, 11, 12, 13, 14]. The motion exhibits three characteristic components: a cyclotron oscillation, a bounce oscillation parallel to **B**, and an $\mathbf{E} \times \mathbf{B}$ drift that leads to a slow magnetron oscillation. A simplified semi-classical picture of a drift-state orbit is shown in Fig. 1. Drift solutions have large densities of states and long lifetimes. Therefore, in collision-rich, strongly magnetized Rydberg-atom ensembles and plasmas they can be quite abundant. The dynamics of these systems exhibits an extraordinary richness, which is shared by systems such as electron/positron plasmas, positronium clouds, and - in solids - electron-hole plasmas and exciton gases.

To study the type of Rydberg atoms sketched in Fig. 1 and their properties in context with cold, magnetized plasma systems, a high-magnetic-field atom and plasma trap is employed that operates at magnetic fields up to six Tesla [15]. The capability of this apparatus to trap ground-state atoms, Rydberg atoms, electrons and ions in close proximity to each other is unique. The trap is employed to prepare clouds of rubidium ground-state atoms laser-cooled to $\approx 150\ \mu$K, as reviewed in in Sec. 2. Rubidium plasmas and / or Rydberg atoms are generated using a pulsed dye laser or a narrow-linewidth continuous-wave laser. Rydberg atoms and plasmas are analyzed using the methods of time resolved electron extraction, state-selective electric-field ionization, magnetic electron imaging and position-sensitive electron detection using a micro-channel-plate detector. Rydberg atoms in long-lived high-angular-momentum states can be magnetically trapped in the high-B trap [16]. The method of Rydberg-atom trapping used in the present work is briefly reviewed in Sec. 3. More recently, electrostatic Rydberg-atom trapping has been demonstrated elsewhere [17].

When combined with a suitable electric potential, the high-B Ioffe-Pritchard atom trap used in the present work can also function as a Ioffe-Penning trap [18, 19, 20]. This type of trap is a modified Penning trap whose inhomogeneous field enables the trapping of neutral particles. In Sec. 4, the stability of the Ioffe-Penning trap used in this work is briefly discussed. Using a nested Ioffe-Penning trap geometry [21], both electron and ion plasmas can be trapped in separated wells. In Sec. 5 recent experiments are described that show the correlated dynamics of electronic and ionic plasma components in a nested Ioffe-Penning trap.

Studies of three-body recombination (TBR) in cold, magnetized plasmas are of fundamental interest [9, 22] and are relevant in ongoing efforts to form and trap antihydrogen [14, 23, 24, 25]. In Sec. 6, first results on recombination of cold rubidium plasmas in a nested Ioffe-Penning trap are presented.

A component of the described research deals with high-resolution spectroscopy of Rydberg atoms in strong magnetic fields. The single-atom dynamics involves several adiabatically conserved degrees of freedom (see Fig. 1), which are weakly coupled by non-adiabatic channel couplings, spin-orbit coupling, and $|m|$-mixing due to transverse electric fields, giving rise to interesting quantum phenomena. In Sec. 7, some recent results on high-resolution spectroscopy performed in the high-B atom trap are discussed. Section 8 concludes the paper.

2 HIGH-MAGNETIC-FIELD ATOM TRAP

The main experimental tool in the described research is a unique high-B atom trap formed by two pairs of superconducting coils. The trap is sketched in Fig. 2 and described in detail in Ref. [15]. The field profile near the trap center is

$$B(\mathbf{r}) \approx B_0 \left[1 + \frac{\beta^2 - \alpha}{2}(x^2 + y^2) + \alpha z^2\right] \tag{2}$$

with a bias field $B_0 < 6$ T and $\alpha, \beta > 0$. The longitudinal field-curvature parameter, $\alpha = 2.2 \times 10^{-3}$ cm^{-2}, is fixed by the coil geometry. The transverse gradient of the racetrack-shaped Ioffe coils, $B_0\beta$, is given by coil geometry and current, and can be as large as 0.75 T/cm.

We mostly operate the system in a Ioffe-Pritchard laser trap mode, in which the field has a local minimum at $\mathbf{r} = 0$ with a depth of a few hundred Gauss in all directions ($\beta^2 > \alpha$ in Eq. 2). A slow beam of ^{85}Rb atoms, generated by a modified magneto-optic trap, enters the high-B region. Atoms are continuously laser-cooled, accumulated and magnetically trapped at the location of minimal magnetic field. The cooling laser is applied from six directions and is tuned approximately one line-width (6 MHz) to the red of the transition ^{85}Rb $5S_{1/2}\ |m_J = 1/2, m_I = 5/2\rangle \rightarrow 5P_{3/2}\ |m_J = 3/2, m_I = 5/2\rangle$ at the trap minimum (*i.e.*, at a magnetic field B_0). In the Ioffe-Pritchard laser trap mode, we can operate the trap at central field values B_0 up to three Tesla. This reflects a 15x improvement over realizations of Ioffe-Pritchard laser traps elsewhere. Since the trap laser ($\lambda = 780$ nm) is up to 42 GHz detuned from the field-free transition frequency, it cannot be locked using standard saturated spectroscopy. It is locked to a home-built

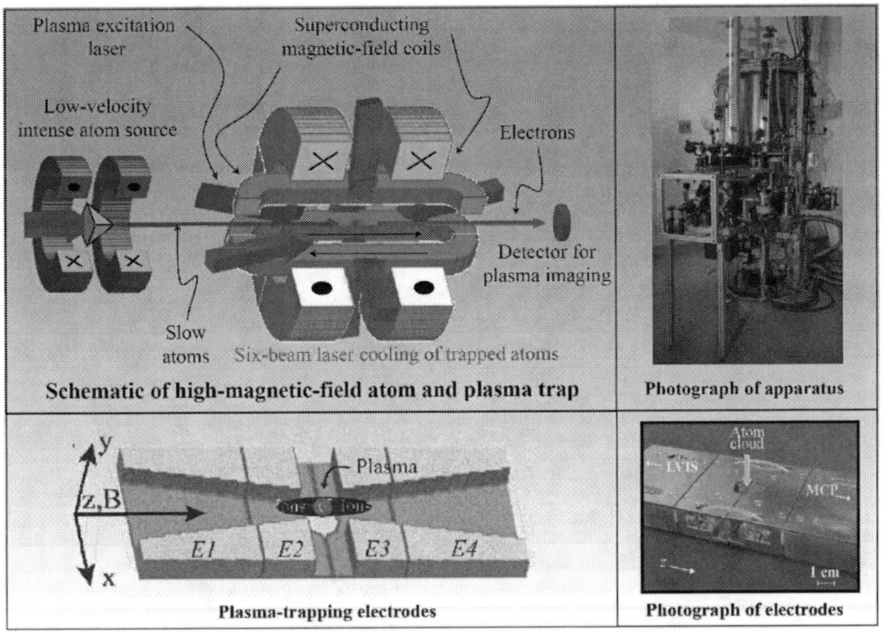

FIGURE 2. Upper left: Sketch of the high-B magnetic trap . Upper right: Photograph of the cryogenic trap setup. Lower left: Sketch of the electrode insert used to manipulate Rydberg atoms and plasmas generated in the trap. Lower right: Photograph of the electrode insert.

pressure-tuned Fabry-Perot interferometer [26]. We trap of order 3×10^7 atoms, at densities of order 10^9 cm^{-3} and temperatures near the Doppler limit (150 μK). An optical image of the trap is shown in Fig. 3. The apparatus can also be used as a pure Ioffe-Pritchard magnetic trap by turning the trap lasers off, after cold atoms have been accumulated. The transfer efficiency from the laser into the magnetic trap is near 100%, and the observed magnetic-trap lifetime is about 80 s.

With the quadrupole coils turned off ($\beta = 0$ in Eq. 2), the trap can be operated at central magnetic fields B_0 up to 6 T. In the case $\beta = 0$, the field still has a minimum in z-direction, but along the transverse axes $\hat{\mathbf{x}}$ and $\hat{\mathbf{y}}$ the curvature of B is negative. While the resultant saddle-point potential does not allow for magnetic trapping, we can still laser-cool atoms near $\mathbf{r} = 0$ in an optical-molasses mode. In this mode, the atoms are also cooled to temperatures of ≈ 150 μK, but there is no restoring force towards the trap center. In this mode, we trap of order 10^6 atoms at densities of order 10^7 cm^{-3}. Using a pulsed loading scheme, these numbers can be increased by about a factor of ten in non-steady-state.

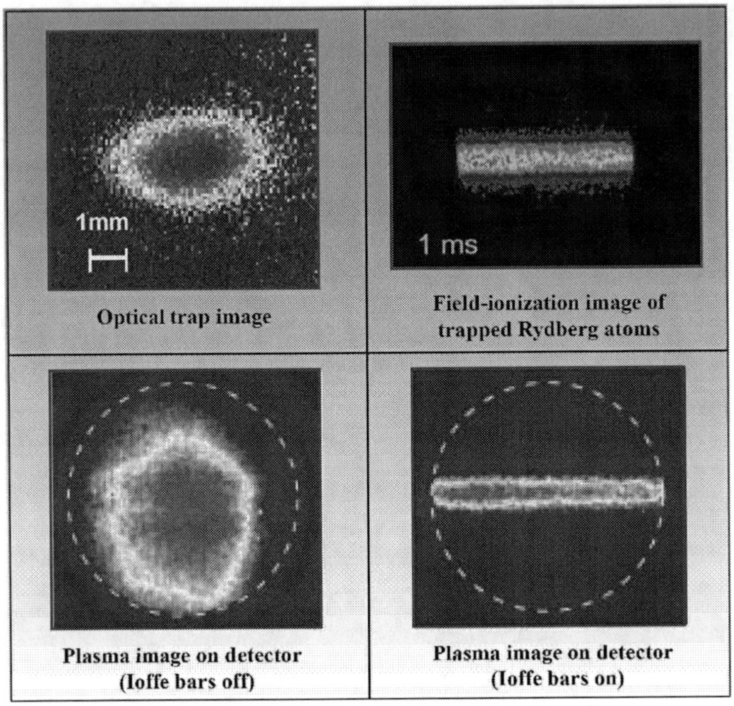

FIGURE 3. Upper left: Optical image of Ioffe-Pritchard laser trap at B=2.9 T. Upper right: Image of trapped Rydberg atoms obtained by electric-field ionization and position-sensitive detection on a micro-channel plate detector. Lower left: Image of an electron plasma stored in a cylindrically symmetric Penning trap. Lower right: Image of an electron plasma stored in a non-symmetric Ioffe-Penning trap.

3 THE TRAPPING OF RYDBERG ATOMS

An interesting area of research follows from the ability of the high-B atom trap to trap Rydberg atoms [16]. More recently, Rydberg atom trapping has also been demonstrated elsewhere using electrostatic fields [17]. Trapped Rydberg atoms have numerous possible applications in the areas of quantum control and quantum information processing [27, 28, 29, 30, 31, 32]. The theory of Rydberg-atom trapping in magnetic quadrupole traps [33, 34] and Ioffe-Pritchard traps [35] - the type of trap used in this work - has been studied in considerable detail. The prospects of magnetic trapping of Rydberg atoms formed by three-body recombination in cold, magnetized plasmas have been discussed in Ref. [22]. Trapping of Rydberg atoms over an extended period of time appears vital in efforts to accumulate anti-hydrogen [23, 36, 37].

Clearly, experimental work on the field of Rydberg-atom trapping is needed to verify theoretical predictions and to lead the way into practical applications. In the present work, we have achieved trapping of Rydberg atoms in three dimensions (3D) for the first time [16]. A typical image of trapped Rydberg atoms, obtained by time-delayed electric-field ionization, field-electron extraction and electron imaging, is shown in Fig. 3. To date, there exist two 3D Rydberg atom traps we know of (ours and one described in Ref. [17]).

To produce trapped Rydberg atoms, we excite dense clouds of ground state atoms collected in the high-B trap into loosely bound diamagnetic Rydberg levels at energies of order -3 cm^{-1}. Due to their high densities and large size, the Rydberg atoms rapidly undergo collision-induced state mixing, leading to the population of long-lived high-$|m|$ levels. These remain magnetically trapped for up to 200 milliseconds. We have observed breathing-mode and sloshing-mode oscillations of the trapped Rydberg atoms; the measured oscillation frequencies have indicated magnetic moments of order -8 Bohr magnetons [38]. The magnetic moment of a Rydberg atom of the type depicted in Fig. 1 is approximately given by the sum of several terms: the cyclotron magnetic moment (trapping), the magnetron magnetic moment (anti-trapping), the electron spin (trapping, for the initial electron spin polarization in the present trap), and the circular motion of the positive charge center in the plane transverse to **B** (trapping). The last term is negligible for rubidium, while for hydrogen it makes a noticeable difference. The trapping of drift-state Rydberg atoms is analyzed in some detail in Ref. [38].

4 Ioffe-Penning trap

Our setup incorporates a Ioffe-Penning trap [18, 19, 20], *i.e.* a trap that is suited for the simultaneous trapping of ions or electrons and neutral atoms. We have demonstrated the trapping stability of our Ioffe-Penning trap by first trapping neutral Rb atoms and subsequently photo-ionizing them, thereby filling the Ioffe-Penning trap "from the inside" with photo-electrons. Electrons have been trapped in the Ioffe-Penning trap for 1/e time scales of up to 50 seconds. Using energy-selective extraction of the electrons from the Penning trap, it was also found that the trapped electrons undergo significant cyclotron cooling, on a time scale of several seconds. Typical electron plasma images obtained from Ioffe-Penning traps loaded with photo-electrons are displayed in Fig. 3. In both displayed cases, the trapped plasmas have an approximately circular cross section in the plane transverse to **B**. The stretching of the plasma image observed in the Ioffe-Penning trap mode is due to distortion of the electron image caused by the return currents of the racetrack-shaped Ioffe coils (see Fig. 2).

To illustrate the Ioffe-Penning trap stability further, in Fig. 4 we show images of electron clouds injected from the outside into Penning- and Ioffe-Penning traps for bias magnetic fields of $B_0 = 1T$. The electrons are injected using an external, heated filament mounted near the axis of the main trap magnet. The loading of thermal electrons emitted from the filament occurs slightly off-axis, leading to the observed circular and elliptical images of the magnetron motion. The elliptical shapes observed in the Ioffe-Penning trap case result from the mentioned distortion of the electron imaging scheme due to the Ioffe coils. The quadrupole magnetic field present in the Ioffe-Penning case is sufficiently strong for atom trapping to be effective. In the study depicted in Fig. 4, the effect of the quadrupole magnetic field on the electron lifetime in the traps is minimal. We note that plasma traps involving magnetic octupole fields rather than quadrupole fields, which are also suitable for the trapping of atoms, have recently been demonstrated [39].

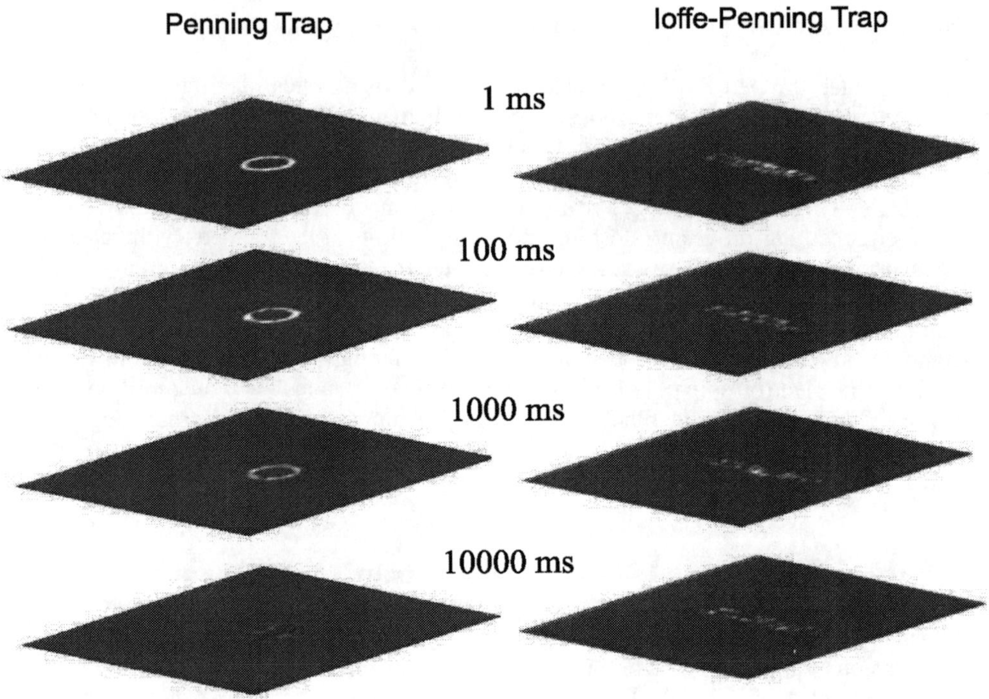

FIGURE 4. Images of electron bunches injected into a Penning trap (left) and a Ioffe-Penning trap (right) extracted after the indicated delay times.

5 Two-component plasma trap

In the past decade, progress in atom cooling and trapping techniques has given birth to the area of Rydberg-atom gases and cold plasmas. Typically, in this research laser-cooled atom clouds are photo-ionized, resulting in plasmas with initial temperatures of the electron and ion components of order 100 K and 1 mK range, respectively [40]. In weak magnetic fields, such systems have been studied experimentally [41, 42, 43, 44, 45] over a wide range of densities and temperatures, and many results have been successfully modeled [44, 45, 46, 47, 48]. In the regime of strong coupling, where the Coulomb interaction energy exceeds the thermal energy, liquid and solid forms of plasmas may be expected. Such plasmas have first been realized in non-neutral, one-component systems of laser-cooled ions [49, 50]. In quasi-neutral, dense and cold plasmas, the ion component can also be strongly coupled [51], leading to observable effects in the ion motion and to spatial correlations [52, 53, 54, 55, 56].

One of the outstanding goals in this field is to reach the regime of strong coupling for both the electronic and ionic components of the plasma. Since the development of spatial correlations may require timescales larger than typical lifetimes of unconfined plasmas, it will be beneficial to trap both plasma components using a strong magnetic field, which

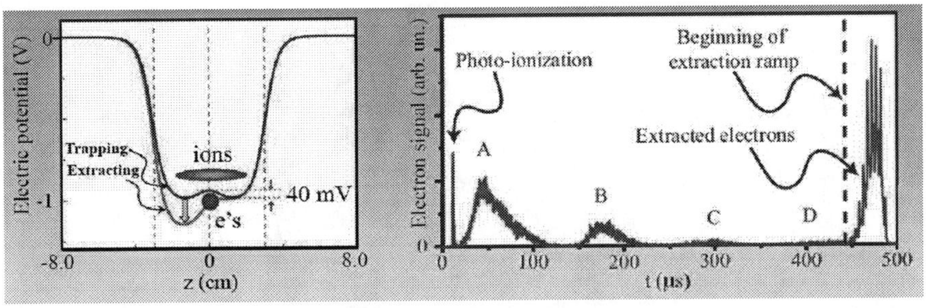

FIGURE 5. Left: Trapping potential (blue curve) along the magnetic-field axis. Electrons are trapped at the center, while ions are trapped in the wider double-well structure. The trap can be distorted (red curve) to extract and measure the electron component of the plasma. Right: Measured electron shake-off signal caused by coupled space-charge oscillations of the electron and ion components of the trapped plasma. The shake-off maxima (A-D) occur at the ion breathing-mode frequency. At 450 μs an electron extraction ramp is applied. The extracted electron signal at $t > 450$ μs is used to determine the electron temperature.

affords long confinement times and cyclotron cooling. The results presented in this Section represent a first step in generating dense, cold, strongly magnetized plasmas.

In recent work, we have excited cold, strongly magnetized two-component plasmas and confined them in a nested Ioffe-Penning trap with a strong bias magnetic field of about three Tesla [57]. The expansion of the plasmas is essentially one-dimensional, leading to confinement times of several milliseconds, which is one to two orders of magnitude longer than that in magnetic-field-free cold-plasma experiments conducted elsewhere. As shown in Fig. 5, we observe a modulation in the effective electron trapping potential, caused by space-charge oscillations of the ionic component. Measurements of the electron temperature, conceptually similar to those in [58], reveal that the electron component undergoes significant cooling of the longitudinal degree of freedom from initially 300 K to < 100 K. The initial electron temperature is consistent with the average potential energy the photo-electrons have in the electric potential of the Penning trap at the time instant of photo-ionization. The observed cooling time constant is of order 1 ms, which is much faster than the expected cyclotron cooling rate (which is of order 300 ms). We are currently considering several possible causes for the electron cooling; the exact origin of this cooling is still under investigation.

6 Recombination experiments

One of the most important atomic-physics processes in cold plasmas is three-body recombination (TBR), which scales in temperature as $T^{-9/2}$ [59, 60, 61] and most likely dominates radiative recombination. While in strong magnetic fields TBR rates are somewhat reduced [9, 22], the process still plays an important role in plasma decay, and it is a key process in efforts to form and trap anti-hydrogen [23, 24, 25].

To study TBR in cold, magnetized rubidium plasmas, we excite the plasmas in shallow two-component traps as discussed in Sec. 5. The bound-state Rydberg-atom population

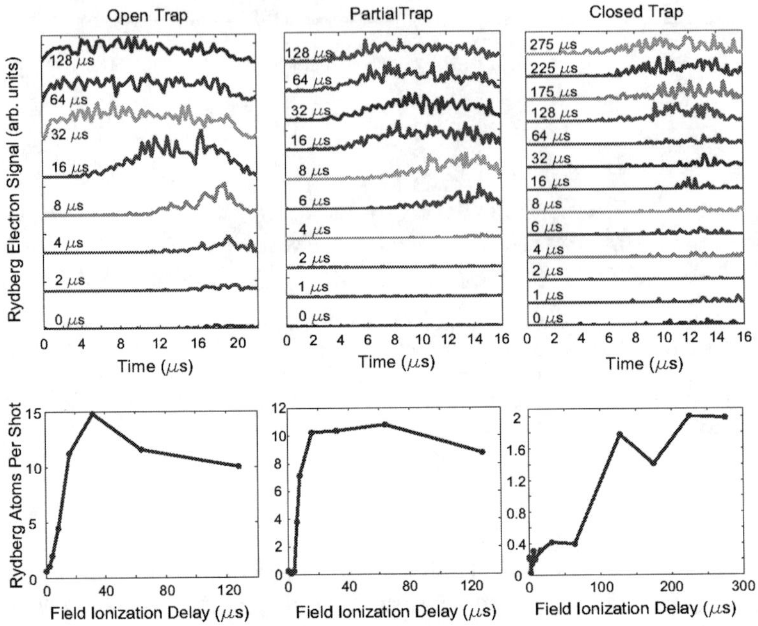

FIGURE 6. Top row: Bound-state field ionization signal for the indicated values of the delay time between plasma excitation and application of the field ionization ramp. The ionization electric field increases linearly from zero to about 200 V/cm over the time interval displayed on the time-axes. Bottom row: Number of recombined atoms per shot vs delay time, obtained by integrating the traces shown in the top row. The electron trap conditions range from symmetric and closed (left column) through partially closed (middle column) to entirely open (right column). The number of initially generated electron/ion pairs is of order 10^5.

that forms via recombination is analyzed using state-selective electric-field ionization. The field-ionization pulse is applied at a variable delay time after plasma generation. The plasma trapping conditions are varied from closed and symmetric (blue curve in Fig. 5) to fully open (red curve). An important experimental issue that has to be addressed in this research is that amplified spontaneous emission (ASE) of pulsed laser sources can photo-excite a significant number of bound-state Rydberg atoms before any recombination has taken place. Those photo-excited Rydberg atoms obviously would distort the measurement, as they could not be distinguished from Rydberg atoms formed by TBR. We have determined that the dye-laser pulses used in our experiment to photo-excite plasmas indeed contain some ASE at frequencies below the photo-ionization threshold of Rb. To eliminate these sub-threshold components of the ASE, we use a spectral filter based on a dispersive prism chain.

Recent results of recombination measurements are shown in Fig. 6. In the case of an open trap, the time window of overlap between ionic and electron components of the plasma is limited to several tens of μs. As a result, we observe considerable build-up of a recombination signal between 0 and 32 μs. The signal develops from the high-electric-field side, corresponding to deeply bound states, and gradually progresses towards the continuum. It appears that at short times any recombined atoms are mostly re-ionized due to ongoing electron-Rydberg-atom collisions. These collisions may also drive a fraction of the recombined atoms into lower-lying states that are de-coupled from further collision-induced dynamics because of small collision cross sections. Any low-lying atoms should rapidly decay into the ground state. As the electron density gradually drops, more recombined Rydberg atoms remain, culminating in a peak of about 15 detected Rydberg atoms per shot at 32 μs (lower left in Fig. 6). Due to detection inefficiencies, this number corresponds to an actual number of about 200 Rydberg atoms, corresponding to of order 1% of the initially ionized atoms. The Rydberg atom yield may be higher, as we only detect Rydberg atoms with binding energies less than $hc \times 80$ cm^{-1}. At times after 32 μs, all electrons have been extracted from the plasma. At later times, the Rydberg-atom population gradually decays, most likely due to electromagnetic decay. Atoms close to the ionization limit decay more slowly, leading to a gradual increase of the relative fraction of loosely bound Rydberg atoms divided by the total Rydberg-atom number. During this evolution, substantial effects caused by Rydberg-Rydberg collisions are unlikely because of the small number of recombined Rydberg atoms.

In closed traps (right column in Fig. 6), we observe very small numbers of recombined Rydberg atoms. The few atoms that are detected have rather high binding energies. These observations are consistent with re-ionization and / or collisional de-excitation of recombined Rydberg atoms due to the persistently dense electron component of the plasma. In closed traps, the small observed Rydberg-atom component increases over the whole investigated range of the delay time (256 μs in Fig. 6). This is attributed to partial separation of the electron and ion components of the plasma in the nested Penning trap, which occurs on a time scale on the order of the ion breathing-mode period (100 μs in Fig. 5). Partially open traps exhibit a recombination behavior that is in-between the extreme cases of fully closed and open traps, as expected (see middle column in Fig. 6).

To estimate the number of recombined Rydberg atoms using theoretical results obtained in Refs. [9, 22], we consider the case that is closest to steady-state, *i.e.* the right column in Fig. 6. We assume an electron and ion number of $N = 3 \times 10^5$, a plasma volume of $V = 10$ mm^3, and an electron temperature of T_e=50 K. The rate at which Rydberg atoms should be created according to [9, 22] is then given by

$$R = 0.07N^3/V^2 v_e b^5 \quad , \tag{3}$$

where the electron velocity $v_e = \sqrt{k_B T_e/m}$ and $b = e^2/4\pi\varepsilon_0 k_B T_e$. For the stated parameters, the Rydberg production rate $R \sim 2000$ s^{-1}, corresponding to of order one atom over 500 μs, or of order 0.1 detected atoms. While this is one order of magnitude less than observed (see Fig. 6, right column), we note that R depends very strongly on both N and T_e. Further, in the experiment we do not consider a steady-state situation. Densities are a function of time, and longitudinal and transverse electron temperatures may differ by as much as about a factor of three [57]. Finally, a large fraction of the detected Rydberg atoms have binding energies lower than what is considered a "successful" recombination in [9, 22]. In view of these shortcomings of the presented comparison between experimentally observed and theoretical recombination rates, the current level of agreement between experiment and theory may seem reasonable. For more accurate comparisons, the experimental system will have to be modeled more accurately in numerical simulations.

An interesting observation is that plasmas that exhibit only transient overlap between electronic an ionic components for rather short times (of order tens of μs) produce about a factor of ten more recombined Rydberg atoms in high-lying states than systems that are closer to steady-state (compare left and right columns in Fig. 6). A qualitative explanation for this behavior may be that the electronic and ionic components become separated when their densities are still very high. The rapid shut-down of electron collisions in the separation process may allow atoms that have recombined into high-lying states to survive. More detailed experiments and simulations will have to clarify this.

7 High resolution spectroscopy

The dynamics of a single Rydberg atom in a strong magnetic field adiabatically separates into several types of motion, as sketched in Fig. 1. This situation gives rise to interesting quantum phenomena that we study using methods of high-resolution spectroscopy.

We have recently observed Landau quantization effects in the photo-, auto- and field-ionization behavior of magnetized Rydberg atoms [62]. The high-B atom trap has been important in obtaining these results because it provides a spin-polarized sample of atoms and therefore a well-defined launch state for the Rydberg excitation. Also, the Rydberg atoms are essentially stationary, allowing for a careful time-resolved study over tens of milliseconds. Perturbations due to motional electric fields, $\mathbf{v} \times \mathbf{B}$, are strongly reduced since the velocity $|\mathbf{v}|$ of the atoms is only of order 0.1 m/s. The energy differences between the observed ionization potentials and the field-free one follow

$$\Delta E_{\mathrm{m,m_s}} = [\frac{1}{2}(m+|m|)+m_{\mathrm{s}}-\frac{1}{2}]\hbar\omega_{\mathrm{c}} \tag{4}$$

where m and m_{s} are magnetic orbital and spin quantum numbers. Due to photo-excitation selection rules that govern the excitation from the intermediate state $5\mathrm{P}_{3/2}\ |m_{\mathrm{J}}=3/2\rangle$, the ionization potentials $\Delta E_{\mathrm{m,m_s=1/2}} = m\hbar\omega_{\mathrm{c}}$ with $m = 0$, 1 or 2 are the most prominent. In the field-ionization data obtained in [62], each ionization potential marks the continuum limit of a field-ionization band described by a relation $W^2 \approx 3.5F$, where W denotes the difference between laser excitation energy and the ionization potential, and F denotes the ionization electric field (in atomic units). Spin-flip dynamics due to inherent LS-coupling and orbital m-mixing lead to additional structures in the observed photo- and field-ionization behavior [62].

In Ref. [62], we used a pulsed dye laser with tens of GHz linewidth, which did not allow us to resolve individual quantum states of the system. Therefore, it was not possible to study the excitation spectra, the decay and the field ionization behavior of individual quantum states, coherent spin-flip dynamics, and electric multi-pole interactions between Rydberg atoms. To enable such studies in the future, a narrow-linewidth Rydberg excitation laser has been added to the experiment (wavelength $\sim$480 nm, linewidth

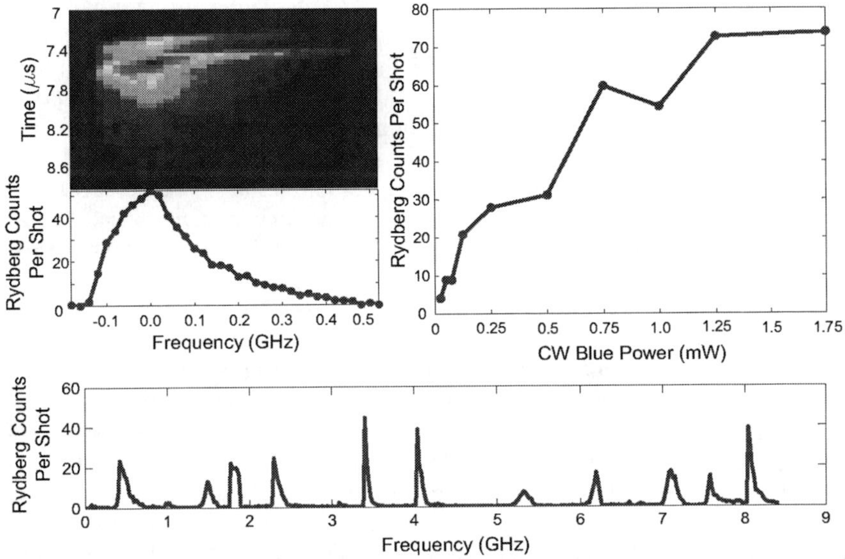

FIGURE 7. Upper left: Diamagnetic Rydberg line at B=2.8 T and energy $-18.4\ \text{cm}^{-1}$ vs the frequency offset of the excitation laser. The false-color field ionization plot is explained in the text. Upper right: Peak count on the Rydberg line vs power of the blue excitation beam. The data exhibit a clear saturation trend. Bottom: Rydberg excitation spectrum starting at -17.6 cm^{-1} vs frequency offset of the excitation laser.

several MHz). To perform a controlled high-resolution scan, the laser is first locked to a pressure-tuned, temperature-stabilized Fabry-Perot interferometer [26]. To perform a scan, the pressure inside the Fabry-Perot is scanned in a well-defined manner, as described in [26]. In Fig. 7 we show first results obtained with this system.

In spectroscopic studies, atoms are excited in a two-step process from $5S_{1/2}$ via $5P_{3/2}$ into diamagnetic Rydberg levels using coincident laser light of about 5 μs duration. The lower-transition pulse is on-resonant with the $5S_{1/2}\ |m_J = 1/2\rangle \rightarrow 5P_{3/2}\ |m_J = 3/2\rangle$ transition (wavelength 780 nm), while the upper-transition pulse (wavelength $\approx$480 nm) is scanned in frequency. We record field ionization traces as a function of the frequency of the upper-transition laser and plot the results in two-dimensional false-color plots. An example of such a plot is shown in the upper left of Fig. 7. There, the frequency is varied over several hundred MHz, and the Rydberg field-ionization signal is shown over a time interval in which the electric field of the field-ionization pulse rises linearly from about 10 V/cm to 40 V/cm. Typically, at fixed frequency the field-ionization signal exhibits peaks at several values of the field-ionization electric field. These fields correspond to different quantum numbers m and m_s of the Rydberg electron. Usually, the quantum numbers corresponding to the strongest field-ionization peak follow from the selection rules for the laser excitation from the intermediate 5P state, while the less strong field-ionization peaks result from mixing due to spin-orbit coupling and m-mixing (which results from stray electric fields and possibly atom-atom interaction). In the example shown in the upper left plot in Fig. 7, the horizontal band at lower electric field (upper band) is due to channels with $m \leq 0$ and $m_s = -1/2$, while the stronger band at higher electric field (lower band) is due to channels with $m \leq 0$ and $m_s = +1/2$. The integrals of the field-ionization traces, *i.e.* integrals along the time axis in plots such as in the upper left of Fig. 7, yield the total Rydberg atom number as a function of laser frequency (*i.e.* an ordinary "spectral line"). Typically, individual spectral lines exhibit an asymmetric structure that is due to the magnetic dipole moment of the excited Rydberg state and the magnetic-field inhomogeneity in the trap (see upper left in Fig. 7). At an energy of about $-18\ \text{cm}^{-1}$ relative to the field-free ionization energy, longer scans reveal diamagnetic Rydberg lines at a density of about one per GHz, as seen in the lower panel of Fig. 7. For most of these lines, we have obtained two-dimensional false-color plots analogous to the one shown in Fig. 7. A complete analysis of these plots is in progress.

Diamagnetic Rydberg states have substantial electric-quadrupole moments. The resultant electrostatic interaction between Rydberg atoms that are close to each other is expected to cause excitation saturation effects, as observed under field-free conditions (see, for instance, Ref. [63]). Measuring the number of Rydberg counts vs power of the excitation laser for a diamagnetic Rydberg state, we observe a clear saturation trend (upper-right plot in Fig. 7). Since the number of detected Rydberg atoms at the highest intensity is only about 70, which is at least a factor of ten lower than the number of atoms that can generated with the pulsed dye laser, optical saturation cannot be the main cause for the observed saturation effect. Therefore, this data may be the first observation of a quadrupole Rydberg excitation blockade. Since the excitation pulse is quite long (5 μs), in future studies we will determine whether other possible causes of saturation play a role, such as level shifts due to ion electric fields.

8 CONCLUSION

In this paper, we have reviewed a high-B atom and plasma trap employed to study laser-excited rubidium Rydberg atoms and cold plasmas in magnetic fields of order three Tesla. Experiments on the correlated dynamics of the electron and ion components of such plasmas in a nested Ioffe-Penning trap have been discussed. The number and the state distribution of Rydberg atoms formed by three-body recombination in these plasmas has been measured under various conditions.

In the future, laboratory studies of strongly magnetized cold plasma could have a broader impact in that they may further our understanding of other artificial or natural plasmas in strong magnetic fields that are less accessible. Those include magnetically confined plasmas in fusion devices and certain astrophysical plasmas (magnetic stars, surface of neutron stars, cores of giant gas planets, etc.). Some of these plasmas are strongly coupled. Three-body recombination in cold plasmas is of high current interest [37, 64, 65, 66], as this process may eventually lead to magnetically stored ground-state anti-hydrogen for fundamental symmetry tests [36]. Insights obtained into the recombination, the trapping and the decay of highly magnetized rubidium Rydberg atoms may become applicable in anti-hydrogen research.

ACKNOWLEDGMENTS

This project is supported by the Atomic, Molecular and Optical Sciences (AMOS) Program of the Chemical Sciences, Geosciences, and Biosciences Division, Office of Basic Energy Sciences, U. S. Department of Energy.

REFERENCES

1. L. D. Landau, and E. M. Lifshitz, *Quantum Mechanics: Non-Relativisitc Theory*, Pergamon Press, 1977, third edn.
2. M. P. Marder, *Condensed Matter Physics*, Wiley-Interscience Publication, 2000.
3. R. F. Stebbings, and F. B. Dunning, *Rydberg States of Atoms and Molecules*, Cambridge University Press, Cambridge, 1983.
4. T. F. Gallagher, *Rydberg Atoms*, Cambridge University Press, Cambridge, 1994.
5. W. Garton, and F. Tomkins, *Astrophys. J.* **158**, 841 (1969).
6. H. Friedrich, and D. Wintgen, *Phys. Rep.* **183**, 37 – 79 (1989), URL `http://dx.doi.org/10.1016/0370-1573(89)90121-X`.
7. M. C. Gutzwiller, *Chaos in classical and quantum mechanics*, Springer, New York, 1990.
8. Q. Wang, and C. Greene, *Phys. Rev. A* **40**, 742 – 51 (1989).
9. M. Glinsky, and T. O'Neil, *Phys. Fluids B* **3**, 1279 – 93 (1991), URL `http://dx.doi.org/10.1063/1.859820`.
10. G. Raithel, and H. Walther, *Phys. Rev. A* **49**, 1646 – 65 (1994), URL `http://dx.doi.org/10.1103/PhysRevA.49.1646`.
11. G. Raithel, and M. Fauth, *J. Phys. B* **28**, 1687 – 706 (1995), URL `http://dx.doi.org/10.1088/0953-4075/28/9/008`.
12. J. Guest, J.-H. Choi, and G. Raithel, *Phys. Rev. A* **68**, 22509 – 1–9 (2003), URL `http://dx.doi.org/10.1103/PhysRevA.68.022509`.
13. J. Guest, and G. Raithel, *Phys. Rev. A* **68**, 52502 – 1–9 (2003), URL `http://dx.doi.org/10.1103/PhysRevA.68.052502`.
14. D. Vrinceanu, B. Granger, R. Parrott, H. Sadeghpour, L. Cederbaum, A. Mody, J. Tan, and G. Gabrielse, *Phys. Rev. Lett.* **92**, 133402 – 1–4 (2004), URL `http://dx.doi.org/10.1103/PhysRevLett.92.133402`.
15. J. Guest, J.-H. Choi, E. Hansis, A. Povilus, and G. Raithel, *Phys. Rev. Lett.* **94**, 073003 – 1–4 (2005), URL `http://dx.doi.org/10.1103/PhysRevLett.94.073003`.
16. J.-H. Choi, J. Guest, A. Povilus, E. Hansis, and G. Raithel, *Phys. Rev. Lett.* **95**, 243001 – 1–4 (2005), URL `http://dx.doi.org/10.1103/PhysRevLett.95.243001`.
17. S. Hogan, and F. Merkt, *Phys. Rev. Lett.* **100**, 043001 – 1–4 (2008), URL `http://dx.doi.org/10.1103/PhysRevLett.100.043001`.
18. T. Squires, P. Yesley, and G. Gabrielse, *Phys. Rev. Lett.* **86**, 5266 – 9 (2001), URL `http://dx.doi.org/10.1103/PhysRevLett.86.5266`.
19. J. Fajans, W. Bertsche, K. Burke, S. Chapman, and D. Van Der Werf, *Phys. Rev. Lett.* **95**, 155001 –1–4 (2005), URL `http://dx.doi.org/10.1103/PhysRevLett.95.155001`.
20. G. Gabrielse *et al.*, *Phys. Rev. Lett.* **98**, 113002 – 1–4 (2007), URL `http://dx.doi.org/10.1103/PhysRevLett.98.113002`.
21. D. Hall, and G. Gabrielse, *Phys. Rev. Lett.* **77**, 1962 – 5 (1996), URL `http://dx.doi.org/10.1103/PhysRevLett.77.1962`.
22. F. Robicheaux, *Phys. Rev. A* **73**, 33401 – 1–5 (2006), URL `http://dx.doi.org/10.1103/PhysRevA.73.033401`.
23. M. Amoretti *et al.*, *Nature* **419**, 456 (2002).
24. G. Gabrielse, N. Bowden, P. Oxley, A. Speck, C. Storry, J. Tan, M. Wessels, D. Grzonka, W. Oelert, G. Schepers, T. Sefzick, J. Walz, H. Pittner, T. Hansch, and E. Hessels, *Physical Review Letters* **89**, 213401 – 1–4 (2002), URL `http://dx.doi.org/10.1103/PhysRevLett.89.213401`.
25. G. Gabrielse, N. Bowden, P. Oxley, A. Speck, C. Storry, J. Tan, M. Wessels, D. Grzonka, W. Oelert, G. Schepers, T. Sefzick, J. Walz, H. Pittner, T. Hansch, and E. Hessels, *Phys. Rev. Lett.* **89**, 233401 –

1–4 (2002), URL http://dx.doi.org/10.1103/PhysRevLett.89.233401.
26. E. Hansis, T. Cubel, J.-H. Choi, J. Guest, and G. Raithel, *Rev. Sci. Instrum.* **76**, 33105 – 1–3 (2005), URL http://dx.doi.org/10.1063/1.1866237.
27. S.-B. Zheng, and G.-C. Guo, *Phys. Rev. Lett.* **85**, 2392 – 5 (2000), URL http://dx.doi.org/10.1103/PhysRevLett.85.2392.
28. D. Jaksch, J. Cirac, P. Zoller, S. Rolston, R. Cote, and M. Lukin, *Phys. Rev. Lett.* **85**, 2208 – 11 (2000), URL http://dx.doi.org/10.1103/PhysRevLett.85.2208.
29. S. Osnaghi, P. Bertet, A. Auffeves, P. Maioli, M. Brune, J. Raimond, and S. Haroche, *Phys. Rev. Lett.* **87**, 037902 – 1–4 (2001), URL http://dx.doi.org/10.1103/PhysRevLett.87.037902.
30. P. Hyafil, J. Mozley, A. Perrin, J. Tailleur, G. Nogues, M. Brune, J. Raimond, and S. Haroche, *Phys. Rev. Lett.* **93**, 103001 – 1–4 (2004), URL http://dx.doi.org/10.1103/PhysRevLett.93.103001.
31. A. Sorensen, C. van der Wal, L. Childress, and M. Lukin, *Phys. Rev. Lett.* **92**, 063601 – 1–4 (2004), URL http://dx.doi.org/10.1103/PhysRevLett.92.063601.
32. J. Mozley, P. Hyafil, G. Nogues, M. Brune, J.-M. Raimond, and S. Haroche, *Eur. Phys. J. D* **35**, 43 – 57 (2005), URL http://dx.doi.org/10.1140/epjd/e2005-00184-7.
33. I. Lesanovsky, J. Schmiedmayer, and P. Schmelcher, *Phys. Rev. A* **70**, 43409 – 1–12 (2004), URL http://dx.doi.org/10.1103/PhysRevA.70.043409.
34. I. Lesanovsky, and P. Schmelcher, *Phys. Rev. Lett.* **95**, 053001 – 1–4 (2005), URL http://dx.doi.org/10.1103/PhysRevLett.95.053001.
35. B. Hezel, I. Lesanovsky, and P. Schmelcher, *Phys. Rev. Lett.* **97**, 223001 – 1–4 (2006), URL http://dx.doi.org/10.1103/PhysRevLett.97.223001.
36. G. Gabrielse, *Fundamental Symmetries*, Plenum, New York, 1987, p. 59 – 75.
37. T. Pohl, H. Sadeghpour, Y. Nagata, and Y. Yamazaki, *Phys. Rev. Lett.* **97**, 213001 – 1–4 (2006), URL http://dx.doi.org/10.1103/PhysRevLett.97.213001.
38. J.-H. Choi, J. Guest, and G. Raithel, *Eur. Phys. J. D* **40**, 19 – 26 (2006), URL http://dx.doi.org/10.1140/epjd/e2006-00086-2.
39. G. Andresen *et al.*, *Phys. Rev. Lett.* **98**, 023402 – 1–4 (2007), URL http://dx.doi.org/10.1103/PhysRevLett.98.023402.
40. T. Killian, S. Kulin, S. Bergeson, L. Orozeo, and S. Rolston, *Phys. Rev. Lett.* **83**, 4776 – 9 (1999), URL http://dx.doi.org/10.1103/PhysRevLett.83.4776.
41. M. Robinson, B. L. Tolra, M. W. Noel, T. Gallagher, and P. Pillet, *Phys. Rev. Lett.* **85**, 4466 – 4469 (2000), URL http://dx.doi.org/10.1103/PhysRevLett.85.4466.
42. C. Simien, Y. Chen, P. Gupta, S. Laha, Y. Martinez, P. Mickelson, S. Nagel, and T. Killian, *Phys. Rev. Lett.* **92**, 143001 – 1–4 (2004), URL http://dx.doi.org/10.1103/PhysRevLett.92.143001.
43. R. Fletcher, X. Zhang, and S. Rolston, *Phys. Rev. Lett.* **96**, 105003 – 1–4 (2006), URL http://dx.doi.org/10.1103/PhysRevLett.96.105003.
44. P. Gupta, S. Laha, C. Simien, H. Gao, J. Castro, T. Killian, and T. Pohl, *Phys. Rev. Lett.* **99**, 075005 – 1–4 (2007), URL http://dx.doi.org/10.1103/PhysRevLett.99.075005.
45. S. Laha, P. Gupta, C. Simien, H. Gao, J. Castro, T. Pohl, and T. Killian, *Phys. Rev. Lett.* **99**, 155001 – 1–4 (2007), URL http://dx.doi.org/10.1103/PhysRevLett.99.155001.
46. S. Mazevet, L. Collins, and J. Kress, *Phys. Rev. Lett.* **88**, 055001 – 1–4 (2002), URL http://dx.doi.org/10.1103/PhysRevLett.88.055001.
47. S. Kuzmin, and T. O'Neil, *Phys. Rev. Lett.* **88**, 065003 – 1–4 (2002), URL http://dx.doi.org/10.1103/PhysRevLett.88.065003.
48. F. Robicheaux, and J. Hanson, *Phys. Rev. Lett.* **88**, 055002 – 1–4 (2002), URL http://dx.doi.org/10.1103/PhysRevLett.88.055002.
49. W. Itano, J. Bollinger, J. Tan, B. Jelenkovic, X.-P. Huang, and D. Wineland, *Science* **279**, 686 – 9 (1998), URL http://dx.doi.org/10.1126/science.279.5351.686.
50. T. Mitchell, J. Bollinger, D. Dubin, X.-P. Huang, W. Itano, and R. Baughman, *Science* **282**, 1290 – 3 (1998), URL http://dx.doi.org/10.1126/science.282.5392.1290.
51. M. Murillo, *Phys. Rev. Lett.* **87**, 115003–1–4 (2001), URL http://dx.doi.org/10.1103/PhysRevLett.87.115003.
52. Y. Chen, C. Simien, S. Laha, P. Gupta, Y. Martinez, P. Mickelson, S. Nagel, and T. Killian, *Phys. Rev.*

Lett. **93**, 265003 – 1–4 (2004),
URL http://dx.doi.org/10.1103/PhysRevLett.93.265003.

53. T. Killian, *Science* **316**, 705 – 8 (2007),
URL http://dx.doi.org/10.1126/science.1130556.
54. T. Pohl, T. Pattard, and J. Rost, *Phys. Rev. Lett.* **92**, 155003 – 1–4 (2004),
URL http://dx.doi.org/10.1103/PhysRevLett.92.155003.
55. T. Pohl, T. Pattard, and J. Rost, *Phys. Rev. Lett.* **94**, 205003 – 1–4 (2005),
URL http://dx.doi.org/10.1103/PhysRevLett.94.205003.
56. M. Murillo, *Phys. Rev. Lett.* **96**, 165001 – 1–4 (2006),
URL http://dx.doi.org/10.1103/PhysRevLett.96.165001.
57. J. H. Choi, B. Knuffman, X. Zhang, A. P. Povilus, and G. Raithel, *Phys. Rev. Lett.* accepted (2008).
58. J. Roberts, C. Fertig, M. Lim, and S. Rolston, *Phys. Rev. Lett.* **92**, 253003 – 1–4 (2004),
URL http://dx.doi.org/10.1103/PhysRevLett.92.253003.
59. T. Killian, M. Lim, S. Kulin, R. Dumke, S. Bergeson, and S. Rolston, *Phys. Rev. Lett.* **86**, 3759 – 62 (2001), URL http://dx.doi.org/10.1103/PhysRevLett.86.3759.
60. T. Gallagher, P. Pillet, M. Robinson, B. Laburthe-Tolra, and M. Noel, *J. Opt. Soc. Am. B* **20**, 1091 – 7 (2003).
61. T. Pohl, and T. Pattard, *J. Phys. A* **39**, 4571 – 7 (2006),
URL http://dx.doi.org/10.1088/0305-4470/39/17/S41.
62. J.-H. Choi, J. Guest, E. Hansis, A. Povilus, and G. Raithel, *Phys. Rev. Lett.* **95**, 253005 – 1–4 (2005),
URL http://dx.doi.org/10.1103/PhysRevLett.95.253005.
63. D. Tong, S. Farooqi, J. Stanojevic, S. Krishnan, Y. Zhang, R. Cote, E. Eyler, and P. Gould, *Phys. Rev. Lett.* **93**, 063001 – 1–4 (2004),
URL http://dx.doi.org/10.1103/PhysRevLett.93.063001.
64. C. Taylor, J. Zhang, and F. Robicheaux, *J. Phys. B* **39**, 4945 – 59 (2006),
URL http://dx.doi.org/10.1088/0953-4075/39/23/012.
65. T. Topcu, and F. Robicheaux, *Phys. Rev. A* **73**, 43405 – 1–10 (2006),
URL http://dx.doi.org/10.1103/PhysRevA.73.043405.
66. A. Wetzels, A. Gurtler, L. Noordam, and F. Robicheaux, *Phys. Rev. A* **73**, 62507 – 1–8 (2006),
URL http://dx.doi.org/10.1103/PhysRevA.73.062507.

Formation of Antihydrogen Rydberg atoms in strong magnetic field traps

T. Pohl and H. R. Sadeghpour

ITAMP, Harvard-Smithsonian Center for Astrophysics, 60 Garden Street, Cambridge MA 02138

Abstract. It is shown that several features of antihydrogen production in nested Penning traps can be described with accurate and efficient Monte Carlo simulations. It is found that cold deeply-bound Rydberg states of antihydrogen ($\bar{\mathrm{H}}$) are produced in three-body capture in the ATRAP experiments and an additional formation mechanism -Rydberg charge transfer-, particular to the nested Penning trap geometry, is responsible for the observed fast (hot) $\bar{\mathrm{H}}$ atoms. Detailed description of the numerical propagation technique for following extreme close encounters is given. An analytic derivation of the power law behavior of the field ionization spectrum is provided.

Keywords: Rydberg antimatter, antihydrogen, non-neutral plasma, three-body capture, charge transfer
PACS: 36.10.-k,34.70.+e,52.20.Hv

I. INTRODUCTION

Since the initial success of the ATHENA [1] and ATRAP [2, 3] collaborations in producing the first neutral form of antimatter (antihydrogen – $\bar{\mathrm{H}}$) in nested Penning traps at CERN, sizable effort has been devoted, both in experiment and in theory, to understanding how these atoms in strong magnetic fields form, how hot (or cold) they are, and what the internal states of such exotic form of matter may be. Subsequent measurements by these groups showed that the produced $\bar{\mathrm{H}}$ atoms are not in thermal equilibrium with the surrounding positron plasma and are hot (fast) atoms [4, 5]. Additionally, ATRAP observations hinted at formation of deeply-bound Rydberg atoms [6]. In the following, we will address in detail, these open questions in $\bar{\mathrm{H}}$ formation in nested Penning traps.

While the numerical simulations have consistently established that the three-body interaction of two positrons (e^+) and an antiproton ($\bar{p}$) in a magnetic field likely forms atoms that are weakly bound [7, 8, 9, 10, 11], a more recent refinement of ATRAP experiments has produced evidence, in the field ionization spectrum, for highly-bound atomic states [6]. The origin and characteristics of such deeply-bound $\bar{\mathrm{H}}$ atoms has only recently been revealed in extensive and precise classical Monte-Carlo simulations of TBC of $\bar{\mathrm{H}}$ atoms [12]. The calculations were conducted on two levels of sophistication. For low- to moderate-fields, at which weakly-bound Rydberg levels are ionized, the guiding-center approximation of an atom in a magnetic field (GCA) was employed, which holds that different degrees of freedom for atomic motion, the cyclotron, magnetron and axial bounce motions, evolve separately in the magnetic field. Since the cyclotron period, $\tau_c = 2\pi m/e$ is the fastest period in the system, followed by the axial bounce period, $\tau_z = 2\pi\sqrt{r_e c^3/\rho^3}$, and the magnetron oscillation period, $\tau_m = (2\pi\omega_c\rho^3)/(r_e c^3)$, the GCA simulations freeze the e^+ cyclotron motion. In the above expressions, e and m are

CP1037, *Cold Antimatter Plasmas and Application to Fundamental Physics,* edited by Y. Kanai and Y. Yamazaki

the charge and the mass in a magnetic field B, r_e and ρ are the classical electron radius and the radial coordinate, assumed to be much larger than the cyclotron radius (circular GCA atom), $r_c = \sqrt{2mk_BT}/(eB)$, and c is the speed of light [13]. Such a picture of strongly-magnetized atoms, breaks down when there is coupling between different degrees of freedom, as for instance, when the atoms are so deeply bound that the Coulomb and magnetic interactions compete *on par*, leading to chaotic electronic motion. In these cases, we employ an efficient and numerically stable regularization scheme in Monte Carlo simulation of TBC for formation of deeply-bound states.

Below, we describe the field ionization of deeply-bound $\bar{\mathrm{H}}$ atoms, and derive an analytic expression for the power-law behavior of surviving atom number with field strength. We will demonstrate the a unique two-step collisional process leads to the formation of such deeply-bound atoms and give a full account of the numerical technique which allows us to account for close binary encounters in the Coulomb field. We account for an additional collision process, which explains the high $\bar{\mathrm{H}}$ axial velocities observed in [4]. This mechanism, the charge exchange of TBC $\bar{\mathrm{H}}$ atoms with $\bar{p}$ ions in the trap, permits the bound e^+ of initially slow $\bar{\mathrm{H}}$ atoms to hop onto the fast $\bar{p}$ ions in the side-wells of the nested Penning trap, thereby producing high velocity $\bar{\mathrm{H}}$ atoms. Detailed comparisons with observations are made.

II. EXPERIMENTAL OBSERVATIONS

A. Plasma confinement and $\bar{\mathrm{H}}$ production

All $\bar{\mathrm{H}}$ production experiments used a nested Penning trap to simultaneously confine e^+ and $\bar{p}$ and to create an overlap region for $\bar{\mathrm{H}}$ formation. A strong homogeneous magnetic field provides confinement of both species radially to the magnetic field axis, while an electrostatic double-well potential axially traps e^+ in the center of the trap and confines $\bar{p}$ in the side-wells.

The ATHENA experiments [1] as well as the initial ATRAP $\bar{\mathrm{H}}$ production scheme [2], were based on injection of energetic $\bar{p}$ which subsequently cooled through collisions with the e^+ . Subsequent ATRAP experiments aimed at optimizing $\bar{\mathrm{H}}$ production, by preparing cold $\bar{p}$ localized in one of the side-wells which were driven through the e^+ cloud by an rf-field. Gentle driving is expected to yield slow $\bar{p}$ in the central reaction region and, hence, to produce slow $\bar{\mathrm{H}}$ atoms. The atoms were detected in an ionization well, located axially outside the trap. Since atoms also drift radially to the magnetic field axis, this methods detects a fraction of the $\bar{\mathrm{H}}$ arriving at the detector within a small solid angle α. This detection scheme, on the other hand, allows for nearly background-free observation of the atoms and has the distinct advantage that pre-ionization fields can be applied to probe internal and translational states of the atoms.

B. Field ionization

In the ATRAP driven experiment [3], a static electric field has been applied in front of the detector that ionizes less deeply bound atoms and prevents them from reaching the detection well. Through varying the strength of the pre-ionization field, and state selective stripping of the bound e^+ , the distribution of internal atomic Rydberg states can be probed. The resulting number of $\bar{H}$ atoms that survive a given electric field, F is shown in Fig. 1. Two main features are evident: (a) at low fields, the surviving atoms follow a simple power-law dependence $N \sim F^{-2}$, and (b) an enhancement of the atom number with respect to this power-law, at larger fields. The origin of both the low-field and the high-field behavior is not immediately obvious, from consideration of either radiative recombination or three-body capture. For example, extrapolating the classical infinite-field calculations [9] overestimates the power-law exponent in the low-field regime [14].

B. Velocity measurement

In 2004, the ATRAP collaboration succeeded in measuring the velocity distribution of such atoms [4]. In this particular experiment, a prestripping oscillating electric field is applied before the detection well, such that the speed of the $\bar{H}$ atoms moving along the magnetic field could be measured by tuning the frequency of the oscillating field.

In arriving at a velocity of the antihydrogen atoms, it was assumed that the number density of observed atoms, with velocity v in an ionizing field F, scaled as $N(v,F) \sim \delta(v-v_0)F^{-3}$. A most probable value of $v_0 = 200$ meV, corresponding to a temperature of $T = 2400$ K, was found to best describe the observed frequency spectrum. This result was a surprise, since the atomic temperature expected from the 4.2 K background temperature is about three orders of magnitude smaller. As we discuss below, an additional mechanism may be at play in forming fast non-thermal $\bar{H}$ from initially slow atoms.

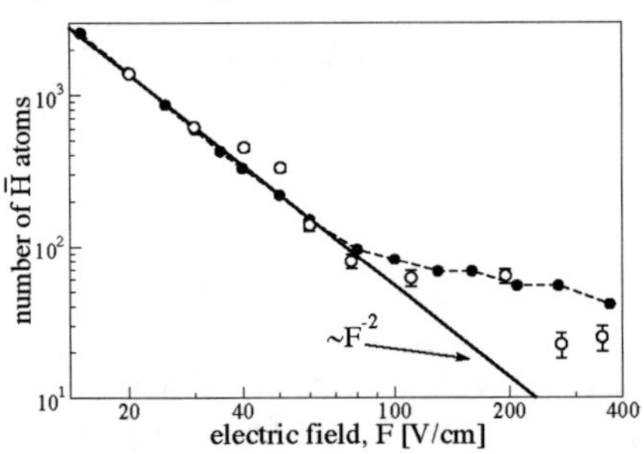

FIGURE 1. Field ionization spectrum of $\bar{H}$ atoms that survive an electric field F. ATRAP measurements [6] (open circles) are compared to exact Monte-Carlo calculations for $kT_{\bar{H}} = 2$ meV (filled circles). The curves have been scaled to experiment at 20 V/cm.

III. CLASSICAL MONTE-CARLO SIMULATIONS

At experimental densities and temperatures, radiative processes and quantum effects are of minor importance, permitting a classical description of $\bar{H}$ through three-body collisions between $\bar{p}$ and e^+ . Although a solution of the classical equations of motion for a two-component plasma system is fairly straightforward, the sheer number of particles ($\sim 10^6$) renders a full scale simulation of the entire trap impractical. However, since $\rho_{\bar{p}}$ in the central trapping region is considerably smaller than the ρ_{e^+}, the $\bar{p}-\bar{p}$ interaction is negligible for the formation process. Exploiting this fact [11, 12], we restrict the simulation to the dynamics of a single $\bar{p}$ traversing the central trap cloud, but repeating the calculations over many cycles. The numerical effort is further reduced by considering e^+ in a finite cubic box, centered around the moving $\bar{p}$. This box is constantly filled with e^+ according to the corresponding equilibrium phase space distribution, with density ρ_{e^+} and temperature T_{e^+}.

After the $\bar{p}$ leave the plasma cloud, the critical ionization field of formed atoms is calculated by slowly ramping up an electric field until a bound e^+ is liberated. Average quantities, such as the corresponding field ionization spectrum are finally calculated from many repetitions of these single $\bar{p}$ simulations. The system is propagated by solving the following classical equations of motion for the $\bar{p}$

$$\dot{\mathbf{R}} = \mathbf{V}\,, M\dot{\mathbf{V}} = \mathbf{F} - m\boldsymbol{\omega}_c \times \mathbf{V} \tag{1}$$

and the e^+ s

$$\dot{\mathbf{r}}^{(i)} = \mathbf{v}^{(i)}\,, \quad m\dot{\mathbf{v}}^{(i)} = \mathbf{f}^{(i)} + m\boldsymbol{\omega}_c \times \mathbf{v}^{(i)}\,, \tag{2}$$

where $\boldsymbol{\omega}_c = (0,0,\omega_c)$ is the e^+ cyclotron frequency, and M, $\mathbf{R}$, $\mathbf{V}$ and m, $\mathbf{r}^{(i)}$, $\mathbf{v}^{(i)}$ denote the mass, the position and the velocity of the $\bar{p}$ and the i-th e^+, respectively. Although the problem appears straightforward, the strong magnetic field renders a numerical solution of the above equations quite challenging. To explore the influence of the magnetic field, we have performed two sets of simulations on different levels of sophistication.

A. Guiding center approximation

On the simplest level, the guiding center description for the e^+ dynamics, i.e. a separation of the e^+ cyclotron motion from the remaining e^+ degrees of freedom can be used. This approximation exploits that for certain strongly magnetized Rydberg states the positronic cyclotron, motion separates from the magnetron and axial bounce motion. If the cyclotron period, is the fastest period in the system, followed by the axial bounce period, and the magnetron oscillation period, ($\tau_c << \tau_z << \tau_m$), the cyclotron dynamics can be decoupled from the remaining degrees of freedom,

$$\begin{aligned} &\dot{\mathbf{r}}^{(i)} = \mathbf{v}^{(i)}\,, \\ &m v_x^{(i)} = -\frac{F_x^{(i)}}{\omega_c}\,, \quad m v_y^{(i)} = \frac{F_y^{(i)}}{\omega_c} \\ &m\dot{v}_x^{(i)} = F_z^{(i)} \end{aligned} \tag{3}$$

where $\mathbf{F}^{(i)}$ is the total Coulomb force on the i-th e^+ due to the central $\bar{p}$ and the remaining e^+. On the other hand the $\bar{p}$ cyclotron period is a factor of 1836 times longer, and must be described exactly according to Eq. 2.

B. Exact propagation of the e^+ motion

To explore the break-down of the GCA, we also follow the exact equations of motion for the e^+ , including its fast cyclotron motion. Compared to the GCA calculations, these simulations are considerably more challenging and computationally intensive; due to two major issues. First, the total simulation time is roughly six orders of magnitude longer than the cyclotron period (from a few picoseconds for the e^+ cyclotron motion to several hundreds of nanoseconds for three-body capture), which when fully resolved, would require a huge number of timesteps. Second, trajectories of deeply-bound non-GCA atoms exhibit frequent close encounters with the $\bar{p}$, making it difficult to fulfill energy conservation over the required long simulation times.

To overcome these difficulties, we used a specially designed integrator which respects the time-symmetry of the underlying Hamiltonian and exploits the disparate dynamical timescales imposed by the Coulomb and magnetic fields, respectively.

Most of the time, the $\bar{p}$ is either freely moving through the plasma, or carrying a loosely bound e^+ , such that the Coulomb interaction barely perturbs the fast cyclotron oscillations. We exploit this fact and employ a symplectic integrator, e.g., in [13], where the system Hamiltonian is split in two parts: the total Coulomb potential energy, and the total kinetic energy, including all magnetic field terms. Since this approach exactly recovers the positronic cyclotron motion, it yields high accuracy even though timesteps are chosen considerably larger than ω_c^{-1}.

For more deeply bound atoms, however, the numerical error drastically increases due to close $e^+ - \bar{p}$ encounters, requiring smaller timesteps. Adjusting the timestep accordingly, however, destroys the symplectic nature of the integrator, and results in a drift of the total energy. To circumvent this problem, we choose the binding energy of the most deeply bound e^+ as a parameter, to switch to a more suitable integrator as the e^+ gets to deeply bound levels to be accurately described by the method described above. We have tested several approaches to follow deeply bound trajectories, ranging from two-body [15] and n-body [16] regularization of the attractive e^+ - $\bar{p}$ interaction, combined with a symplectic integrator based on a triply-split Hamiltonian, to an adapted version of the time-transformed leap-frog (TTL) scheme introduced in [17]. While all of these methods gave an accurate description of the trajectories, the latter approach is conceptually more simple and was found to yield somewhat higher efficiency than the more complex regularization procedures.

The underlying idea of the TTL scheme is to introduce a scaled time $\tau = \Omega(\mathbf{R},\mathbf{r}^{(1)},\ldots)t$, but treat the scaling factor $W = \Omega$, as an independent variable. The resulting equations of motion are split into two parts

$$\frac{d}{d\tau}\begin{pmatrix}\mathbf{r}^{(i)}\\ t\\ \mathbf{v}^{(i)}\\ W\end{pmatrix} = \begin{pmatrix}\mathbf{v}/W\\ 1/W\\ (\boldsymbol{\omega}_c \times \mathbf{v}^{(i)})/W\\ 0\end{pmatrix} + \begin{pmatrix}\mathbf{0}\\ 0\\ \mathbf{f}^{(i)}/m\Omega\\ (\mathbf{v}^{(i)}/\Omega)\cdot\partial\Omega/\partial\mathbf{r}^{(i)}\end{pmatrix} \tag{4}$$

such that each part can be separately solved analytically, from which we construct a fifth order leap-frog-like integration scheme. For the scaling factor, we choose

$$\Omega = \left(\sum_i \frac{1}{\left| \mathbf{R} - \mathbf{r}^{(i)} \right|} \right)^{-1}, \tag{5}$$

which has the striking advantage that the corresponding integrator yields the exact solution for the two-body Kepler problem, in the limit of very deeply bound trajectories, where the magnetic field and the interaction with the surrounding e^+ can be regarded as a negligible perturbation. The combined integrator tracts the exact e^+ dynamics in both relevant limiting cases, i.e. for a dominating magnetic or antiprotonic Coulomb field. This ensures a highly efficient and at the same time accurate determination of the solutions of the many-body dynamics, allowing for accumulation of sufficient statistics for comparison with experiments. The solutions are themselves a valuable consistency test of the quality of the GCA.

IV. CALCULATIONS AND COMPARISON WITH OBSERVATIONS

A. Field ionization spectrum

1. Numerical results

The results of the GCA simulations, shown in Fig. 1, nicely agree with the observed low-field ionization spectrum. This suggests that the produced atoms which survive low ionizing fields are likely guiding center atoms. Indeed, the vast majority of atoms in this regime are found to exhibit the GCA hierarchy of timescales discussed above. At higher ionization fields, however, the GCA prediction for the field ionization spectrum deviates from the measured atom yield, indicating a possible break-down of the GCA approximation for more strongly bound atoms.

Our full simulations confirm the validity of the GCA for low ionization fields, but consistent with observation, deviate from it at higher ionization fields. The good agreement between the results of our full-scale calculations and the experiment, suggests that the observed deviation from the low-field power-law behavior is indeed a manifestation of deeply-bound non-GCA atoms.

2. Two-step mechanism of weakly bound states

A typical evolution of the minimal binding energy leading to a deeply bound atom is shown in Fig. 2. In accordance with previous studies [9], we find that the atomic states are changed by so-called "replacement" collisions (vertical lines in Fig.2), where the initially bound e^+ is replaced by an incident e^+ , but with a different binding energy. The e^+ is initially captured into a weakly bound orbit at a few $k_B T_{e^+}$. Subsequent thermal collisions at energies above $E_b \sim 4 k_B T_{e^+}$ — the bottleneck energy- drive the distribution of binding energies $W(E)$ into fast equilibrium with the free plasma e^+ ,

$$W^{(eq)} \propto E^{-7/2} \exp\left(\frac{E}{k_B T_{e^+}}\right) . \tag{6}$$

Below E_b, atoms are predominantly driven down to deeper binding and reionization is greatly suppressed [9], such that $W(E)$ strongly deviates from Eq. (6). Guided by our numerical simulations, we can determine the form of $W(E)$ below E_b. The striking feature in Fig. 2 is the prominence of a *single* replacement collision at later times, which abruptly brings the atom into a tightly-bound orbit, far removed from the thermal equilibrium. Due to the small size of the resulting atom, collisions are unlikely to significantly modify the binding energy distribution and whose dynamics can be understood in the two-step picture as an initial capture, followed by one single de-exciting collision. The rate for de-excitation at an energy E is proportional to $K(E_i,E) \propto (-E)^{-E_b/k_B T}$.

Once captured into a highly excited state with energy E_i above the bottleneck, the final energy distribution after a de-exciting event to a deeply bound state with energy $E = E_f$ is thus proportional to the corresponding de-excitation rate $K(E_i,E_f)$, which can

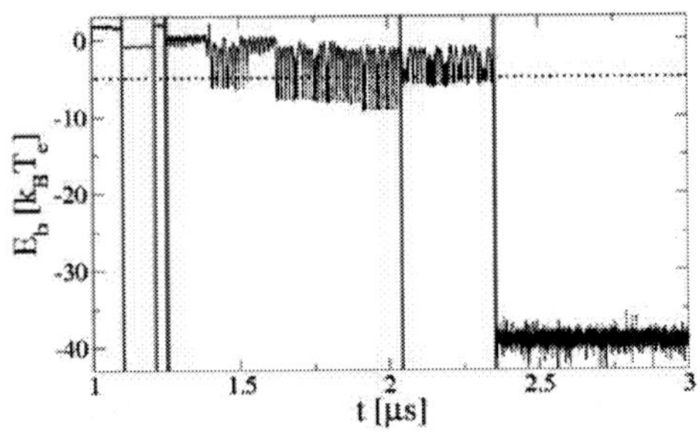

FIGURE 2. Time evolution of the minimal atomic binding, illustrating the two-step formation process. The e^+ is initially captured near the kinetic bottleneck as indicated by the horizontal dotted line. During this phase the atom undergoes a number of replacement collisions (vertical lines) until the e^+ is eventually driven down abruptly to very deep binding ($\sim -40 k_B T_{e^+}$), by a single collision near $t \sim 2.3\,\mu$s..

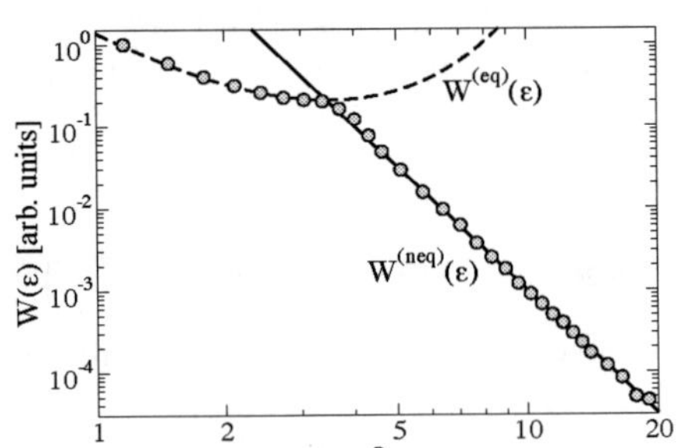

FIGURE 3. Numerically calculated binding energy distribution (dots) compared with the equilibrium distribution W_{eq} according to Eq.(6) (dashed line) and the power-law dependence Eq.(8) with $\varepsilon_b = 4$ (solid line) arising from the two-step capture.

be obtained from [7, 9]

$$K(E_\mathrm{i},E_\mathrm{f}) = R(E_\mathrm{f},E_\mathrm{i})W_\mathrm{eq}(E_\mathrm{i})/\rho_{\mathrm{e}^+} \tag{7}$$

where $R(E_\mathrm{f},E_\mathrm{i})$ is the transition kernel for a transition between the binding energies E_i and E_f. In the limits of $B=0$ and $B \longrightarrow \infty$, the energy dependence of $R(E_\mathrm{f},E_\mathrm{i})$ is well described by $R(E_\mathrm{f},E_\mathrm{i}) \propto e^{-\varepsilon_i}(-E_f)^{-(\varepsilon_\mathrm{b}+1)}$ [7, 9], where ε_i and ε_b denote the initial and bottleneck energies in units of $k_\mathrm{B}T_{e^+}$. The non-equilibrium binding energy distribution W_neq below the bottleneck hence satisfies the relation

$$W_\mathrm{neq}(E)dE \sim K(E_\mathrm{i},E)dE \propto E^{-(\varepsilon_\mathrm{b})}dE\ . \tag{8}$$

Our numerical binding energy distribution agrees well in Fig. 3 with this relation and the equilibrium distribution, Eq. (6), in the high and low energy regimes, respectively. Fitting Eq. (8) to the deep-binding energy tail yields $\varepsilon_\mathrm{b}=4$, in nice accord with previous findings for the field-free [7] and infinite-field [9] cases.

To compare with the measured field ionization spectrum, Eq. (8) should be transformed into the corresponding distribution of maximal ionization fields, F. A one-to-one correspondence between E and F is only possible in the GCA case. As shown in [18], the maximum ionization field and binding energy of GCA atoms scale as $F \sim \rho/(\rho^2+z_\mathrm{m}^2)^{3/2}$ and $E_\mathrm{b} \propto 1/\sqrt{\rho^2+z_\mathrm{m}^2}$, respectively, where z_m denotes the amplitude of the axial bounce motion. Exploiting these relations, the ionization field distribution can be obtained in few straightforward steps. We first transform Eq. (8) to the spherical coordinates with the radius determined by $r=1/E$ and subsequently to the cylindrical coordinates (ρ,z_m) with $r^2=\rho^2+z_\mathrm{m}^2$, i.e.

$$W_\mathrm{neq}(E)dE \sim r^2 r^{\varepsilon_\mathrm{b}-3}dr \sim \rho\left(\rho^2+z_\mathrm{m}^2\right)^{\frac{\varepsilon_\mathrm{b}-3}{2}} dz_\mathrm{m} d\rho\,. \tag{9}$$

Transforming the arguments in the right hand side to $x=(\rho/r)^{3/2}$ and $F=\rho/r^3$ and integrating over x yields the ionization field distribution

$$P(F)dF \sim \int_0^1 \frac{x^{\frac{\varepsilon_\mathrm{b}+2}{3}}}{\sqrt{x^{2/3}-x^2}} dx F^{-\frac{\varepsilon_\mathrm{b}+2}{2}} dF\ . \tag{10}$$

The integral over x is well defined, so that upon integrating $P(F)$ over the ionization field F, we obtain the total number of atoms surviving a given electric field

$$N(F) \propto \int_F^\infty \tilde{F}^{-\frac{\varepsilon+2}{2}} d\tilde{F} \propto F^{-\varepsilon_\mathrm{b}/2}\ . \tag{11}$$

which for $\varepsilon_\mathrm{b}=4$ gives $N(F) \propto F^{-2}$, as observed in the experiment [6] and in agreement with our numerical simulations.

3. Deeply-bound Rydberg states

The exact numerical scheme, outlined above, is used to search for the TBC formation of deeply-bound $\bar{\mathrm{H}}$ atoms in magnetic fields. Several such classes of atoms are shown in Fig. 4, highlighting the differences in how these atoms ionize in experiments. Most GCA atoms form tightly in the transverse direction, but are highly elongated in the axial direction, making them susceptible to field ionization. We caution that in most theoretical treatments involving GCA [9, 13], the atoms are assume dto have negligible axial amplitude and are considered to be circular GCA atoms. The more deeply-bound atoms have, on the other hand, tight axial and transverse confinement, and although often exhibit chaotic behavior, fill a nearly spherical phase-space volume. These atoms survive large ionizing fields, resulting in a departure from the F^{-2} power-law behavior, as shown in Fig. 1.

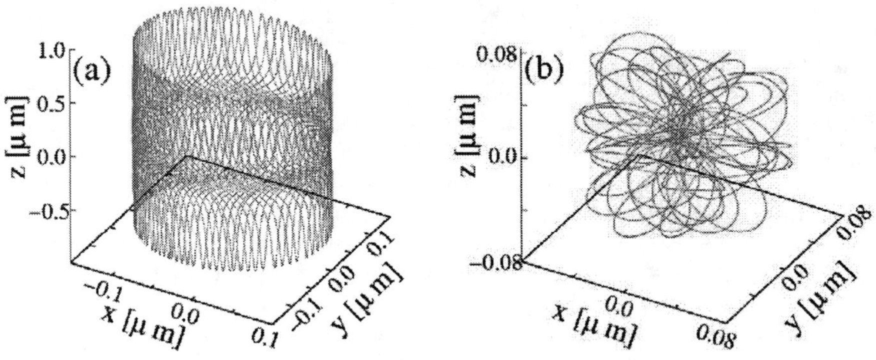

FIGURE 4. Trajectories for $\bar{\text{H}}$ atoms formed in TBC are shown depicting an elongated GCA atom (a) and a chaotic deeply-bound atom (b). From Ref. [12].

B. Velocity spectrum

The driven $\bar{\text{H}}$ production technique [3] anticipated formation of atoms with nearly thermal axial velocities, an expectation that in a later ATRAP experiment [4] was shown not to be the case. In this velocity measurement [4], the speed of $\bar{\text{H}}$ atoms was measured by adding a time-varying electric field to the existing static field, acting as a velocity selective barrier. By monitoring the fraction of atoms that passed through this oscillating field as a function of the field frequency, the ATRAP collaboration was able to deduce the characteristic velocity of the formed atoms. A monoenergetic velocity distribution $\varphi(v) = \delta(v - v_0)$ was assumed. The extracted characteristic velocity of 200 meV was found to be much larger than the thermal velocity of 0.3 meV (4.2 K).

Below, we show how charge-exchange collisions with fast $\bar{p}$ in the side wells of the nested Penning trap provide a self-consistent explanation for the observed, fast non-thermal $\bar{\text{H}}$ atoms. The TBC formed $\bar{\text{H}}$ atoms in passing through the plasma interact with the fast $\bar{p}$ ions in the side wells of the nested Penning trap; the bound positrons hop from the slow $\bar{\text{H}}$ atoms to fast $\bar{p}$ "projectiles" forming fast $\bar{\text{H}}$ atoms [12].

1. Rydberg charge exchange cross section

The Rydberg charge-exchange [19] in a magnetic field has received only little attention. Bradenbrink *et al.*[20] studied the electron transfer from Rydberg atoms in collision with singly-charged ions in a laboratory magnetic field. The initial state was chosen from a stationary microcanonical ensemble corresponding to a fixed binding energy.

Here, we calculate the charge exchange cross section for each microscopic internal and translational atomic state, produced by the preceding Monte-Carlo simulation to describe the initial formation of $\bar{\mathrm{H}}$. By firing a large number $\bar{p}$ of varying incident energy onto the TBC $\bar{\mathrm{H}}$ atoms, we obtain the cross section for charge exchange for each microscopic initial state. The initial and final states of the $\bar{\mathrm{H}}$ atoms are labeled by the maximum surviving fields, F_{i} and F_{f}, such that the state-selective cross section $\sigma_{\mathrm{cx}}(F_{\mathrm{i}},F_{\mathrm{f}};v_{\bar{H}},v_{\bar{p}})$ for charge exchange between an incident $\bar{p}$ with velocity $v_{\bar{p}}$ and an $\bar{\mathrm{H}}$ with velocity $v_{\bar{H}}$ ionizing at a maximum field F_{i} can be defined. The resulting $\bar{\mathrm{H}}$ atom moving at $v_{\bar{p}}$ will ionize at a maximum field F_{f}. Since the charge exchange depends only on the relative velocity between the projectile and the initial TBC atom, and $v_{\bar{p}} \gg v_{\bar{H}}$, the resulting cross sections are nearly independent of $v_{\bar{H}}$.

In Fig. 6, we show the average total capture cross section $\bar{\sigma}_{\mathrm{cx}}(F_{\mathrm{i}},v_{\bar{p}}) = \int \sigma_{\mathrm{cx}}(F_{\mathrm{i}},F_{\mathrm{f}};v_{\bar{p}})dF_{\mathrm{f}}$ as a function of $v_{\bar{p}}$ for different F_{i}. As expected, the cross sections are very large and quickly drop as the relative collision velocity increases. The cross section maxima are due to a matching of Rydberg e^+ and $\bar{p}$ velocities [19].

2. Charge exchange influence on detection

The probability for charge exchange as the initially slow $\bar{\mathrm{H}}$ atoms move through the $\bar{p}$ plasma in the side wells is obtained from

$$P_{\mathrm{cx}}(F_{\mathrm{i}},F_{\mathrm{f}};v_{\bar{p}},v_{\bar{H}}) = 1 - \exp\left(-\sigma_{ex} n_{\bar{p}} d \frac{v_{\bar{p}}}{v_{\bar{H}}}\right) , \tag{12}$$

where $n_{\bar{p}}$ and d denotes the density and the length of the $\bar{p}$ plasma in the side wells of the nested Penning trap and the $\bar{\mathrm{H}}$ travel time through the $\bar{p}$ cloud is $d/v_{\bar{H}}$.

Despite the large cross sections, Eq. (12) yields small exchange probabilities for typical experimental values. The ATRAP experiments, however, only detect atoms within a small opening angle $2\pi\alpha$ along the magnetic field direction, such that the corresponding detection efficiency $\kappa(v,\alpha)$ strongly depends on the atomic velocity v and hence can strongly be influenced by charge exchange collisions. Assuming free motion of the produced atoms, the transverse thermal diffusion gives a detection efficiency,

$$\kappa(v,\alpha) = 1 - \exp\left(-\frac{mv^2}{2k_B T_\perp}\frac{\alpha}{\pi}\right) , \tag{13}$$

where we assumed a transverse $\bar{\mathrm{H}}$ temperature ($T_\perp = T_{e^+}$). Fig. 5 shows the detection efficiency κ as a function of the $\bar{\mathrm{H}}$ kinetic energy. Fast $\bar{\mathrm{H}}$ produced by charge exchange are detected with practically 100% efficiency while κ quickly drops as the $\bar{\mathrm{H}}$ velocity decreases.

With $N(F)$ being the number of initially formed H̄ that survive a given field F, the fraction of detected slow H̄ atoms that do not charge exchange, after traversing the side well, is given by

$$f_{\rm ncx}(F,v) = N(F)\kappa(\alpha,v)\phi_{\bar{H}}(v)\int \phi_{\bar{p}}(v_{\bar{p}})\left[1-P_{\rm cx}(F,F_{\rm f};v_{\bar{p}},v)\right]dF_{\rm f}d^3v_{\bar{p}}\,, \qquad (14)$$

while the fraction of detected fast H̄ produced by charge exchange collisions is obtained from

$$f_{\rm cx}(F,v) = \phi_{\bar{p}}(v)\int N(F_{\rm i})\phi_{\bar{H}}(v_{\bar{H}})P_{\rm cx}(F_{\rm i},F;v,v_{\bar{H}})dF_{\rm i}d^3v_{\bar{H}}\,, \qquad (15)$$

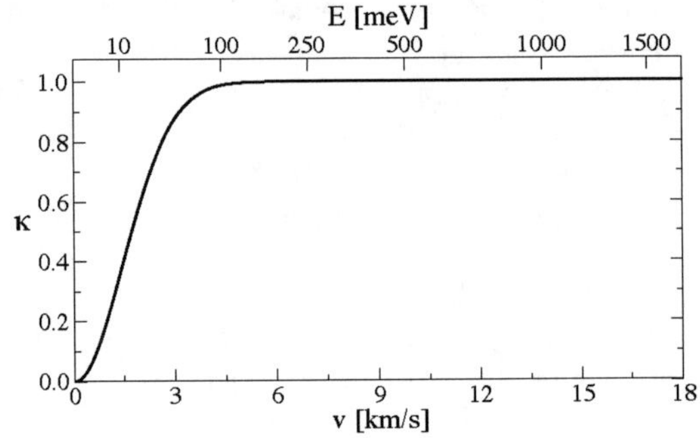

FIGURE 5. Detection efficiency $\kappa(v,\alpha)$ as a function of the velocity v of the detected H̄ atom, and for $\alpha = 4\pi/250$.

3. Charge-exchange influence on H̄ *field ionization*

Let us first examine the influence of charge exchange collisions on the field ionization spectrum discussed previously. The average number of slow and fast atoms – those that undergo charge exchange, as a function of the initial H̄ velocity, is obtained from

$$\bar{f}_{\rm ncx} = \kappa(\alpha,v_{\bar{H}})\left[1-\int \phi_{\bar{p}}(v_{\bar{p}})P_{\rm cx}(F_{\rm i},F_{\rm f};v_{\bar{p}},v_{\bar{H}})\,dF_{\rm f}d^3v_{\bar{p}}\right]\,, \qquad (16)$$

and

$$\bar{f}_{\rm cx} = \int \kappa(\alpha,v_{\bar{p}})\phi_{\bar{p}}(v_{\bar{p}})P_{\rm cx}(F_{\rm i},F_{\rm f};v_{\bar{p}},v_{\bar{H}})\,dF_{\rm f}d^3v_{\bar{p}}\,, \qquad (17)$$

The resulting fraction of fast atoms, $\bar{f}_{\rm cx}/(\bar{f}_{\rm cx}+\bar{f}_{\rm ncx})$, is shown in Fig. 6(b) as a function of the kinetic energy $E_{\bar{H}} = Mv'^2/2$ of the slow initially formed atoms, for different ionizing electric fields. The fraction is nearly independent of the electric field, such that the field ionization spectrum remains unaffected by charge exchange collisions. In other words, our findings discussed in the preceding sections also apply when charge-exchange collisions are included. This point is illustrated in Fig. 7, where we show the field spectrum for atoms surviving an electric field before and after charge exchange.

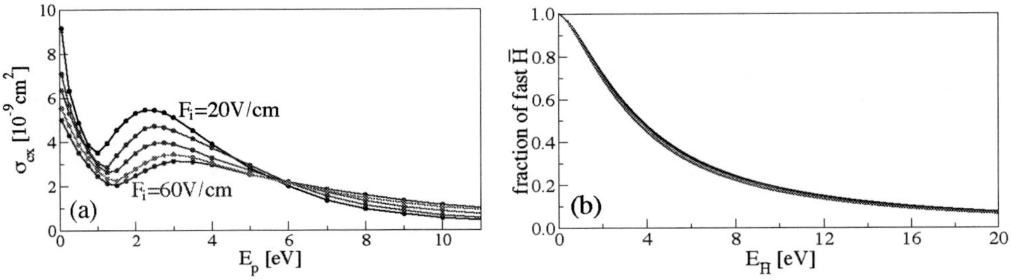

FIGURE 6. e^+ exchange cross section from 1 meV $\bar{H}$ atoms as a function of $\bar{p}$ axial energy, for different ionization fields in the range, 20 – 60 V/cm.

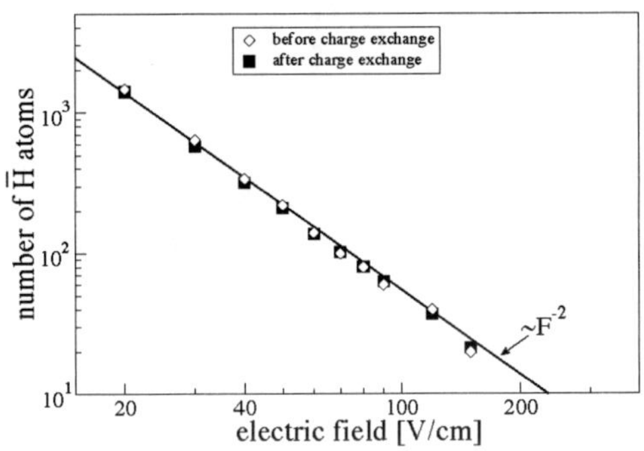

FIGURE 7. Comparison of the field ionization spectrum, calculated within the GCA, before (open diamonds) and after (filled squares) charge exchange.

4. Charge exchange effect on $\bar{H}$ *velocity distribution*

The velocity distribution of the atoms that reach the detector is obtained by adding the contributions from the slow (fast) atoms that do not (do) charge exchange, as given by Eqs. (14) and (15) and integrate over the ionization field

$$\phi(v) = \int f_{\text{ncx}}(F,v) + f_{\text{cx}}(F,v)dF \tag{18}$$

The exact initial velocity distributions $\phi_{\bar{p}}$ and $\phi_{\bar{H}}$ are not known. Here, we assume Gaussian distributions with temperatures $T_{\bar{H}}$ and $T_{\bar{p}}$, respectively.

The characteristic energy of the high-energy $\bar{p}$ is determined by the depth of the side-wells, which is about 10 eV. The exact form of the $\bar{p}$ axial velocity distribution is not important for obtaining the $\bar{H}$ final velocity distribution, since the charge exchange cross sections (Fig. 6) decrease quickly above projectile energies above several eV. The temperature $T_{\bar{H}}$ of the initially formed $\bar{H}$, on the other hand, is determined by comparing the resulting field ionization spectrum with the experimental observation. As shown in Fig.8, our calculated spectra agree with the experimental data for temperatures ranging from 1 meV to 10 meV – considerably smaller values than one would extract by including charge exchange collisions.

The effect of charge exchange collisions is further illustrated in Fig. 8b, where we show the calculated velocity distribution of $\bar{H}$ atoms arriving at the detector. The low-velocity peaks are due to the slow, initially formed, $\bar{H}$ atoms that do not charge exchange, i. e. $\phi_{\bar{H}}$. The extended plateau at larger velocities arises from charge exchange with the fast $\bar{p}$ atoms in the side wells. The latter produces the high-frequency tail in the observed field spectrum.

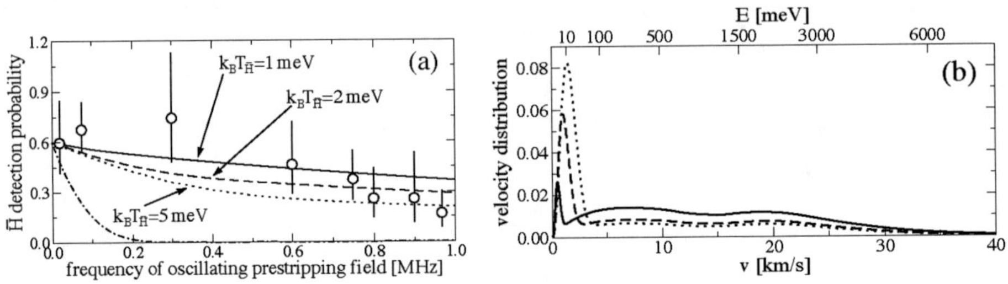

FIGURE 8. (a) Proposed charge exchange mechanism produces a frequency dependent field ionization spectrum consistent with experimental data [4] (circles). The spectra have been calculated for $k_B T_{\bar{p}} = 8$eV and $k_B T_{\bar{H}} = 1$meV (solid line), $k_B T_{\bar{H}} = 2$ meV (dashed line) and $k_B T_{\bar{H}} = 5$meV (dotted line). The dot-dashed line shows the spectrum that results by neglecting e^+ charge transfer for $kT_{\bar{H}} = 2$meV ($T_{\bar{H}} = 30$K). (b) Corresponding velocity distributions after charge exchange. From Ref. [12]

V. SUMMARY

We present classical Monte Carlo simulations that explain several observations of previous ATRAP $\bar{H}$ experiments. It is found that the formation of guiding center atoms at low ionizing fields is responsible for the F^{-2} dependence of observed field ionization spectrum, allowing for an analytical verification of the numerical results. Deviations from this power-law at higher ionization fields, occurring in the Monte-Carlo simulations and observed experimentally, can be attributed to the formation of non-guiding center atoms. It is further demonstrated that an auxiliary formation mechanism, particular to nested Penning traps, is responsible for producing fast non-thermal $\bar{H}$ atoms. This mechanism, the charge exchange with fast $\bar{p}$ projectiles in the side wells, yields field ionization spectra in agreement with the experiment and suggest that $\bar{H}$ atoms much slower than detected are produced.

VI. ACKNOWLEDGEMENT

This work was supported by NSF through a grant for ITAMP at Harvard University and Smithsonian Astrophysical Observatory.

REFERENCES

1. M. Amoretti *et al.*, Nature **419**, 456 (2002).

2. G. Gabrielse *et al*., Phys. Rev. Lett. **89**, 213401 (2002).
3. G. Gabrielse *et al*., Phys. Rev. Lett. **89**, 233401 (2002).
4. G. Gabrielse *et al*., Phys. Rev. Lett. **93**, 073401 (2004).
5. M. Amoretti *et al*., Phys. Lett. B **583**, 59(2004).
6. G. Gabrielse, Ad. At. Mol. Opt. Phys. **50**, 155(2005).
7. P. Mansbach and J. Keck, Phys. Rev. **181**, 275(1969).
8. F. Robicheaux and J. D. Hanson, Phys. Rev. A **69**, 010701(R) (2004).
9. M. Glinsky and T. O'Neil, Phys. Fluids **B 3**, 1279(1991)
10. S. X. Hu, D. Vrinceanu, S. Mazevet, and L. A. Collins Phys. Rev. Lett. **95**, 163402 (2005).
11. F. Robicheaux, Phys. Rev. A **70**, 022510 (2004).
12. T. Pohl, H. R. Sadeghpour, and G. Gabrielse, Phys. Rev. Lett. **97**, 143401 (2006).
13. D. Vrinceanu *et al*., Phys. Rev. Lett. **92**, 133402 (2004).
14. F. Driscoll, Phys. Rev. Lett. **92**, 149303 (2004).
15. G. D. Quinlan and L. Hernquist, New Astron. **2**, 533(1997).
16. C. D. Heggie, Celest. Mech. **10**, 217(1974).
17. S. Mikkola and S. Aarseth, Celest. Mech. & Dyn. Astron. **84**, 343(2002).
18. S.G. Kuzmin and T.M. O'Neil Phys. Plas. **12**, 012101 (2005).
19. D. M. Homan *et al*., Phys. Rev. A **58**, 4565 (1998); K. B. MacAdam *et al*., Phys. Rev. Lett. **75**, 1723 (1995); V. N. Ostrovsky, J. Phys. B: **28**, 3901 (1995).
20. S. Bradenbrink *et al*., J. Phys. B: **30**, L161 (1997).

Particle Physics Aspects of Antihydrogen Studies with ALPHA at CERN

M.C. Fujiwara[a], G. B. Andresen[b], W. Bertsche[c], P.D. Bowe[b], C.C. Bray[d], E. Butler[c], C. L. Cesar[e], S. Chapman[d], M. Charlton[c], J. Fajans[d], R. Funakoshi[f], D.R. Gill[a], J.S. Hangst[b], W.N. Hardy[g], R.S. Hayano[f], M.E. Hayden[h], A.J. Humphries[c], R. Hydomako[i], M.J. Jenkins[c], L.V. Jørgensen[c], L. Kurchaninov[a], W. Lai[a], R. Lambo[e], N. Madsen[c], P. Nolan[j], K. Olchanski[a], A. Olin[a], A. Povilus[d], P. Pusa[j], F. Robicheaux[k], E. Sarid[l], S. Seif El Nasr[g], D.M. Silveira[e], J.W. Storey[a], R.I. Thompson[i], D.P. van der Werf[c], L. Wasilenko[a], J.S. Wurtele[d], and Y. Yamazaki[m]
(ALPHA Collaboration)

[a]TRIUMF, 4004 Wesbrook Mall, Vancouver, BC, V6T 2A3, Canada
[b]Department of Physics and Astronomy, Aarhus University, DK-8000 Aarhus C, Denmark
[c]Department of Physics, Swansea University, Swansea SA2 8PP, United Kingdom
[d]Department of Physics, University of California at Berkeley, Berkeley, CA 94720-7300, USA
[e]Instituto de Fisica, Universidade Federal do Rio de Janeiro, Rio de Janeiro 21941-972, Brazil
[f]Department of Physics, University of Tokyo, Tokyo 113-0033, Japan
[g]Department of Physics and Astronomy, University of British Columbia, Vancouver BC, Canada
[h]Department of Physics, Simon Fraser University, Burnaby BC, V5A 1S6, Canada
[i]Department of Physics and Astronomy, University of Calgary, Calgary AB, T2N 1N4, Canada
[j]Department of Physics, University of Liverpool, Liverpool L69 7ZE, United Kingdom
[l]Department of Physics, Auburn University, Auburn, AL 36849-5311, USA
Atomic Physics Laboratory, RIKEN, Saitama 351-0198, Japan

Abstract. We discuss aspects of antihydrogen studies, that relate to particle physics ideas and techniques, within the context of the ALPHA experiment at CERN's Antiproton Decelerator facility. We review the fundamental physics motivations for antihydrogen studies, and their potential physics reach. We argue that initial spectroscopy measurements, once antihydrogen is trapped, could provide competitive tests of CPT, possibly probing physics at the Planck Scale. We discuss some of the particle detection techniques used in ALPHA. Preliminary results from commissioning studies of a partial system of the ALPHA Si vertex detector are presented, the results of which highlight the power of annihilation vertex detection capability in antihydrogen studies.

Keywords: Cold antihydrogen, fundamental symmetries, particle detection, Si vertex detector
PACS: Replace 36.10.-k, 52.27.Jt, 39.10.+j

INTRODUCTION

One of the prime goals of antihydrogen research is to study symmetry between matter and antimatter. Given that atomic hydrogen is one of the best studied systems in physics, a comparison of hydrogen and antihydrogen properties belongs to the class

CP1037, *Cold Antimatter Plasmas and Application to Fundamental Physics,* edited by Y. Kanai and Y. Yamazaki

of experiments that provide the foundations of modern physics. Cold antihydrogen atoms were first produced by the ATHENA [1] and ATRAP [2] experiments in 2002 at CERN's Antiproton Decelerator (AD), currently the world's only source of low energy antiprotons. These anti-atoms, while nearly at rest, were not trapped and hence annihilated on the apparatus walls shortly after production. The next major step in the field is stable trapping of antihydrogen atoms, and this is the short-term goal of ALPHA, the successor of ATHENA.

The ultimate goal of performing fundamental tests is ambitious: numerous technical developments are required in order to produce, detect, trap, cool and interrogate anti-atoms. This requires us to drive progress in atomic, plasma, and ion trap physics in regimes previously unexplored, even with matter particles. Since its start in 2006, ALPHA has been making aggressive progress, already producing several publications [3,4,5,6,7]. See Ref. [8] for a review of the early results.

In this article, we will discuss particle physics aspects of antihydrogen studies, particularly focusing on the physics motivations and particle detection techniques.

FUNDAMENTAL PHYSICS MOTIVATIONS

In this section, we discuss the scientific case for antihydrogen research in general and ALPHA in particular, from the viewpoint of fundamental subatomic physics. Plasma and trap physics related interests in antihydrogen studies are not described here, but see, e.g., Ref. [9] for discussions of related phenomena.

CPT symmetry

Our belief in CPT invariance is largely theoretical, based on the remarkable success of quantum field theory. Whether CPT is in fact an exact symmetry is a question that should be tested by experiment. Recall that Nature has given us a list of symmetries that are fundamental, yet broken: parity, time-reversal, electroweak, chiral, and perhaps supersymmetry. Precision comparisons of antihydrogen with hydrogen atoms could provide some of the most stringent direct tests of CPT, with the possibility of probing the energy scale beyond the Planck scale.

The CPT theorem [10] guarantees the invariance of CPT in quantum field theories in a flat space time, with mild assumptions including Lorentz invariance, locality, unitary and the spin statistics connection. These assumptions may not be valid in quantum gravity, string theory, or theories with large extra dimensions.

CPT violation could also have implications in cosmology. Baryon asymmetry in the Universe is usually associated with the famous Sakharov condition: (1) Baryon number violation, (2) C and CP violation, and (3) that these occur out of thermal equilibrium. It is, however, possible to generate baryon number excess without (2) and (3), if CPT is violated [11,12]. In fact, an $O(10^{-6})$ difference in top and anti-top quark mass can generate the observed baryon asymmetry [13]. Note that the (anti)top quark mass is currently only known to the 1% level. Another important, yet challenging, goal is a test of the gravitational interaction between matter and antimatter, for which there exist no direct experimental data. Such measurements will test the Weak Equivalence Principle of general relativity.

TABLE 1. Direct tests of CPT via particle-antiparticle comparisons [14].

Particle, CPT quantity	Relative precision	Δm in energy
e- e+ mass	0.8×10^{-8}	4×10^{-12} GeV
$K^0\overline{K^0}$ mass※	10^{-18}※	5×10^{-19} GeV※
$p\overline{p}$ mass	2×10^{-9}	2×10^{-9} GeV
e- e+ g-2	2×10^{-9}	
μ- μ+ g-2	0.7×10^{-6}	
$p\overline{p}$ q/m	0.9×10^{-10}	
$p\overline{p}$ magnetic moment	3×10^{-3}	

※See text for discussion of the kaon test.

A theory recently proposed by Kostelecky and co-workers, the so-called Standard Model Extension (SME) [15] has generated much attention with regard to CPT and Lorentz violation tests. According to the SME, a large parameter space for CPT violation can be explored by searching for Lorentz violation signals with normal matter systems, and impressive progress is being made in this area [16]. Direct comparisons of matter-antimatter systems, on the other hand, give model-independent tests of CPT invariance. Table 1 shows some of the existing direct tests of CPT, with emphasis on the mass measurements.

It should be noted that meson-antimeson systems ($q\overline{q}$ - $\overline{q}q$) are inherently different from baryon-antibaryon (qqq - $\overline{qqq}$) or lepton-antilepton ($l-\overline{l}$) systems, and could have different patterns of symmetry violation. Given the fundamental importance of CPT symmetry, it should be tested in all particle sectors. Furthermore the interpretation of the $K^0-\overline{K^0}$ results as a test of CPT is controversial. The impressive quoted precision is derived with some theoretical assumptions, and Kobayashi and Sanda [17] have put forward an argument that the CPT violating interaction is tested only at much reduced level (currently ~10^{-5} [18]). Bigi has also questioned the quoted relative precision of 10^{-18} for similar reasons [19].

In addition to the relative precision, one can consider the absolute energy precision as an alternative figure of merit. In precision measurements, typical energy corrections from short distance effects may have the form [20]:

$$\Delta Energy \sim \frac{m^{n+1}}{\Lambda_{NP}^{\ n}} \qquad (1)$$

where n depends on the theory and is typically > 0, and m is the characteristic energy scale of the system, while Λ_{NP} is the high energy scale applicable to the CPT violating new physics.[i] To give a scale, for n = 1, m ~ GeV, and $\Lambda_{NP} \sim m_{\rm Pl}$ (the Planck scale ~ 10^{19} GeV), we may expect the CPT violating energy shift at $\Delta Energy$ ~ 10^{-19} GeV level. The $K^0-\overline{K^0}$ comparison, taken at face value, is constrained only to ~ 100 kHz in frequency or 5 x 10^{-19} GeV in terms of the absolute energy scale. In this respect,

[i] Since direct CPT tests challenge the entire framework of (effective) field theory, one has to be cautious in applying the language of field theory as in Eq. 1.

antihydrogen experiments could compete favorably with the kaon and other direct tests of CPT, the details of which now follow.

Physics Reach with Antihydrogen Laser Spectroscopy

As previously mentioned, atomic hydrogen is one of the most precisely studied simple systems in physics. The two photon transition between the 1s and 2s states is an ideal system for precision spectroscopy, due to its narrow intrinsic line width ($\Delta\nu/\nu = 4 \times 10^{-16}$), as well as the possibility for cancellation of the 1st order Doppler broadening when counter propagating photons are employed (the well-established "Doppler-free spectroscopy" techniques). Further, spectral lines collected at sufficiently high signal-to-noise level can be fit to known spectral shapes, allowing determination of transition line frequencies to well below natural linewidths. In principle, if this approach could be used to "split" the line by a factor of 500, precisions approaching 10^{-18} would be achievable. Obviously, there are many technical challenges before this ultimate precision can be realized. Nonetheless, progress in the measurement of this transition in atomic hydrogen has been remarkable, and the current relative precision is at a level of 2×10^{-14} with an uncertainty of ~50 Hz [21].

While the ultimate precision of 10^{-18} may become possible one day, what precision can we achieve with antihydrogen in the short term? Clearly, because of the scarcity of the anti-atoms, the technological bottle-neck is in achieving stable trapping. In 1993 Hänsch and Zimmermann estimated that 1000 trapped antihydrogen atoms would be required to achieve 10^{-12} precision in 1s-2s spectroscopy [22]. In fact, this level of line width has been observed with trapped hydrogen atoms by a member of the ALPHA collaboration in 1996 [23]. With the progress in producing continuous Lyman alpha lasers, Walz et al. [24] argued that even fewer atoms would be needed to make precision spectroscopy by using the shelving scheme, i.e., repeating many times the transitions of the same atom. The precision at this level would be limited in part by the Zeeman effect in the trapping magnetic field. If laser cooling, which has been demonstrated for trapped hydrogen atoms [25], can be applied to antihydrogen, it would dramatically reduce the influence of the Zeeman effect, hence improving the precision. Introduction of a high power laser into an antihydrogen production apparatus has already been achieved [26], hence the technical bottle-neck for antihydrogen laser cooling is the stable trapping of antihydrogen. Many other developments relevant for antihydrogen spectroscopy are being reported in the literature. (For example, 2 photon laser cooling [27], coil-gun slowing [28], single atom cooling [29], Rydberg trapping [30], to name just a few).

An initial 10^{-12} measurement of the 1s-2s transition would be already competitive with other CPT tests involving leptons and baryons (Table 1). In particular, it would constitute a 4 order of magnitude improvement in the equality of the electron and the positron mass[ii]. In terms of absolute precision, $\Delta\nu \sim$ few kHz corresponds to an energy sensitivity of $\Delta E \sim 10^{-20}$ GeV.

Table 2 illustrates the physics reach of the initial spectroscopy measurement with trapped antihydrogen. These initial measurements could already reach the Planck suppressed $1/M_{pl}$ sensitivity. Further improvements are envisioned with existing and emerging atomic techniques.

TABLE 2. Expected precision of initial 1s- 2s measurement and the physics reach

Relative precision $\Delta\nu/\nu$	Frequency precision $\Delta\nu$	Absolute energy sensitivity ΔE	CPT quantity	Possible improvement factor
10^{-12}	few 10^3 Hz	10^{-20} GeV	e- e+ mass ratio[ii]	10^4

Antihydrogen Hyperfine Splitting

We are developing a technique for preliminary microwave spectroscopy of antihydrogen hyperfine splitting in a high magnetic field, based on the present ALPHA apparatus [8]. Hyperfine measurement in a low field has been considered by the ASACUSA experiment [31]. The hyperfine fine splitting of atomic hydrogen, the famous 21 cm line, is very well measured to the 7 x 10^{-13} level with an absolute precision of 1 mHz. The magnetic moment of antiprotons is currently only known to 0.3%. The deviation of the (anti)proton magnetic moment from the Dirac value reflects its internal structure, and hence its value is possibly more sensitive to short-distance effects than 1s-2s transitions. In addition, *within the SME model,* the sensitivity to the CPT violating effects are enhanced by $O(1/\alpha^2) \sim 10^4$, compared to the 1s-2s case. Hyperfine splitting can therefore provide a powerful complementary test of CPT.

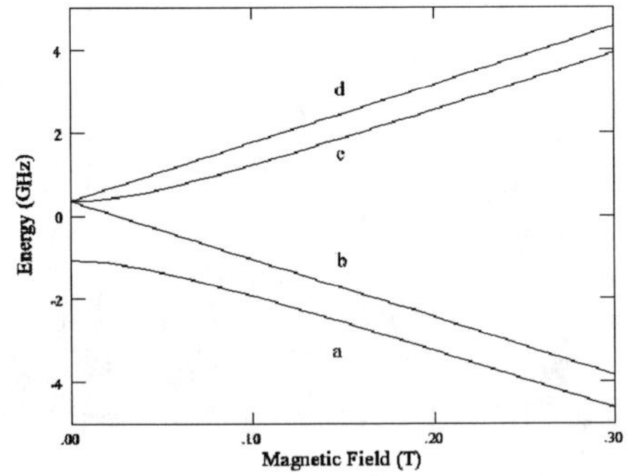

FIGURE 1. Breit-Rabi diagram for energy levels of (anti)hydrogen atoms in a magnetic field.

The energy levels of (anti)hydrogen atoms in a magnetic field, given by the Breit-Rabi formula, are shown in Fig. 1. Microwave transitions from low field seeking, trappable

---

[ii] The charge equality $|q_e| = |q_{e+}| = |q_p| = |q_{\bar{p}}|$ is assumed to derive the e- e+ mass ratio from the hydrogen-antihydrogen comparison of the 1s-2s level. Alternatively, limits on the charges can be obtained.

states (c, d in Fig. 1) to high field seeking un-trappable states (a, b) would lead to ejection of the antihydrogen atom from the trap. The transition can be detected, with nearly 100% efficiency, via annihilation signals on the trap wall measured by ALPHA's position sensitive vertex detector (see below). An onset of this signal is expected to occur when the frequency of an applied microwave is varied such that the anti-atoms come into resonance. Each transition frequency is sensitive to a different combination of fundamental parameters. For example, in the high-field limit, the transitions between the states (d→ a) and (c→b) are given by [8]:

$$\nu_{ad} = \frac{\Delta\nu_H}{2} + \frac{\mu_B g_{e+}}{h} B, \text{ and}$$
$$\nu_{bc} = -\frac{\Delta\nu_H}{2} + \frac{\mu_B g_{e+}}{h} B. \qquad (2)$$

Here $\Delta\nu_H$ is the antihydrogen hyperfine splitting at zero field, μ_B the Bohr magneton, and g_{e+} the positron g-factor in the antihydrogen atom. The difference $(\nu_{ad} - \nu_{bc}) = \Delta\nu_H$ will give a measure of the antihydrogen hyperfine splitting, which is proportional in leading order to the antiproton magnetic moment. The sum $(\nu_{ad} + \nu_{bc})$ will give the value of the positron bound state g-factor. The resonant microwave transitions as in Eq. (2) may be one of the first spectroscopic signatures of trapped antihydrogen. Given that the antiproton magnetic moment is currently known only to 0.3%, a measurement of $\Delta\nu_H$ with an initial precision of 0.1% would already constitute a significant CPT test of the magnetic properties of an anti-baryon. A 10^{-4} measurement would reach an absolute energy sensitivity of $\sim 10^{-18}$ GeV, competitive with the neutral kaon test. Dramatic improvements beyond these proof-of-principle measurements can be envisioned, when combined with existing and emerging cooling and trapping techniques.

Antimatter Gravity

Finally, the gravitational influence on antimatter could be observable with antihydrogen atoms laser cooled to mK.[iii] For a vertical trap, the antihydrogen density would have the form $\sim \exp(-M\bar{g}h/kT)$, where M is the antihydrogen gravitational mass, $\bar{g}$ the gravitational acceleration for antimatter and T the antihydrogen temperature [32]. For $T = 1$ mK, we have a typical height of $h \sim 85$ cm. Hence with a few mK antihydrogen, we would know at least the sign $\bar{g}$ of in a reasonably sized apparatus. Other measurements of antimatter gravity have been proposed at the AD [33].

PARTICLE DETECTION IN ALPHA

In this section, we discuss some of the particle detection systems used in the ALPHA experiment. There are two distinct kinds of particle detection that are relevant. One is *counting* of antiproton annihilation events, and the other is *imaging* of antiproton

[iii] The Doppler limit is 2.4 mK, and the recoil limit 1.3 mK, for Lyman alpha cooling.

annihilations. A unique feature of antiparticle studies (as opposed to normal particles) is the high efficiency for its detection via annihilations, which allows sensitive studies of important processes such as antihydrogen production and antiproton losses due to non-uniform magnetic fields.

External Annihilation Detector

The annihilation counting technique is often used to count the number of trapped antiprotons. When trapped antiprotons are released to collide with the apparatus walls, an annihilation event produces several charged particles (mostly pions). The counting detector usually consists of a material which emits light when a particle goes through it, and a photon sensor which converts the light into an electrical signal. The most common combination is a plastic scintillator, coupled with a photo-multiplier tube (PMT). Standard PMTs, however, are susceptible to magnetic fields. In a Penning trap environment, there is an unavoidable stray field of up to a few 100 G. For the external annihilation detector for ALPHA, we use triple shields for our PMTs with mu-metal and iron shielding. Such a PMT is coupled to a relatively large scintillator (60 cm x 40 cm x 1 cm), placed outside the main solenoid magnet. Two scintillator-PMT systems are used as a coincidence pair, which reduces noise. The background is mainly due to cosmic rays, typically 20 Hz per coincidence pair. For our first run in 2006 we had 2 such pairs. The coverage was doubled for the 2007 run. We are currently constructing 2 more pairs, making the total to be 6 coincidence pairs, or 12 scintillator-PMT systems. Fig. 2 (left) illustrates an example of early physics results obtained in 2006 with this external annihilation detector.

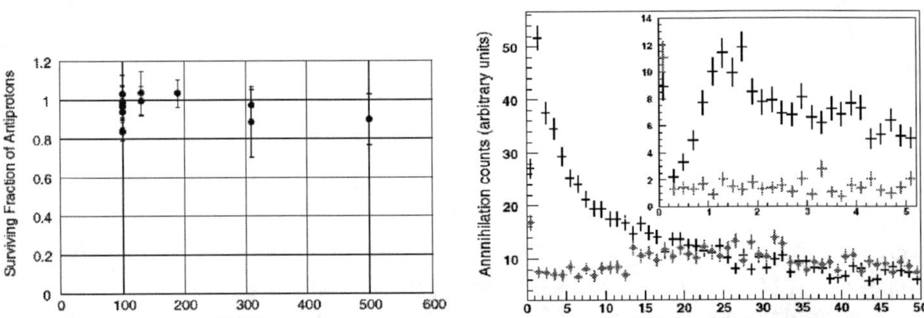

FIGURE 2. (Left) Examples of physics results obtained with the external annihilation detector, indicating survival of antiprotons in an octupole field [4]. (Right) Results with the internal annihilation detector, demonstrating antihydrogen production at 1 T field [5].

Internal Annihilation Detector

With the background level of tens of Hz, the external annihilation detector is not well suited for detecting small signals. The antihydrogen annihilation signal during the mixing of antiprotons and positrons is one such example. In this case, it would be preferable to place the detector inside the bore of the solenoid magnet, close to the

Penning trap walls, in order to increase the signal to background. The strong magnetic field and the space restrictions make it difficult to use a PMT as a photon sensor. The Si strip detector we are developing (see below) will work in such an environment, but delays in the delivery by the Si manufacturer forced us to develop an alternative internal annihilation detection system on a very short time scale.
Our initial attempt was based on a design similar to that used for the electromagnetic calorimeter [34] developed for the KOPIO rare kaon decay experiment. The system consisted of a scintillator with extruded holes through which wave-shifting fibers were run to bring the photons out. Unfortunately, the light yield from the KOPIO system was not sufficient for our application. Instead, we developed a system which coupled a scintillator directly to an avalanche photodiode (APD), as shown in Fig. 3. The main volume of the scintillator has a dimension of 30 cm x 7 cm x 2.5 cm. One of the sides of the volume is extended and tapered to match the 8 mm x 8 mm active area of the APD. Both the main volume and the tapered volume are machined out of the same piece of scintillator material. A low noise pre-amplifier has been developed to read out the APD signal. At a typical operating voltage of 1.8 kV, the APD has a gain of about 500. Fig. 2 (right) shows physics data taken in 2006 using an initial system of four scintillator-APD modules.

The internal detector system was substantially improved for the 2007 run. The detector assembly was redesigned and rebuilt, which allowed for a better noise insulation as well as improved mechanical strength. The new system allows temperature control of the APD to 0.1 degree C using a water cooling system, which stabilizes the gain of the APD. A total of up to 14 scintillator-APD modules surrounded the Penning traps to provide a high solid angle coverage. Physics results obtained with this new system in the 2007 run were recently published in Ref. [6]. It has also been used to search for trapped antihydrogen as described in [35]

FIGURE 3. (Left) Internal annihilation detector assembly. The tapered scintillator is directly coupled to an APD. (Right) Installation of the internal detector into the ALPHA solenoid magnet.

Antiproton Beam Detectors

ALPHA's beam detection systems use techniques similar to those developed for ATHENA, but adopted specifically for the ALPHA application. The spatial profile of the antiproton beam is measured with a segmented Si detector of a small thickness (60 μm). The detector, operated at a few 10's of Kelvin, is biased typically at -75 V, 10 times higher than the depletion voltage. This is necessary since a pulse of 10^7 antiprotons generates a large number of electron-hole pairs in Si, some of which recombine unless there is a large electric field to sweep out the charge carriers. We are currently investigating the possibility of a new beam detector based on a CVD diamond film, with even smaller thickness and improved spatial resolution.

The antiproton beam intensity is measured with a scintillator coupled to a hybrid photodiode, a system similar to that discussed in [36].

Si Vertex Detector

One of the important design features of the ALPHA apparatus is the capability to imagine annihilations by means of a Si vertex detector. As demonstrated in ATHENA [37,38], this is an extremely powerful tool, not only for detecting antihydrogen, but also for diagnosing related plasma processes. Also, for the initial detection of trapped antihydrogen and the subsequent spectroscopy experiments, it will probably necessary to search for very rare events in the presence of backgrounds such as cosmic rays. The high background rejection capability from a position sensitive detector is likely be important for the first detection of trapped antihydrogen and its spectroscopy. This is why we have made a great effort to incorporate annihilation imaging capability compatible with a neutral atom trap. See [39] for conceptual discussions of the ALPHA detector.

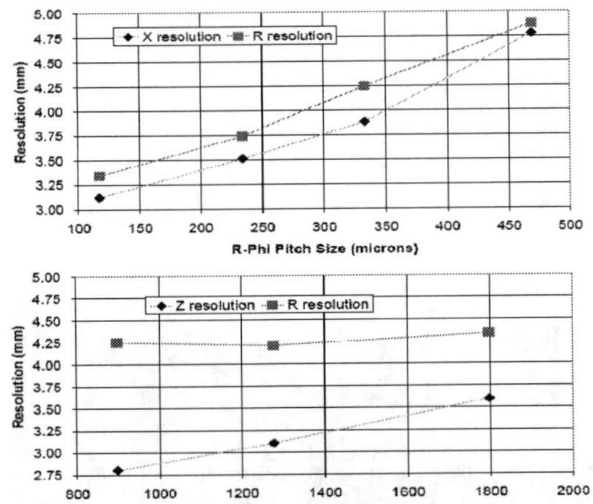

FIGURE 4. Monte Carlo simulation studies of the expected resolutions for reconstructed vertices as a function of R-phi and Z pitch sizes. See the text for details [40].

Since the early studies [39], detailed Monte Carlo simulations have been performed in order to determine the design parameters of the ALPHA vertex detector (see, e.g, Fig. 4) [40]. Our Si module has a dimension of 230 mm x 60 mm, and is double-sided. Strip pitches are 900 μm and 230 μm along the z- (along the solenoid field) and phi-directions, respectively. A total of 60 such modules will be used to surround the trapping region in 3 layers. Our simulations show that it is essential to have 3 layers of

Si detectors in order to have a useful reconstruction capability, as opposed to the ATHENA detector, which had only 2 layers. The main difference in ALPHA is the increased scattering material thickness and the larger distance from the trap walls, both dictated by the requirements for the neutral (anti-atom) trap.

The data acquisition system to read out and record a total of 30,000 channels of the ALPHA Si detector features a custom-made high density flash ADC (VF48), and also a custom-made trigger and control module (TTC). Si sensors and front-end ASIC chips are connected via Front-end Repeater Cards to the ADCs. The MIDAS software framework controls the readout sequence via a VME controller.

Si Detector Commissioning

In the final 3 weeks of the 2007 run, we commissioned the first Si modules in the antihydrogen experiment environment at the AD. Two groups of 3 layer Si modules (see Fig. 5), constituting 1/10 of the full detector system, were placed in the trapping region of the ALPHA apparatus. We would have been satisfied with simply detecting a few pion tracks with the detector, but we have managed to reconstruct antiproton annihilation vertices (despite the low efficiency due to the small coverage), and obtain some physics results during our commissioning. Figure 5 illustrates examples of cosmic events, as well as antiproton annihilation events, both detected by the Si detector and reconstructed via a vertex analysis software routine.

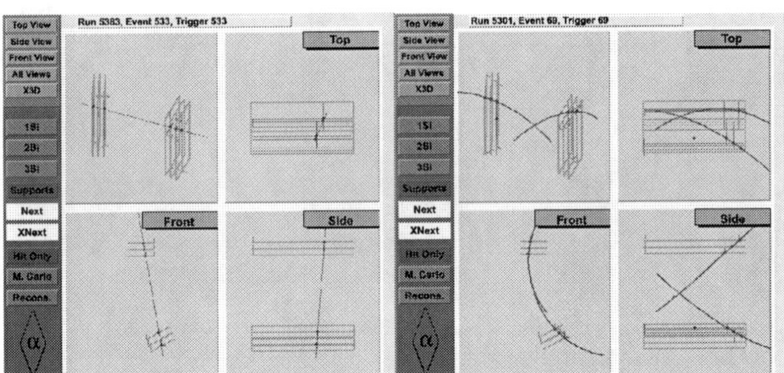

FIGURE 5. Reconstructed tracks from a cosmic ray (B = 0 T) and antiproton annihilation (1 T).

The axial (along the solenoid field) distribution of antiproton annihilations is shown in Fig. 6, with the Penning trap potential well moved to different positions. The annihilations follow the trap wells as expected, demonstrating the basic functionality of the entire chain of the Si detector system, starting from the Si sensors, the front-end electronics, the data acquisition system, and the vertex reconstruction routine (including the calibration and the alignment).

Figure 7 shows the axial distribution of antiproton annihilation vertices during mixing of antiprotons and positrons in a uniform 1 T field, a condition similar to that for Fig.

2 (Right). A peak at the position of the positron plasma is characteristic of antihydrogen annihilations (c.f. Ref. [41]), corroborating the production of antihydrogen at 1 T, which we reported earlier in Ref. [5].

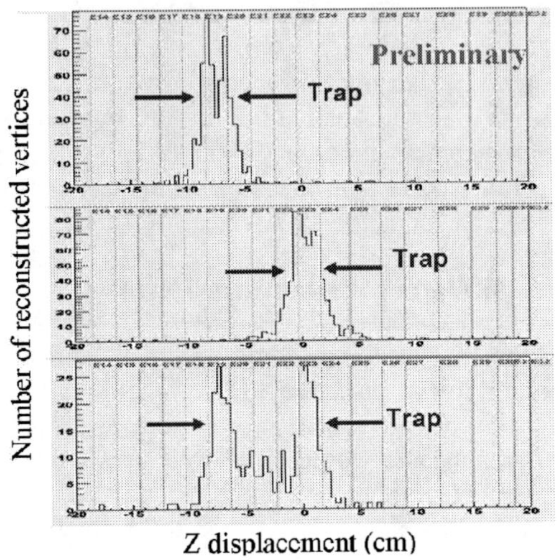

FIGURE 6. Axial position of the reconstructed annihilation vertices, when the Penning trap potential walls (indicated with arrows) are moved. Vertical lines indicate the position of the trap electrodes.

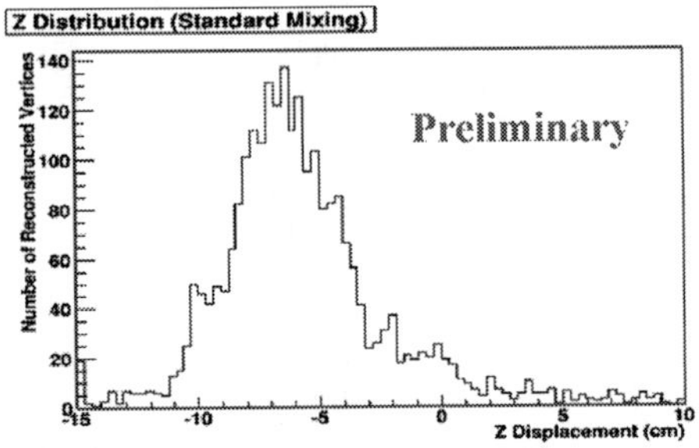

FIGURE 7. Axial distribution of the annihilation vertices during antiproton-positron mixing, indicating the production of antihydrogen. The positron plasma is centered near -6 cm.

New physics information, although still preliminary, is obtained with the partial detector during our study of antiproton dynamics in an octupole magnetic field. The axial distribution of antiproton annihilation shown in Fig. 8 (Left) is indicative of the so-called ballistic loss of antiprotons in an octupole field [6]. The larger the axial excursion of the antiproton in an octupole field, the more likely it is that the trajectory of the particle coincides with the trap wall. This leads to the observed concentration of annihilations at both edges of the trapping potential.

Furthermore, the ballistic loss model predicts that antiproton losses are azimuthally peaked four-folds at trap edges, and that these four peaks are shifted by 45 degrees between the edges [6]. The Preliminary z-phi distributions shown in Fig. 8 (Right) provides an indication of such peaks. Note that due to the limited solid angle coverage of the Si modules, the reconstruction efficiency is not uniform. It should be stressed that Figs. 8, while consistent with the ballistic model predictions, are not definitive proof for it, and further studies with a full detector are required. Nonetheless, these examples highlight the power of the annihilation imaging detector, even with 1/10 of the full system, when studying antiproton processes in a new condition. We are currently preparing to install more than 1/2 of the full system for the 2008 run. It is hoped that the full detector system will be completed by the end of 2008.

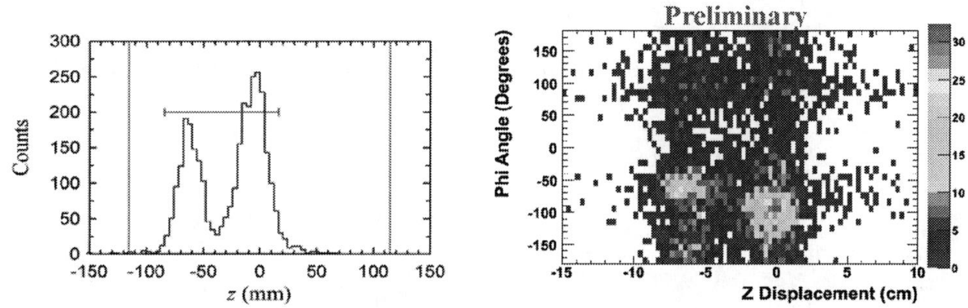

FIGURE 8. (Left) Axial distribution of antiproton annihilations when the octupole field is ramped up. Reproduced from Ref. [6]. (Right) The azimuthal, as well as the axial annihilation distributions for the same data. Both plots are consistent with the ballistic model of antiproton loss in an octupole field.

SUMMARY

In this article we have discussed some aspects of the ALPHA experiment related to particle physics techniques and ideas. We reviewed the fundamental physics case for antihydrogen studies and their potential physics reach, emphasizing the possible sensitivity to Planck scale phenomena once antihydrogen trapping is achieved. We have also discussed some of the particle detection techniques used in ALPHA. Preliminary results for a commissioning run of a partial system of a Si vertex detector were presented, which highlights the power of an annihilation vertex detection capability in antihydrogen studies.

ACKNOWLEDGMENTS

We gratefully acknowledge the technical support by TRIUMF and the University of Liverpool for detector development, design and construction. In particular, we are indebted to Pierre Amaudruz, Daryl Bishop, Zlatko Bjelic, Travis Howland, Kevin Langton, and Peter Vincent of TRIUMF, Jean-Pierre Martin of Montreal, and David Seddon, Jim Thornhill, and David Wells of the University of Liverpool. We also thank Pablo Genoa, Lawrence Posada, and Alberto Rotondi for their help in the early stages of Si simulations. This work is supported in part by by CNPq, FINEP (Brazil), NSERC, NRC/TRIUMF (Canada), FNU (Denmark), ISF (Israel), MEXT, RIKEN (Japan), EPSRC and Leverhulme Trust (UK) and DOE (USA).

REFERENCES

1. M. Amoretti et al. (ATHENA Collaboration), Nature (London) **419**, 456 (2002).
2. G. Gabrielse et al. (ATRAP Collaboration), Phys. Rev. Lett. **89**, 213401 (2002).
3. W. Bertsche et al. (ALPHA Collaboration), Nucl. Instrum. Methods A **566**, 746 (2006).
4. G. Andresen et al. (ALPHA Collaboration), Phys. Rev. Lett. **98**, 023402 (2007).
5. G. Andresen et al. (ALPHA Collaboration), J. Phys. B **41**, 011001 (2008).
6. G. Andresen et al. (ALPHA Collaboration), Phys. Plasmas **15**, 032107 (2008).
7. G. Andresen et al. (ALPHA Collaboration), Phys. Rev. Lett. (in press).
8. M.C. Fujiwara et al. (ALPHA Collaboration), Hyperfine Interact. **172**, 81 (2006).
9. C. Surko, these proceedings.
10. G. Lüders, Ann. Phys. **2**, 1 (1957).
11. A.D. Dolgov, Y.B. Zeldovich, Rev. Mod. Phys. **53**, 1 (1981).
12. For recent theory, see e.g. H. Davoudiasl et al. Phys. Rev. Lett. **93**, 201301 (2004); S. M. Carroll et al., Phys. Rev. D **73**, 103515 (2006); P.A. Bolokhov et al., Phys. Rev. D **74**, 123517 (2006).
13. A.D. Dolgov, Phys. Rep. **222**, 309 (1992).
14. Particle Data Group, J. Phys. G **33**, 1 (2006) with some updates.
15. E.g. D. Colladay, V.A. Kostelecky, Phys. Rev. D **55**, 6760 (1997); **58**, 116002 (1998); Proc. of 4th Meeting on CPT and Lorentz Symmetry, CPT07 (ed. by V.A. Kostelecky).
16. See e.g., V. A. Kostelecky, N. Russell, arxiv:hep-ph/0801.0287; M. Pospelov and M. Romalis, Phys. Today, 57, **40** (2004); S. Reinhardt, G. Gwinner et al., Nature Phys. **3**, 861 (2007); G.W. Bennet et al., Phys. Rev. Lett. **100**, 09106 (2008).
17. M. Kobayashi and A.I. Sanda, Phys. Rev. Lett. **69**, 3139 (1992).
18. Y. Takeuchi, S.Y. Tsai, Int. J. Mod. Phys. A**18**, 1551 (2003): arxiv:hep-ph/0208148.
19. I.I. Bigi, Nucl. Phys. A**692**, 227c (2001).
20. M. Pospelov, Hyperfine Interact. **172**, 63 (2006).
21. M. Niering et al., Phys. Rev. Lett. **84**, 5496 (2000).
22. T.W. Hänsch and C. Zimmermann, Hyperfine Interact. **76**, 47 (1993).
23. C.L. Cesar et al., Phys. Rev. Lett. **77**, 255 (1996).
24. J. Walz, A. Pahl, K.S.E. Eikema and T.W. Hänsch, Hyperfine Interact. **127**, 167 (2000)
25. I.D. Setija et al., Phys. Rev. Lett. **70**, 2257 (1993).
26. M. Amoretti et al. (ATHENA Collaboration), Phys. Rev. Lett. **97**, 213401 (2006).
27. D. Kielpinski, Phys. Rev. A **73**, 063407 (2006).
28. E. Narevicius et al., Phys. Rev. Lett. **100**, 093003 (2008).
29. G.N. Price et al., Phys. Rev. Lett. **100**, 093004 (2008).
30. S.D. Hogan and F. Merkt, Phys. Rev. Lett. **100**, 043001 (2008).
31. E. Widmann et al., Nucl. Instrum. Methods B **214**, 31 (2004).
32. W.D. Phillips et al., Hyperfine Interact. **76**, 265 (1993).
33. G. Testera et al. (AEGIS Collaboration), Proposal to CERN 2007; P. Perez et al., (Antimatter Free Fall exp.), Letter of Intent to CERN 2007.
34. T. Numao et al., in *Scintillating Crystals*, KEK Proceedings **2004-4**, p. 179 (2004).
35. G. Andresen et al. (ALPHA Collaboration), these proceedings.
36. M.C. Fujiwara and M. Marchesotti, Nucl. Instrum. Methods A **484**, 162 (2002).
37. C. Regenfus, Nucl. Instrum. Methods A **501**, 65 (2003).
38. M.C. Fujiwara et al. (ATHENA Collaboration), Phys. Rev. Lett. **92**, 065005 (2004).
39. M.C. Fujiwara, AIP Conf. Proc. **793**, 111 (2005): arxiv:hep-ex/0507082.
40. M.C. Fujiwara et al., ALPHA Detector Simulation Report, unpublished (2005)
41. N. Madsen et al. (ATHENA Collaboration), Phys. Rev. Lett. **94**, 033403 (2005).

A Particle Physicist Looks At Global Warming

John Eades

University of Tokyo, c/o Div PH, CERN, 1211 Geneva 23

Abstract: Specialised literature on global warming is not easy to understand. The media usually try to present this complex subject to the public in an attention-getting way, short on numbers and quantities and far removed from the content of the original source material. Since this kind of treatment is hardly going to satisfy physics professionals working in other fields, we owe it to ourselves, as well as to those who may ask our opinion on these matters, to inform ourselves appropriately. I shall therefore try to look at this topic from a particle physicist's point of view with special reference to the many points of contact with our own discipline.

Keywords: Global Warming, Climate Change, Greenhouse Gases
PACS: 01.30.Rr , 92.70.Aa , 92.70.-j, 92.30.Np

INTRODUCTION

Whereas one might hope after 20 years of scientific and political debate to see the outlines of an action plan, climate policy has instead stalled in a mind-numbing blame game in which governments, corporations and advocacy groups argue over who is most at fault for past emissions, who should cut future emissions, and who should pay for it all—while the collective gaze of the global public glazes over in boredom and incomprehension. Paul Roberts, in '*The End of Oil*' [1].

In 2007, the UN Intergovernmental Panel on Climate Change (IPCC) published its *Fourth Assessment Report* [2]. This received wide coverage in newspapers, magazines and on TV, much of it sympathetic to the panel's conclusions. The consequence has been an unusual level of public acceptance of the reality of Anthropogenic Global Warming (AGW).

Nevertheless, a well-organised and vociferous body of dissenters has been able use the media to call the report's credibility and conclusions into question. It includes many individuals not normally recognized as authorities on scientific matters, most of whom tend to make the same mistakes and to have the same illusions and misunderstandings about the nature and practice of science. We first look at what some of the better known AGW doubters have to say.

Author and film producer Michael Crichton asks in a Jan 17 2003 lecture at Caltech [3] why anyone should believe climate forecasts for a century ahead when tomorrow's weather forecast often turns out to be wrong. Here he makes the elementary mistake of confusing weather with climate. Weather forecasts are day by

CP1037, *Cold Antimatter Plasmas and Application to Fundamental Physics,* edited by Y. Kanai and Y. Yamazaki

day (often hour by hour) predictions of air temperature, wind speed etc. for thousands of locations over the globe, while climate forecasting deals only with global averages of such values over an interval of ten years or more, 50-100 years hence. Crichton goes on to complain, as do other doubters, about the often-quoted consensus about AGW among scientists:

'There is no such thing as consensus science. If it's consensus, it isn't science. If it's science, it isn't consensus. Period. Consensus is invoked only in situations where the science is not solid enough. ...Nobody says the consensus of scientists agrees that $E = mc^2$*... '*

Of course it is not the case that we decided that $E = mc^2$ by voting on it. Nor however is it true that the consensus on this question among the people in this room is seriously going to undermine the scientific basis of Special Relativity.

Australian Geology Professor Robert Carter observes [4] in a talk to The Melbourne Rotary Club, 15/06/05 that *'AGW assertions are a symptom of a disease called Hansenism ... Hansenist climate hysteria is driven by relentless, ideological, pseudo-scientific drivel, most of which issues from green political activists and their supporters. Some scientists, too, contribute to the Hansenist alarmism. Such environmental lobbying and emotional propaganda (is) a greater threat to our society and way of life than communism....'*

The Hansen in question is the director of NASA's Goddard Institute for Space Studies. Carter goes on to say that *'.... the probable effect of human-injected carbon dioxide is a miniscule 0.12% of the greenhouse warming, that is a temperature rise of 0.036 degrees ...*99.88% of the *greenhouse effect has nothing to do with carbon dioxide emissions from human activity.'*

The surprising thing here is that while most critics claim that it is quite impossible to attach any degree of certainty to AGW figures, Carter confidently works one of them out to a thousandth of a Kelvin on the basis of a back of an envelope calculation.

Vaclav Klaus, Czech Republic President informs us on his website [5] that: *'Environmentalism ..is becoming a new dominant ideology, if not a religion. Its main weapon is raising the alarm and predicting the human life endangering climate change based on man-made global warming. The recent awarding of Nobel Prize to the main apostle of this hypothesis was the last straw because by this these ideas were elevated to the pedestal of "holy and sacred" uncriticisable truths.'*

Moreover, he continues : *'It became politically correct to caricature us, who dare to speak about it, (for) talking about things they do not understand and are not experts on. People like me ... do not try to better measure global temperature or to present alternative scenarios of the future global climate fluctuations....* They need not do it *because the climate change debate is basically not about science; it is about ideology.*

It is not about global temperature; it is about the concept of human society. It is not about scientific ecology; it is about environmentalism, which is a new anti-individualistic, pseudo-collectivistic ideology based on putting nature and environment and their supposed protection and preservation before and above freedom.'

Klaus thus conveniently exempts himself from the burden of proof, which in climate science as in particle physics is always on the claimant.

John Coleman, USA TV weatherman believes (*Daily Telegraph, 11/11/2007*) that: '*Some dastardly scientists with environmental and political motives manipulated long term scientific data to create an illusion of rapid global warming. ...Their friends in government steered huge research grants their way ... And I am telling you Global Warming is a non-event, a manufactured crisis. ..Natural cycles and drifts in climate are as much if not more responsible for climate changes under way.*'

Coleman, who has never presented any evidence in support of these claims, now (March 2008) suggests that the abovementioned AGW apostle be sued for fraud.

On the other hand, the carpet bombing of newspaper readers with stories sympathetic to the anthropogenic origin of recent global warming will hardly satisfy physicists either. Here are a few typical examples :

The *Daily Telegraph* (*'Britain's no 1 quality newspaper') 6 June 2007* reports under the headline *'Scientist Implicates Worms in Global Warming'* that *'Jim Frederickson, the research director at the Composting Association has called for data on worms and composting to be re-examined after a German study found that worms produce greenhouse gases 290 times more potent than carbon dioxide... '*

The Associated Press notes (*'Ecofashion Wins on Runway', 9 Jan 2008)* that *'In a workshop in (San Francisco's) Mission District, Ally Beran's team of fashion designers is sprawled out over buttons and spools of thread, hoping to stem global warming by stitching new outfits from thrift store finds....As with the movement for locally harvested food, ecofashion's devotees seek to lower their toll on the earth by buying clothes made of recycled materials and sustainably harvested, homegrown fibers.'*

The *Japan Times 18 Feb 2008* tells us that *'The world needs to find smarter ways to feed a rising population while cutting emissions of laughing gas, a widely forgotten greenhouse gas that is stoked by the use of fertilizer... Nitrous Oxide, best-known for its mirth-producing qualities, is 310 times more powerful at trapping heat that than* CO_2 *... The University of East Anglia said nitrous oxide ... is no laughing matter.'*

Nitrous Oxide is a greenhouse gas (GHG), and it is indeed necessary to understand its effect better, while directly or indirectly all waste adds CO_2 to the atmosphere unnecessarily. The point I wish to make here is not that the research described is not worth reporting nor that efforts to reduce waste are unimportant, but that the form of these stories trivializes and decontextualizes issues that need to be addressed seriously. It is not then surprising that the global public's gaze is glazing over in boredom and incomprehension.

The basic problem seems to me that the media and the scientific community have different ideas of what constitutes explanation. Media people aim to tell an interesting story; for them a good explanation is one that gives equal time to people who have controversial views (journalistic 'balance'), says what may happen, not what has been established and with what probability, and grabs the lay reader's attention with snappily-worded news products, remote from the original scientific report, and short on numbers and quantities. Scientists, consider that a good explanation is necessary as well as sufficient (Occam's razor [6]), introduces only testable, *i.e.* falsifiable hypotheses à la Karl Popper, and clearly distinguishes between number and quantity.

QUESTIONS TO BE ASKED

In navigating through the maze of specialized literature we can usefully start by seeking answers to a number of questions we might typically ask about work in our own field.

What Needs to be Explained ?

The first question to be asked is whether there is in fact a phenomenon in need of explanation. Figure 1 shows the measured air temperature at the surface of the globe from 1850-2006. Reliable but sparse meteorological data are available from the 1800s. They include indirect measures calibrated against modern ones but also high quality numerical data such as those of the Radcliffe Observatory of the University of Oxford, which has provided a continuous daily record since 1814.

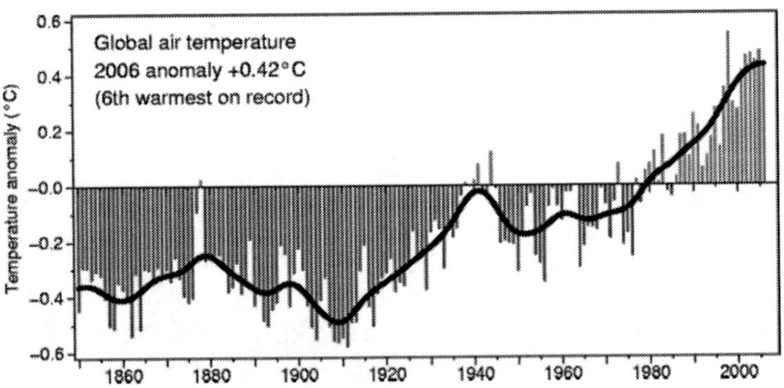

FIGURE 1. Measured global temperature over 160 years relative to 1980 value [7].

Over the last several decades, data on temperature etc. have been collected by the World Meteorological Organization (WMO) from some 17000 identical land weather stations and ten meteorological satellites, as well as from many research ships on data gathering cruises. The overall change in the 20th C was +0.85 K. Since this small number is a gigantic quantity seen against the yardstick of the 4.5 K *peak-to-*

peak swings between ice ages and interglacial periods an explanation is clearly necessary. If it is also to be sufficient however, it must, whether it includes anthropogenic effects or not, explain not just the overall change but the relatively constant periods before 1910 and from 1940-1970. Only then can we expect to make reasonable estimates of future temperatures and other climatic features.

What are the Radiative Characteristics of Earth and Sun ?

Absorption and emission of radiation can be fruitfully discussed in terms of the Black Body concept.

<u>The Earth:</u> Since we can see the earth from space, and black bodies absorb all the radiation falling on them by definition, Earth is clearly not a black body absorber. However its departure from black body absorption can be quantified by its albedo, or reflective power, $\alpha = \alpha(\lambda)$, λ being the wavelength of the incident solar radiation. Likewise $\varepsilon = \varepsilon(\lambda)$, the 'emissivity', can take care of Earth's departure from black body radiative properties, for which ε would be one for all wavelengths. There is almost no overlap in wavelength between the the incident solar radiation and the earth's IR radiation, emitted in the 5-50μ range. Both α and ε can for many purposes be replaced by an average over their respective wavelength intervals. A decrease or increase in either α or ε will warm the earth up or cool it down respectively.

<u>The Sun:</u> The sun on the other hand is quite well represented as a black body.

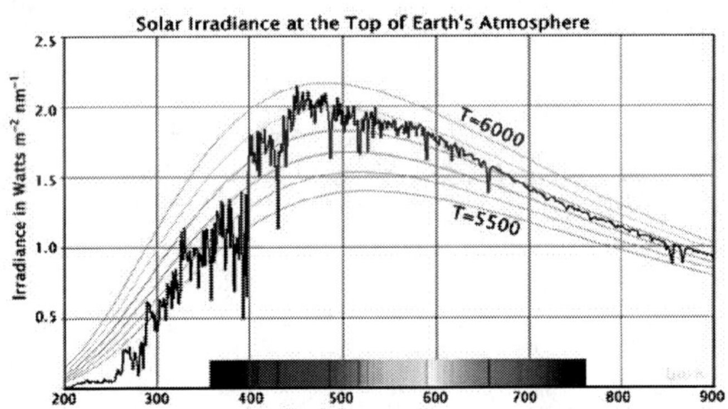

FIGURE 2. Solar emission spectrum, with superimposed black body curves.

The Total Solar Irradiance I_S is measured to be 1367 Wm^{-2} at the top of the atmosphere in a plane perpendicular to the direction of the sun, with a measured spectral distribution function that follows quite well the Planck distribution:

$$dI_S(\lambda,T_S)/d\lambda = (2hc^2/\lambda^5)/(\exp h\nu/kT_S - 1) \quad W\,m^{-3}sr^{-1}. \tag{1}$$

Putting I_S equal to the Planck integrated power density

$$I_S = \int_\lambda dI_S(\lambda,T_S)/d\lambda\ d\lambda = \sigma T_S^4 \quad Wm^{-2} \tag{2}$$

reduced to the earth's orbit, we get the value 5780 K for the sun's equivalent black body temperature.

What is the Expected mean Equilibrium Temperature of the Earth ?

We can now calculate the expected equilibrium temperature of the earth, T_E, by equating the power input from the sun to its own power output. This can only be in the form of radiation as convective transfer of gas into space is negligible, and there can be no heat loss by conduction. Allowing for the fact that the earth intercepts a disc of solar radiation but emits into a 4π solid angle we obtain :

$$I_S \pi r_E^2 (1-\alpha) = 4\pi r_E^2\ \varepsilon\sigma T_E^4 \text{ with } \sigma = 5.67\ 10^{-8}\ Wm^{-2}\ K^{-4},\ I_s = 1367\ Wm^{-2}. \tag{3}$$

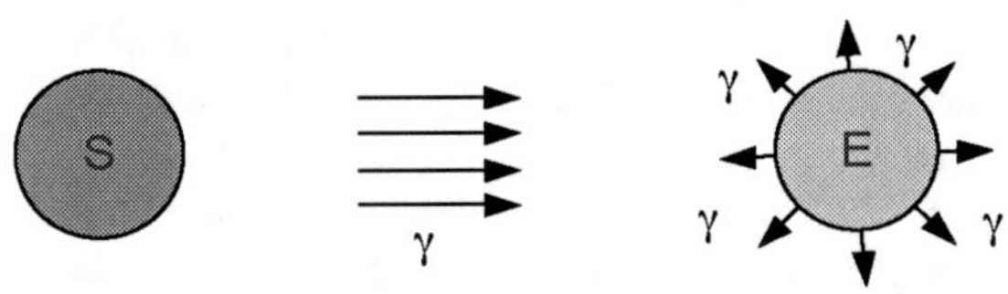

FIGURE 3. Determination of the expected equilibrium temperature of Earth, T_E.

Black body and other values are tabulated below. Here we neglect the contribution to the power balance of 75 mWm^{-2} from the earth's molten interior. The 278 K black body value is not so far from the measured one of 288 K, but this is purely fortuitous. The 256 K value is determined solely by the absorbed part of the solar power density - 239 Wm^{-2} given the measured value of α. Exactly this power must be re-radiated if the earth is to remain in equilibrium. The emissivity ε then represents the effect of the atmosphere, which owing to the presence of H_2O, CO_2 , CH_4, N_2O and traces of other GHGs, absorbs the 5-50μ IR radiation characteristic of these temperatures , heats up, and reradiates both up and down. The upward part goes toward space, the downward part toward the surface, thereby raising its temperature from 256 K to the observed value of 288 K.

TABLE 1. T_E for various values of albedo and emissivity.

α	ε	T_E
0 (BB absorption)	1 (BB emission)	278 K
0.3 (measured)	1 (BB emission)	256 K
0.3	0.613	288 K (measured)

Were there no GHGs the atmosphere would be perfectly transparent, the surface would remain at 256 K and we would simultaneously freeze and starve to death. The whole question of AGW is thus about the extent to which human activity is changing atmospheric GHG concentrations from preindustrial levels, and consequently slowly changing the equilibrium temperature of the planet. Since this depends only on I_S, α and ε , any effect, anthropogenic or otherwise, that changes T_E must do so by its effect on these parameters. So although the surface temperature will surely change if ε does, we must allow for the possibility that I_s and α may also change at the same time. Earthly effects cannot change I_S, but changes in α may occur with both anthropogenic and natural earthly origins.

Has I_S changed during the 20th C ?

Differentiation of equation 3 gives the sensitivity dI_S/dT_E of T_E to changes in I_s as $20\ Wm^{-2}K^{-1}$. Its value has been monitored very systematically by satellite probes since 1978. Before that date, measurements relied on more or less indirect measurements like counting sunspots, the number of which depends on I_S.

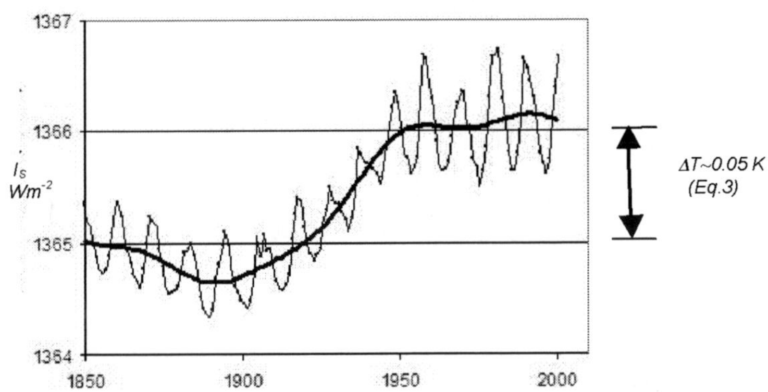

FIGURE 4. Variations in Total Solar Irradiance through the 20th C.

Values [8] for the 20th C are displayed in Figure 4 and Table 2. An important point to remember here is that the thermal inertia of the oceans is much too great for the planet to follow the 11 year cyclic variation closely. It seems safe to conclude that although variation in the TSI may well have been responsible for some of the observed 1910-1950 temperature rise, it has not contributed to the change observed in T_E since 1975. This must therefore come from a change in α, ε or both.

TABLE 2. 20th C variations in the Total Solar irradiance

Period	dI_S/I_S	dT_E
1900-1950	0.1%	0.07K
1950-2005	~ 0%	0 K
Over 11 yr solar cycle	~ 0.05%	± 0.02 K

Has α changed during the 20th C ?

From equation 3, the sensitivity $d\alpha/dT_E$ of the earth's equilibrium temperature to α is - 0.01/K. One of the most effective known agents for increasing α (reflecting solar radiation) and therefore cooling the earth is the creation of aerosols in the troposphere and stratosphere by ejecta from volcanic eruptions. The most famous of these was the eruption of the Indonesian volcano Tambora in 1815. This produced a worldwide drop in average temperature of ~ 1 K in 1816, which consequently became known as the year without a summer. More recently the Pinatubo eruption ejected 20 Mt of SO_2, and is estimated to have reduced T_E by ~ 0.3 K.

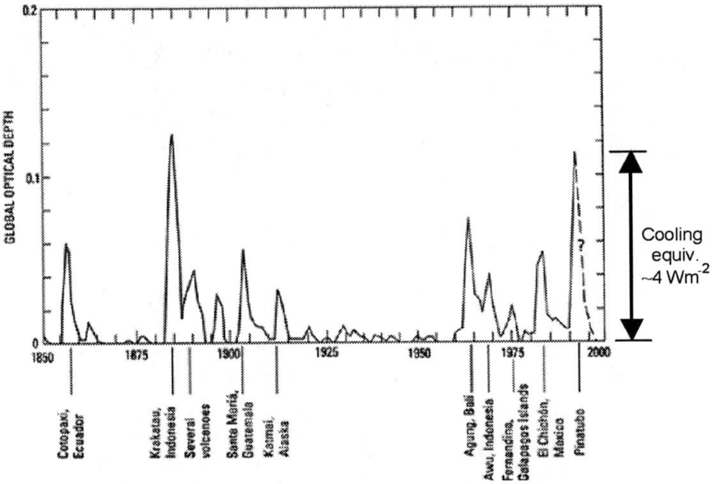

FIGURE 5. Change [9] in atmospheric optical depth (*ln* incident/transmitted radiation) from volcanic eruptions1850-2000, and estimated atmospheric cooling effect in Wm^{-2}.

Volcanic aerosols can persist in the stratosphere for a year or more, and their cooling effect can be estimated from the change they induce in the visible region atmospheric transmissivity. This has been measured by satellite since the late 1970s but was estimated indirectly before then. The important feature in Figure 5 is the 1935-1955 volcanically quiet period, when reduced cooling relative to the adjacent periods may have contributed to the relatively high temperatures then measured.

Other natural aerosols include forest and grassland fires and sea-spray. Starting around 1900, anthropogenic aerosols such as sulphates and soot from fossil fuels (and later in the 20th C airliner contrails) began to have an increasingly important effect (presently estimated at 25% - 50%) on α. All these effects together have contributed to the 20th C temperature record of Figure 1 in some complex way. Anthropogenic aerosols, unlike volcanic ones, normally have lifetimes of days or weeks, while fossil CO_2 stabilization is delayed by many years. This means that their cooling effect can partly explain the relative 1950-1970 constancy of T_E. When, however, around 1970 we started, for good reasons, to clean up our environmental act we began to remove this compensating effect on fossil CO_2 warming.

Have GHG Levels Changed, And If So, With What Effect ?

By burning fossil fuel we transfer carbon from under the ground to the atmosphere in the form of CO_2, where its effect will be to reduce ε and therefore increase T_E. As noted above the warming is not instantaneous. The ocean and other sinks currently reabsorb some 50% of the annual fossil CO_2, but these reabsorption processes continue only until a new atmospheric equilibrium has been reached at a higher temperature. Unless we put the carbon back underground therefore, some part of the temperature change will be permanent. Table 3 shows how the relative abundance of various GHGs has changed since the preindustrial era and in recent years.

TABLE 3. Global abundances key GH gases, WMO GHG monitor network.

	CO_2 ppm	CH_4 ppb	N_2O ppb
2006	381.2	1782	320.1
2006 wrt 1750	+36%	+155%	+19%
2005-2006	+2.0	-1	+0.8
Last 10 y mean	1.93	2.4	0.76

The effect on ε of a given quantity of GHG depends not only on how much was released, but on its interaction with other GHGs and atmospheric components. So although the lifetime in the atmosphere of a molecule of added GHG is well defined, its effect on ε is not simple. This is measured by a dimensionless parameter called the Global Warming Potential per molecule relative to CO_2 (Table 4) which depends on the time horizon over which the reabsorption process has continued. Doubters often quote such figures to imply that fossil CO_2 is unimportant with respect to CH_4 and N_2O.

TABLE 4. GWP (Global Warming Potential) of GHGs per molecule wrt CO_2

Molecule	Lifetime	GWP 20 y	GWP 100 y	GWP 500 y
CH_4	12y	72	25	7.6
N_2O	114y	310	298	153
SF_6	3200y	16300	22800	32600

Since there are in the atmosphere 200 CO_2 molecules for every CH_4 molecule and 1200 for every N_2O molecule, this is quite wrong. N_2O is the *wormogenic* or laughing gas referred to in media reports quoted above. Fortunately the world economy does not depend on SF_6, but O_3, present in minute quantities, has an important effect.

Why do GHGs (ppm-ppb) reduce ε while N_2/O_2 (78%-21%) do not ?

Although IR photons do not have enough energy to induce electronic transitions in elements or molecules, they can excite molecular rotational and/ or vibrational transitions. The important difference between N_2/O_2 and CO_2/CH_4 is then that the former are homopolar molecules, the symmetry of which ensures that there is no change of electric dipole moment when a photon excites a transition between two vibrational or rotational states. The GHGs on the other hand can bend or vibrate asymmetrically, and therefore present an oscillating dipole moment to the incoming photons. Consequently their photon absorption cross-section can approach a few percent of the geometric one at resonant wavelengths, and if these lie near the Planck distribution maximum of the earth's radiation, the GHG will absorb an appreciable fraction of the IR power even at low number densities.

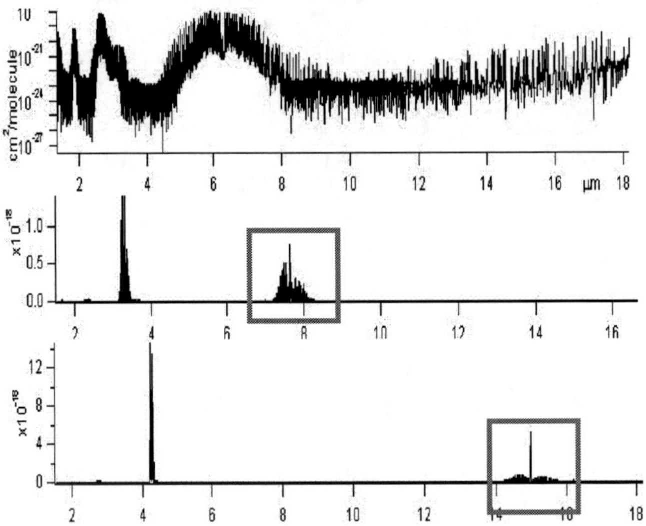

FIGURE 6. Absorption photon cross-sections [10] of H_2O (top) CH_4 (centre) and CO_2 (bottom) in cm^2, λ=1- 20μ.

Photon absorption cross-sections for H_2O, CH_4 and CO_2 in Figure 6 show bands of closely spaced lines. For CH_4 and CO_2, vibrational bands with cross-section of order 10^{-18} cm^2 lie near the emission peak for 250-300 K. For H_2O the spectrum extends over the whole 5-50μ range within which surface and atmospheric IR is emitted. The cross-sections are however some four orders of magnitude smaller over most of this range than those of CH_4 and CO_2, which explains why approximately 2 % of atmospheric water vapour plays a role in global warming not that much more important than the 200-400 ppm of CO_2.

At present GHG gas surface concentrations, ~385 ppm CO_2 and ~1782 ppb CH_4, the absorption length of photons with wavelengths in the corresponding vibrational bands averages to only a few metres for CO_2 and a few hundred metres for CH_4. This means that the earth can only cool off at these wavelengths from altitudes above which the GHG number density is low enough to allow IR photons to escape into space. Away from these λ values, the earth can radiate efficiently to space from ground level.

How do GHGs Work to Reduce ε ?

Figure 7 shows a typical earth emission spectrum seen by a low orbit satellite. The effect of the GHGs in the bands noted in Figure 6 and of O_3 is evident. It is instructive to interpret this as follows. As the satellite's monochromator scans over wavelength it 'sees' a series of black bodies at various temperatures. In regions where the black body temperature is 288 K, it is clearly seeing the surface of the planet, while in the absorption bands it sees colder black bodies. These black bodies must be at high altitudes since, as we saw above, radiation emitted from the surface and from within the atmosphere is all absorbed by GHGs at low altitudes. This is quite consistent with the known lapse rate, or rate of cooling of the atmosphere with altitude.

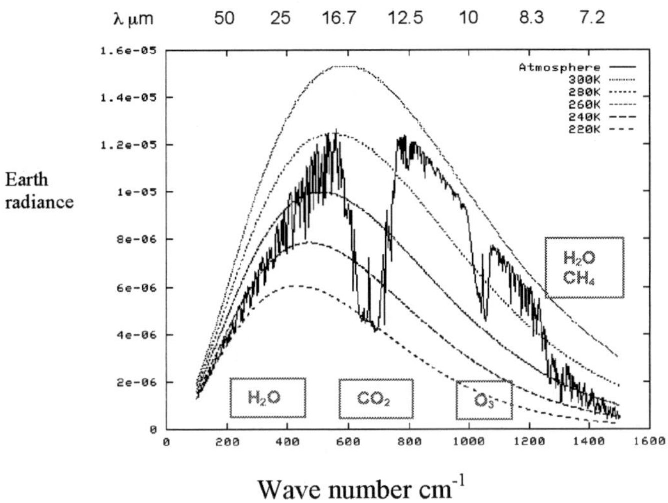

FIGURE 7. Typical earth emission spectrum seen from a low orbit satellite over a temperate latitude.

If now the CO_2 concentration increases because of increasing fossil fuel consumption, the 'bite' at 14-16 μm will get deeper, and radiation from it will originate at a higher, cooler level. The total area under the curve, averaged seasonally and globally, must nevertheless remain equal to that of a black body at 256 K, this being set by I_S and α alone. The global surface temperature T_E must then increase from 288 K in order to maintain this value. Simple back of the envelope calculations can be done following this line of argument, to give [11] $\Delta T_E = 1.2$ K for a doubling of the 1800 CO_2 level. This is surely an underestimate, since a number of positive feedback mechanisms occur to increase T_E. These include the contribution from the higher water vapour content of the atmosphere that will result from the initial 1.2 K change. More accurate estimates from the GCMs discussed below give a value closer to 2.5 K.

Low-orbit satellite data like that depicted in Figure 7 are for a given latitude and longitude, at the time the satellite passed over that location. An enormous increase in our understanding of global warming would result from a satellite located at the Lagrange point L1, which is about 1.6 million km from earth and constitutes a kind of gravitational trap for satellites. From L1, the earth's apparent size is the same as that of the moon from earth. Such a satellite would be able integrate the spectrum over the entire sunlit hemisphere of the planet, as well as the corresponding albedo spectrum and the total solar irradiance. The key parameters I_S , α and ε could thus be surveyed continuously over many years, thereby answering many remaining questions about climate change. This satellite (the Deep Space Climate Observatory, DSCOVR) does exist [12] but has remained in NASA mothballs since 2001 on cost-cutting grounds.

CLIMATIC DATA PAST AND PRESENT

Sources, Sinks and Reservoirs of CO_2

Most figures quoted in connection with CO_2 quantities in the literature refer to its carbon content, which makes it easier to compare the effect of different fuels such as

TABLE5. Magnitudes of sinks and sources of CO_2

Useful conversion factor :	1 *ppm* CO_2 ~ 2 *GtC*
Ocean reservoir :	35000 *GtC*
Ocean recycling	90 $GtCy^{-1}$
Plant reservoir :	600 *GtC*
Plant (biosphere) decay	~60 $GtCy^{-1}$
Plant growth	~62 $GtCy^{-1}$
Fossil fuel burning	8 $GtCy^{-1}$
Fossil fuel CO_2 sinks	~3.5 $GtCy^{-1}$
Added annually to atmosphere	~4 *GtC* ~ 2 *ppm*
Volcanic CO_2 production	0.08 $GtCy^{-1}$
Heat input from sun	239 Wm^{-2}
Heat flow from earth core	75 mWm^{-2}
Direct Heat flow from fossil fuel burning	25 mWm^{-2}

coal methane, LNG and oil. For the mass of CO_2 gas multiply by 3.7. The equivalence of 1 ppm of CO_2 to 2 GtC (7.4 $GtCO_2$) is worth remembering, as is the fact that some 10% of the plant reservoir is recycled each year.

The 2 $GtCy^{-1}$ excess of growth over decay meant that until the industrial era, we were slightly ahead of the game, as additional CO_2 from forest fires and other occasional sources were in the long run soaked up by biomass. We no longer have this insurance policy. Normal annual volcanic CO_2 represents only about 1% of the annual fossil fuel source, and can, like the heat flow from the earth's core and direct heat from fossil fuel burning, be safely neglected.

Ancient Climate and Modern Physics

Thermometric records of temperature go back a few hundred years. Although the length of a column of mercury is a convenient, easily calibrated measure of added heat, it is not the only one. Until fairly recently non-thermometric measures included only dendrochronology (tree-ring growth), historical records, pollen counts etc. These nevertheless informative archives of climatic history have now been powerfully supplemented by a large number of techniques from modern physics [13]. Thus air bubbles in deep-bore ice cores go back hundreds of thousands of years, and archive CO_2 concentration directly. Most physical processes fractionate to some extent, at a rate that usually depends on temperature and other environmental factors. Stable H, O, and C isotope fractions turn out to be especially useful. Thus the D/H ratio in ice core water measures air temperature at the time the water precipitated, with sensitivity almost $10^{-3}\,K^{-1}$ for Antarctic cores [14], while the C^{13}/C^{12} ratio in the atmospheric CO_2 has changed by 1.5 per mil since 1800. Plant carbon discriminates against C^{13} - its level is less than that of atmospheric carbon from which it comes. This is a smoking gun for the fossil origin of postindustrial CO_2 were one needed. Moreover C^{14} is absent from fossil carbon. Other archives include Be^{10} and C^{14} for cosmic ray history and the neutron count for more recent cosmic radiation intensity.

Changes of T_E and CO_2 over Millennia and Centuries

The results of these and other studies are summarized in Figure 8, which charts atmospheric CO_2 and T_E from 400 ky BP [14] and T_E from 1800 y BP [15]. The temperature change at the poles has been well established to be about twice the value

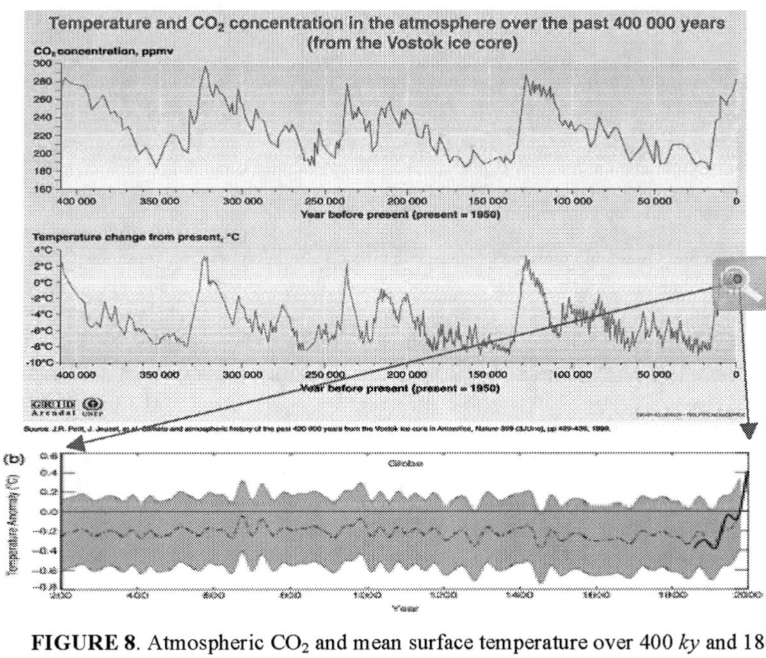

FIGURE 8. Atmospheric CO_2 and mean surface temperature over 400 *ky* and 1800 *y*.

over the whole globe, a feature also seen in recent temperature trends. Before our industrial era, the CO_2 level was never more than 300 *ppm* at any time in at least the last half million years. Doubters frequently point out that on careful inspection the

CO_2 record lags rather than leads the temperature record by several centuries. This is to be expected since entry into and exit from ice ages are now well explained by Milankovic cycles, in which secular variations of the earth's eccentricity, axial tilt and perihelion longitude produce small changes in insolation and consequently in the earth's average temperature. This original small adjustment of the global thermostat evaporates additional water vapour and CO_2 from the ocean, resulting in a positive feedback loop which amplifies the change in temperature with delay of several hundred years set by the thermal inertia of the ocean.

Doubters also claim that the last 2000 years included periods when the climate was much warmer than it is at present, and that recent changes cannot therefore be due to fossil fuel burning. Recent studies [16] have shown that the so-called 'Medieval Warm Period' was a regional and not a global phenomemon, restricted to northern Europe, and absent even from the northern hemispheric average. The same is true for the so-called 'Little Ice Age'. Neither feature is evident in Figure 8 (lower panel).

FROM QUALITATIVE TO QUANTITATIVE CALCULATIONS

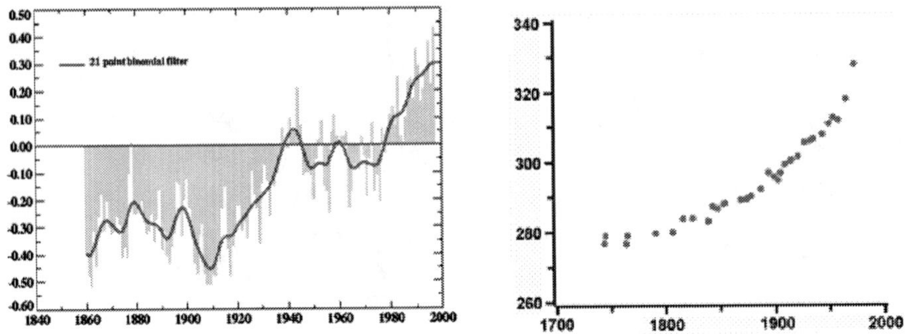

Figure 9. L: ΔT_E wrt 1961-1990 mean; R: Ice core CO_2 *ppmv*

The arguments presented above to explain the details of Figure 1, reproduced in Figure 9, have the most part been qualitative. We saw that a small increase in I_s from 1910-1945, coupled with reduced cooling from volcanic aerosols 1935-1955, probably accounted for some of the 1910-1945 increase in T_E. From 1960 onwards fossil fuel CO_2 increased sharply, with cooling from the associated aerosols partly compensating the GHG effect of CO_2. Finally, since 1970 the rate of fossil fuel consumption has risen even more dramatically while at the same time we started to clean up our environmental act, removing aerosol pollutants and so reducing this compensation. The correlation between CO_2 and T_E seen in Figure 9 suggests an underlying cause and effect relationship. However these numbers represent incommensurate quantities, and their correlation means nothing until they can be linked by calculation. This is the purpose of the so-called General Circulation Models.

General Circulation Models (GCMs) of the Atmosphere

Equation 3 can be thought of as a very simple model of the atmosphere. At the other end of the scale, there are about ten independent GCMs now in use worldwide, of which the UK Met Office Hadley Centre version [17] is recognized as one of the most complete and reliable. GCMs are the counterparts of, and are at least a complex as, the Standard Model in particle physics. In them [11] the atmosphere and ocean are represented by a 3-dimensional lattice of cells, within which measured values for pressure, temperature, velocity, humidity (for the atmosphere), salinity (for the ocean) etc., can be specified as initial conditions. Dynamical processes are represented in each cell by the Navier-Stokes equations (Newton's second law for fluids) :

$$D\boldsymbol{u}/Dt = -1/\rho\ \nabla p - g\boldsymbol{k} + \eta/\rho\ \nabla^2 \boldsymbol{u}\ , \tag{4}$$

where $D/Dt = \partial/\partial t + \mathbf{u}\cdot\nabla$, ρ is the cell density, η its viscosity, $\mathbf{k}$ its unit vertical and $\mathbf{u}$ and p are its velocity and pressure respectively. Physical processes, such as the absorption, emission and reflection of solar radiation (considered above), evaporation and condensation, also drive the global heat engine and are included with the dynamical ones as parametrizations. The equations are then integrated numerically under the constraints supplied by hydrostatic and continuity equations, energy conservation, and the equations of state. While the Hadley GCM currently employs about a million grid points, more powerful computers and more complex programmes allow an ever finer lattice. Better understanding of atmospheric and oceanic processes is also being incorporated into GCMs, so that smaller uncertainties and more accurate predictions can be expected in future.

Some Results from GCM Calculations

As in particle physics, GCMs are tested in terms of their success in postdicting known phenomena, in this case the climatic history of the 20th C. They are then used to predict the climatic development of the 21st C and form the basis of the reports of the IPCC. Figure 10 (top) shows the 20th century history as measured and as simulated by GCMs with only natural climatic influences, or 'forcings' included. It is clear that natural factors, as described above, while they may be necessary, do not provide a sufficient explanation of the 20th C temperature record from 1970, the period during which fossil fuel use increased dramatically throughout the world. In the lower panel, the effect of fossil fuel burning is added to these natural factors, resulting in an explanation that is now both necessary and sufficient. Moreover the fact that adding the anthropogenic term also produces a better fit *before* the crucial post 1970 period is in itself a significant indication of the validity of the calculations.

The Fourth Assessment Report (FAR) of the IPCC makes [2] extensive predictions of the likely consequences of continued Global Warming, including sea-level rise, precipitation and regional forecasts. I reproduce here only the development of the global mean surface temperature T_E. While the past is well defined, the future depends on the assumptions made about how the world reacts to the present situation, if it reacts at all, and how the population evolves. Figure 11 shows what can be expected

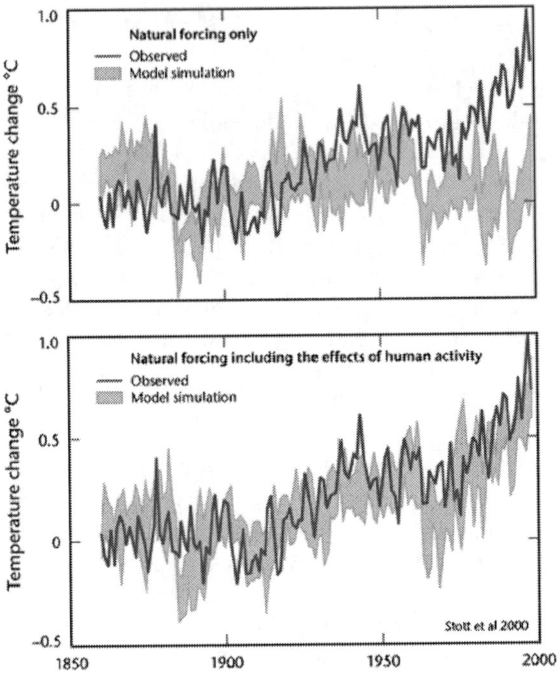

Figure 10. Temperature record [18] of the 20th C, measured (continuous curve) and as calculated from the Hadlley GCM. The error band is for one standard deviation.

for some of these 'emissions scenarios'. Perhaps the most interesting ones to physicists are those which define limits. The lowest curve represents one of these extremes, the so-called *Constant Composition Commitment*, under which CO_2

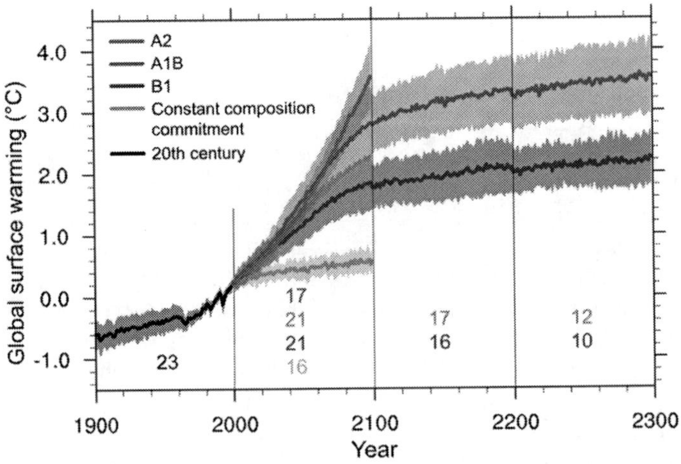

Figure 11. ΔT_E predictions w.r.t. 1980-1989 [2]. Numerals show the number of GCM runs included.

concentrations are stabilized at the 2000 level. In this case we are already committed to a rate of increase of T_E of about 0.1 K per decade until 2020, due mainly, as mentioned above, to the thermal inertia of the oceans. It is of course a completely unattainable goal. The uppermost curve (A2) corresponds to CO_2 stabilization at a level 1250 ppmv, a little over four times the preindustrial level. This is qualitatively characterized by the IPCC as *self-reliance and preservation of local identities*. From a physicist's point of view, it is more interesting to see it as one in which the post-industrial increase in T_E is equal to the one that occurred between the last ice age and the present interglacial period.

OBJECTIONS

I began by quoting pronouncements by various objectors to the claim that the recent increase in global temperature is due to the burning of fossil fuels. Although none of them raised serious scientific arguments, it would be unusual if in a contentious field like science, no-one was able to find better arguments than these.

Freeman Dyson

Freeman Dyson is one of the few AGW critics whose arguments might carry considerable weight among particle physicists. These have most recently been expressed in the January 2008 issue of Physics World [19] :

(All) the fuss about global warming is grossly exaggerated......There is no doubt that parts of the world are getting warmer, but the warming is not global. I am not saying that the warming does not cause problems. Obviously it does .. (and we) should be trying to understand it better... One of the main causes of warming **is** *the increase of CO_2 in the atmosphere resulting from our burning of fossil fuels .....*

The average global temperature being perfectly well defined statistically , it is difficult to understand Dyson's objection about the warming not being global, as no-one claims anyway that every point on the earth has warmed up equally. More importantly, he parts company with many objectors by agreeing that warming does exist, is at least partly due to fossil fuel burning and requires action. He just claims that it is no big deal when put in perspective:

.... BUT consider the half of the land area of the earth...that supports vegetation...The increase in thickness of the biomass that would result if all the carbon that we are emitting by burning fossil fuels were absorbed and converted into biomass (over this area) is 0.01 inches per year...

Dyson's solution is as follows:

I conclude from this calculation that the problem of carbon dioxide in the atmosphere is a problem of land management, not a problem of meteorology ..We do not know whether intelligent land-management could increase the growth of the topsoil reservoir by four billion tons of carbon per year, the amount needed to stop the increase of carbon dioxide in the atmosphere. All that we can say for sure is that this is a theoretical possibility and ought to be seriously explored.

Dyson suggests that this be achieved by what is called no-till farming. Only 0.01 inches, 0.25 mm, one has to remark. This figure is equivalent to increasing the biomass of entire planet by something of the order of 1% *each year*, the implication being that we do not then need to reduce consumption or switch to renewables. In this case the fraction will increase each year in proportion to the annual world consumption. This would be a gigantic undertaking, at least as challenging as getting people to reduce fossil fuel consumption or switch to renewables. An appropriate distinction between number and quantity seems to be missing here.

H. Svensmark

H. Svensmark (Danish Space Research Institute, Copenhagen) suggests that although I_S , the sun's power, has been constant since the 1950s, indirect solar effects may occur to increase T_E, so that we have either no need of AGW to explain its increase, or that its contribution is smaller. The effect suggested involves the effect of cosmic rays on the atmosphere: a) A long term rise (>>11y) in the solar magnetic field over the 20th C would deflect more galactic cosmic rays from the earth, b) Cosmic rays can produce Cloud Condensation Nuclei (CCN), so that fewer cosmic rays means fewer clouds, c) Fewer clouds result in net surface warming.

The reason water vapour is a powerful net GHG, is because at high altitudes its warming effect is greater than the cooling it exerts by reflecting solar radiation and increasing the albedo. Fewer clouds should therefore normally result in less warming. At low altitudes this situation can be reversed, and this is claimed for the hypothesized clouds. The links in the above chain of reasoning are not well-sustained by quantitative data, but the CERN SPSC has approved [20] an experiment which uses a particle beam as an artificial source of cosmic rays to investigate CCN formation.

However, the very existence of the necessary downward trend in the cosmic ray flux has also been questioned. We remarked already that the ocean's thermal inertia ensures that Earth does not respond to the 11 year solar cycle. Any longer term trend is difficult to discern over the few cycles included in the period (1970-2007) in question. Lockwood and Fröhlich (2007) have studied neutron counts of cosmic radiation over the 20th C and find [21] that if any such a trend, if indeed it exists, was in the wrong direction for most of this time. If Dyson appears not to distinguish number and quantity adequately, Svensmark's idea seems a somewhat contrived example of the multiplication of entities beyond necessity.

SUMMARY

As working scientists, we are in a position to provide ourselves with a better explanation of the scientific basis of the contentious topic of Anthropogenic Global warming than is available in the media. A review of the field from a particle physicist's standpoint reveals several important points. First the claim that fossil fuel burning is responsible for a large part of the Global Warming of the last 50 years is not just handwaving. In particular, the Total Solar Irradiance has not changed during the 50 y that fossil fuel use rapidly increased, apart from the normal 11 year fluctuation. Other climatic influences, whether natural or anthropogenic must act via their effect on the earth's albedo and emissivity. Known natural factors are all taken into account into GCM calculations, and turn out to be necessary but insufficient to explain 20th C warming; adding the anthropogenic contribution provides the missing sufficiency.

Swings in Earth's temperature over the last 100 ky happened long before dawn of civilization, which of course depends absolutely on stable climatic conditions and indeed owes its existence to them. Previous swings associated with entry into and exit from ice ages are now well explained by Milankovic cycles, and occurred at times when no, or few people were around to experience the consequences.

We should listen to the few doubters of AGW who have something of importance to say, but ask for hard evidence and apply the same critical stance we use in our own work. Finally we owe it to those whose *'gaze has glazed over with boredom and incomprehension'* to point out not just the common misconceptions promoted by so-called AGW skeptics - confusing weather and climate, number and quantity, regional and global climate, necessity and sufficiency - but also that media coverage sympathetic to the IPCCs conclusions often presents only a caricature of the serious research behind them.

REFERENCES

1. Paul Roberts, *The End of Oil*, London: Bloomsbury, 2005, pp. 122-123.
2. IPCC, *Climate Change 2007: The Physical Science Basis, Summary for Policymakers,* www.ipcc.ch
3. See Crichton's official website: www.michaelcrichton.com/index.html
4 www.rotaryclubofmelbourne.org.au; Text: http://members.iinet.net.au/~glrmc/new_page_9241.htm
5. *The Other Side of Global Warming Alarmism*, 7.11.2007, Vaclav Klaus: www.klaus.cz
6. William of Occam (14th C) : *Entia non sunt multiplicanda praeter necessitatem – Entities should not be multiplied beyond necessity.*
7. Data from Climatic Research Institute, University of East Anglia, Norwich.
8. J. Lean, J. Beer and R. Bradley, *"Reconstruction of Solar Irradiance since 1610: Implications for Climate Change "*, Geophys. Res. Letts., 22, 3195-3198, 1995; updated in ref 17 to 2003.
9. Sato et al. ***Stratospheric Aerosol Optical Thickness***: J. Geophys. Res., 98, pp. 22987-22994
10. These and many other spectra are available at http://vpl.astro.washington.edu/spectra/
11. J. Houghton, *"Global Warming"*, Rep. Prog. Phys. 68 , 2005, 1343-1403
12. Scripps Institution of Oceanography, UCSD: cloud.ucsd.edu/dscovr/mission_summary.html
13. P. Ghosh and W..A. Brand, *"Global Isotope Ratio Mass Spectrometry in Global Climate Change Research"*, Int. J. Mass Spect. 228, 2003, pp 1-33
14. J.R. Petit et al., *"Climate and Atmospheric history of the Last 420000 years from the Vostok Ice Core in Antarctica"*, Nature 399, 1999, pp. 429-436.
15. M.E. Mann and P.D. Jones, *"Global Surface temperature over the past two Millenia"*,Geophys. Res. Lett.30, 1820.
16. R.S. Bradley, M.K. Hughes and H. F. Diaz, , *"Climate in Medieval Time"*, *Science* 3022, 2003, No. 5644, pp. 404-405.
17. Hadley Centre December 2005: *"Climate Change and the Greenhouse Effect"*, www.metoffice.gov.uk/research/hadleycentre/,
18. P.A. Stott et al., *"External control of 20th Century Temperature Variations by Natural and Anthropogenic Forcings"*, Science, 15, pp. 2133-2177, 2000.

19. F. Dyson, *Many Colored Glass: Reflections on the place of Life in the Universe*, Barbour Lectures, University of Virginia Press 2007, and *Physics World*, Jan 2008,
20. CERN Report number SPSC-M-721 ; CERN-SPSC-2004-023
21. M. Lockwood and C. Fröhlich, *"Recent Oppositely Directed Trends in Solar Climate Forcings and the Global Mean Surface Air Temperature"*, Proc. R. Soc. A, doi:10.1098/rspa.2007.1880

First Attempts at Antihydrogen Trapping in ALPHA

G.B. Andresen*, W. Bertsche†, P.D. Bowe*, C.C. Bray**, E. Butler†, C.L. Cesar‡, S. Chapman**, M. Charlton†, J. Fajans**, M.C. Fujiwara§, R. Funakoshi¶, D.R. Gill§, J.S. Hangst*, W.N. Hardy‖, R.S. Hayano¶, M.E. Hayden††, A.J. Humphries†, R. Hydomako‡‡, M.J. Jenkins†, L.V. Jørgensen†, L. Kurchaninov§, R. Lambo‡, N. Madsen†, P. Nolan§§, K. Olchanski§, A. Olin§, R.D. Page§§, A. Povilus**, P. Pusa§§, F. Robicheaux¶¶, E. Sarid***, S. Seif El Nasr‖, D.M. Silveira‡, J.W. Storey§, R.I. Thompson‡‡, D.P. van der Werf†, J.S. Wurtele** and Y. Yamazaki†††

*Department of Physics and Astronomy, Aarhus University, DK-8000 Aarhus C, Denmark
†Department of Physics, Swansea University, Swansea SA2 8PP, United Kingdom
**Department of Physics, University of California at Berkeley, Berkeley, CA 94720-7300, USA
‡Instituto de Física, Universidade Federal do Rio de Janeiro, Rio de Janeiro 21941-972, Brazil
§TRIUMF, 4004 Wesbrook Mall, Vancouver BC, Canada V6T 2A3
¶Department of Physics, University of Tokyo, Tokyo 113-0033, Japan
‖Department of Physics and Astronomy, University of British Columbia, Vancouver BC, Canada V6T 1Z4
††Department of Physics, Simon Fraser University, Burnaby BC, Canada V5A 1S6
‡‡Department of Physics and Astronomy, University of Calgary, Calgary AB, Canada T2N 1N4
§§Department of Physics, University of Liverpool, Liverpool L69 7ZE, United Kingdom
¶¶Department of Physics, Auburn University, Auburn, AL 36849-5311, USA
***Department of Physics, NRCN-Nuclear Research Center Negev, Beer Sheva, IL-84190, Israel
†††Atomic Physics Laboratory, RIKEN, Saitama 351-0198, Japan

Abstract. The ALPHA apparatus is designed to produce and trap antihydrogen atoms. The device comprises a multifunction Penning trap and a superconducting, neutral atom trap having a minimum-B configuration. The atom trap features an octupole magnet for transverse confinement and solenoidal mirror coils for longitudinal confinement. The magnetic trap employs a fast shutdown system to maximize the probability of detecting the annihilation of released antihydrogen. In this article we describe the first attempts to observe antihydrogen trapping.

Keywords: antihydrogen, antiprotons
PACS: 36.10.?k, 34.80.Lx, 52.20.Hv

INTRODUCTION

Antihydrogen is of fundamental interest for use in precision tests of CPT symmetry and in investigations of antimatter gravitation. Following the first synthesis of antihydrogen from trapped antimatter plasmas in 2002 [1, 2], the recent focus of the two operational antihydrogen experiments at CERN's Antiproton Decelerator (AD) [3] has been on trapping of the produced anti-atoms. Early experiments in the second-generation antihydrogen devices (ALPHA and ATRAP2) addressed the longevity of trapped plasmas

CP1037, *Cold Antimatter Plasmas and Application to Fundamental Physics*, edited by Y. Kanai and Y. Yamazaki in combined Penning/neutral atom trapping fields [4, 5]. In this article we demonstrate the complete sequence of manipulations comprising an antihydrogen trapping experiment in the ALPHA device, and we discuss the results of the first effort to trap neutral antimatter.

THE ALPHA APPARATUS

The ALPHA apparatus is schematically depicted in Figure 1. Antiprotons from the AD are dynamically captured and electron-cooled in a Penning trap (3 T solenoidal field, 5 keV well depth). Typically, 3×10^7 extracted antiprotons from the AD result in 40000 captured and cooled antiprotons in the catching trap. The catching trap includes a "rotating wall" electric field system which can be used either to expand the electron cloud radius for maximizing antiproton capture efficiency or to reduce the antiproton cloud radius to maximize the probability of antihydrogen formation and trapping [6, 7].

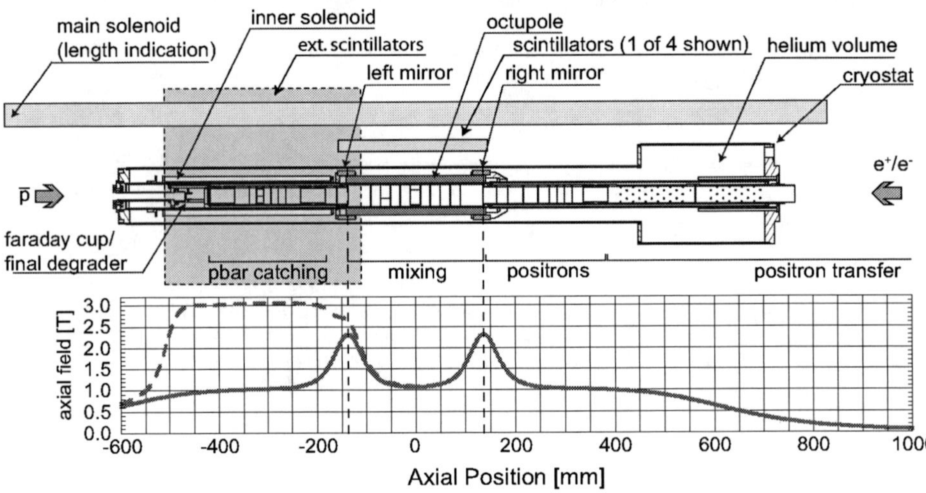

FIGURE 1. Schematic diagram of the ALPHA apparatus. The graph shows the on-axis longitudinal magnetic field due to the solenoids and mirror coils. The blue (red) curve is the field with (without) the inner solenoid energized.

Positrons from a Surko-type accumulator [8] are loaded into the mixing region of the device, trapped dynamically, and allowed to cool by cyclotron radiation. The solenoidal field in the mixing region is only 1 T, in order that the trap depth of the neutral atom trap surrounding this region is as large as reasonably possible [9, 10]. ALPHA generates the necessary two-region solenoid field using an outer 1 T, warm bore magnet, and an inner 2 T winding on the Penning trap vacuum chamber. All Penning traps in the ALPHA device are cooled through contact with the liquid helium system for the superconducting magnet coils comprising the neutral atom trap and inner solenoid (Figure 1). Typical positron plasmas for the mixing experiments described here contained 30-50 million positrons.

The mixing region is surrounded by a multipole trap for confining neutral anti-atoms. The transverse confinement is by an eight-layer octupole winding, and the longitudinal confinement is by solenoidal mirror coils generating a peak field of 2 T [9]. The maximum achievable neutral trap depth, in temperature units, is about 0.7 K for ground state antihydrogen. Slightly smaller depths were employed here.

The ALPHA design features the capabilities for fast ramp-up of the neutral atom trapping coils and for very fast shutdown of the entire system. The fast ramp-up is desirable so that the antiprotons and positrons can be accumulated and manipulated into proximity without the perturbing, transverse fields present in the experimental volume. The fast shutdown feature is incorporated to optimize the detection of trapped antihydrogen atoms by quickly turning off the trapping fields and looking for antiproton annihilations in the detectors surrounding the trapping region (Figure 1). Since trapping an anti-atom in the first place is a potentially rare event, it is desirable to maximize the signal to noise ratio of the annihilation detector by dumping the trap in a narrow time span. The ALPHA magnets feature a fast energy extraction system that employs an isolated gate bipolar transistor (IGBT) current switch [11] to dump the excitation current into a resistive load. This same circuit also functions as a part of the quench protection system for the magnets.

Figure 2 shows the voltage across the octupole during a fast shutdown from 950 A. (Note that the maximum design current is 1100 A.) The e-folding time is 9.6 ms. The resistance in the circuit is about 330 mΩ, and the magnet inductance is 3.2 mH. We have no direct measurement of the field decay in the superconducting magnet. The mirror coils have a slightly faster e-folding time of 8.3 ms.

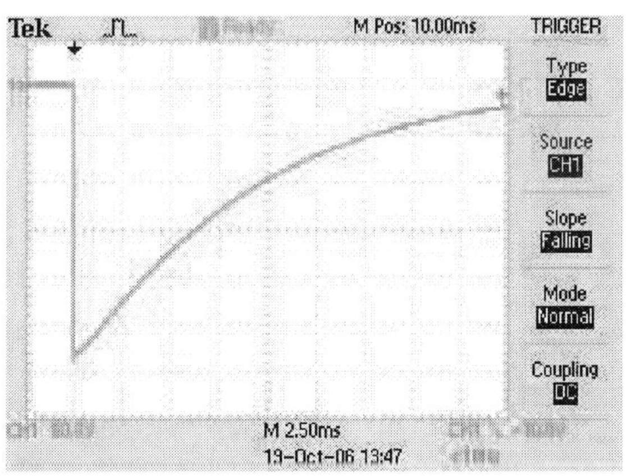

FIGURE 2. Voltage induced across the octupole magnet during a fast shutdown event. See text for details.

For the measurements described in the following section, the apparatus was equipped with scintillation detectors read out by avalanche photodiodes (APD) [12]. The detectors were placed inside the outer solenoid and adjacent to the mixing trap (Figure 1). The geometry of the detector assembly is shown in Figure 3. Four of the seven positions contained scintillation detectors, and two contained three-layer, position sensitive, silicon detectors. An event was registered if two or more of the scintillation detectors fired in coincidence (100 ns window). The solid angle subtended by the scintillation detectors was about 50% of 4π, with respect to the center of the mixing region. The background rate for the scintillation detectors in this coincidence mode is about 2 Hz.

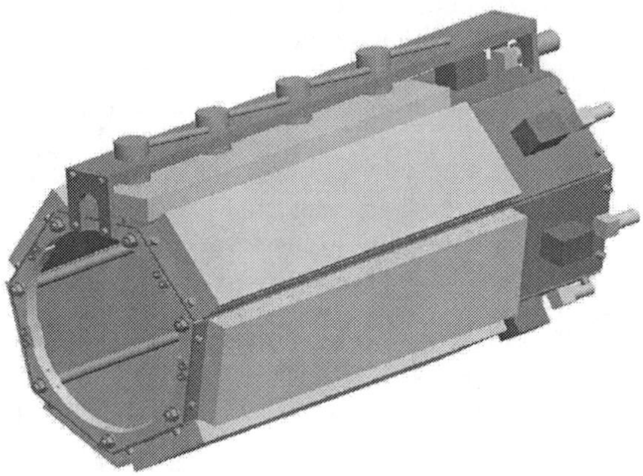

FIGURE 3. Mounting geometry for the scintillators surrounding the mixing region. For the measurements reported here, three of the seven modules were replaced with two layers of position sensitive silicon strip detector. The small detectors (orange) are CsI crystals for looking at positron annihilations. A similar, seven-segment scintillation detector surrounds the degrader foil.

THE TRAPPING EXPERIMENT

The trapping experiment begins with the transverse octupole turned off, but the mirror coils energized. Antiprotons from the AD are accumulated for up to eight AD cycles. The AD cycle time was about 70 s. The resulting antiproton stack of roughly 3×10^5 particles was transferred to the mixing region after dynamic removal of the cooling electrons. Positrons accumulated during the antiproton accumulation were transfered to a potential well adjacent to the antiprotons. With both species of particle in the mixing region, the octupole field was ramped up from 0 to 900 A in 45 s. The antiprotons are then injected into the positron plasma using the nested Penning trap [13] configuration as employed for the first antihydrogen production in ATHENA [1] .

The antiprotons are cooled by collisions with the positron cloud, and this mixing/cooling process is monitored by recording scintillation events in the scintillators surrounding the mixing region. Two other sets of scintillators, one external to the main solenoid and one adjacent to the degrader, are also sensitive to antiproton annihilations, but with much smaller solid angle and efficiency. The time development of the scintillation signal during mixing in the combined trap is shown in Figure 4. For comparison, the scintillation signal for the same procedure, but without positrons, is also shown (in red). The scintillation trigger rate evolves in qualitatively the same fashion as is observed in previous antihydrogen production cycles in ATHENA [14] and in ALPHA [10] without the presence of the neutral atom trapping fields. There is an initial rise in annihilations as the antiprotons cool and interact with the positrons, followed by a slow decay of the annihilation signal. These annihilation events are presumably related to neutral antihydrogen production, with the neutrals annihilating on the Penning trap wall. Another possibility is that antihydrogen is field-ionized at a radius at which antiprotons follow field lines to the trap wall and annihilate (see reference [7]). Without positrons, there is no evidence of antiproton annihilation, except for the immediate loss associated with the injection manipulations.

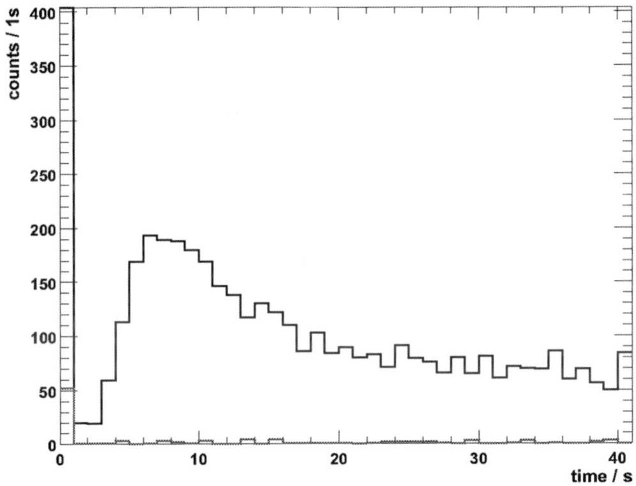

FIGURE 4. Scintillation triggers versus time after the start of mixing for normal mixing (black) and for the same sequence but without positrons present (red). The octupole was at 900 A, corresponding to a trap depth of about 0.5 K, for these measurements.

Further evidence for antiproton-positron interaction is obtained by releasing the remaining antiprotons from the trap after the end of the mixing cycle. This is done by lowering the confining potentials in a controlled way, such that particles with higher energies escape the trap first. The two side wells of the nested trap ("left" and "right") can be emptied separately, see Figure 5. Antiprotons which lie above the energy of the central positron well emerge in the left dump, which occurs first. Antiprotons that are cooled by the positrons but have not formed antihydrogen emerge later in time as the electrodes are ramped down. The results of these energy dumps are shown in Figure 6.

With no positrons present, antiprotons remain at the energy at which they are injected into the nested well, and they are all released early in the left dump. When positrons are present, the antiprotons can cool to the level of the positron cloud, and interactions between the two can yield antihydrogen. Thus with positrons present, we also observe an annihilation signal in the right dump, and the left dump antiprotons are shifted to lower energy. No antiprotons were present in the right well after mixing without positrons. Note that it is also possible for antihydrogen atoms to field-ionize at the edges of the nested well, leading to additional accumulation of antiprotons in each side well.

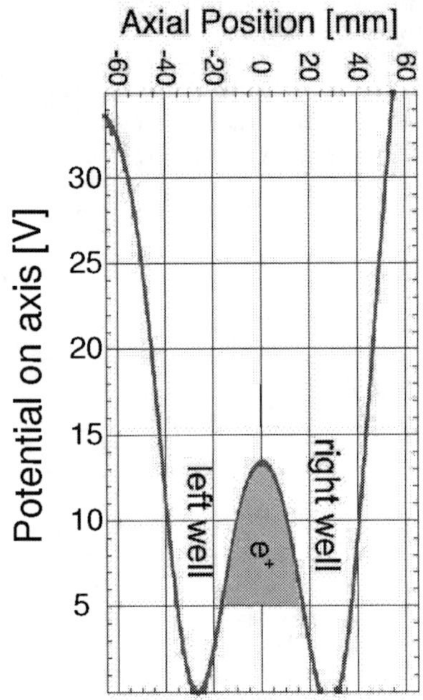

FIGURE 5. Potential on the axis of the nested well used for positron-antiproton mixing. The antiprotons are injected into the nested well with a potential of about 23 V on this diagram. The blue shaded region represents the approximate spacecharge potential of the positron cloud.

The data in Figures 4 and 6 are consistent with antihydrogen production in the combined trap. At the end of each mixing cycle, the mixing region was emptied of charged particles using applied electric fields. To look for evidence for trapped antihydrogen, the neutral trap was then rapidly de-energized, and the signal from the trap scintillators was scrutinized for evidence of annihilations from released antihydrogen atoms. The integrated number of particles mixed using this sequence during the 2007 run was about 2.3 million antiprotons and 5.6×10^9 positrons.

With this data sample, we observed no evidence for annihilations above the background level in the 10 ms time window immediately following the trap turn-off. There was thus no indication that antihydrogen atoms had been trapped and released. As we do not yet know the absolute antihydrogen production rate, we are unable to put an upper

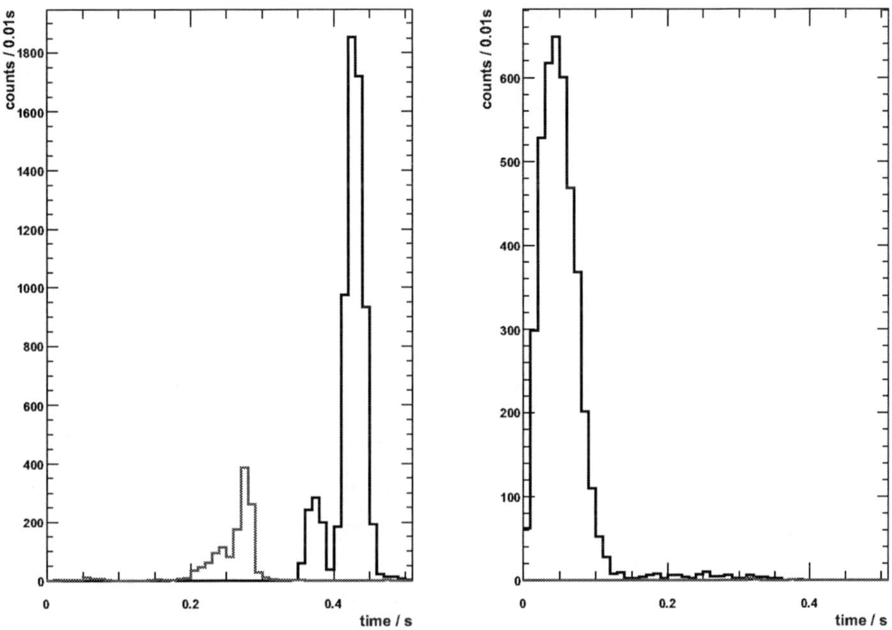

FIGURE 6. Antiproton annihilation signals observed while ramping down the left potential wall (left panel) followed by the right potential wall (right panel), after the mixing cycle was completed. The black (red) curves are with (without) positrons. The voltage ramp-down rate is about 70 V/s. This signal is for a single mixing cycle with eight AD shots.

limit on trapping probability.

In addition to the obvious statistical limitations of the data sample, there are several possible explanations for the lack of a trapping signal. It is of course possible that no antihydrogen is being produced in the combined trap, and that the annihilation signal observed in Figure 4 is just due to loss of antiprotons from the trap. This is highly unlikely, due to the corroborating evidence of antiproton cooling, but we have not yet systematically ruled out this scenario. A second possibility is that the antihydrogen produced is too energetic to be trapped in the 0.5 K well depth employed here. This is quite likely, as it was shown in both ATHENA [16] and ATRAP [17] that antihydrogen formation in this mixing scheme takes place before the antiprotons come into equilibrium with the positron cloud, which is presumably in equilibrium with its cryogenic surroundings. There is in fact no evidence to date for cryogenic antihydrogen production by any method in any device, and there are no direct measurements that show that the positrons in fact cool to the 4.2 K temperature of the apparatus. A third possibility is that some antihydrogen is trapped initially in an excited internal state having a large magnetic moment, but escapes the trap during decay to more tightly bound levels. We currently have no way of testing this hypothesis, although some of us have addressed this question theoretically [15]. Other loss mechanisms leading to trapping lifetimes much shorter than the time scale of the experimental manipulations used here can also not be ruled out at this early stage.

SUMMARY AND OUTLOOK

We described the first attempt at trapping antihydrogen in a combined Penning/neutral atom trap. No evidence for antihydrogen trapping has been observed yet, but all of the necessary hardware operations for realistic trapping attempts have been successfully demonstrated. A likely explanation for our result is that the three-body production in the "normal" mixing technique yields antihydrogen that is too energetic to be trapped, but much work remains to be done before abandoning the high-rate antihydrogen production characteristic of this technique. The 2008 experimental program in ALPHA will focus on careful manipulations and mixing techniques designed to minimize the temperature of the produced antihydrogen atoms. In these endeavours, we expect to benefit both from the novel diagnostic techniques reported elsewhere in these proceedings [7] and from the ALPHA imaging silicon detector [12], expected to be installed in 2008.

ACKNOWLEDGMENTS

This work was supported by CNPq, FINEP (Brazil), ISF (Israel), MEXT (Japan), FNU (Denmark), NSERC, NRC/TRIUMF (Canada), DOE (USA), EPSRC and the Leverhulme Trust (UK) and HELEN/ALFA-EC.

REFERENCES

1. M. Amoretti *et al.*, Nature **419**, 456 (2002).
2. G. Gabrielse *et al.*, Phys. Rev. Lett. **89**, 213401 (2002).
3. S. Maury, Hyperfine Int. **109**, 43 (1997).
4. G. Andresen *et al.*, Phys.Rev.Lett. **98**, 023402 (2007).
5. G. Gabrielse *et al.*, Phys. Rev.Lett. **98**, 113002 (2007).
6. G. Andresen *et al.*, to be published in Phys.Rev.Lett. (2008).
7. G. Andresen *et al.*, *Antiproton compression and radial measurements*, these Proceedings.
8. T. J. Murphy and C. M. Surko, Phys. Rev. A **46**, 5696 (1992).
9. W. Bertsche *et al.*, Nucl. Inst. Meth. Phys. Res. A **566**, 746 (2006).
10. G. Andresen *et al.*, J. Phys. B: At. Mol. Opt. Phys. **41**, (2008).
11. This system was developed in consultation with G. Ganetis of Brookhaven National Laboratory.
12. M.C. Fujiwara *et al.*, *Particle Physics Aspects of Antihydrogen Studies with ALPHA at CERN*, these Proceedings.
13. G. Gabrielse, *et al.*, Phys. Lett. A **129**, 38 (1988).
14. M. Amoretti *et al.*, Phys. Lett. B **578**, 23 (2004).
15. C.L. Taylor, Jingjing Zhang and F. Robicheaux, J. Phys. B **39**, 4945 (2006).
16. N. Madsen *et al.*, Phys. Rev. Lett. **94**, 033403 (2005).
17. G. Gabrielse *et al.*, Phys. Rev. Lett. **93**, 073401 (2004).

Anti-hydrogen production with positron beam ion trap

Takahisa Itahashi

Department of Physics, Graduate School of Science, Osaka University
1-1 Machikaneyama, Toyonaka, Osaka 560-0043, Japan

Abstract. In low-energy antiproton physics, it is advantageous to be able to manipulate anti-particles as freely as normal particles. A robust production and storage system for high-quality positrons and antiprotons would be a substantial advance for the development of anti-matter science. The idea of electron beam ion trap could be applied for storage of anti-particle when the electron beam could be replaced by the positron beam. The bright positron beam would be brought about using synchrotron radiation source with a superconducting wiggler. The new scheme for production of anti-particles is proposed by using new accelerator technologies.

Keywords: EBIT, Positron, Decelerator
PACS: 29.25.-t; 29.27.-a; 36.10.-k

I. INTRODUCTION

Since the discovery of weak-bosons, an extensive high-energy collision experiment between protons and antiprotons has been undertaken at the Tevatron in Fermilab. Also, it is expected that weak-bosons will be a powerful tool for investigating new physics of Higgs bosons in LHC experiments at CERN. Formation and storage is a crucial issue for investigating anti-particles. Existing technologies for producing antiprotons and positrons have advanced over thirty years in the top-quark search and in B-physics. Combining the two accelerator technologies of storage and collision would be a significant further advance for accelerator science. However, a more sophisticated method is required than that applied at CERN-SPS, Tevatron and other existing accelerators [1]. An underdeveloped subject for anti-matter study is the formation and storage of low-energy and low-emittance positrons and anti-protons, which is quite different from high-energy physics experiments. To this end, a study has been initiated to examine anti-hydrogen formation process using a limiting number of positrons and antiprotons as the primary constituent of anti-matter synthesis. This paper describes low-energy positron production, storage and anti-hydrogen beam generation.

CP1037, *Cold Antimatter Plasmas and Application to Fundamental Physics,* edited by Y. Kanai and Y. Yamazaki

II. POSITRON PRODUCTION

As far as in a high energy positron and electron collision experiment, luminosity improvements for e+ and e- collider achieved at KEK and SLAC have set a state of the art benchmark for positron sources. However in a low energy positron experiment there are still undeveloped subjects concerning with its production, deceleration and storage. There are three methods for low-energy and high-intensity positron production: bremsstrahlung of accelerated electrons with accelerator, beta decay of radioactive nuclei and pair creation of high-energy photons. Two of these methods, using a high-energy electron beam and using a beta-radioactive source, are in practical use for material science. Positron beams and spectroscopic studies using positron beams are powerful, established techniques for atomic physics, condensed matter physics, surface chemistry and microelectronics and semiconductor technology. In addition, positron and antiproton plasma studies of CPT-theorem have played an important role in anti-hydrogen formation [2].

(i) Positron Production with Nuclear Beta Decay

The technique of positron emission from beta decay has been developed in nuclear physics and other applied fields such as PET for medical diagnosis of diseases and positron annihilation spectroscopy in material science. Beta radioactive nuclei can be produced by several nuclear reactions such as (n,γ), (n,p) and (p,n). The (n,γ) reaction uses high-flux neutrons produced by a nuclear reactor while the (p,n) reaction requires a proton of energy 10 MeV over the reaction threshold energy. The maximum expected positron intensity from radioactive nuclei produced by the (n,γ) reaction is 8×10^{16} e^+/sec, though the short lifetime of beta radioactivity limits its enrichment. For the (p,n) reaction, there are several nuclear reactions which can produce positron emitting nuclei of intensity around 6×10^{11} e^+/sec using a compact cyclotron [3]. A recently proposed method from plasma physics offers the possibility of producing mA positron beams. High-energy protons produced by the ^{3}He + d reaction inside a plasma could generate a subsequent (p,n) reaction and produce a positron beam of hundreds of μA. The expected e^+ intensity is 6×10^{14} e^+/sec [4]. The use of intense beta radioactive sources as positron emitters would be suitable for high duty factor or DC operation positron storage or post acceleration.

(ii) Positron Production with an Electromagnetic Interaction

A currently used, general method for positron production is through electron bombardment of a target with a linear accelerator at energy of about 50 MeV. A more energetic and high power electron beam of a few hundreds of MeV to a few GeV could produce a large number of positrons via an electromagnetic shower process. Such a production system would consist of a linear accelerator, production target (converter), collection solenoid and positron accelerator. In general, the production rate for positron production should be roughly proportional to the incident electron energy. For a 65-MeV linac, the obtainable positron flux is 10^8 e^+/sec. Recent work on B-physics at SLAC and KEK has led to remarkable developments in bright positron sources. At energy of GeV, the electron to positron conversion rate should be about a few 10^{-2}, giving an estimated positron flux of 10^{16} e^+/sec.

(iii) Positron Production with a Wiggler

It is expected that the demand for low-energy bright positrons will be greater than can be met by high-energy laboratories and therefore a more effective source of low-energy positron production is required. A possible source is high-energy photons produced by synchrotron radiation source facilities such as SPRING-8. The wiggler proposed for the SPRING-8 storage ring has a magnetic field of 10 T and can be used to produce hard synchrotron radiation with photon energy greater than 1 MeV, the threshold energy for electromagnetic pair creation. The expected positron production for a field of 10 T and a photon flux of 7.1×10^{14} photon/sec is 7.4×10^{12} e^+/sec [5].

III. ANTI-HYDROGEN PRODUCTION WITH A POSITRON TRAP

We propose the storage of a large number of positrons in the drift space of a tubular-type apparatus similar to the Tubular Electron Beam/String Ion Source (TEBIS) in a reflex mode of operation, using a positron beam instead of an electron beam as used in Electron Beam Ion Source(EBIS) or TEBIS for a heavy ion source [6]. Applying the general theory of cooling for positrons due to synchrotron radiation in the usual magnetic structure of EBIS, the estimated cooling time is around 300 ms for an initial positron energy of E_e = 1–5 eV to 0.5 meV. The proposed trap, called Tubular Electron String Ion Source (TESIS), could store as many as 10^{10} positrons. Instead of TESIS a detailed assessment of the positron string trap (PST) has already been given [6]. Various scenarios are expected for TESIS, from virtual cathode formation causing positron heating, to positron "crystallization". We also anticipate that cooling to the cryogenic temperatures of a considerable number of positrons confined in the positron string might be possible. Based on the expected number of cooled positrons, we can estimate the average rate of anti-hydrogen production when the positron trap is installed in a storage ring. The installation of the facility is assumed to be at a positive voltage of around 1–5 kV, as a positron and an antiproton have the same velocity in the storage ring. For 10^{11} stored anti-protons becomes, the resulting rate of anti-hydrogen production is 2×10^7 based on the production rate of the ATHENA facility [7].

IV. PROSPECTIVE SCHEME FOR ANTI-HYDROGEN PRODUCTION

The latest experiment at RIKEN notices that studies of positron accumulation in ultra high vacuum using both high-density electron plasma and an ion cloud as an energy absorber have achieved high positron accumulation rates, e.g., 3.6×10^2

e^+/s/mCi from a radioactive ^{22}Na source [8]. A further scheme has been proposed, more complicated than that used for radioactive sources. The scheme was originally proposed for nuclear astrophysics to investigate fusion reactions between bare nuclei [9]. It is based on the positron string potential scheme of EBIT and TEBIS for positron storage and cooling, described in section III, applied to anti-hydrogen production. The expected number density of cooled positrons in a string mode trap should be comparable to that in other traps in a reasonable time interval. Low-temperature positrons with a cryogenic temperature of T_{e+}= 0.5 meV can be stored to form a dense positron string in a Penning trap for an injection of 10^{10} e^+/s positrons of a few eV from one side of the trap in a synchrotron facility. At the other side of the trap, a decelerated pulsed antiproton beam can then be injected. The trap potentials should be chosen to give convenient center of mass energy between the antiprotons and positrons to form anti-hydrogen inside TEBIS.

In addition, we propose a convenient method of simultaneous production and deceleration of anti-protons that can be achieved with the FFAG accelerator [10]. Recent progress on the FFAG accelerator suggests that storage and phase space rotation should be possible using a 6-sector ring constructed for muon lepton flavor violation experiments [11]. FFAG has the great advantage of large horizontal (40,000 π mm-mrad) and vertical (6000 π mm-mrad) acceptance than other accelerators. Such large acceptances are suitable for the acceleration or deceleration of secondary particles such as muons and antiprotons produced by GeV protons.

IV. CONCLUSION

From the experimental point of view for fundamental physics, it is certain that the direction so far developed at three experiments ASACUSA, ATHENA and ATRAP [12] the antiproton decelerator (AD) of the antiproton facility would be more advanced in a dedicated facility such as FLAIR at GSI. In order to realize the project entitled "From the LEAR to the FLAIR", a new accelerator complex and related technologies will play an important role for anti-matter science [13, 14]. As an option of technological development the storage of positron by using the same technology as it has been well studied for heavy ion production has been proposed for anti-matter physics. In addition, we have presented the FFAG accelerator as a new and high qualified scheme of positrons decelerator. The production, collection, acceleration or deceleration, cooling and storage proposed here might be new ideas for the development of anti-matter physics. In order to make progress in overcoming the various limitations, it would be greatly advantageous to bring in new insight from the interdisciplinary regions of science. New technologies for manipulating anti-particles such as positrons and antiprotons will be crucial in the next half century.

ACKNOWLEDGMENTS

This work was supported by the Grant-in-Aid of Scientific Research, Ministry of Education, Culture and Sports, No. 19540292.

REFERENCES

1. H. Schopper, CERN 89-02 ISSN 0007-8328, ISBN 92-9083-007-7.
2. For a review see: R. S. Hayano and E. Widman (ed), *Nucl. Instr. and Meth. B* **214** (2004).
3. R. Chehab, CERN 89-05 (1989) pp. 105–134.
4. J. Dawson, proceedings of the workshop "Critical issues in the development of new linear colliders" Madison-Wisconsin; and August 1986 CAS, CERN 89-05, pp. 105–134.
5. A. Ando et.al, *J. Synchrotron Radiation* **5,** 360–362 (1998).
6. E. D. Donets, E. E. Donets, E. M. Syresin, T. Itahashi and A. E. Dubinov, *Rev. of Sci. Instrum.*, **75,** 1563–1565 (2004).
7. M. C. Fujiwara et. al, *Nucl. Instr. and Meth. B* **214**, 11–16 (2004).
8. N. Ohshima, T. M. Kojima, M. Niigaki, A. Mohri, K. Komaki and Y. Yamazaki, *Phys. Rev. Lett.* **93,** 195001-4 (2004).
9. T. Itahashi, N. Kudomi, E. D. Donets and E. E. Donets, *Rev. Sci. Instrum.*, **73**, 667–669 (2002).
10. Y. Mori, *J. Phys. G* **29**, 1527 (2003).
11. Y. Kuno, *Nucl. Phys. Proc. Suppl.* **155**, 53–57 (2006).
12. S.Muury, CERN/PS 99-50 (1999).
13. Conceptional Design Report of GSI "*An International Accelerator Facility for Beams of Ions and Antiprotons*".
14. W. F. Henning, *Nucl. Instr. and Meth. B* **214**, 191–195 (2004).

Cryogenic Particle Accumulation In ATRAP And The First Antihydrogen Production Within A Magnetic Gradient Trap For Neutral Antimatter

C. H. Storry[a], A. Carew[a] D. Comeau[a], E. A. Hessels[a], M. Weel[a], M. C. George[a,b], D. Grzonka[b], W. Oelert[b], T. Sefzick[b], Z. Zhang[b], G. Gabrielse[c], P. Larochelle[c], D. LeSage[c], B. Levitt[c], W. S. Kolthammer[c], R. McConnell[c], P. Richerme[c], J. Wrubel[c], A. Speck[d], F. Markert[e], F. Nillius[e], M. Scheid[e] and J. Walz[e]

[a]York University, Department of Physics and Astronomy, Toronto, Ontario M3J 1P3, Canada
[b]IKP, Forschungszentrum Jülich GmbH, 52425 Jülich, Germany
[c]Department of Physics, Harvard University, Cambridge, Massachusetts 02138, USA
[d]Rowland Institute at Harvard, Harvard University, Cambridge, Massachusetts 02142, USA
[e]Institut fur Physik, Johannes Gutenberg-Universität, D-55099 Mainz, Germany

Abstract. ATRAP has made many important improvements since CERN's Antiproton Decelerator (AD) was restarted in 2006. These include substantial increases in the number of positrons (e+) and antiprotons (Pbars) used to make antihydrogen (Hbar) atoms, a new technique for loading electrons (e-) that are used to cool Pbars and e+, implementation of a completely new, larger and more robust apparatus in our second experimental zone and the inclusion of a quadrupole Ioffe trap intended to trap the coldest Hbar atoms produced. Using this new apparatus we have produced large numbers of Hbar atoms within a Penning trap that is located within this quadrupole Ioffe trap using a new technique which shows promise for producing even colder atoms. These observed Hbar atoms resolve a debate about whether positrons and antiprotons can be brought together to form atoms within the divergent magnetic fields of a quadrupole Ioffe trap.

Keywords: Positrons, Antiprotons, Antihydrogen, Buffer gas positron accumulator, Ioffe trap, Trapping, CPT test, CERN, antiproton decelerator (AD), Photoelectrons, Cryogenic Penning trap, Solid neon moderator.
PACS: 37.10.De, 36.10.-k, 41.20.-q, 41.90.+e.

2 GENERATIONS OF ANTIHYDROGEN PRODUCTION APPARATUS

The ATRAP collaboration has continued its aggressive progress toward its goal of producing cold Hbar atoms that can be confined and laser cooled within a neutral particle trap and studied with the tools of laser spectroscopy for comparison to hydrogen [1]. The goal of these experiments is to compare the spectra of these simplest matter and antimatter atoms for a sensitive test of CPT and Lorentz

CP1037, *Cold Antimatter Plasmas and Application to Fundamental Physics,* edited by Y. Kanai and Y. Yamazaki

invariance with leptons and baryons [2]. Prior to 2006 ATRAP experiments were done in a liquid helium cooled Penning trap mounted within the bore of a superconducting solenoid having an inner diameter of 10cm and providing a 5.4 Tesla magnetic field for radial confinement of charged particles [Fig 1]. In this first generation apparatus large numbers of cold e+ and Pbars were brought together within the 0.6cm radius Penning trap electrodes to form Hbar atoms in a variety of ways [3-5]. For these experiments, Pbars were accumulated and electron cooled using techniques developed by the ATRAP and TRAP collaboration from which if was formed and now used by all Hbar experiments [6]. Pbars, produced in collisions at CERN, are slowed to 5 MeV and axially compressed in the AD. Typically 30 million Pbars are ejected from the AD in a 100nS pulse and enter the Penning trap from below. Some of the Pbars are slowed further within a beryllium degrader foil at the bottom of the trap and exit the degrader with keV energies. Pbars exiting the degrader with a kinetic energy of less than 3keV are repelled by the electric field produced from the applied -3kV potential on an electrode part way up the Penning trap. After the pulse of Pbars has passed through the degrader, but before it returns to the degrader, a -3kV potential is applied to the degrader, thus Pbars with less than 3keV become trapped within this long Penning trap well. These Pbars are cooled further by collisions with e-, previously loaded into the trap from a field emission point. Additional pulses of Pbars from the AD are stacked and electron cooled until the desired number of particles is accumulated. The e- are ejected from the Pbars by briefly applying an electric field that accelerates the particles. The field is applied for long enough that the electrons leave the trap but the much heavier Pbars have not. The result is a dense cloud of cold Pbars to be used for Hbar production.

The e+ for these experiments are loaded directly into the sealed Penning trap enclosure from a ^{22}Na radioactive source lowered into the high field region of the solenoid [7]. Relativistic e+ are slowed within a thin single crystal, W (100), tungsten moderator at the top of the Penning trap. As the e+ exit the moderator very often an e- is also ejected. Highly excited positronium atoms (Ps) are produced from these particles by applying an electric field applied at the surface of the moderator which is used to match the velocities of the e- and e+. Some positrons are not moderated on this single pass, exit the moderator with high energy, and enter a second, thicker tungsten crystal, W (110), part way down the Penning trap. Many of these e+ are moderated within this thick crystal, ejected from its surface, are accelerated back toward the top of the apparatus where they re-enter the thin moderator and have a second opportunity to make Ps as described. These highly excited Ps atoms are easily ionized by the fields of a Penning trap well and the e+ are accumulated for Hbar production. In addition to producing Hbar atoms in this apparatus from these particles some details of the atoms produced have been studied. ATRAP has measured the excited state distribution of the atoms produced [3] and the velocity of the atoms detected [8].

Following these successes there was a 1 year break in Pbar production at CERN in 2005, while the technical staff was devoted to construction of the LHC, scheduled to begin operation in 2008. Since then ATRAP has dedicated itself to the construction and implementation of a completely new, larger and more robust apparatus for Hbar

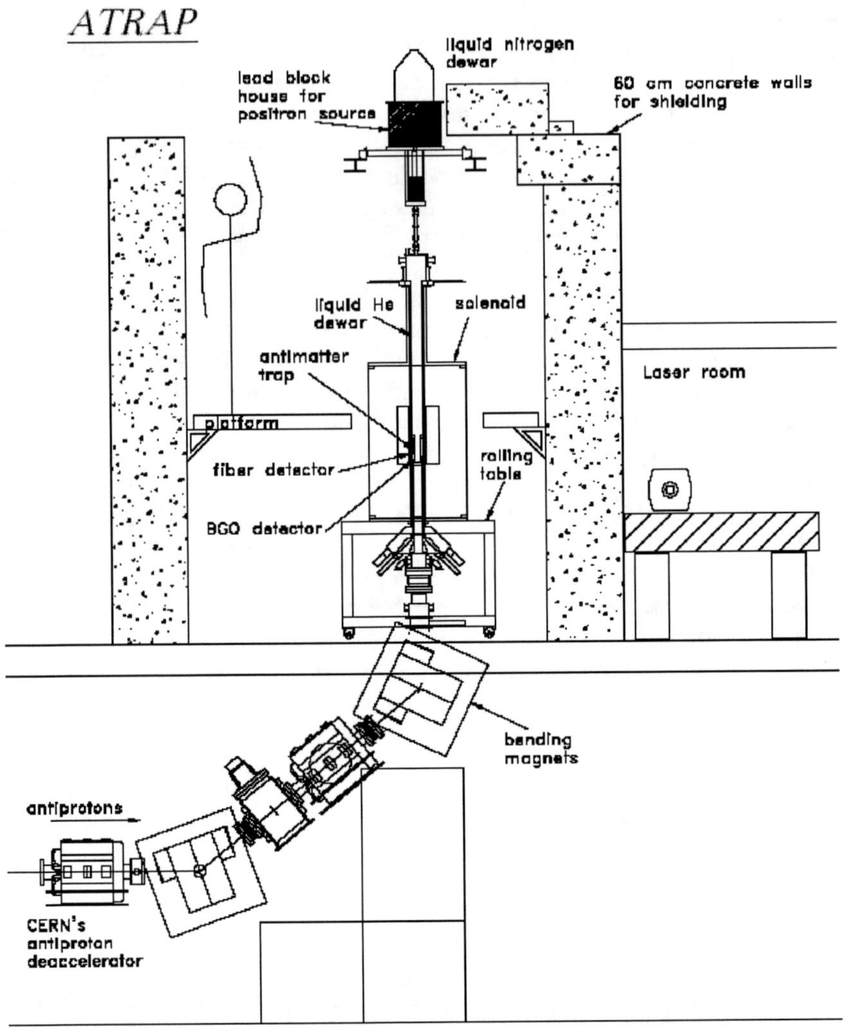

FIGURE 1. Schematic representation of the first generation ATRAP apparatus used to produce antihydrogen. The major components of the apparatus are shown including some elements of the antiproton beam line, the 10 cm bore superconducting solenoid and some of the detectors. It is also evident from the figure that the e+ system used in these experiments is extremely compact.

experiments. This next generation apparatus is housed in a second experimental zone containing a much larger superconducting solenoid, shown schematically in figure 2, having and inner diameter of 50 cm and producing magnetic fields of up to 3 Tesla for the radial confinement of charged particles. Two layers of plastic scintillating paddles surrounding the magnet in combination with 10 layers of scintillating fibers within the bore of the magnet are used to detect charged pions ejected when Pbars annihilated in Hbar experiments. The 2 layers of scintillating paddles are assembled from 24 separate pieces and there are 1732 scintillating fibers. This high degree of granularity allows reduction of background counts from other CERN experiments and from cosmic rays and is designed to allow position sensitive detection of the Pbars that annihilate in the Penning trap. The scintillating fibers are assembled in 3 separate groups, such that the inner 2 groups can be removed for the inclusion of a superconducting Ioffe quadrupole trap.

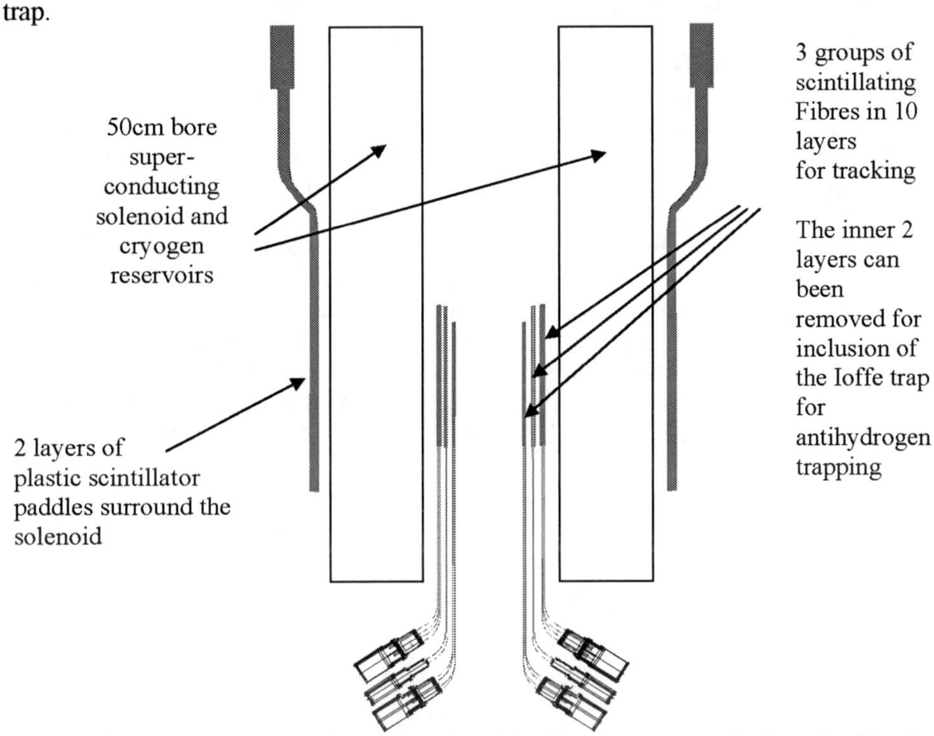

FIGURE 2. Figure 2: Schematic representation of the 50cm bore superconducting solenoid and scintillation detectors for detection of antiproton and positron annihilations that houses the cryogenic Penning trap.

The Ioffe trap shown in figure 3 is intended for trapping of Hbar atoms such that the small number of atoms produced can be studied for extended periods of time. The trap consists of 4 racetrack windings that produce a strong radial gradient in the magnetic field and pinch coils at both ends that produce an axial gradient and close the ends of the magnetic confinement field. Although a radial gradient in the magnetic field is required for trapping of the Hbar atoms, it poses new challenges for charge particle confinement. This radial gradient destroys the cylindrical symmetry of the uniform axial magnetic field produced by the solenoid and that is desired for charged particle confinement in a Penning trap.

The electric fields for charged particle confinement in this Penning trap are produced by potentials applied to a stack of 45 cylindrical electrodes [Fig 3]. These electrodes are 3 times larger than in the previous apparatus and have an inner diameter of 3.6 cm. This larger diameter allows for much larger plasmas and for a much greater number of particles to be stored. In a test apparatus with these large electrodes ATRAP has shown that plasmas of up to 1 billion e- can be confined [9].

In that apparatus ATRAP developed a new technique for accumulating electrons in a cryogenic Penning trap for Hbar experiments. This new technique is now used to accumulate the electrons required to cool the Pbars as described above and to efficiently trap and cool e+. Briefly; 248 nm UV laser pulses from a commercial KrF eximer laser source pass through a sapphire window at the top of the Penning trap enclosure. The later surface of the window is coated with a thin layer of gold and its potential controlled with an external voltage source. Alternately the laser light can enter through a hole at the top of the trap and are incident on a gold plated degrader foil that closes the bottom of the Penning trap. The energy of these laser photons is higher than the work function which binds the e- to the gold; hence, photoelectrons are ejected from the surface with little kinetic energy. The kinetic energy of the photoelectrons can be retarded or increased by changing the applied potential on the gold surface. To increase the number of e- that leave the surface they are accelerated so as to reduce the space charge from the e- near the gold. The e- travel some distance and are then turned back around by a negative potential farther along the electrode stack. Before the e- return to the gold a potential barrier is raised near it trapping the e- in a similar way that Pbars are captured and described above. This new technique now used for loading the e- for Pbar and e+ cooling in ATRAP has a number of advantages over the techniques previously used. Loading of field emission point e- decreases over time, the e- are typically much higher energy of order 1000 eV when first loaded and hence require more time to cool to cryogenic temperatures and to compress radially.

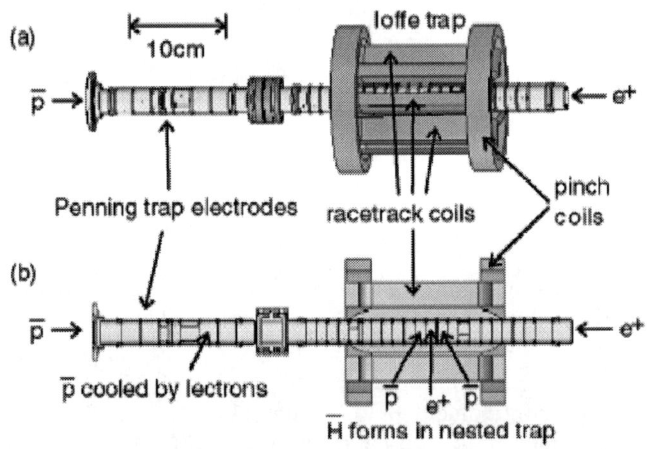

FIGURE 3. Penning trap electrodes and Ioffe trap for antihydrogen trapping.

The e+ for Hbar production in ATRAP experiments are now loaded at a greatly increased rate. In the previous generation of ATRAP experiments e+ were loaded at a rate of <1000 per second using the techniques described above. Recently ATRAP has constructed and implemented a solid neon moderator [10] and buffer gas cooled e+ accumulator [11] [Fig. 4]. Using this newly included technique for e+ accumulation ATRAP is able to load particles at a rate which is 1000 times higher for a given radioactive source strength. In this technique the radioactive source and a copper electrode in front of it are cooled using a mechanical refrigerator to cryogenic temperatures of 4K. Neon gas, released into the vacuum chamber, is frozen onto the window of the source and to the surfaces of the electrode and acts as a moderator for the high energy e+ emitted upon decay of the ^{22}Na. A small electric field applied near the source pulls them out of the solid neon and accelerates them toward a room temperature Penning trap system. Magnetic fields are used to guide them around a dog leg in the vacuum apparatus and out of the lead enclosure protecting the users from the harmful radiation of the source.

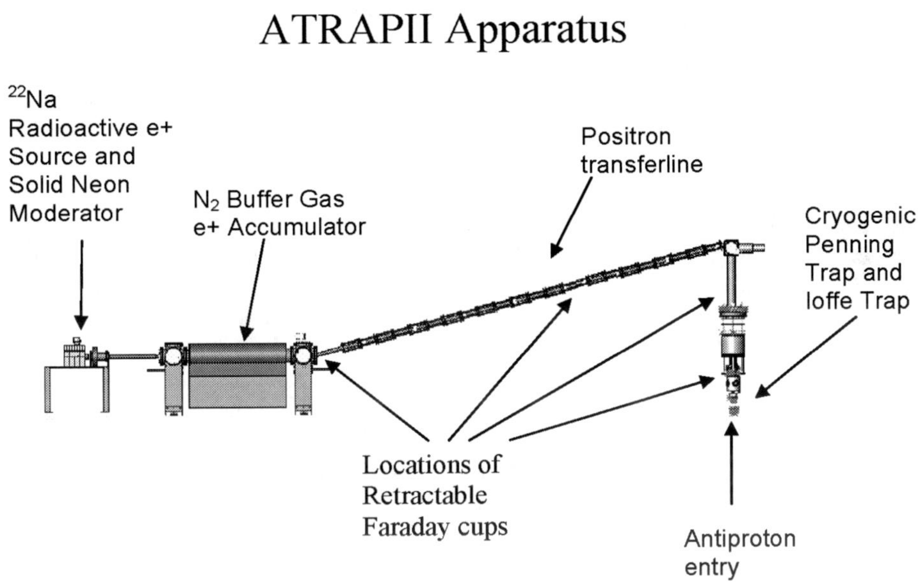

FIGURE 4. Schematic representation of some of the other major components of the second generation ATRAP apparatus. Positrons are emitted from a radioactive ^{22}Na source cooled to cryogenic temperatures by a mechanical fridge. They are moderated in a layer of frozen neon deposited onto the ^{22}Na capsule and an electrode in front of it. Moderated positrons are guided into the buffer gas accumulator by magnetic fields where they are accumulated and compressed radially by a rotating electric field. Bunches of positrons are ejected to the right in the figure with a kinetic energy of order 10eV and guided to a position above the cryogenic Penning trap by the 5 meter long transferline shown. Retractable Faraday cups are used to optimize the positron transfer into the cryogenic trap. A superconducting Ioffe trap for antihydrogen confinement is located within the 50cm bore superconducting solenoid not shown here.

These e+ with a kinetic energy of 10eV travel a distance of 3 meters and enter a large diameter copper solenoid providing a magnetic field of 0.15 Tesla for radial confinement in the Penning trap it houses. The e+ are slowed within an applied Penning trap well by collisions with nitrogen gas which is injected into the second electrode of the accumulator. Upon colliding with nitrogen molecules e+ are slowed and become trapped in the 2 meter long electric potential well produced by the potentials applied to the first and last electrode of this Penning trap. To avoid fast annihilation of the e+ with the e- of the nitrogen gas the minimum potential in the Penning trap is positioned 1.5 meters after the nitrogen input location where the pressure is greatly reduced. This axial gradient in the nitrogen pressure is produced by allowing most of the gas to exit the third electrode through radial holes 30cm after the gas input. Additional reduction in the nitrogen pressure where the e+ accumulate is achieved by the combination of large cryopumps at either end of the accumulator and by increasing the diameter, and hence molecular diffusion rate, though the electrodes near the output end of the accumulator. With this combination we are able to store the e+ accumulated for 100's of seconds in the lowest potential of the accumulator. This allows for many transits back and forth in the accumulator before annihilation, sufficient time to allow further collisions with the nitrogen molecules and to allow the e+ to cool to room temperature. Once the e+ are trapped here a rotating electric field is used to compress the cloud of e+ radially to the central axis of the trap [12]. This rotating field is achieved by applying a constant radio frequency of 0.5 MHz with phases shifted by 90 degrees consecutively to the 4 quadrants of the electrode near the e+ accumulation site. Typically e+ are accumulated and compressed radially for 50 seconds, the potentials are then adjusted so the e+ well bottom is +10V relative to earth ground. The e+ are then ejected toward the cryogenic Penning trap by quickly lowering the last electrode potential so that an electric field accelerates them to 10eV and directs them out of the accumulator.

Due to spatial constraints in the AD hall, this 5 meter long apparatus ends at a location 5 meters to the side of the vertical superconducting magnet housing the Penning trap for Hbar production. e+ are transported to a location 1.5 meters above the top of the superconducting solenoid and from there follow the central magnetic field lines down into the cryogenic Penning trap. To reach this location, the e+ are transported through a vacuum tube with an inside diameter of 4cm. This transferline uses a solenoidal magnetic field of 0.02 Tesla applied along its axis by copper windings to guide them along the tube. The fields from the magnets of the AD which change through the AD cycle, from other experiments in the AD hall and from the large fringing magnetic field of ATRAP's superconducting solenoid tend to give a transverse component to the guiding field and direct the e+ into the vacuum tube wall. To cancel these undesirable components of the field, computer controled dipole electromagnets along the entire length of the transferline are included to cancel both radial components along the transferline. In spite of these challenges, the 15 degree upward bend that the e+ take as they enter the transferline and the 105 degree downward bend that they take above the superconducting solenoid, nearly 100% of the e+ that are accumulated are transported to a location just above the superconducting solenoid as measured using a Faraday cup and charge sensitive amplifier. This is only one of a series of retractable Faraday cups along the transferline indicated in figure 4 which are used while optimizing the 95 adjustable electromagnets between the accumulator and the superconducting solenoid. As e+ travel down into the superconducting solenoid they are compressed radially by the increasing magnetic field strength and enter into the Penning trap through a 1.5cm long, 1mm diameter tube that closes the top of the Penning trap enclosure. Over half of the accumulated e+ make it through this tube as indicated by the charge deposited on the degrader at the bottom of the cryogenic Penning trap and also by the annihilation signal from the scintillating fibers which can be used to detect the gamma photons emitted when e+ annihilate within the Penning trap. To trap the e+, the potential on an electrode part way down the Penning trap is set to a positive potential which reflects them back up. Using a similar technique to that which is used to trap Pbars, the potential on an electrode near the top of the trap is quickly increased before the e+ can make it back out. A dense could of photoelectrons

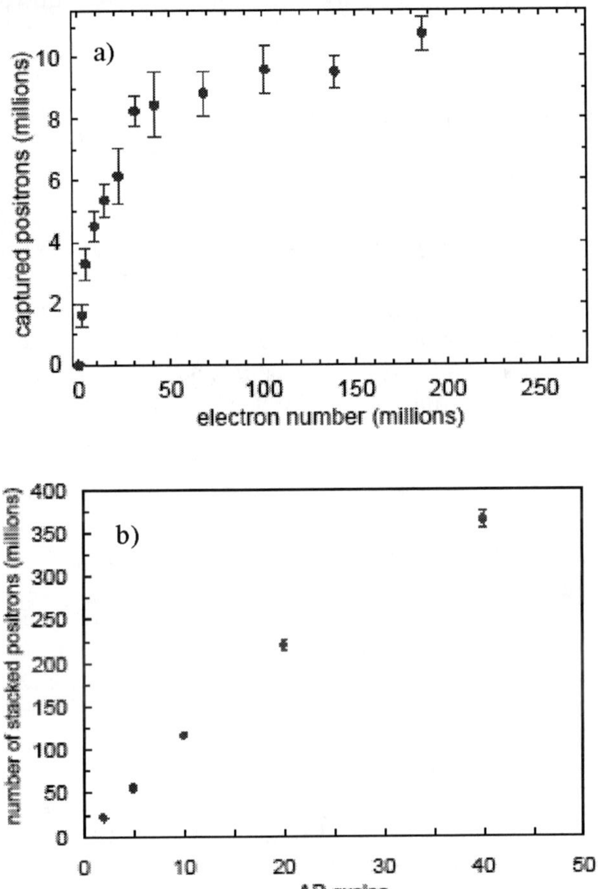

FIGURE 5. a) Positron capture and cooling is aided by photoelectrons loaded into the Penning trap prior to positron transfer. 150million e- are used in Hbar production experiments within the Ioffe trap. b) The number of e+ stacked into the cryogenic Penning trap grows almost linearly with additional pulses of positrons transferred.

previously loaded into the Penning trap helps to efficiently cool and trap these e+ [Fig 5]. Typically 10 million e+ are loaded in 2 pulses during a 100 second cycle of the AD, stacking and e- cooling for many AD cycles has allowed ATRAP to accumulate nearly 400 million e+ using this technique [Fig 5].

For trapping of the largest fraction of the cold Hbar atoms formed, the deepest possible Ioffe trap is desired. Since the bias field from the superconducting solenoid reduces the depth of the Ioffe trap, recent ATRAP experiments are done in a bias magnetic field of 1 Tesla and a radial Ioffe trap gradient of 93Tesla/m. There was some controversy as to whether large numbers of Pbars and e+ could be confined in these fields and if Pbars and e+ could be made to interact to form Hbar atoms. After the AD resumed operation in 2006, ATRAP demonstrated that indeed charged particles could be confined in the combined fields of the solenoid and Ioffe trap [13], a critical step toward the eventual goal of trapping Hbar atoms for spectroscopic measurements.

ANTIHYDROGEN PRODUCTION IN AN IOFFE TRAP FOR NEUTRAL ANTIMATTER CONFINEMENT

With the success of the described experiments, ATRAP has gone on to confirm that the charged constituents can be made to interact and indeed large numbers of Hbar atoms can be produced within the fields of the Ioffe trap [14]. Some new challenges have been overcome to achieve this important milestone as described above. Further; e- and e+ cool at a slower rate in the reduced 1 Tesla bias magnetic field required to ensure a significant trap depth for Hbar atoms. To overcome this hurdle and to ensure that the charged particles are cooled to cryogenic temperatures, much larger numbers of e- are used to cool the Pbars than were used in the previous generation of experiments. In addition we include a number of 5 to 10 minute periods in our experimental procedure to allow the e- and e+ to cool.

In preparation for Hbar production experiments in the Ioffe trap, e- are loaded and allowed to cool to cryogenic temperatures over a 10 minute period into 2 Penning trap wells. 150 million e- are loaded into a well in the upper part of the electrode stack and used to efficiently capture and cool the e+. 400 million e- are loaded in a separate well near the bottom of the electrode stack and are used for e- cooling of the Pbars. During the next 15 minutes the e+ and Pbars are simultaneously accumulated using the techniques described above. Typically 0.2 million cold Pbars and 60 million cold e+ are accumulated for these experiments although much larger numbers of both species have been accumulated in this apparatus. The particles are then adiabatically transferred to a positioned near the center of the de-energized Ioffe trap for Hbar production. The Pbars are then injected into a nested trap used to induce particle interactions and form Hbar. The Hbar atoms that form as the e+ initially cool the Pbars are discarded, and the scintillating detectors record how many Pbars are lost. The currents in the Ioffe trap are then ramped up over a period of 14 minutes and a detection well [Fig. 6] is then created near the nested Penning trap. Pbars and e+ are then made to interact and form cold Hbar atoms by slowly lowering the depth of the central e+ well over an 11 minute period. The Hbar atoms that form in the energized Ioffe trap and are detected in these experiments are those which travel to the detection well and are ionized by the electric field as they enter the well. The Pbars from ionized Hbar atoms are trapped in this well for later detection. The currents in the Ioffe trap are then reduced to zero over a period of 1 minute. Finally, the trapped Pbars from ionized Hbar atoms are ejected from the detection well by lowering one end of the well and annihilate upon contact with the electrode at the end of the trap. The scintillating paddles and fibers detect the charged pions emitted upon annihilation of the Pbars and indicate the number of Hbar atoms produced and ionized. Experiments are done at a variety of Ioffe trap depths and also with only the pinch coils charged. The number of Hbar atoms detected in these experiments is plotted in Figure 7

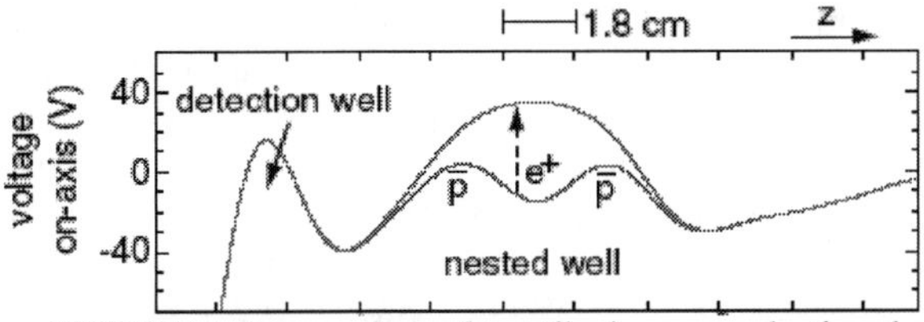

FIGURE 6. A new scheme for antihydrogen production shows great promise for producing slow, trappable antihydrogen. Electric potentials on the axis of the trap for antihydrogen production are shown. The depth of the e+ well in the center of the Penning trap is slowly decreased over an 11 minute period, providing long interaction times for Hbar production.

These experiments demonstrate the first production of Hbar atoms within a Penning-Ioffe trap and open the way to observing trapped Hbar atoms and optimizing their production. The fastest release of trapped Hbar atoms from this apparatus comes by ramping down the radial quadrupole field in 1 minute, limited by internal quench-protection diodes in the Ioffe trap. A signal equal to our detector background used here would be produced if 120 trapped Hbar atoms per second are released while this quadrupole field of the Ioffe trap is ramp down.

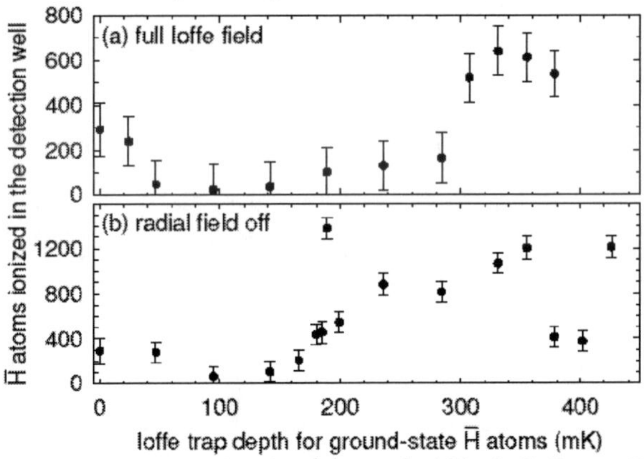

FIGURE 7: Antihydrogen atoms are produced and detected within the Ioffe trap. The number of Hbar atoms ionized in the detection well per trial within the Ioffe trap (a) and with only the axial field applied (b), as a function of the trap depth. The temperature units on the horizontal axis are for the low-field seeking ground state atoms with the number of atoms normalized to 100,000 Pbars used in the production.

In conclusion, ATRAP has commissioned a completely new set of apparatus for Hbar experiments with a wide range of improvements over the earlier generation. These include improved detectors for annihilation products from Pbars and e+, a faster and more robust technique for loading of photoelectrons used to cool charged antiparticles, enhanced Pbar loading in spite of the reduced bias magnetic field in the larger superconducting solenoid. A 1000 fold improvement in the rate of e+ accumulation in the cryogenic Penning trap has also been achieved, as compared to the most recent ATRAP experiments prior to the AD's one year shut down. This improved rate is achieved by the inclusion of a solid neon moderated, buffer gas cooled e+ accumulator and in face of the challenges that result from transporting the e+ over the complicated path from the accumulator to the cryogenic Penning trap apparatus. For the first time, the ATRAP apparatus includes a hole to allow entry of charged particles and laser light while maintaining the exquisite vacuum required for antimatter experiments and shown to be less than 5 x 10^{-17} torr in the earlier systems [15] that used a sealed Penning trap enclosure. ATRAP has provided the first demonstration that Hbar atoms can be produced within a magnetic gradient trap required for antimatter confinement in tests of CPT and Lorentz invariance. The interaction of Pbars and e+ to form Hbar atoms is prolonged with a new method that shows great promise for producing lower energy Hbar atoms. The first trapped Hbar atoms may soon be detected by the inclusion of a new magnetic gradient trap, now being developed by ATRAP.

ACKNOWLEDGMENTS

We are grateful to CERN for to delivering large numbers of slow Pbars from its unique Antiproton Decelerator. This work was supported by the NSF and AFOSR of the U.S., the BMBF, DFG, and Jülich Laboratory of Germany, along with the NSERC, CRC, CFI, and OIT of Canada.

REFERENCES

1. G. Gabrielse, in Fundamental Symmetries, edited by P. Bloch, P. Pavlopoulos, and R. Klapisch (Plenum, New York, 1987), pp. 59–75.
2. R. Bluhm, V. A. Kostelecky, and N. Russell, Phys. Rev. Lett. **82**, 2254 (1999).
3. G. Gabrielse, N. S. Bowden, P. Oxley, A. Speck, C. H. Storry, J. N. Tan, M. Wessels, D. Grzonka, W. Oelert, G. Schepers, T. Sefzick, J.Walz, H. Pittner, T.W. Hänsch, and E. A. Hessels, Phys. Rev. Lett. **89**, 213401 (2002).
4. G. Gabrielse, N. S. Bowden, P. Oxley, A. Speck, C. H. Storry, J. N. Tan, M. Wessels, D. Grzonka, W. Oelert, G. Schepers, T. Sefzick, J.Walz, H. Pittner, T.W. Hänsch, and E. A. Hessels, Phys. Rev. Lett. **89**, 233401 (2002).
5. C. H. Storry, A. Speck, D. LeSage, N. Guise, G. Gabrielse, D. Grzonka, W. Oelert, G. Schepers, T. Sefzick, H. Pittner, M. Herrmann, J.Walz, T.W. Hänsch, D. Comeau, and E. A. Hessels, , Phys. Rev. Lett. **93**, 263401 (2004).
6. G. Gabrielse, N.S. Bowden, P. Oxley, A. Speck, C.H. Storry, J.N. Tan, M. Wessels, D. Grzonka, W. Oelert, G. Schepers, T. Sefzick, J.Walz, H. Pittner, T.W. Hänsch, and E.A. Hessels, Phys. Lett. B, **548** (2002) pp. 140–145.
7. J. Estrada, T. Roach, J. N. Tan, P. Yesley and G. Gabrielse, Phys. Rev. Lett. **84**, 859 (2000).
8. G. Gabrielse, A. Speck, C. H. Storry, D. LeSage, N. Guise, D. Grzonka, W. Oelert, G. Schepers, T. Sefzick, H. Pittner, J.Walz, T.W. Hänsch, D. Comeau, and E. A. Hessels, Phys. Rev. Lett. **93**, 073401 (2004).
9. B. Levitt, G. Gabrielse, P. Larochelle, D. LeSage, W.S. Kolthammer, R. McConnell, J. Wrubel, A. Speck, D. Grzonka, W. Oelert, T. Sefzick, Z. Zhang, D. Comeau, M.C. George, E.A. Hessels, C.H. Storry, M. Weel and J. Walz, Phys. Lett. B, **656** (2007), pp. 25-29.

10. R.G. Greaves and C.M. Surko, Can. J. Phys., **74** (1996), pp. 7-8.
11. T. J. Murphy and C. M. Surko, Phys. Rev. A, **46** (1992), pp. 5696-5705.
12. R. G. Greaves and C. M. Surko, Phys. Rev. Lett. **85**, (2000), pp. 1883-1886.
13. G. Gabrielse, P. Larochelle, D. LeSage, B. Levitt, W. S. Kolthammer, I. Kuljanishvili, R. McConnell, J. Wrubel, F. M. Esser, H. Glückler, D. Grzonka, G. Hansen, S. Martin, W. Oelert, J. Schillings, M. Schmitt, T. Sefzick, H. Soltner, Z. Zhang, D. Comeau, M. C. George, E. A. Hessels, C. H. Storry, M. Weel, A. Speck, F. Nillius, J. Walz, and T.W. Hänsch, Phys. Rev. Lett. **98**,113002 (2007).
14. G. Gabrielse, P. Larochelle, D. LeSage, B. Levitt, W.S. Kolthammer, R. McConnell, P. Richerme, J. Wrubel, A. Speck, M.C. George, D. Grzonka, W. Oelert, T. Sefzick, Z. Zhang, A. Carew, D. Comeau, E.A. Hessels, C.H. Storry, M. Weel, and J. Walz, Phys. Rev. Lett. **100**, 113001 (2008).
15. G. Gabrielse, X. Fei, L. A. Orozco, R. L. Tjoelker, J. Haas, H. Kalinowsky, T. A. Trainor, and W. Kells, Phys. Rev. Lett. **63**, 1360 (1989).

Solid-state continuous Lyman-alpha source for laser-cooling of antihydrogen

Jochen Walz, Thomas Beyer, Daniel Kolbe, Frank Markert, Andreas Müllers, and Martin Scheid

Institut für Physik, Johannes Gutenberg-Universität Mainz, Staudinger Weg 7, D-55128 Mainz

Abstract. Cooling antihydrogen atoms is important for future experiments both to test the fundamental CPT symmetry by high-resolution laser spectroscopy and also to measure the gravitational acceleration of antimatter. Laser-cooling of antihydrogen can be done on the strong 1 S–2 P transition at the wavelength of Lyman-alpha (121.6 nm). Ongoing work to set up a solid-state continuous-wave laser source at Lyman-alpha is described.

Keywords: antihydrogen, laser-cooling, Lyman-alpha
PACS: 36.20.-k, 37.10.De, 42.65.Ky

INTRODUCTION

Future high-resolution laser-spectroscopy of antihydrogen in a magnetic trap can provide very stringent tests of the fundamental symmetry between matter and antimatter (CPT symmetry) [1]. The two-photon transition from the 1 S ground state to the 2 S metastable excited state is an excellent candidate for such precision experiments because it has a natural linewidth of only 1.6 Hz at a transition frequency of 2 466 THz. The absolute frequency of this transition has already been measured in ordinary hydrogen with enormous precision [2, 3]. However, the 1S–2S ($F = 1, m_F = 1 \quad F = 1, m_F = 1$) transition frequency has a residual dependence on the magnetic field of 186 kHz per Tesla. This will broaden and shift the spectral line of antihydrogen atoms in a magnetic trap [4, 5]. It will thus be very important to cool antihydrogen atoms, thereby reducing their spatial spread in the inhomogeneous magnetic field of the trap. Laser cooling of ordinary hydrogen atoms in a magnetic trap to the milli-Kelvin temperature range has been demonstrated [6]. A pulsed laser source has been used to drive the the strong Lyman-alpha transition at 121.6 nm wavelength from the 1S ground state to the 2P excited state.

In addition to testing CPT there is also the intriguing prospect to measure the gravitational acceleration of antimatter for the first time using antihydrogen atoms [7]. The thermal motion of antihydrogen is a critical factor in this type of experiments and laser-cooling at Lyman-alpha to milli-Kelvin temperatures will help a lot. Ultimately, ultracold temperatures in the sub-milli-Kelvin range are desirable for practical experiments. These temperatures are beyond standard laser-cooling limits for (anti-)hydrogen. Novel cooling schemes for ultracold temperatures have been proposed [8, 9] and experiments have started along these lines [10, 11].

At Mainz we are meanwhile setting up a new continuous-wave coherent Lyman-alpha source which is based on solid-state laser-systems. This contribution discusses the status of the new Lyman-alpha source.

CP1037, *Cold Antimatter Plasmas and Application to Fundamental Physics,* edited by Y. Kanai and Y. Yamazaki

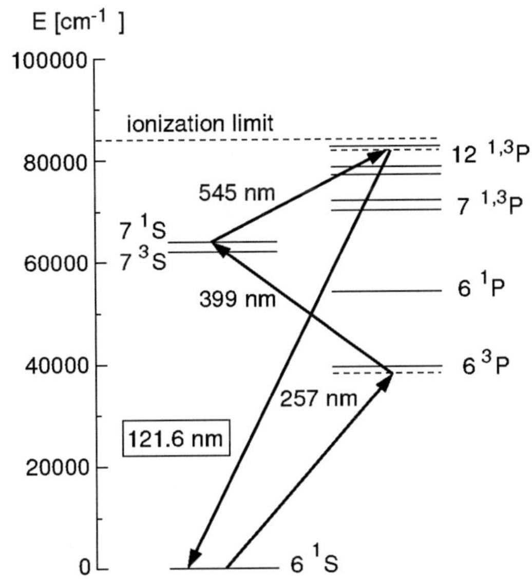

FIGURE 1. Four-wave mixing scheme to produce continuous laser radiation at Lyman-alpha in mercury vapor.

CONTINUOUS-WAVE COHERENT LYMAN-ALPHA GENERATION

Producing coherent radiation at 121.6 nm (Lyman-alpha) is a challenge as there are no tunable lasers and nonlinear frequency-doubling crystals available for that spectral region. Sum-frequency generation of several incident laser beams utilizing the nonlinear susceptibility of atomic vapors and gases is commonly used to produce coherent radiation in the VUV. Four-wave sum-frequency mixing produces the sum-frequency of three fundamental colors [12] and has been employed to generate *pulsed* laser radiation at Lyman-alpha, typically using Krypton gas [13, 14, 15, 16, 17].

Continuous coherent radiation at Lyman-alpha can have distinct advantages for laser-cooling of antihydrogen. An important difference, however, is that the power levels of continuous fundamental beams are many orders of magnitude lower than the peak powers typically used in pulsed Lyman-alpha generation. Continuous Lyman-alpha generation therefore uses resonances and near-resonances in the nonlinear optical medium.

Figure 1 shows the scheme for continuous Lyman-alpha generation in mercury vapor. Four-wave sum-frequency mixing of three fundamental beams is used [18, 19, 20]. The first fundamental beam at 257 nm is on the long wavelength side of the transition 6s $^1S_0$ 6p $^3P_1$ in mercury at 253.7 nm. This beam is obtained by frequency doubling the radiation from a single-mode Ar^+-laser. A nonlinear optical crystal (β-barium borate, BBO) placed in an enhancement cavity produces up to 900 mW of power at 257 nm. The second fundamental beam at 399 nm establishes an exact two-photon resonance with the 7 $^1S_0$ state. This beam is obtained by frequency-doubling the radiation from a

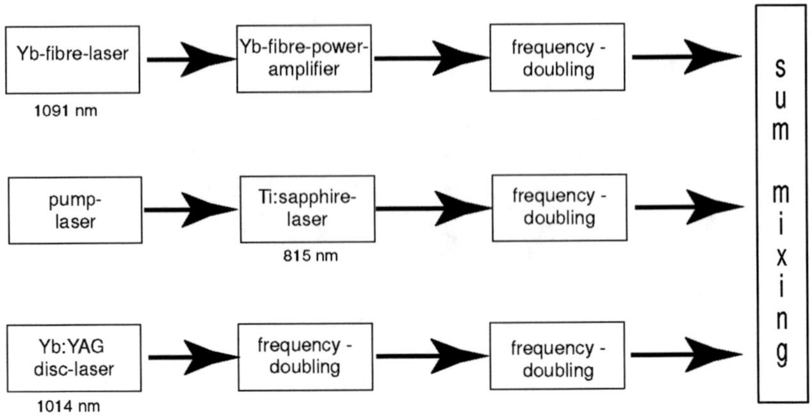

FIGURE 2. Laser-system overview for continuous Lyman-alpha generation using solid-state lasers.

titanium-sapphire laser. A nonlinear optical crystal (lithium borate, LBO) placed in an enhancement cavity produces up to 920 mW of power at 399 nm. The third fundamental light field at 545 nm comes from a dye-laser which operates on rhodamine 110 and produces up to 1.7 W. The wavelength of this third fundamental beam is chosen such that the sum-frequency is at Lyman-alpha. Bound states in mercury such as 11p $^1P_1$ and 12p $^1P_1$ contribute significantly to the non-linear susceptibility [21]. The three fundamental beams of parallel and linear polarization were focused to a beam waist of 30 μm in the middle of a 15 mm long zone of mercury vapor at 20–30 Torr. The setup has generated up to 20 nW of continuous laser radiation at Lyman-alpha and has been used to demonstrate 1S 2P spectroscopy of atomic hydrogen at natural linewidth (about 100 MHz) [19].

SOLID-STATE LYMAN-ALPHA SOURCE

The first continuous coherent Lyman-alpha source as described in the previous section used to employ up to three large-frame Argon-ion lasers. Two of them were used as pump-lasers for the titanium-sapphire laser and for the the dye-laser. The third Argon-ion laser had an intracavity etalon for single-frequency operation which allowed frequency-doubling in an external enhancement cavity. These Argon-ion lasers tended to limit the reliability of Lyman-alpha generation.

We are therefore now setting up a new continuous-wave Lyman-alpha source which is based on solid-state laser systems. Again, four-wave sum-frequency mixing of three fundamental beams in mercury vapor will be used. Figure 2 gives an overview of the laser systems. The first fundamental beam is at 254 nm, close to the transition 6s $^1S_0$ 6p $^3P_1$ in mercury. This beam is produced by frequency-quadrupling the radiation of an Yb:YAG disk-laser at 1014 nm in two subsequent enhancement cavities. Output powers of more than 700 mW in the ultraviolet have been generated in this way. The system is described in [22]. The second fundamental beam establishes an exact two-photon resonance with the 7 $^1S_0$ state in mercury and is now at 408 nm wavelength.

This beam will come from frequency-doubling the radiation of a titanium:sapphire laser in an enhancement cavity. A commercial diode-pumped solid-state laser at 533 nm will be used to pump the titanium:sapphire laser. The third fundamental beam at 545 nm is generated by frequency-doubling the radiation of an Yb fiber-laser at 1091 nm in an enhancement cavity. This system has produced output powers of up to 4 W in the yellow-green [23].

Compared to the previous Lyman-alpha source [18, 19, 20] the new setup has the additional advantage of being tunable at the one-photon height. We plan to explore the potential of the close 6s $^1S_0$ 6p $^3P_1$ mercury resonance for enhanced Lyman-alpha generation.

ACKNOWLEDGMENTS

We greatfully acknowledge support by DFG and BMBF.

REFERENCES

1. R. Bluhm, V. A. Kostelecký, and N. Russell, *Phys. Rev. Lett.* **82**, 2254–2257 (1999).
2. M. Niering, R. Holzwarth, J. Reichert, P. Pokasov, T. Udem, M. Weitz, T. W. Hänsch, P. Lemonde, G. Santarelli, M. Abgrall, P. Laurent, C. Salomon, and A. Clairon, *Phys. Rev. Lett.* **84**, 5496–5499 (2000).
3. M. Fischer, N. Kolachevsky, M. Zimmermann, R. Holzwarth, T. Udem, T. W. Hänsch, M. Abgrall, J. Grünert, I. Maksimovic, S. Bize, H. Marion, F. P. D. Santos, P. Lemonde, G. Santarelli, P. Laurent, A. Clairon, C. Salomon, M. Haas, U. D. Jentschura, and C. H. Keitel, *Phys. Rev. Lett.* **92**, 230802–1–4 (2004).
4. T. W. Hänsch, and C. Zimmermann, *Hyp. Int.* **76**, 47–57 (1993).
5. C. L. Cesar, *Phys. Rev. A* **64**, 023418(5) (2001).
6. I. D. Setija, H. G. C. Werij, O. J. Luiten, M. W. Reynolds, T. W. Hijmans, and J. T. M. Walraven, *Phys. Rev. Lett.* **70**, 2257–2260 (1993).
7. G. Gabrielse, *Hyperfine Interact.* **44**, 349–356 (1988).
8. J. Walz, and T. W. Hänsch, *Gen. Rel. Grav.* **36**, 561–570 (2004).
9. A. Kellerbauer, and J. Walz, *New J. Phys.* **8**, 1–9 (2006).
10. P. Pérez *et al.*, these proceedings (2008).
11. A. Kellerbauer *et al.*, these proceedings (2008).
12. C. R. Vidal, "Four-Wave Frequency Mixing in Gases," in *Tunable Lasers*, edited by L. F. Mollenauer, J. C. White, and C. R. Pollock, Springer, Berlin, 1992, vol. 59 of *Topics in Applied Physics*, chap. 3, pp. 57–113, 2 edn.
13. R. Mahon, T. J. McIlrath, and D. W. Koopman, *Appl. Phys. Lett.* **33**, 305–307 (1978).
14. D. Cotter, *Opt. Commun.* **31**, 397–400 (1979).
15. R. Wallenstein, *Opt. Commun.* **33**, 119–122 (1980).
16. L. Cabaret, C. Delsart, and C. Blondel, *Opt. Commun.* **61**, 116–119 (1987).
17. J. P. Marangos, N. Shen, H. Ma, M. H. R. Hutchinson, and J. P. Connerade, *J. Opt. Soc. Am. B* **7**, 1254–1259 (1990).
18. K. S. E. Eikema, J. Walz, and T. W. Hänsch, *Phys. Rev. Lett.* **83**, 3828–3831 (1999).
19. K. S. E. Eikema, J. Walz, and T. W. Hänsch, *Phys. Rev. Lett.* **86**, 5679–5682 (2001).
20. A. Pahl, P. Fendel, B. R. Henrich, J. Walz, T. W. Hänsch, and K. S. E. Eikema, *Laser Physics* **15**, 46–54 (2005).
21. A. V. Smith, and W. J. Alford, *J. Opt. Soc. Am. B* **4**, 1765–1770 (1987).
22. M. Scheid, F. Markert, J. Walz, J. Wang, M. Kirchner, and T. W. Hänsch, *Opt. Lett.* **32**, 955–957 (2007).
23. F. Markert, M. Scheid, D. Kolbe, and J. Walz, *Optics Express* **15**, 14476–14481 (2007).

Antihydrogen production

Evandro Lodi Rizzini[*,†], Luca Venturelli[*,†] and Nicola Zurlo[*,†]

[*] *Dipartimento di Chimica e Fisica per l'Ingegneria e per i Materiali, Università di Brescia, 25133 Brescia, Italy*
[†] *Istituto Nazionale di Fisica Nucleare, Gruppo Collegato di Brescia, 25133 Brescia, Italy*

Abstract. Antihydrogen production in ATHENA is analyzed more carefully. The most important peculiarities of the different experimental situations are discussed. The protonium production via the first matter-antimatter chemical reaction is commented too.

Keywords: Antihydrogen, Antiproton, Positron, Protonium
PACS: 36.10.-k, 34.80.Lx, 52.20.Hv

INTRODUCTION

In 2002, ATHENA and ATRAP demonstrated that cold antihydrogen atoms ($\bar{\mathrm{H}}$) can be abundantly produced by mixing antiprotons ($\bar{p}$) and positrons (e^+) in a nested Penning trap [1][2], see Fig. 1.

Subsequent studies showed that i) the $\bar{\mathrm{H}}$ production occurs at a much higher rate than the radiative ($e^+ + \bar{p} \rightarrow \bar{\mathrm{H}} + h\nu$) rate indicating the importance of the three-body process ($e^+ + e^+ + \bar{p} \rightarrow \bar{\mathrm{H}} + e^+$) (ATHENA[3]), ii) antihydrogen atoms are produced even when the positrons are heated to $T_e^+ \sim 300$ K, whereas the simple three-body assumption predicts a diminishing rate at high temperatures (ATHENA[4]), iii) antihydrogen atoms are in large principal quantum numbers ($n \sim 50$, ATRAP[5]), and iv) antihydrogen atoms are not cold enough for confinement in the "minimum-B" neutral atom traps (ATRAP[6], ATHENA[7]).

Utilizing the unique "three-dimensional annihilation imaging" capability of the ATHENA antihydrogen apparatus[8], see Figs. 2,3, i.e. the ability to localize the $\bar{p}$ annihilation position, we also established that v) the radial loss of antiprotons on the trap wall is localized to small spots (hot spots), strongly breaking the azimuthal symmetry expected for an ideal trap[9], and that vi) metastable antiprotonic hydrogen atoms ($\bar{p}p$) are produced, as byproducts of antihydrogen, following the interaction of antiprotons with the molecular ion H_2^+ [10],[11], see Figs. 4,5.

These studies revealed that the behaviors of $\bar{p}$s and e^+s in a Penning trap are much more complex than were naively assumed, and their detailed studies are essential to the success of antihydrogen trapping and spectroscopy.

The purpose of this presentation is to discuss how we can disentangle antiproton annihilations in the ATHENA trap due to 1) $\bar{\mathrm{H}}$ and 2) $\bar{p}p$ from 3) "hot spot", see Fig. 6, which can all occur simultaneously, but with different dependence on time, position, plasma temperature, positron density, positron shape and antiproton injection energy. We also address the question whether or not the "hot spots" are produced in a unique mechanism.

CP1037, *Cold Antimatter Plasmas and Application to Fundamental Physics,* edited by Y. Kanai and Y. Yamazaki

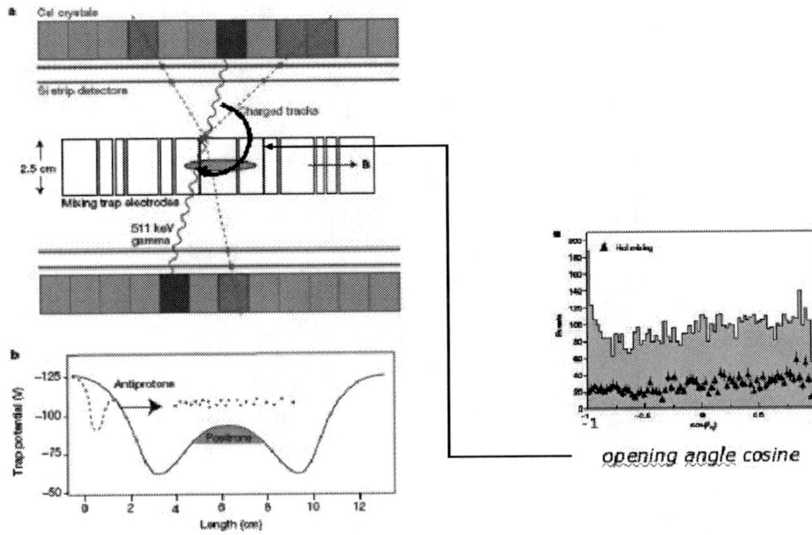

FIGURE 1. Central part of the ATHENA apparatus and trapping potential. **a**, Schematic diagram, in axial section, of the ATHENA mixing trap and antihydrogen detector. The cylindrical electrodes and the position of positron cloud (blue ellipse) are shown. A typical antihydrogen annihilation into three charged pions and two back-to-back 511-keV photons is also shown. The arrow indicates the direction of the magnetic field. The detector active volume is 16 cm long and has inner and outer diameters of 7.5 cm and 14 cm, respectively. **b**, The trapping potential is plotted against lenght along the trap. The dashed line is the potential immediately before antiproton transfer. The solid line is the potential during mixing.

HOT SPOTS IN $\bar{p}+e^+$ COLD MIXING SAMPLES

In Ref.[9], it is reported that some antiprotons in the ATHENA trap (without being mixed with e^+) annihilate on well-localized spots (hot spots), see Fig.6, but "hot spots" do occur in normal $\bar{p}-e^+$ mixing cycles.

Figure 7 is for the "Cold-mixing 2004" data set for the $\bar{p}\ e^+$ mixing cycles, showing the $r-z$ vertex distribution. We clearly observe "hot spots", $\bar{H}$ and $\bar{p}-p$ superposition. "Hot spots" are clearly observable on the wall ($r \gtrsim 0.7$ cm) at $z \sim -4, -1.5, +2$ and $+4$ cm.

The time distribution of the annihilation events for the same data set, for the first half a second, is shown in Fig. 8. Such distributions can be compared with the "trigger" time distributions illustrated in the ATHENA pubblications. Is the prominent peak found in Fig. 8 at $t < 40$ ms an indication of $\bar{\mathrm{H}}$ production immediately after the start of $\bar{p}-\mathrm{e}^+$ mixing cycle?

In order to examine the nature of this peak, we present in Fig. 9 the time evolution of the $z-$distributions of annihilation vertex positions. These distributions clearly show that the early-time behavior is dominated by hot spots. In the first time slice ($0 < t < 3$ ms) the hot spot is at $z = -4$ cm, in the second time slice ($3 < t < 8$ ms), additional hot spots appear at $z = -2$ as well as at $z = 4$. Even after the spike in the "trigger" distribution,

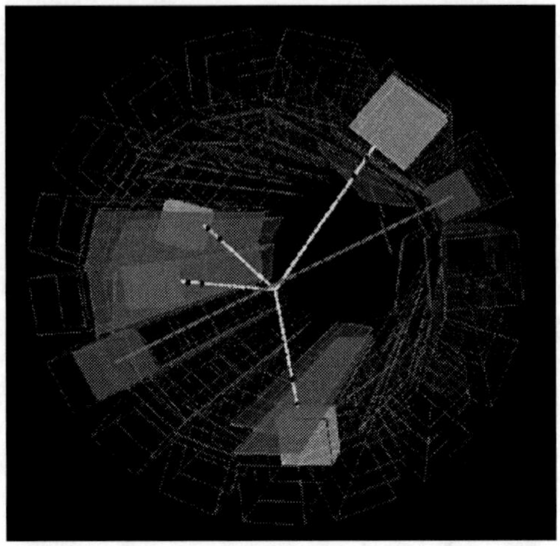

FIGURE 2. ATHENA detector allows reconstruction of the $\bar{p}$ annihilation point, as we see in the following picture (that refers to a simulated antihydrogen event). But not all the annihilations seen inside the trap are from antihydrogen!!

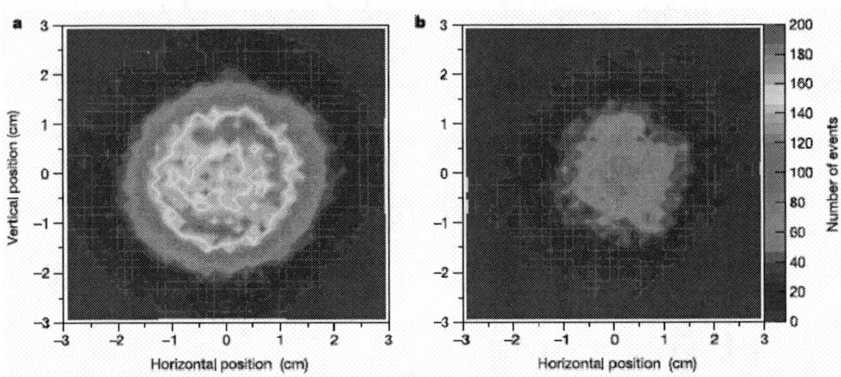

FIGURE 3. **a** on wall $\bar{H}$ annihilation **b** $\bar{p}p$ (protonium) annihilation

$40 < t < 100$ ms, the z−distribution is still dominated by hot spots, while the $\bar{\mathrm{H}}$ signal which we expected centered $z = 0$ cm, is almost absent. The situation becomes different after $t = 100$ ms (bottom-left and right of Fig. 9) where there is a clear peak around $z = 0$, which is presumably due to antihydrogen production.

These observations make it clear that it is difficult to discuss the $\bar{\mathrm{H}}$ production/annihilation rates based on the 'trigger' time distribution alone.

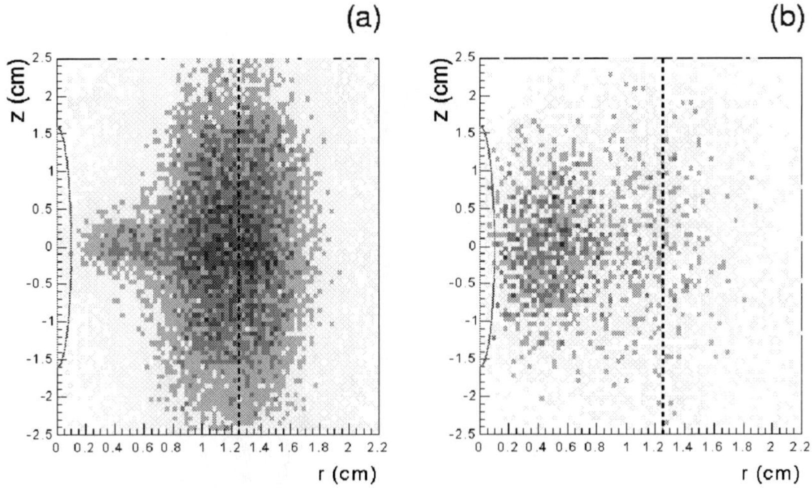

FIGURE 4. If we look at the projection of the reconstructed vertices on a plane (r,z) containing the trap axis (z), we see that this "background" around the positron's cloud has a clearly recognizable structure. The width of this structure along the axial direction increases with the temperature.

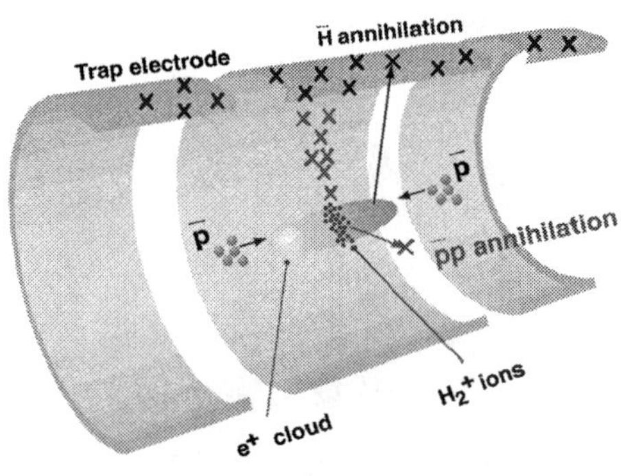

FIGURE 5. Overall picture of what happens inside the ATHENA trap at 15 K.

DETAILED EXAMINATION OF HOTSPOT-H̄-$\bar{p}p$ COMPOSITION

The "background" due to hot spots occurring in $\bar{p}-e^+$ mixing cycles has already been observed by the ATHENA collaboration, and hence in papers such as [3],[4] , those backgrounds were removed by selecting the events in the central region of $|z| < 2$ cm.

The $rz-$distribution presented in Fig. 7 however indicates that there are hot spots even within $|z| < 2$, and a more tighter cut of $|z| < 1$ cm is necessary to select hotspot-free events.

As shown schematically in the top of Fig. 1, when we mixed $\bar{p}$s with e^+s, we

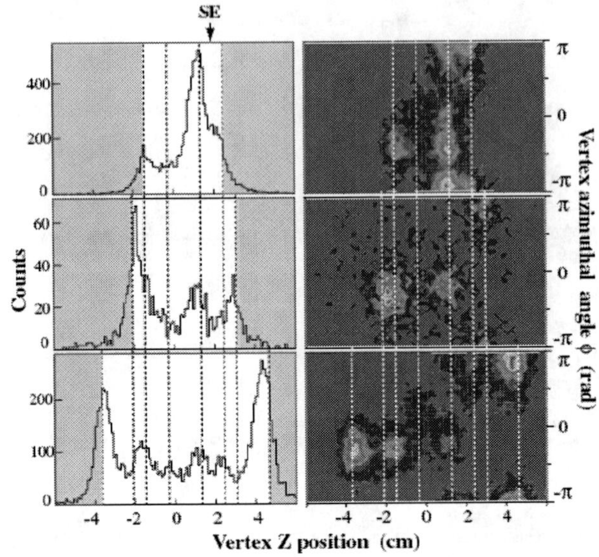

FIGURE 6. "Hot spots" distribution in different situations with only antiprotons in the trap.

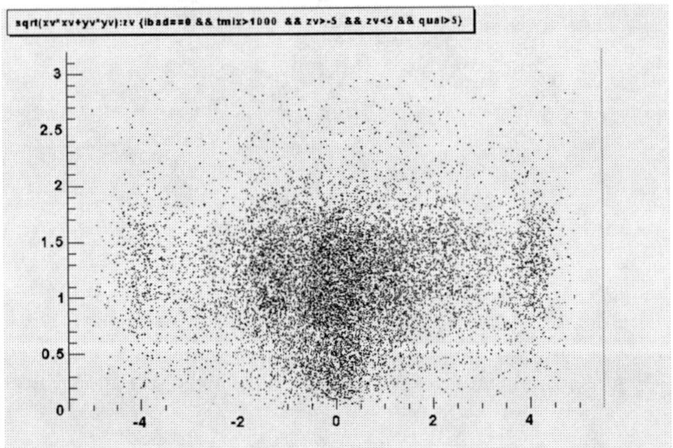

FIGURE 7. 2004 data taking: $r-z$ scatter plot of $\bar{p}$ annihilations on the wall or near the central axis.

initially hold $\sim 10^8$ positrons at the center of the nested Penning trap, in the region of $-1.6 < z < 1.6$ cm, and $\sim 10^4$ $\bar{p}$s in a potential well situated at around $z \sim -6$ cm; at $t = 0$, the $\bar{p}$-holding well was opened, and antiprotons were injected into the e^+ cloud.

The injection energy must be different: "high"$\approx$ 30 eV or "low"$\approx$ 5 eV. Moreover, after around half of the mixing time we can "reheat" the antiproton in the two lateral

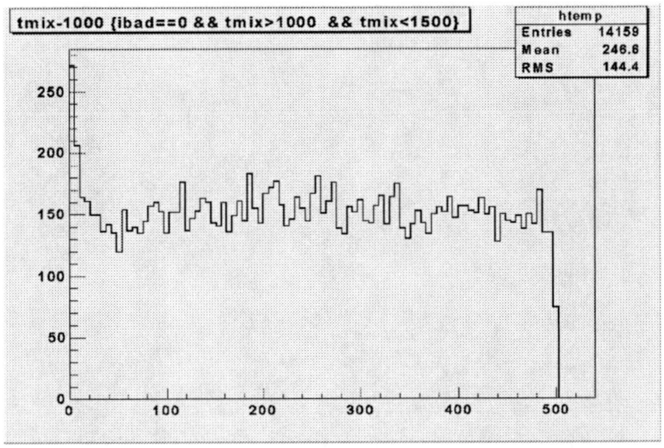

FIGURE 8. 2004 cold mixing data: the time distribution of the annihilation events in the first half a second.

wells by diminishing the deep of the wells. By so doing a lot of $\bar{p}$ are reinjected into the positron cloud and we observe a fast increasing in the $\bar{H}$ production rate, see Fig.10.

For the reheating at 40 s after the start of the mixing cycle (i.e. $t = 1$ s) we can compare the (x,y) distribution in the central trap region ($0 < z < 1$ cm). In Fig.11 we observe $\bar{H}$ annihilations on the wall in the first 2 s in the left figure, the $\bar{p}p$ annihilations near the centre being quite absent. Quite different is the situation after the reheating, figure 11 right, were the $\bar{p}p$ annihilation appears dominant.

The radial distributions in Fig.12 confirm the relevant presence of $\bar{p}p$ annihilation for $r < 0.7$ cm, right distribution, in respect to the one before the reheating as in Fig. 12 left.

Moreover, after the reheating a clear hot-spot appears around $z = -1$ cm in the z distribution, Fig.13 bottom, not present before the reheating as in the upper part of Figure 13. Such a hot spot is clearly observed in Fig.14 right.

In Table 1 we report the $\bar{H}$ production before and after reheating for such 2003 events sample.

TABLE 1. Antihydrogen production for different samples and in different temporal ranges

Temporal range	**Events per mixing cycle (with $r > 0.7$cm)**	**Sample**
1 s< t < 3 s	147	global cold mixing 2003
1 s< t < 3 s	141	cold mixing with reheating
39 s< t < 41 s	14	cold mixing with reheating
41 s< t < 43 s	18	global cold mixing 2003
41 s< t < 43 s	83	cold mixing with reheating

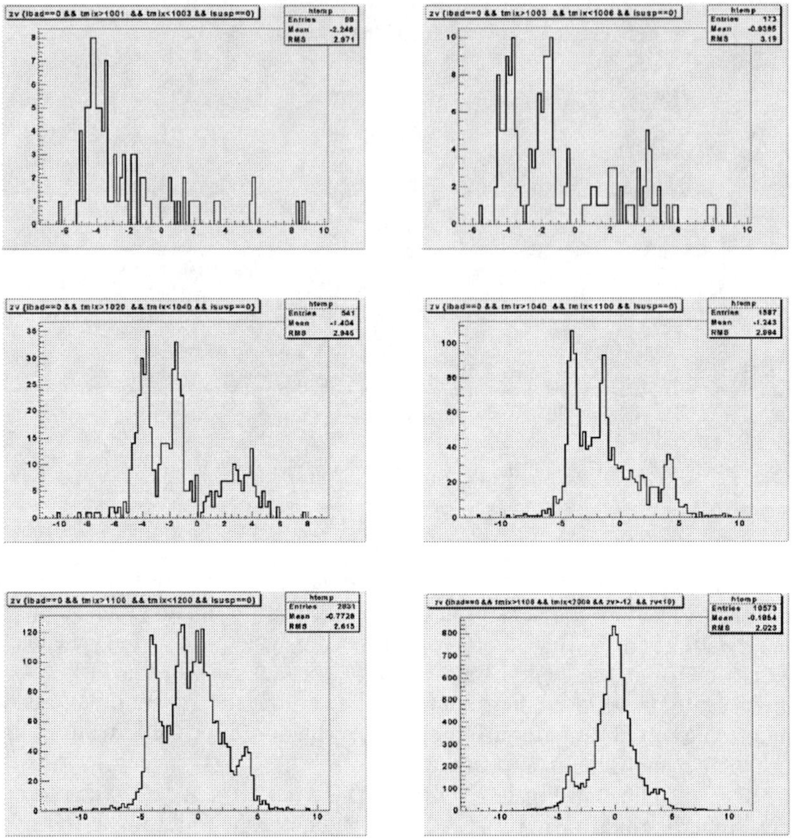

FIGURE 9. 2004 cold mixing data. See text for the content of the different samples.

Is it possible that low $\bar{p}$'s injection energies mimics the reheating conditions, i.e. the $\bar{H}$ and $\bar{p}p$ productions? As it is observed in Fig.15 bottom, the "low" $\bar{p}$'s injection energy presents in the first 4 seconds a situation quite different from the normal "high" injection energy, with an higher $\bar{p}p$ annihilation, i.e. protonium formation, at $r \lesssim 0.7$ cm.

In the so-called "electron through positron" mixing scheme we don't observe protonium formation. Clear hot-spot are present at least around $z = 4.5$ cm, see Fig.16.

Moreover, protonium formation is copious at any mixing time if we warm up the positrons cloud at temperature around 10,000 K, see Fig.17. It is very interesting to observe how change the presence of $\bar{H}$ annihilations on the wall and the $\bar{p}p$ annihilations in the central part of the trap along the e^+ cloud as we change the e^+ cloud radius: compressed ≈ 0.2 mm, normal ≈ 1 mm, expanded ≈ 3 mm see Figs. 18,19,20. In the 3 time intervals ($0 \div 7$ s; $7 \div 24$ s; $24 \div 60$ s) we observe a clear dominance of $\bar{H}$ production in the first time interval for the "compressed" mode, Fig.18 whereas the $\bar{p}p$ dominate

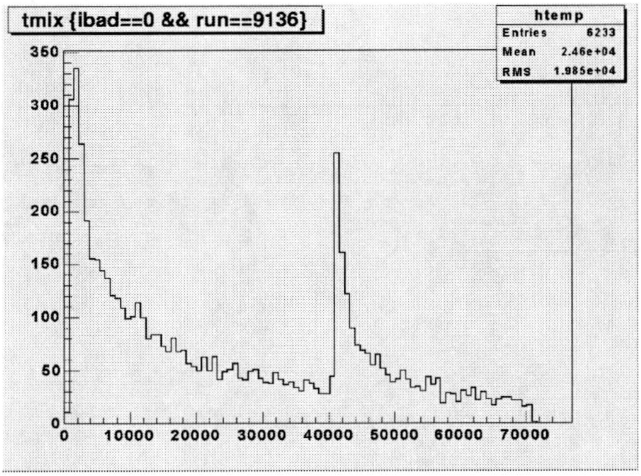

FIGURE 10. Antiproton's "reheating" after 40 s from the beginning of mixing, see text

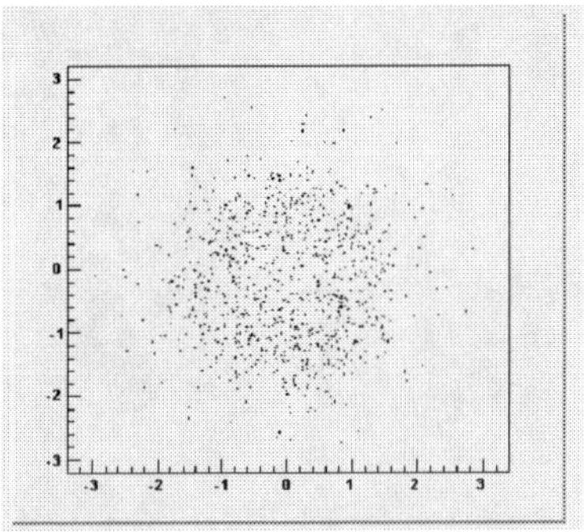

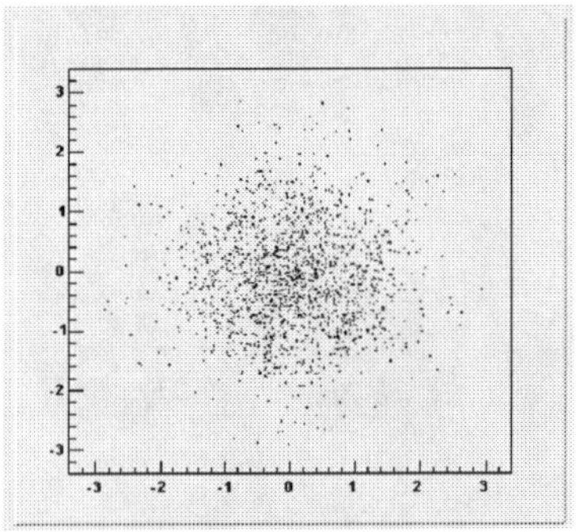

FIGURE 11. 2003 cold mixing data with reheating, (x,y) plane. Annihilations in the central trap region (0 cm$< z < 1$ cm) in the first two seconds, left, and after "reheating", right picture.

the second half of mixing time in such a mode. In the "expanded" situation the $\bar{H}$ production is the most important even in such second half of the mixing time, see Fig.20. In conclusion, a deeper analysis of the ATHENA data is very useful to disantangle the three different antiprotons annihilations we can observe in the ATHENA apparatus.

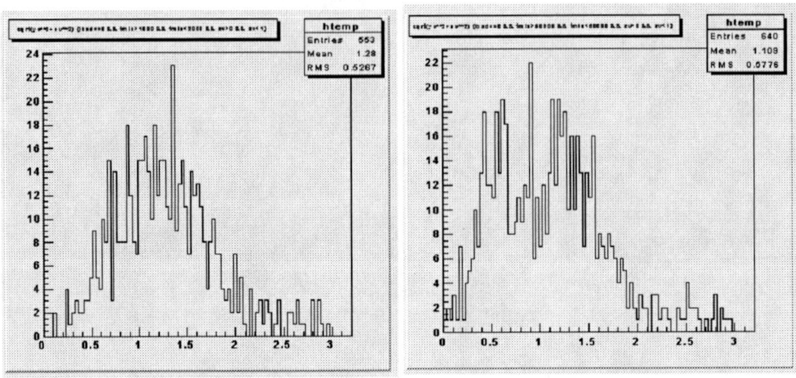

FIGURE 12. The same as in Fig.11 but for r distribution

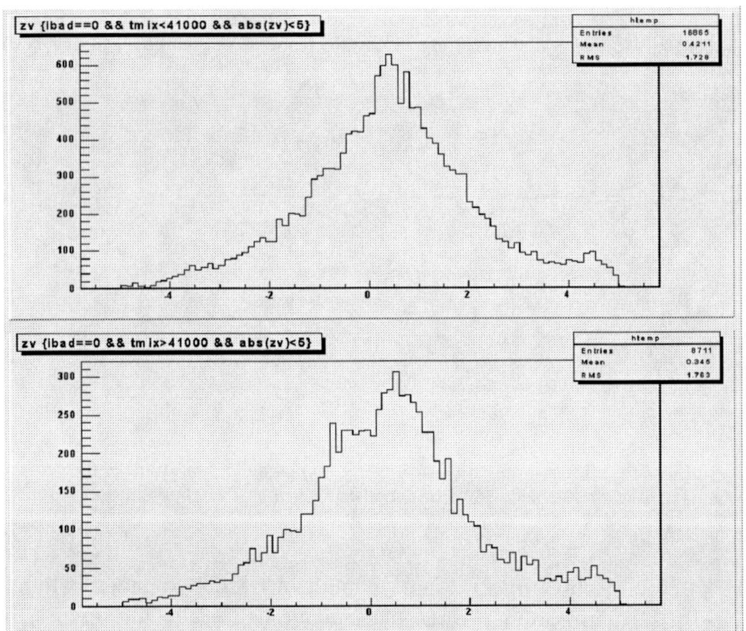

FIGURE 13. Cold mix 2003 sample as in Fig.11 and 12 but for the on wall $\bar{p}$'s annihilations for |z|<5 cm.

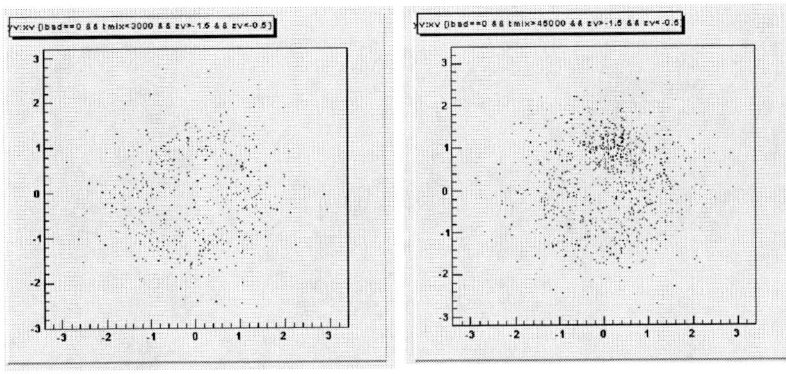

FIGURE 14. (x,y) distribution for the $\bar{p}$ annihilation in Fig.13 with -1.5 cm < z < 0.5 cm. A clear hotspot is present around $y \approx 1$ cm, $x \approx 0.5$ cm

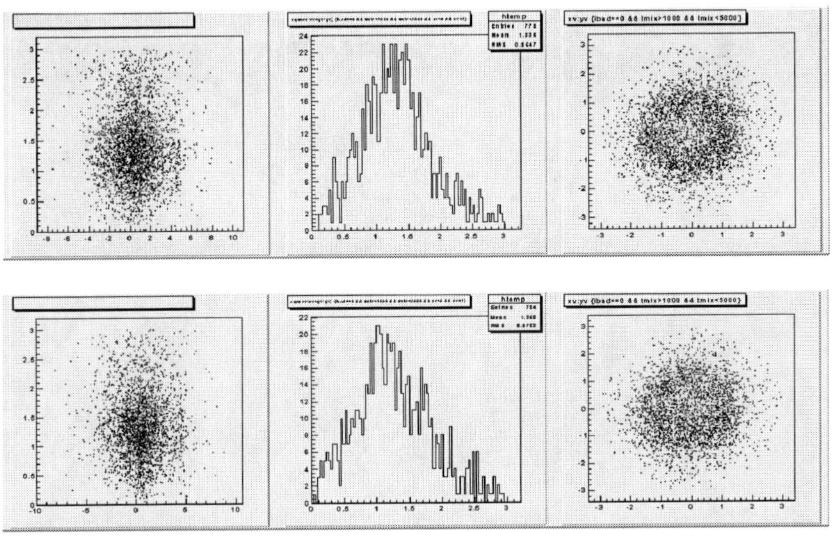

FIGURE 15. In the upper distributions we observe the $r - z$ scatter plot, the radial distribution for 0 cm <z<1 cm and the (x,y) distribution for $\bar{p}$'s annihilations after injection at "high" energy; bottom, the same distributions but after "low" energy injection

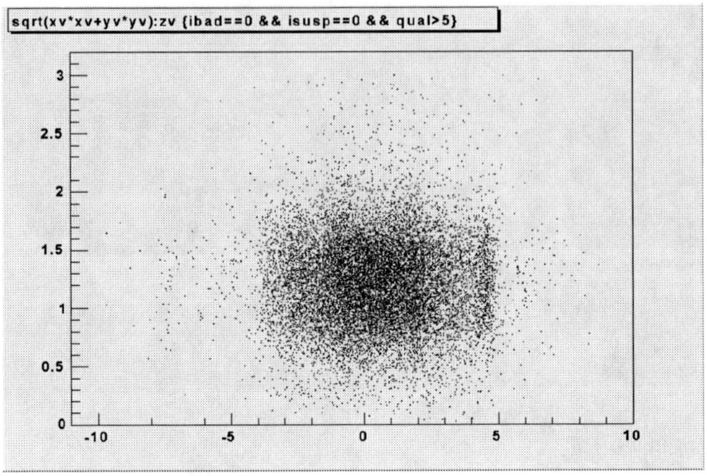

FIGURE 16. E.T.P.: Electron through positron mode for the transfer of electrons from the gun to the "catching trap".

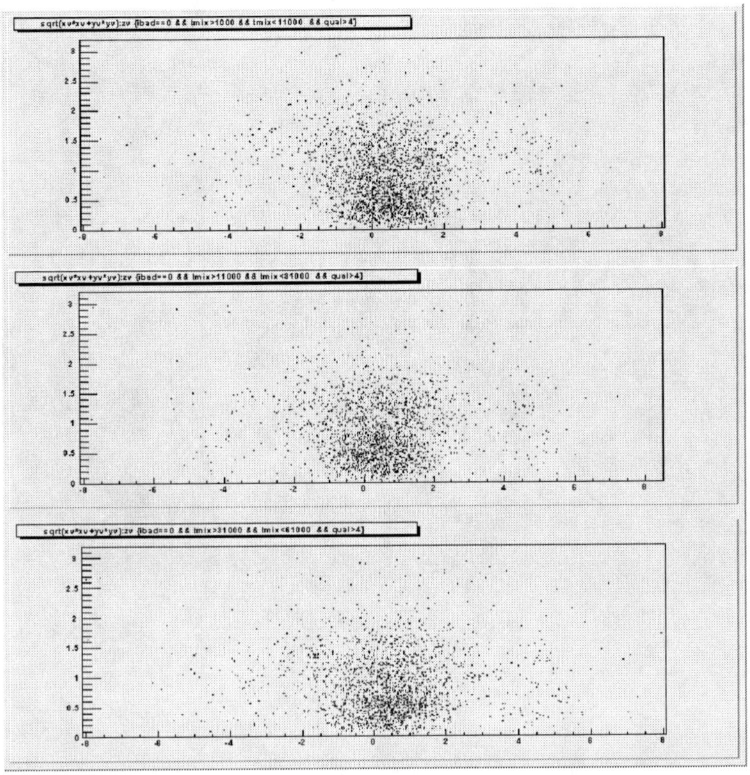

FIGURE 17. Hot mixing: positron cloud is warmed up to $\approx 10,000$ k. Upper picture, the first 10 s; central picture, 20 s time interval after the first 10s; lower picture, the final 30 s mixing.

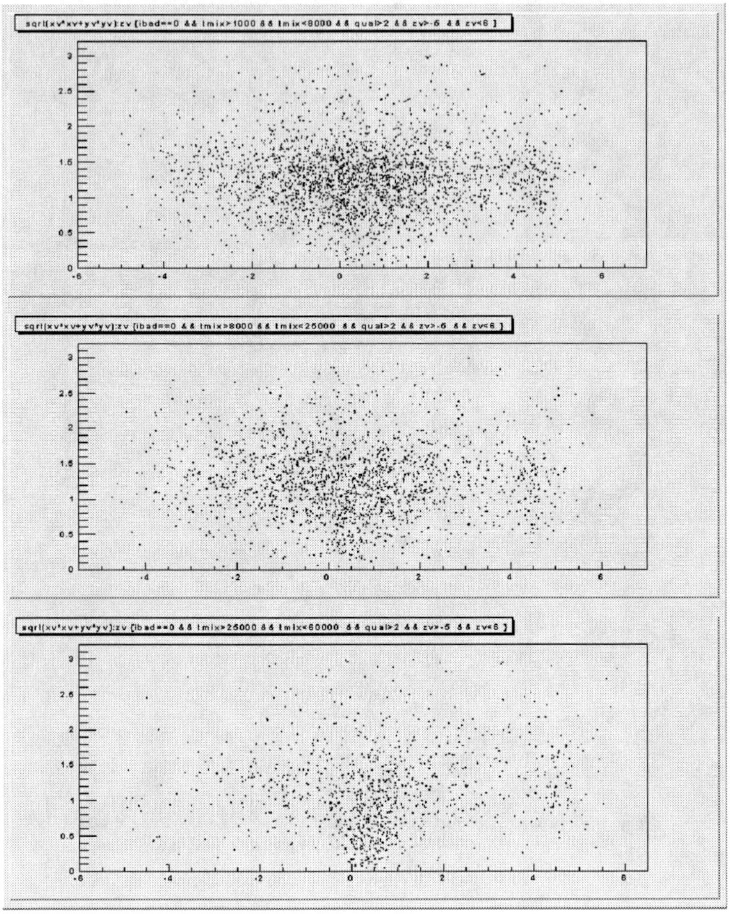

FIGURE 18. $\bar{H}$ production by "compressed" positrons cloud. Upper picture: the first 7 s; central picture: 7 s < t < 24 s; bottom picture: 24 s < t < 60 s mix

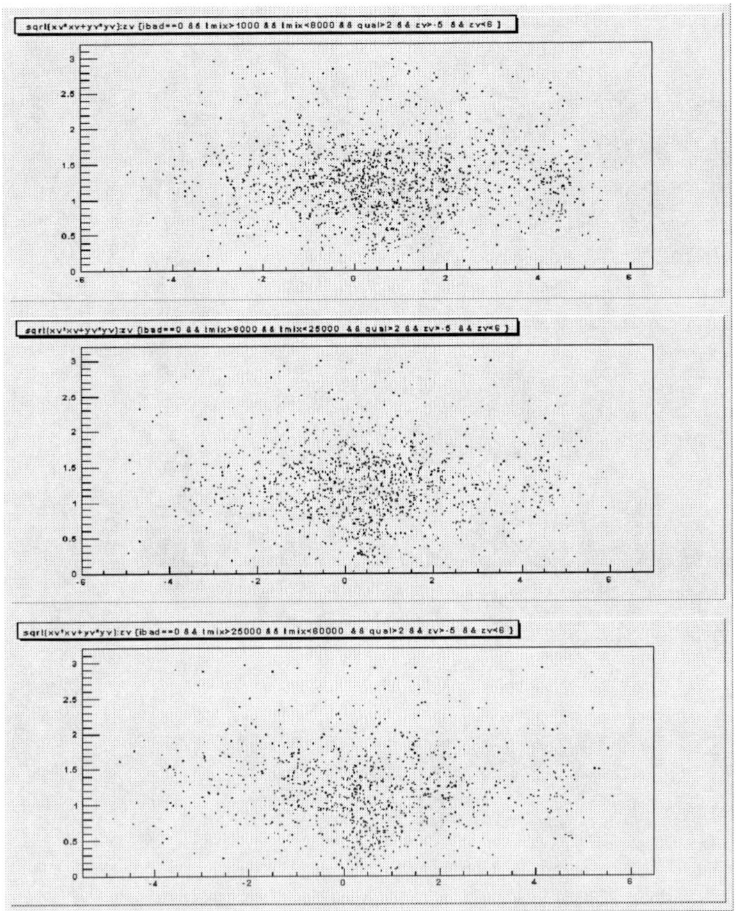

FIGURE 19. $\bar{H}$ production by "standard" positron cloud. The time intervals as in Fig.18.

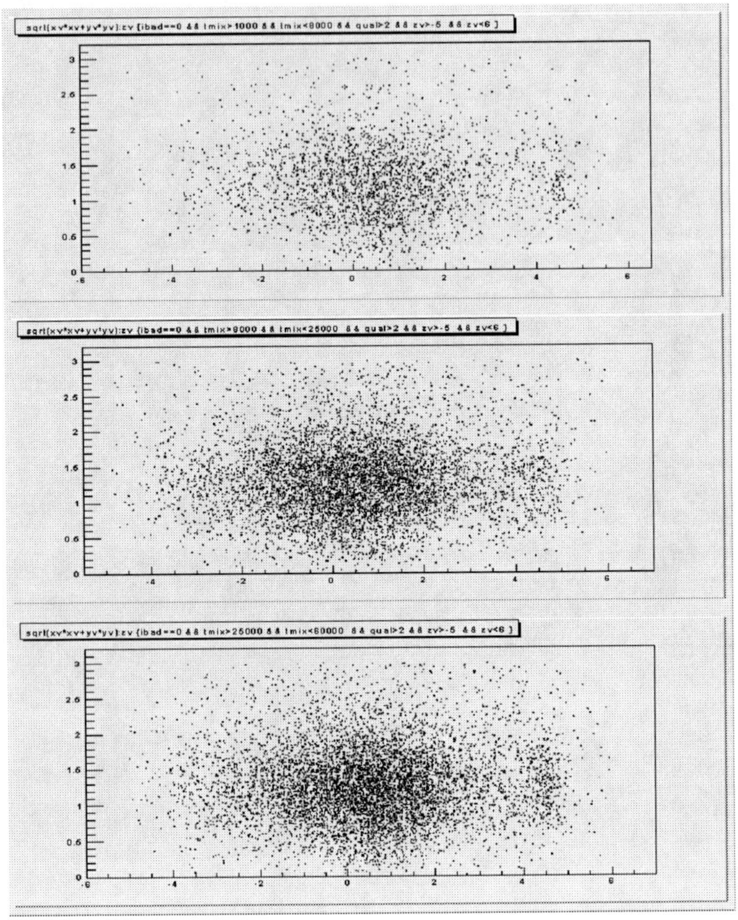

FIGURE 20. $\bar{H}$ production by "expanded" positrons cloud. The time intervals as in Fig.18.

REFERENCES

1. M. Amoretti *et al.*, *Nature* **419**, 456 (2002).
2. G. Gabrielse *et al.*, Phys. Rev. Lett. **89**, 213401 (2002).
3. M. Amoretti *et al.*, Phys. Lett. B **578**, 23 (2004).
4. M. Amoretti *et al.*, Phys. Lett. B **583**, 59 (2004).
5. G. Gabrielse *et al.*, Phys. Rev. Lett. **89**, 233401 (2002).
6. G. Gabrielse *et al.*, Phys. Rev. Lett. **93**, 073401 (2004)
7. N. Madsen *et al.*, Phys. Rev. Lett. **94**, 033403 (2005).
8. M. Amoretti *et al.*, Nucl. Instrum. Meth. A **518**, 679 (2004).
9. M. C. Fujiwara *et al.*, Phys. Rev. Lett. **92**, 065005 (2004).
10. N. Zurlo *et al.*, Phys. Rev. Lett. **97**, 153401 (2006).
11. E. Lodi Rizzini *et al.*, ChemPhysChem **8**, 1145 (2007)

A New Time-Dependent Scattering Theory: Application to the Capture of Antiprotons by Hydrogen Atoms and Helium Atoms

X. M. Tong[a,b], K. Hino[b,c], and N. Toshima[a]

[a]*Institute of Materials Science, Graduate School of Pure and Applied Sciences, University of Tsukuba, 1-1-1 Tennodai, Tsukuba, Ibaraki 305-8573, Japan*
[b]*Center for Computational Sciences, University of Tsukuba, 1-1-1 Tennodai, Tsukuba, Ibaraki 305-8577, Japan*
[c]*Institute of Frontier Science, Graduate School of Pure and Applied Sciences, University of Tsukuba, 1-1-1 Tennodai, Tsukuba, Ibaraki 305-8573, Japan*

Abstract. We present a theoretical method for Coulomb three-body rearrangement collisions solving a Chew-Goldberger-type integral equation directly. The scattering boundary condition is automatically satisfied by adiabatically switching on the interaction between the projectile and target. Hence the outgoing wave function is obtained without the tedious procedure of adjusting the total wave function in the asymptotic region. All the dynamical information can be derived from the scattering wave function obtained on pseudo-spectral grids numerically. Using this method, we obtained the state-specified capture cross sections when antiprotons collide with hydrogen atoms or helium atoms. Differing from the capture processes of antiprotons by hydrogen atoms, the anomalous bumpy structures are revealed in the total angular momentum dependent capture cross sections by helium atoms. Further analysis shows that the bumps arise from the partial channel closing due to the removal of the energy degeneracy in the antiprotonic helium atom. The ejected electron energy distributions are also provided for the comparison with future experiments.

Keywords: Coulomb three-body rearrangement, time-dependent scattering theory, antiprotons
PACS numbers: 34.70.+e, 36.10.-k, 25.43.+t

INTRODUCTION

Coulomb three-body rearrangement collisions have been studied extensively as one of the most fundamental dynamical processes in atomic physics. They serve as a prototype of general ion-atom collisions and electron-atom collisions. In high energy region, the perturbative treatments, such as the Born approximation, are generally applicable and the theoretical study of them is not so difficult. As the collision energy decreases higher-order effects become important and we are obliged to use nonperturbative approaches. For heavy negative charged particles captured by atoms, the number of capture channels increases dramatically due to the huge mass difference between electron and negative charged particles. Thus, the computational time required by the nonperturbative approaches, e.g., the close-coupling method, inevitably increases dramatically as discussed by Esry and Sadeghpour [1]. It is very difficult to solve the scattering equations

CP1037, *Cold Antimatter Plasmas and Application to Fundamental Physics,* edited by Y. Kanai and Y. Yamazaki

without reducing the antiproton mass artificially. Protonium formation cross sections were calculated by scaling down the proton and antiproton masses to 20 a.u. Hesse *et al.* [2] used the diabatization technique to calculate the protonium formation but they were also obliged to use scaled proton and antiproton masses of 100 a.u. Their formation cross sections are peaked at $n = 7$, which is much lower than the estimated value of $n = 30$ for the protonium with the real mass, 1836 a.u. There is no definite way to convert the n-distribution of the scaled-mass calculations to that of the real system. Cohen [3] calculated the protonium formation cross section by the classical trajectory Monte Carlo (CTMC) method. The applicability of the classical mechanics at such low energies is not clear, especially for the motion of the electron. Besides ambiguity occurs in extracting the discrete quantum numbers n and ℓ from the classical continuous quantities.

Sakimoto [4] studied the formation process by the time-dependent wavepacket method but he did not obtain the state-specified formation cross section due to the numerical difficulty. He calculated the formation probability by subtracting the sum of elastic and excitation probabilities from unity. In order to localize the wave packet in a sufficiently small region, one has to superpose a bunch of states belonging to different collision energies so that keeping high accuracy throughout the collision is not easy. Ovchinnikov and Macek [5] applied the advanced adiabatic approach and presented n-dependent formation cross sections. Although their sophisticated method may be accurate qualitatively for the relevant process, it is not easy to estimate the quantitative reliability for the minute information such as the n-distribution. Yamanaka and Ichimura [6] studied the protonium formation by solving Faddeev equations only involved the ejected electron in the s-wave states. To the best of our knowledge, reliable state-specified formation cross sections have not been obtained yet from nonperturbative quantal calculations apart from our recent works [7, 8].

As mentioned above, the traditional coupled-channel method for the time-independent differential Schrödinger equation is not practically applicable to the process. To solve the problem differently, we start from the Chew and Goldberger integral representation of the scattering equation. The advantage of this method is that we transfer the complicated boundary conditions into a simple initial condition and convert the time-independent scattering theory into the time-dependent scattering theory. The theoretical details will be presented in the following section, followed by the applications of the theory to the captures of antiprotons by hydrogen atoms and helium atoms.

THEORETICAL METHOD

General time-dependent formalism

The time-independent Schrödinger equation of a Coulomb three-body system in the center-of-mass frame is written as (atomic units $\hbar = m_e = e = 1$ are used hereafter unless otherwise stated)

$$H|\Psi> = E|\Psi>, \tag{1}$$

with

$$H = -\frac{1}{2\mu_e}\nabla_{\mathbf{r}}^2 - \frac{1}{2\mu_{\bar{p}}}\nabla_{\mathbf{R}}^2 + V_{n,e}(\mathbf{r}) + V_{n,\bar{p}}(\mathbf{R}) + V_{\bar{p},e}(\mathbf{r},\mathbf{R}) \tag{2}$$

Here $E = E_c + E_{1s}$ is the total energy of the collision system in the center-of-mass frame with E_c the incident energy of the projectile and E_{1s} the ground state energy of target atom. $\mathbf{r}$ and $\mathbf{R}$ are respectively the distances from the nucleus to the electron and from the center-of-mass of the atom to the projectile. $V_{n,e}(\mathbf{r})$, $V_{n,\bar{p}}(\mathbf{R})$, and $V_{\bar{p},e}(\mathbf{r},\mathbf{R})$ are the interactions of the nucleus and the electron, the nucleus and the antiproton, and the electron and the antiproton, respectively. μ_e is the reduced mass of the nucleus and the electron and $\mu_{\bar{p}} = M_{\bar{p}}(m_e + M_n)/(M_n + m_e + M_{\bar{p}})$. M_n, m_e and $M_{\bar{p}}$ are the masses of nucleus, electron and antiproton, respectively. The Hamiltonian can be recast into two forms, before and after the collision, for rearrangement processes,

$$H_i|\Psi_i> \;=\; E|\Psi_i>, \tag{3}$$

$$H_f|\Psi_f> \;=\; E|\Psi_f>, \tag{4}$$

and the corresponding interactions are,

$$V_i = H - H_i, \tag{5}$$

$$V_f = H - H_f. \tag{6}$$

The solutions of Eq. (1) satisfying a proper boundary condition can be conveniently expressed utilizing an integral expression;

$$\begin{aligned}\Psi^+(t) &= \Psi_i - i\int_{-\infty}^{t} e^{-i(H-E)(t-t')} f(t')V_i\Psi_i dt' \\ &= \Psi_i - i\int_{-\infty}^{t} e^{iE(t-t')}U(t,t')f(t')V_i\Psi_i dt'.\end{aligned} \tag{7}$$

The time-dependent wave function $\Psi^+(t)$ satisfies Eq. (3) as $t \to -\infty$, and it also satisfies Eq. (1) at $t = 0$ if $f(-\infty) = 0$, $f(0) = 1$ and $f(t)$ switches on adiabatically [9], i.e., $\frac{d}{dt}f(t) \to 0$. A choice of $f(t) = \exp(-\eta|t|)$ gives Chew-Goldberger formal solution at $t = 0$ as

$$\Psi^+ = \Psi_i + \frac{1}{E - H + i\eta}V_i\Psi_i, \tag{8}$$

where η is a positive infinitesimal number. We integrate Eq. (7) propagating the time-dependent part instead of solving Eq. (1) or Eq. (8). $f(t)$ can be any function so long as it is switched on adiabatically. For instance, $f(t) = \exp(-t^2/\tau^2)$ can be used for $\tau \to \infty$. Once we get $\Psi^+(t=0)$ by propagating Eq.(7), we can derive all dynamical information of the collision system from it.

Numerical procedures

We have introduced a general theory for solving scattering problems. Coulomb three-body systems cover a variety of physical processes. The characteristic of each system depends strongly on the combination of the masses and the charges of the constituting particles. We focus on dealing with heavy negatively-charged particles, antiprotons, colliding with hydrogen atoms as examples to show how to propagate Eq. (7) effectively and accurately. The initial wave function Ψ_i of the system is the product of the ground state wave function of the target atom and the wave function of the projectile;

$$\Psi_i(\mathbf{r},\mathbf{R}) = \psi_{1s}(\mathbf{r})\psi_c(\mathbf{k}_0,\mathbf{R}), \tag{9}$$

where $\mathbf{k}_0$ is the initial momentum of the projectile in the center-of-mass frame. $\psi_{1s}(\mathbf{r})$ is the ground state wave function of the Hamiltonian $H(r) = -\frac{1}{2\mu_e}\nabla^2_{\mathbf{r}} + V_{n,e}(\mathbf{r})$ and $\psi_c(\mathbf{k}_0,\mathbf{R})$ is the wave function of the Hamiltonian $H = -\frac{1}{2\mu_{\bar{p}}}\nabla^2_{\mathbf{R}} + V_{n,\bar{p}}(\mathbf{R}) - \frac{1}{R}$ with the incoming wave boundary condition. The Hamiltonian of the initial channel and the corresponding interaction are

$$H_i = -\frac{1}{2\mu_e}\nabla^2_{\mathbf{r}} + V_{n,e}(\mathbf{r}) - \frac{1}{2\mu_{\bar{p}}}\nabla^2_{\mathbf{R}} + V_{n,\bar{p}}(\mathbf{R}) - \frac{1}{R}, \tag{10}$$

$$V_i = V_{\bar{p},e}(\mathbf{r},\mathbf{R}) + \frac{1}{R}. \tag{11}$$

Since the total angular momentum L, its component M onto the incident beam direction, and the parity are good quantum numbers, we can expand the time-dependent wave function as

$$\Psi(t) = \sum_{LM}\Psi_{LM}(\mathbf{r},\mathbf{R},t) = \sum_{LM}\sum_{\alpha}F^{\alpha}(r,R,t)\Omega^{\alpha}_{LM}(l_e,l), \tag{12}$$

with

$$\Omega^{\alpha}_{LM} = \sum_{m_e,m} < l_e m_e l m | LM > Y_{l_e m_e}(\hat{r})Y_{lm}(\hat{R}). \tag{13}$$

Here α represents the angular momentum of the electron l_e, the angular momentum of the antiproton l, and the parity collectively. Each $\Psi_{LM}(\mathbf{r},\mathbf{R},t)$ can be propagated independently. The corresponding radial part of the total Hamiltonian for a given LM is written as

$$\begin{aligned} < \Omega^{\alpha}_{LM}|H|\Omega^{\alpha'}_{LM} > &= \left(-\frac{1}{2\mu_{\bar{p}}}\frac{\partial^2}{\partial R^2} + \frac{l(l+1)}{2\mu_{\bar{p}}R^2} + V_{n,\bar{p}}(\mathbf{R})\right)\delta_{\alpha,\alpha'} \\ &+ \left(-\frac{1}{2\mu_e}\frac{\partial^2}{\partial r^2} + \frac{l_e(l_e+1)}{2\mu_e r^2} + V_{n,e}(\mathbf{r})\right)\delta_{\alpha,\alpha'} \\ &+ < \Omega^{\alpha}_{LM}|V_{\bar{p},e}(\mathbf{r},\mathbf{R})|\Omega^{\alpha'}_{LM} > \\ &= \mathbf{H}^{\alpha}(R) + \mathbf{H}^{\alpha}(r) + \mathbf{V}^{\alpha,\alpha'}(r,R). \end{aligned} \tag{14}$$

After the partition, the time propagation from t to $t+\Delta t$ can be performed by the split-operator method in the energy representation [10, 11] as

$$U(t+\Delta t,t) = e^{-i\mathbf{V}(r,R)\Delta t/2}e^{-i\mathbf{H}(R)\Delta t}e^{-i\mathbf{H}(r)\Delta t}e^{-i\mathbf{V}(r,R)\Delta t/2} + O(\Delta t^3).$$

We first propagate the wave function a half-time step in $\mathbf{V}(r,R)$, and then a full-time step in $\mathbf{H}(r)$ and $\mathbf{H}(R)$, and finally another half-time step in $\mathbf{V}(r,R)$. The advantage of this time-propagation scheme is that each operator is diagonal in the other two dimensions. $\mathbf{V}(r,R)$ couples only different channels $\alpha \neq \alpha'$ and $\mathbf{H}(R)$ and $\mathbf{H}(r)$ are, on the other hand, diagonal for (α,α'). These characters make the time propagation computationally efficient.

For three-body rearrangement collisions, we also have to deal with continuum wave functions. In the direct integration method on grid points, the whole coordinate space is confined in a finite box. We have to filter out the outgoing waves at the boundary in order to eliminate the unphysical reflection. For this purpose, we add an optical potential near the boundary to absorb the outgoing flux. The optical potential [12] we used is of the form of

$$V_{opt}(\varepsilon,r) = \begin{cases} 0 & \text{for } r < r_c \\ -iV_0\left(\frac{r-r_c}{r_{max}-r_c}\right) & \text{for } r \geq r_c \end{cases}, \tag{15}$$

with

$$2\varepsilon^{1/2}/A < V_0 < \varepsilon^{3/2}A, \tag{16}$$

$$A = \sqrt{2\mu}(r_{max} - r_c). \tag{17}$$

Here ε is the energy of the basis wave function and r_{max} and r_c are the position of the boundary in the radial direction and the position to start the filtering, respectively. μ is the reduced mass associated with coordinate r. The optical potential added to Hamiltonian $\mathbf{H}(r)$ is

$$V_{opt}(r) = \sum_i V_{opt}(\varepsilon_i,r)|\psi_i><\psi_i| \quad \text{for} \quad \varepsilon_i \geq 0, \tag{18}$$

where $\{\varepsilon_i\}$ and $\{\psi_i\}$ are the eigen energies and eigen wave functions of $\mathbf{H}(r)$. The $V_{opt}(r)$ depends on the grid structure but it does not depend on the collision energy. It is a universal optical potential which can be used to absorb the ejected electron in a broad energy range. Similarly we add an optical potential to $\mathbf{H}(R)$. After a long time propagation from $t=-\infty$ to $t=0$, in the interaction region, we get the scattering wave function $\Psi^+(t=0)$, which satisfies the proper boundary condition. The final wave function corresponding to the capture to the state of a principal quantum number n and an angular momentum l is expressed as

$$\Psi_f^\alpha = j_{l_e}(k_{nl}r)\psi_{nl}(R)\Omega_{LM}^\alpha \tag{19}$$

where k_{nl} is the ejected electron momentum associated with the formation of an exotic atom in the nl state. The T-matrix element is given by

$$T_{nl}^\alpha = <\Psi_f^\alpha|V_f|\Psi_{LM}^+>, \tag{20}$$

and the state specified capture cross section is readily obtained as

$$\sigma_{nl}^\alpha(L) = \frac{4\mu_{\bar{p}}\,\mu_e\,k_{nl}}{k_0}\sum_\alpha |T_{nl}^\alpha|^2. \tag{21}$$

We introduce two other ways to calculate the capture cross sections. The wave function $\Psi^{+}_{LM}(0)$ can be further recast into

$$< \Omega^{\alpha}_{LM}|\Psi^{+}_{LM} >=< \Omega^{\alpha}_{LM}|\Psi_i > + \sum a^{\alpha}_{nl_e}\psi_{nl}(r)F^{\alpha}(R) + \sum_{nl} b^{\alpha}_{nl}\psi_{nl}(R)f^{\alpha}_{nl}(r). \tag{22}$$

The first term represents the incident channel, the second term stands for the excitation and elastic channel, and the third term is the capture channel. The asymptotic form of the ejected electron wave function is

$$f^{\alpha}_{nl}(r) = \frac{e^{ik_{nl}r}}{r} \quad \text{for} \quad r \to \infty. \tag{23}$$

Define

$$G^{\alpha}_{nl}(r) = \int \psi^{*}_{nl}(R) < \Omega^{\alpha}_{LM}|\Psi^{+} > dR, \tag{24}$$

and we obtain b^{α}_{nl} as

$$\frac{k_{nl}}{\mu_e}|b^{\alpha}_{nl}|^2 = \Re\left\{-\frac{i}{\mu_e}(rG^{\alpha}_{nl}(r))^{*}\frac{\partial}{\partial r}(rG^{\alpha}_{nl}(r))\right\}_{r_c}, \tag{25}$$

where $\Re$ denotes the real part of a complex number. The definition of the capture cross sections gives

$$\sigma^{\alpha}_{nl}(L) = \frac{\mu_{\bar{p}}\, k_{nl}}{\mu_e\, k_0}\sum_{\alpha}|b^{\alpha}_{nl}|^2. \tag{26}$$

Suppose that we calculate b^{α}_{nl} in Eq. (25) at each time t and propagate the wave function of Eq. (7) with $f(t) = \exp(-t^2/\tau^2)$ from $-\infty$ to ∞, then the averaged ejected electron current is

$$\frac{k_{nl}}{\mu_e}|\bar{b}^{\alpha}_{nl}|^2 = \frac{\sqrt{2}}{\sqrt{\pi}\tau}\int_{-\infty}^{\infty}\frac{k_{nl}}{\mu_e}|b^{\alpha}_{nl}(t)|^2 dt. \tag{27}$$

The capture cross section can be calculated in the third way as

$$\sigma^{\alpha}_{nl}(L) = \frac{\mu_{\bar{p}}\, k_{nl}}{\mu_e\, k_0}\sum_{\alpha}|\bar{b}^{\alpha}_{nl}|^2. \tag{28}$$

Eq. (28) corresponds to the procedure that we collect all the ejected electrons associated with capture to the nl states in the whole collision process. If the wave function $\Psi^{+}(0)$ is converged satisfactorily, the three expressions, Eqs. (21), (26) and (28), should give the same results. In the following calculations, we always check the cross sections calculated by the above three expressions to make sure that the results are converged. We also change the simulation parameters, such as the size of the interaction region, the number of the partial wave, the number of the grid point, and the propagation time, and so on, to check the numerical stability. The extensive convergence test was presented in our previous work [8].

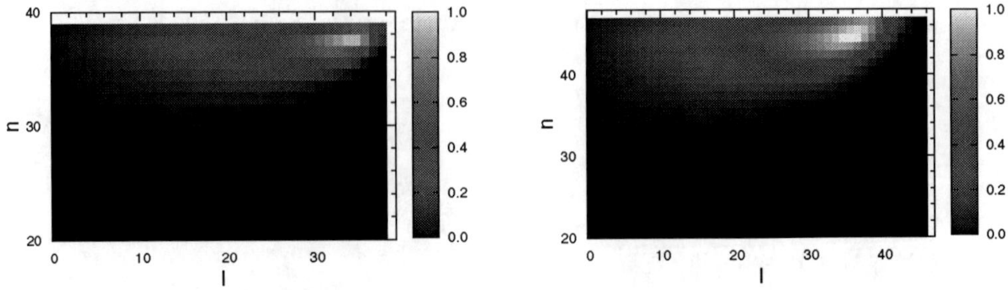

FIGURE 1. The normalized n,ℓ-dependent protonium formation cross sections at 5.44 eV (left) and 8.16 eV (right) incident energies.

RESULTS AND DISCUSSION

Based on the above theoretical method, we can investigate the dynamical processes involved in a Coulomb three-body problem or many-body problem which can be reduced to a three-body problem approximately. Following, we will present our studies on the antiproton captured by hydrogen atom, which stands for a Coulomb three-body rearrangement process and the antiproton captured by helium atom, which stands for a coulomb three-body rearrangement reduced from many-body system by frozen one of the two electrons in the helium atom.

Antiprotons captured by hydrogen atoms

By the present method, we can obtain all the detailed scattering information, including the state-specified capture cross-sections. Figure 1 shows a 2-dimensional plot of the state specified (n,ℓ dependent) cross sections $\sigma_{n\ell} = \sum_{LM} \sigma_{LM}^{n\ell}$ at 5.44 and 8.16 eV incident energies. For the lower incident energy, the protonium atoms are mainly formed in the states with the angular momentum $\ell \sim 32$ and the principal quantum number $n \sim 37$. The formation cross sections drop quick as the n decreases. The cross sections are of a broad distribution as a function of the ℓ. As the incident energy increases from 5.44 to 8.16 eV, the protonium atoms are mainly formed in the states with $\ell \sim 36$ and $n \sim 44$. Similar to the lower incident energy case, the probability of formation in the lower n decreases quickly. Though the 2-dimensional image can give a global feature well, minute information is not easy to be presented from them. We give ℓ-dependent cross sections ($\sigma_\ell = \sum_n \sigma_{n\ell}$) and n-dependent cross sections ($\sigma_n = \sum_\ell \sigma_{n\ell}$) separately in Fig. 2. The ℓ-distribution of the formation cross sections increases smoothly as ℓ increases, reaches a peak position and then drops quickly as ℓ increases further. For the higher incident energy, the distribution is getting broader. The n-distribution shows a sharp increase and decrease around the peak position, which means the antiproton is mainly captures in the high n region. The n distribution of the formation cross sections shows large energy dependence at variance with the simple estimate that, based on the energy matching consideration between the initial and final bound states, the protonium

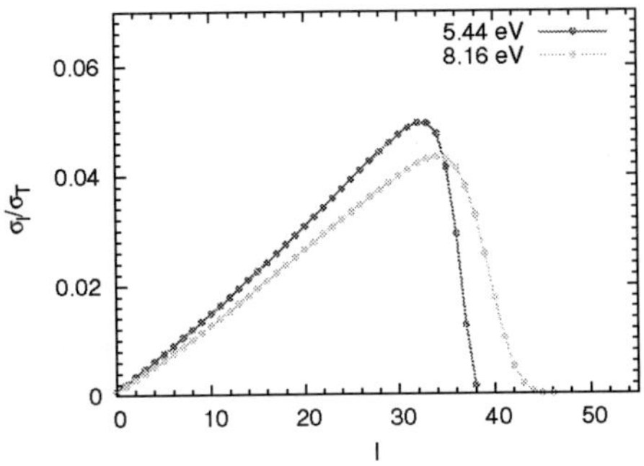

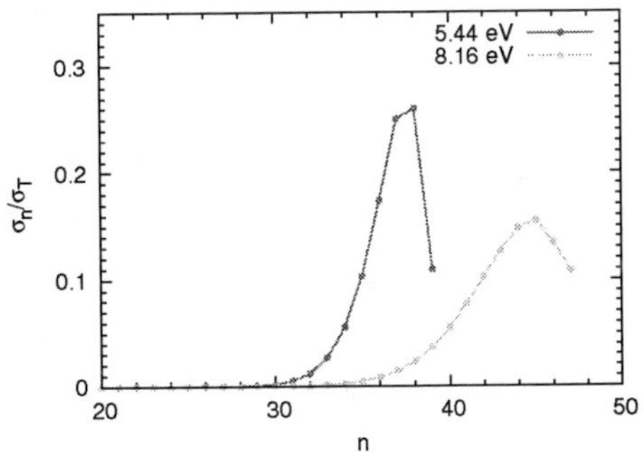

FIGURE 2. The ℓ (left) and n (right) dependent protonium formation cross sections at 5.44 and 8.16 eV incident energies.

is mainly formed in the states with $n \sim \sqrt{\mu} \sim 30$ where μ is the reduced mass of the protonium. This prediction holds only for the lowest incident energy. As the incident energy increases, the n-distribution becomes broader and the peak position moves to higher n.

From our present calculation, we obtain all the detailed capture cross sections $\sigma^L_{n\ell}$. In Fig. 3 we show the ejected electron energy distribution, given by,

$$P(\varepsilon) = \frac{d\sigma}{d\varepsilon} = \sum_{L,n,\ell} \delta(\varepsilon - \varepsilon_{Ln\ell})\sigma^L_{n\ell}. \tag{29}$$

Here $\varepsilon_{Ln\ell}$ is the ejected electron energy associated with the antiproton captured to the (n,ℓ) states with the total angular momentum L. In Fig. 3, we see that the ejected electrons mainly distribute in the low energy region ($< 10eV$) and peak around 1 eV. The thin curves show the non-isotropic distribution of the ejected electron, which is much smaller than the total ejected electron energy distribution. Thus, we can conclude that the ejected electron is almost isotropically distributed in the center of mass frame.

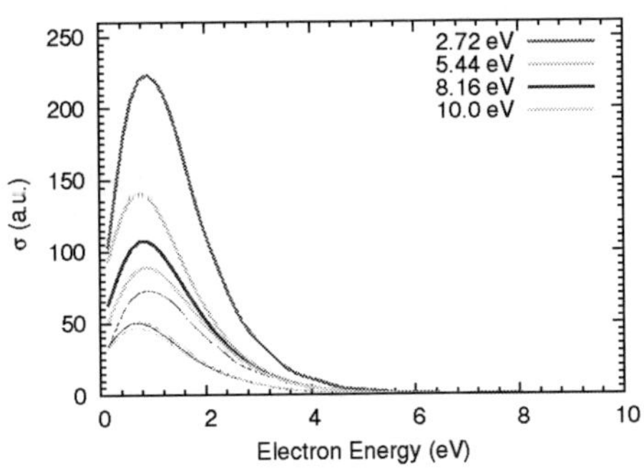

FIGURE 3. The total ejected electron energy distributions (thick curves) and the non-isotropic components (thin curves).

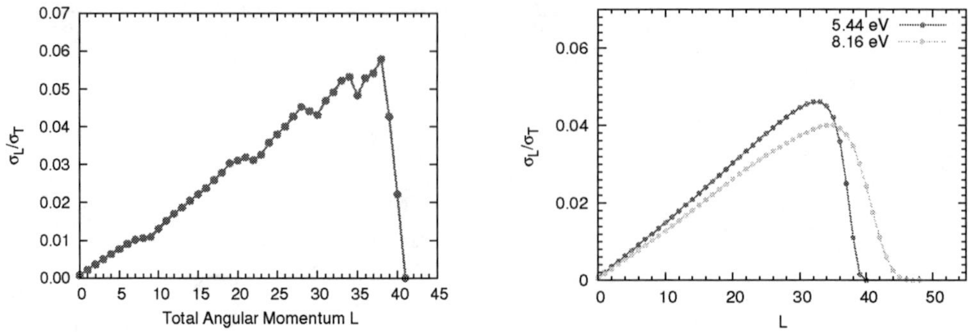

FIGURE 4. The L-dependent capture cross sections of antiprotons by helium atoms at 8 eV (left) and by hydrogen atoms (right).

Antiprotons captured by helium atoms

The capture process of antiprotons by helium atoms is a Coulomb four-body problem. So far, we can not solve the problem non-perturbatively even with the largest supercomputer. Thus, we have to solve the problem approximately. To make the problem fit to the above theoretical method, we freeze one of the two electrons in the helium atoms and use a model potential [13] to describe the active electron interaction with the helium ion and a adiabatic potential to describe the antiproton interaction with the helium ion. With this approximation, we can obtain the state-specified capture cross sections.

Figure 4 shows the L-dependent capture cross sections. To compare with, we also plot the corresponding cross sections when the antiproton captured hydrogen atoms. For the capture by hydrogen atoms, as shown on the right side of Fig. 4, the capture cross sections increase smoothly as L increases, reach a peak at L = 30 and then drop quickly as L increases further. The peak position moves to the higher L as the incident energy increases. For the capture by helium atoms, as shown on the left side of Fig. 4, the capture cross sections first increase in an irregular manner, reach the maximum at $L = 38$, and then, drop sharply. This global feature is similar to the capture of antiprotons by hydrogen atoms. Unlike the hydrogen case, however, the present cross sections show the anomalous bumpy structures. It is notable that the present calculations are more difficult than those of the hydrogen case due to the larger reduced mass μ_p and the larger effective charge of helium. To check the numerical accuracy, the simulation conditions, such as the number of grids, the size of the space surrounding the whole system, and so on, have been varied. As the result, it is convinced that the anomalies manifested in Fig. 4 are insensitive to the changes of these parameters. Thus, the anomalies are ascribable to a certain physics, not due to the numerical inaccuracy.

Although the capture process of the antiprotons by hydrogen atoms and helium atoms are similar to each other within the present approximation, the energy structures of protonium ($\bar{p}p$) and antiprotonic helium ($\bar{p}He^+$) are different. Replacing the Coulomb potential in the former by the effective potential in the latter, the energy degeneracy assigned by n and ℓ is removed, where n and ℓ are the principle and angular momentum quantum numbers, respectively. Figure 5 shows the energy structures of antiprotonic

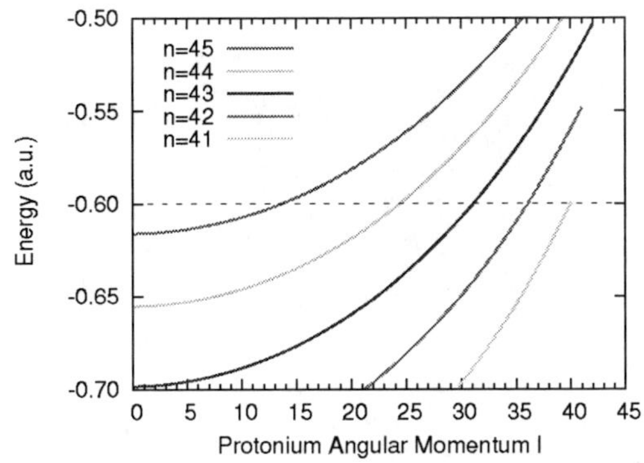

FIGURE 5. The energy levels of antiprotonic helium. The horizontal dashed line is the total energy for 8 eV incident energy.

helium. Apparently, for a given n, the energy increases as the ℓ increases. Following the Hund's rule, a lower angular momentum state can penetrate into the deeper inner region than a higher angular momentum state, resulting in a lower binding energy for the lower ℓ state with the same n.

The same n and different l states form a channel (the energy curve in Fig. 5). For instance, channel 44 includes all the quantum states with $n = 44$ and different ℓ's. The energy curve for a given channel crosses the total energy (dashed straight line in Fig. 5). The cross over positions are in good agreement with the dip positions in the total L-dependent capture cross sections in Fig. 4. Let us focus on channel 44 as an example to show the original of the dips. With 8 eV incident energy, the antiprotons can be captured to $n = 44$ and $\ell \leq 25$ due to the energy conservation. This means that this channel is open for the capture to $\ell \leq 24$ states and is closed for the capture to $\ell \geq 25$ states. This fact would allow us to speculate that the manifestation of the anomaly around $L \sim l = 25$ is attributed to the channel closing of the states with $l \geq 25$. This speculation is more clarified by comparing the situations between the cases of the hydrogen and helium targets. In the former, all channels with $n = 44$ and different ℓ's contribute to the capture cross sections because of the energy degeneracy, and the capture cross sections as a function of L show no anomaly. In contrast to this, in the latter, the states with $n = 44$ and $l > 25$ do not contribute to the capture cross sections, leading to the dip. In a similar way, it is showed that the next dip corresponds to the closing of channel 43 at L=30, and so on. The similar phenomena are also observed at other incident energies with the change of the dip positions and hence this anomaly is considered a general phenomenon.

Figure 6 shows the state-specified capture cross sections, where the gradated trace is depicted in the (n, ℓ) plane. The antiprotons are mostly captured to the region around $n \sim 42$ and $\ell \sim 38$ states, which is much higher than that of the hydrogen case. The detailed results of antiprotons captured by the helium atoms will be present elsewhere.

To summarize, we presented a new time-dependent scattering theory, which can be used to study general Coulomb three-body rearrangement processes. The key procedure is that we transfer the very complicated boundary conditions into a simple initial condition. Using this method, we obtained the state-specified capture cross sections when

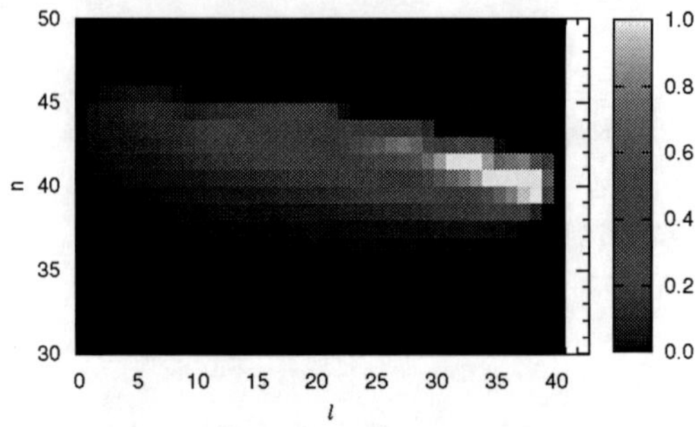

FIGURE 6. The state-specified capture cross sections normalized to the total capture cross section of 8.0 eV incident energy for antiprotons captured by helium atoms.

antiprotons collide with hydrogen atoms or helium atoms. Differing from the capture processes of antiprotons by hydrogen atoms, the anomalous bumpy structures are revealed in the total angular momentum dependent capture cross sections by helium. Further analysis shows that the bumps arise from the partial channel closing due to the removal of the energy degeneracy in the antiprotonic helium atom. Ejected electron energy distributions are also provided for the comparison with future experiments.

ACKNOWLEDGMENTS

This work was supported by Grant-in-Aids for Scientific Research (B) and (C) from the Japan Society for the Promotion of Science.

REFERENCES

1. Esry, B. D., and Sadeghpour, H. R., *Phys. Rev. A*, **67**, 012704 (2003).
2. Hesse, M., Le, A. T., and Lin, C. D., *Phys. Rev. A*, **69**, 052712 (2004).
3. Cohen, J. S., *Phys. Rev. A*, **36**, 2024–2030 (1987).
4. Sakimoto, K., *J. Phys. B*, **35**, 997–1014 (2002).
5. Ovchinnikov, S. Y., and Macek, J. H., *Phys. Rev. A*, **71**, 052717 (2005).
6. Yamanaka, N., and Ichimura, A., *Phys. Rev. A*, **74**, 012503 (2006).
7. Tong, X. M., Hino, K., and Toshima, N., *Phys. Rev. Lett.*, **97**, 243202 (2006).
8. Tong, X. M., Shirahama, T., Hino, K., and Toshima, N., *Phys. Rev. A*, **75**, 052711 (2007).
9. Avron, J. E., and Elgart, A., *Phys. Rev. A*, **58**, 4300–4306 (1998).
10. Tong, X. M., and Chu, S. I., *Chem. Phys.*, **217**, 119–130 (1997).
11. Telnov, D. A., and Chu, S.-I., *Phys. Rev. A*, **59**, 2864 (1999).
12. Perie, J., and Jolicard, G., *J. Phys. B*, **26**, 4491–4502 (1993).
13. Tong, X. M., and Lin, C. D., *J. Phys. B*, **38**, 2593–2600 (2005).

Atomic Collision Experiments with Ultra-Low-Energy Antiprotons

Hiroyuki A. Torii*, Yugo Nagata*,†, Hiroshi Imao†, Victor L. Varentsov**, Naofumi Kuroda*, Masahiro Shibata†, Koremitsu Ogata*, Hiroshi Toyoda*, Takuya Shimoyama*, Yoshinori Enomoto*, Hiroyuki Higaki‡, Yasuyuki Kanai†, Akihiro Mohri† and Yasunori Yamazaki*,†

*Institute of Physics, University of Tokyo, 3-8-1 Komaba, Meguro-ku, Tokyo 153-8902, Japan
†Atomic Lab., RIKEN, 2-1 Hirosawa, Wako-shi, Saitama 351-0198, Japan
**V. G. Khlopin Radium Institute, 2-nd Murinsky Ave. 28, 194021 St. Petersburg, Russia
‡Graduate School of Advanced Sciences of Matter, Hiroshima University, 1-3-1 Kagamiyama, Higashi-Hiroshima, Hiroshima, 739-8530, Japan

Abstract. We have developed techniques to decelerate, cool and confine antiprotons in vacuo with an electromagnetic trap, for production of a Monoenergetic Ultra-Slow Antiproton Source for High-precision Investigation (MUSASHI) over the last several years. The ultra-slow antiproton beam which can now be extracted stably has opened up the possibility to study ionization and atomic capture processes between an antiproton and an atom at an unprecedented low energy under the single-collision condition for the first time. The collision energy can be tuned from 10 eV to 1 keV either by varying the beam transport energy or by biasing the voltages at the collision region.

Since the number of available antiprotons is very much limited, the reaction probability must be maximized in order to make best use of them. We have prepared a powerful supersonic helium gas jet with a density of 3×10^{12} atoms/cm^3 to be crossed with the antiproton beam. For rigorous identification of particles (e^-, $\bar{p}$ and $(\bar{p}A^+)^0$) needed for reduction of huge background signals, we developed a detection system with two microchannel plates each with a delay-line two-dimensional position sensitive detector, and a box of scintillator plates. A set of electrodes and coils were placed near the collision point to guide electromagnetically the electrons perpendicular to the antiproton beam. The reaction events will be recognized by an electron signal followed by an antiproton annihilation with an appropriate time of flight. Our design and strategy of the experiment are discussed.

Keywords: antiproton, ionization, atomic capture, antiprotonic atoms, supersonic gas jet, Antiproton Decelerator (AD), RFQD (Radio Frequency Quadrupole Decelerator), Multi-Ring electrode Trap (MRT), ultra-slow antiproton beam.
PACS: 34.90.+q, 7.77.Ka, 36.10.Gv, 47.40.Ki, 29.40.Gx, 52.27.Jt

INTRODUCTION

Antiproton is a unique probe for atomic physics researches. Having the same mass as the proton but with opposite charge, It behaves as a "heavy electron" or as a "negative nucleus", and gives a new test ground for studies of atomic collision dynamics. In fact, it is a "theoreticians' ideal projectile" because lack of electron capture processes avoids complication in theoretical treatments. In addition to various CTMC (Classical Trajectory Monte Carlo) calculations hitherto available [1], recently semi-quantum treatments have become possible [2, 3] (see Fig. 1). Yet so far no experiment was possible at energies less than around 5 keV [4, 5, 6].

CP1037, *Cold Antimatter Plasmas and Application to Fundamental Physics,* edited by Y. Kanai and Y. Yamazaki

Experimentally, antiprotons can be produced at accelerator facilities, but only at a huge energy in the GeV range. The Antiproton Decelerator (AD) [7] at CERN, Geneva, is a unique facility which provides cooled low-energy antiprotons with MeV energies. There, the ASACUSA collaboration has been doing researches on "Atomic Spectroscopy And Collisions Using Slow Antiprotons" [8].

Along with a wide scientific program such as laser and microwave spectroscopy of metastable antiprotonic helium atom [9, 10, 11, 12], nuclear collision experiment with antiprotons etc. in the collaboration, our subgroup, the MUSASHI group deals with atomic collision dynamics involving ultra-low-energy antiprotons. In the experimental realization, an ultra-slow antiproton beam at a well-defined energy in the eV to keV range must be prepared for antiproton–atom cross-beam experiment. We have developed a system consisting of an RFQ Decelerator, an electromagnetic trap and an extraction beam line, and have recently succeeded in production of ultra-low-energy monoenergetic antiproton beams. These beams are crossed with an atomic gas-jet target, to study ionization and atomic capture processes of an antiproton projectile against various atoms "A",

$$\overline{p} + \mathrm{A} \rightarrow \overline{p} + \mathrm{A}^+ + e^-$$

$$\overline{p} + \mathrm{A} \rightarrow (\overline{p}\mathrm{A}^+)^0 + e^-$$

for the first time under the single-collision condition at very low energies. This study also reveals the properties of exotic antiprotonic atoms formed in vacuo.

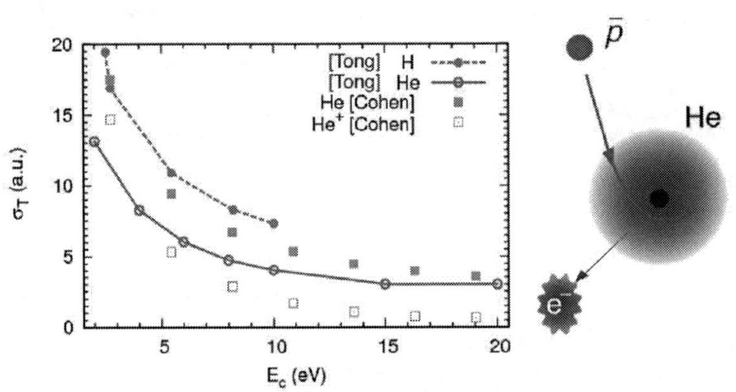

FIGURE 1. Theoretically calculated total cross section of the antiproton colliding with hydrogen and helium atoms [1, 2]. The antiproton is captured into atomic orbits to form an antiprotonic atom below the ionization threshold (13.6 eV for hydrogen and 24.6 eV for helium atom).

DECELERATION OF ANTIPROTON BEAM

At CERN, antiprotons are produced by collision of protons at 26 GeV/c with an Ir target via the reaction $p + p \rightarrow p + p + p + \overline{p}$. A part of them, 5×10^7 in number, are collected at 3.6 GeV/c and stored in the AD ring [7], which are then cooled via electron cooling

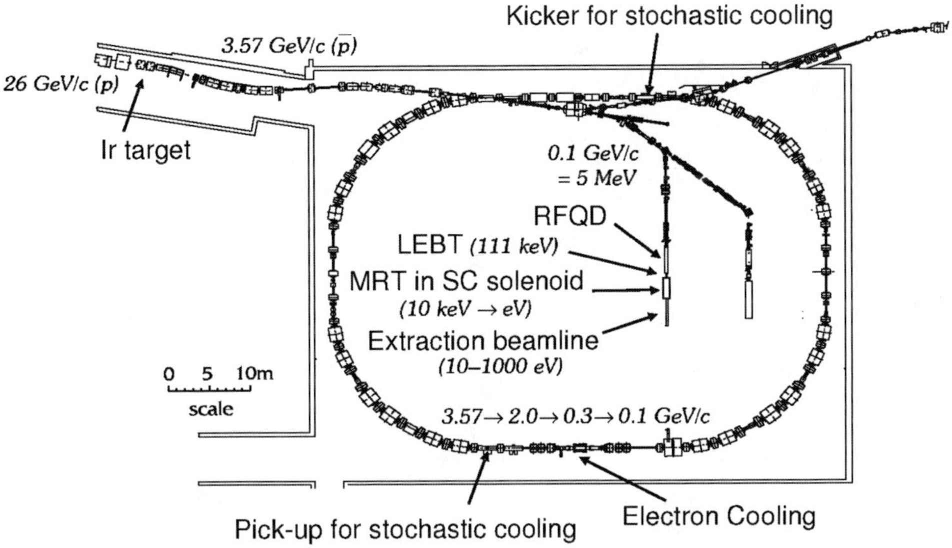

FIGURE 2. Layout of the Antiproton Decelerator (AD) at CERN.

and stochastic cooling techniques [13, 14] and decelerated down to a momentum of 100 MeV/c or 5.3 MeV in kinetic energy, before they are extracted to experimental zones as a pulsed beam of typically 3×10^7 antiprotons in a bunch of 100–200 ns (see Fig. 2).

Thus produced antiproton beam, though already lower in energy by 3 orders of magnitude than at production, needs further deceleration for them to be captured in vacuo electrostatically. In order to decelerate antiprotons efficiently, the ASACUSA collaboration [8] developed a Radio Frequency Quadrupole Decelerator (RFQD) [15] in collaboration with the CERN PS group. An RF wave was applied to the cavities in such a way as to decelerate microbunches of the antiproton beam at 5.3 MeV down to 63 keV with an efficiency of 30%. Since the RF cavities can be biased $\pm$ 60 keV, the output antiproton energy can be varied 10–120 keV.

The output beam from the RFQD was injected into an eletromagnetic trap in a strong magnetic field of 2.5 T produced by a superconducting solenoid. The beam was focused to a diameter of 3–4 mm on a thin Mylar i.e. PET (polyethylene terephthalate) double-layered foil of 90 μg/cm^2 for each layer [16], used to isolate ultrahigh vacuum of 10^{-12} mbar inside the trap.

Taking into account the energy loss in the foil, we set the bias voltage of the RFQD to 110 kV so as to maximize the number of antiprotons captured, with their transverse energy less than 10 keV after penetration through the foil [17].

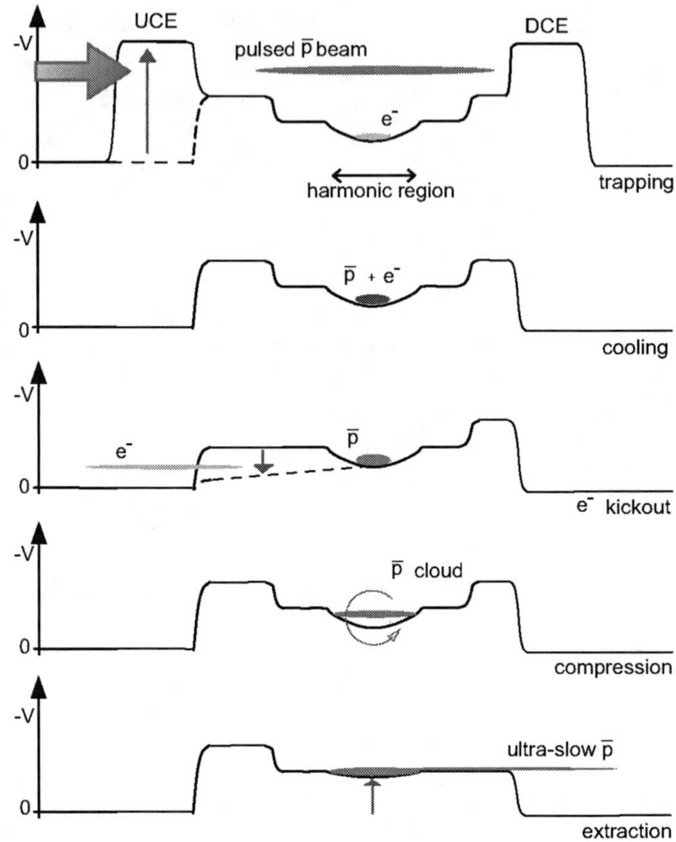

FIGURE 3. Sequential procedures of antiproton capture, cooling and extraction.

CONFINEMENT AND COOLING

Antiprotons were then captured and confined in the trap. We used a Multi-Ring Trap (MRT) [18] consisting of 14 cylindrical electrodes placed coaxially along the magnetic field line. A favorable feature of the MRT compared with a normal Penning trap and a trap with a well-type potential is that a harmonic electric potential can be prepared in a wide region near the trap axis by application of appropriate voltage on each electrode, which enables extremely stable confinement of a large number of charged particles.

Figure 3 shows sequential steps for antiproton capture, cooling and extraction. The pulse of incident antiprotons were reflected backward at the DCE electrode floated at -10 kV. By the time the pulse returned after its round trip of typically 500 ns back to the UCE, the trap was closed by a fast switch which biased the UCE to -10 kV, confining a major part of the antiprotons. The antiprotons were then cooled by a plasma of typically 3×10^8 electrons preloaded in the harmonic potential. Antiprotons lost their energy by transferring it to electrons, while the heated electrons cooled by themselves by emission of synchrotron radiation in the magnetic field of 2.5 T, until the antiprotons were trapped in the bottom of the harmonic potential of 50 V depth.

In order to diagnose cooling process of the antiproton in the trap, we prepared a set of track detectors to know the position and time of antiproton annihilation [19]. Two scintillator bars of 2 m in length, with a rectangular cross section of 4(H) $\times$ 6(V) cm^2, were placed in the same plane and parallel to the trap axis. Passage of charged particles such as pions, or electron-positron showers converted from gamma rays originating in the decay of neutral pions, was detected and the position of the crossing points calculated from the time-of-flight difference of the scintillation light arriving at photomultiplier tubes (PMTs) at both ends of the bars. Tracking back the reconstructed particle trajectory onto the trap axis, antiproton annihilation can be detected with a position resolution of 20 cm and with a detection efficiency of $\varepsilon \sim 5\%$, which agreed with our simulation using GEANT trajectory calculation code [20].

Integrating the signal counts from the track detectors over the trapping period, we can estimate the total number of trapped antiprotons. With this trap, we successfully confined 1.2×10^6 cooled antiprotons until the end of our trap cycle of 1–5 minutes [21]. Typically 1 million antiprotons were trapped for each AD shot. We then accumulated antiprotons for several AD shots. This technique of "stacking" also worked fine, and we trapped 1.2×10^7 antiprotons simultaneously for stacking of 12 AD shots, the largest number of antiprotons ever accumulated.

RADIAL COMPRESSION OF ANTIPROTON CLOUD

After the antiprotons have been cooled by the electrons, we then opened one side of the potential for 550 ns to selectively release electrons: lighter and thus faster electrons escaped within this short period, while heavier and much slower antiprotons remained inside. This release of electron was necessary for efficient extraction of antiprotons. The antiproton cloud was then given torque by a rotating electric field to be compressed radially [22]. For this purpose, one of the electrodes was segmented azimuthally into four parts, and an RF voltage was applied to each segment with a phase difference of $\pi/2$ next to each other.

The antiprotons were then released from the trapping potential as it was gradually shallowed, and were extracted as an ultra-slow continuous beam of 10–500 eV. Since the antiprotons tend to expand in radial direction when they follow the strongly diverging magnetic field line, it was essential that the antiproton cloud be well compressed radially in the trap. Precise alignment of the magnetic field axis, electric trap axis, and the beam transport axes was also important in efficient extraction; a misalignment of even a few milliradians was enough to prevent effective extraction.

For the study of radial compression of the antiproton cloud, the profile of the antiproton beam was monitored by a microchannel plate (MCP) with a two-demensional position sensitive detector (PSD), placed 1.75 m downstream from the center of the MRT. The annihilation position along the beam line was determined by another set of track detectors installed parallel to the beam line. An additional plastic scintillator plate near the MCP-PSD identified the antiprotons by detecting passage of annihilation products

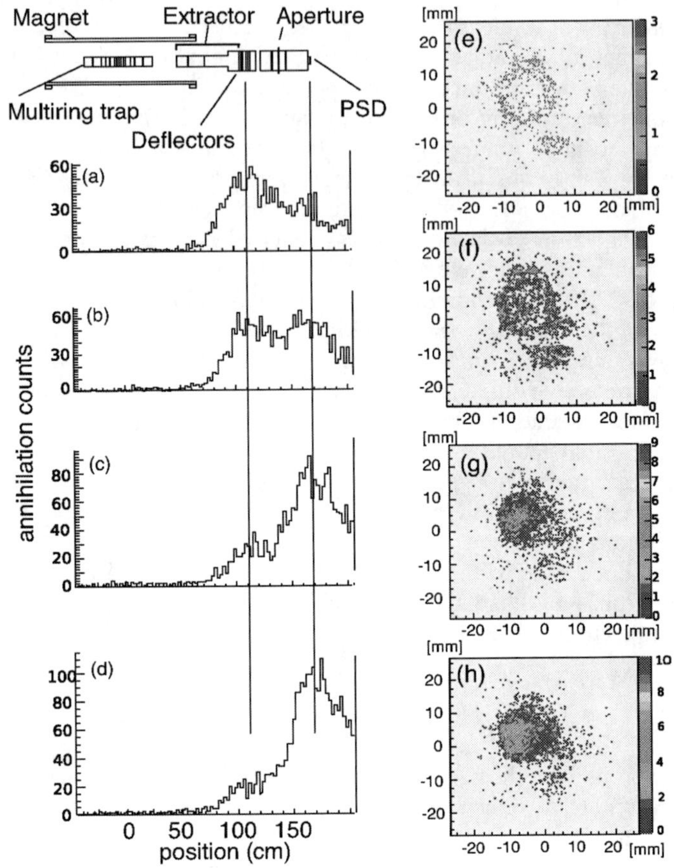

FIGURE 4. (a)–(d) The annihilation position of extracted antiprotons along the transport beam line for the rotation time of $t_r = 0$, 60, 120, and 200 s. (e)–(h) The MCP-PSD images of extracted antiprotons for $t_r = 0$, 60, 120, and 200 s. The rotation frequency was $f = 247$ kHz here, but the compression worked similarly effectively for a wide range of frequencies.

such as energetic pions.

Figures 4(a)–4(d) show the annihilation distribution along the transport beam line. When no rotating field was applied [Fig. 4(a)], most of the antiprotons annihilated around the extractor electrodes at the exit of the superconducting solenoid. Figures 4(b)–4(d) correspond to the cases where a rotating electric field with its frequency $f = 247$ kHz was applied for $t_r = 60$, 120, and 200 s, respectively. The annihilation peak around the extractors diminished while the peak around the MCP-PSD increased in size as t_r became longer. The corresponding 2D images on the MCP-PSD are shown in Figs. 4(e)–4(h). Without the rotating field [case (a)], the image is dim with low intensity and shows a depopulated region (hollow) in its center. This hollow gradually fills up as the field is applied, as in the cases (b), (c), and (d). The antiproton cloud was therefore

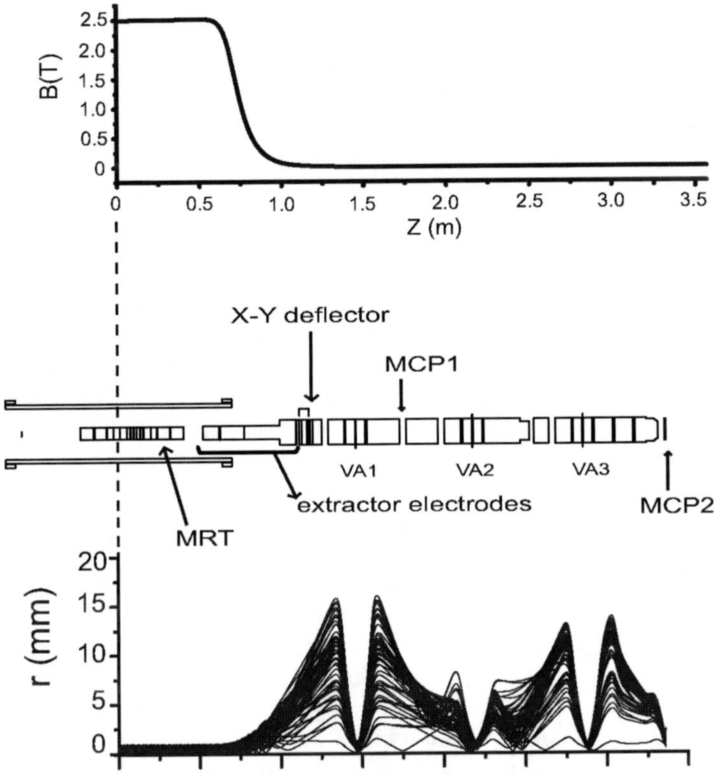

FIGURE 5. Schematic of the extraction beam line for ultra-slow antiproton beams. The beams are focused 3 times at positions of variable apertures (VAs), to comply with the requirement of effective differential pumping (bottom). Magnetic field strength is also shown in the top figure.

effectively being compressed by the rotating field. Comparison with our beam transport simulation demonstrates that the major part of the antiproton cloud was compressed from a radius of 3.4 mm to 0.25 mm in the case of $t_r = 200$ s [22]. The compressed cloud remained stable in the MRT for at least a few minutes.

EXTRACTION AND BEAM TRANSPORT

The extraction beam line was designed to transport antiproton beams over a length of 3 m, at variable energies ranging from 10 to 1000 eV. The antiproton beams were refocused three times by sets of Einzel lenses at the position of apertures, as shown in Fig. 5. These variable apertures of diameter 4–10 mm allow differential pumping of 6 orders of magnitude along the beam line [23], which was necessary to keep the trap region at an extremely high vacuum better than 10^{-12} mbar so as to avoid antiproton annihilation, while the end of the beam line is to be exposed to atomic or molecular gas jets of upto 10^{-6} mbar.

The MRT, the superconducting solenoid and the transport line for the ultra-slow beam are jointly known as "MUSASHI", or the Monoenergetic Ultra-Slow Antiproton Source for High-precision Investigations. MUSASHI opens a new research field ranging from atomic physics to nuclear physics [24, 25, 26, 27], including our near-future project of antihydrogen synthesis in a cusp trap [28].

ATOMIC COLLISION EXPERIMENTS

With this unique beam, atomic formation and ionization processes by very-low-energy antiprotons can now be studied under single collision conditions for the first time.

Since the number of available antiprotons is very much limited, the reaction probability must be maximized in order to make best use of them, and therefore a maximum possible density of the atomic gas jets need to be prepared to be crossed with the antiproton beams. Then the number of antiprotonic atoms formed is given by the formula

$$N_{\overline{p}A^+} = \sigma n_A L N_{\overline{p}},$$

where σ is the formation cross section, n_A the number density of the atomic target, L the interaction length, and $N_{\overline{p}}$ is the number of antiprotons.

The reaction cross sections are naturally expected to be of the order of 10^{-16} cm^2 [1]. Suppose a beam with 10^5 antiprotons is available every shot per several minutes, we will obtain 30 antiprotonic atoms for a gas density of 3×10^{12} atoms/cm^3 at an interaction length of 1 cm. The reaction probability is of the order of 10^{-3}–10^{-4}. Rigorous particle identification is necessary in order to assure enough statistical significance for the reaction events to be detected and distinguished from background events.

ATOMIC GAS-JET TARGET

At our early stage of development, several ideas for realization of the gas target was considered [29]. Usage of effusive gas out of micro-capillary array was also considered, but was rejected in favor of supersonic gas targets for two main reasons: 1) the gas jet should be well-collimated and 2) the background vacuum should be kept at a level below 10^{-6} mbar.

Helium atom was chosen for our first target, for technical simplicity and because of the fact that the lifetime of antiprotonic helium atoms are already known from our spectroscopic measurements [30]. Atomic and molecular hydrogen targets are also interesting and simple from theoretical point of view, but we shall consider them later in our next step, because preparation of these hydrogen targets confront experimental difficulties including special safety precautions.

Figure 6 shows a schematic design of our apparatus for the supersonic gas-jet target [31]. It has 5 chamber stages in order to achieve good jet collimation and efficient pumping. A supersonic gas jet emerging from a 0.1-mm-diameter nozzle was skimmed by two stages of conically shaped skimmers with a hole diameter of 0.6 mm and 2.2 mm.

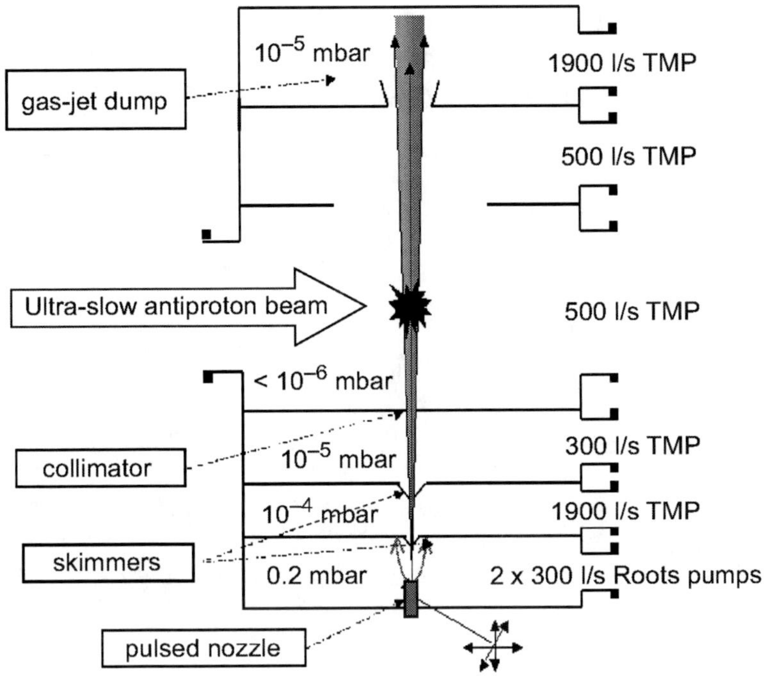

FIGURE 6. Schematic design of our supersonic gas-jet system.

The stagnation pressure before the nozzle was optimized at 25 bar at room temperature, for production of a powerful jet with a large Mach number of 35.8 according to our gas-jet fluid-dynamics simulation. A Pitot tube was used in advance to monitor the pressure profile of the gas jet, in order to determine the optimum distance between the nozzle and the first skimmer.

The nozzle had a poppet valve just in front of it, operated by a small solenoidal magnet. Open and close of the gas-jet flow was controlled by an external electric switch, and the equilibrium of the gas flow was reached within a small fraction of a second after the switch was turned on. This pulsed valve allows economical usage of the gas jet which is turned on only during the extraction period of the antiproton beam in a whole cycle of antiproton capture, cooling, trapping and extraction, and hence reduction of the total gas load. The nozzle and the valve was mounted on a three-dimensionally movable stage, so as to adjust the position and maximize the gas-jet density.

Most of the gas out of the nozzle was skimmed away and was differentially pumped. Two powerful 300 l/s Roots pumps were employed to cope with the great gas load at 0.2 mbar at the first stage, while the other chamber stages with better vacuum than 10^{-4} mbar were pumped by turbo-molecular pumps (TMPs) with a large pumping speed of typically 500 l/s. The gas jet was then collimated to form a target of about 1 cm diameter at the cross point in the main chamber. The jet was then collected at the dump stages with an over-90% efficiency by a powerful TMP with a pumping speed of 1900 l/s.

The vacuum in the main chamber was kept to a level of less than 10^{-6} mbar.

ANTIPROTON BEAM TRAJECTORY

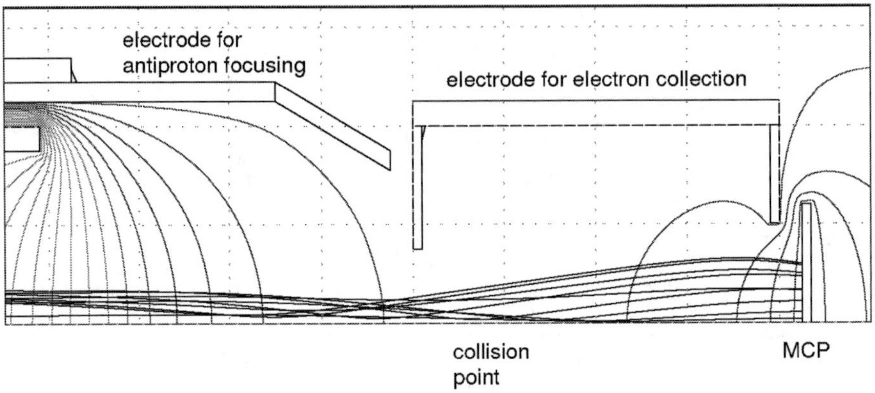

FIGURE 7. Simulated antiproton trajectoris at 250 eV at the last stage of the beam line for our first version of the detector design. The antiproton is well focused near the cross point to overlap the gas-jet target while the beam divergence is kept small enough for the beam to be detected within the sensitive area with a 4-cm diameter of the MCP-PSD placed 6.5 cm downstream.

Designing the beam transport of ultra-slow antiprotons is not a trivial task at this very low energy. The antiproton beam needs to go through 3 aperture stages along the 3-m extraction beam line before it is focused at the collision point [23]. The last focusing has to fulfill two contradictory conditions: the beam spot must be squeezed to the small size of the gas jet, yet the focusing has to be weak enough to keep the beam divergence to a small value so that all the antiprotons and antiprotonic atoms should be collected at the sensitive area of an MCP-PSD downstream. Figure 7 shows a solution to the simulation of antiproton trajectories at 250 eV, from the last stage of our beam line through the collision point against the gas-jet target, until the MCP downstream. The MCP is biased at $+470$ eV to draw in the negative particles. The beam size is squeezed to a diameter of 8 mm at the waist, to ensure a good overlap with the gas jet.

DETECTION AND DATA ACQUISITION SYSTEMS

Antiprotons are extracted as a beam in a slow-extraction mode. This continuous beam allows event-by-event data acquisition associated with each single antiproton extracted. For particle detection, an MCP-PSD was placed inside the gas-jet chamber, downstream of the cross point along the antiproton beam path. The above-mentioned good vacuum level in the main chamber was required also for proper operation of the MCP without a risk of discharge. The MCP-PSD was surrounded by a box of scintillator plates for identification of antiproton annihilation by detecting passage of its annihilation products.

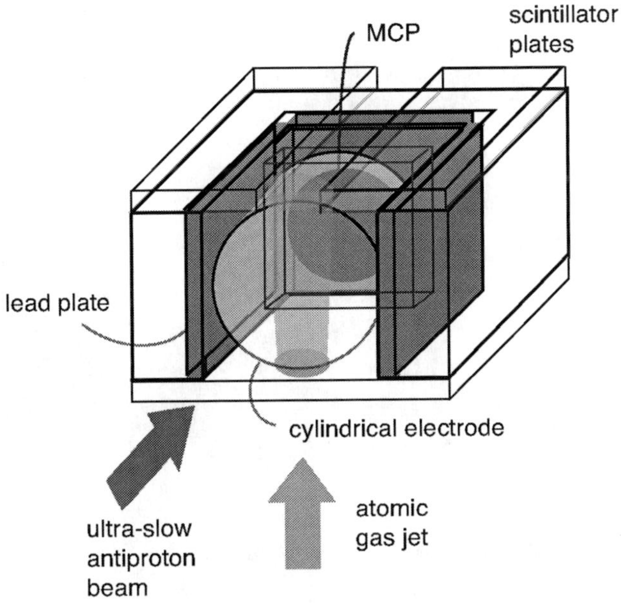

FIGURE 8. Schematic setup of the first version of detectors surrounding the collision point of the antiproton beam and the gas-jet target.

In our first attempt, we used a single MCP-PSD. An antiproton impacts a helium atom and ionizes it, emitting an electron, which is then guided by a cylindrical electrode toward the MCP with an efficiency larger than 90%. Depending on the impact energy, the antiproton either travels downstream at a lower energy or is captured to form an antiprotonic helium atom which flies at an even lower speed toward downstream. The MCP was used to detect any of these particles: an electron, a bare antiproton, or an antiprotonic atom; while the scintillation counter with lead plates in front detected passage of high-energy particles such as pions, originating from antiproton annihilation (see Fig. 8).

The key point in this experiment is the rigorous identification of particles. As was mentioned above, the reaction probability is less than 0.1%. So that one single proper reaction event must be well identified out of the rest 999 antiproton annihilation events without reaction. The solid angle coverage of the scintillation counter was nearly 90% seen from the MCP. Antiproton annihilation is usually accompanied by emission of 3 charged pions on average and 2 neutral pions. A neutral pion decays into 2 gamma rays, which are converted to electromagnetic showers of electrons and positrons in the lead plates, and together with charged pions, they can be detected by the scintillation counter. This multiplicity of the annihilation products helps to achieve an extremely high detection efficiency of 99.7% for the detection of antiproton annihilation [32]. Thus, an electron signal is identified by an MCP hit and no scintillator hit, while an antiproton signal is identified by the coincidence of both MCP and scintillator hits.

Our experiments with this setup proved the highly efficient discrimination between electrons and antiprotons. However, the configuration above turned out not to be effective enough to disentangle signals by electrons emitted in the ionization reaction from quite a number of unidentified background signals.

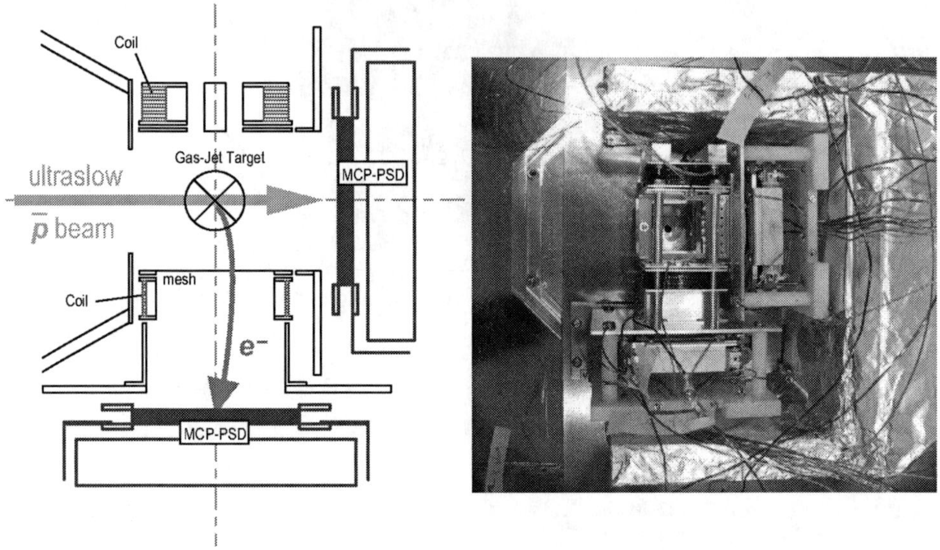

FIGURE 9. Schematic setup (left) and a photo image (right) of the modified version of detectors surrounding the collision point of the antiproton beam and the gas-jet target.

Therefore, we decided to modify the setup by employing one more MCP-PSD placed sideways of the antiproton beam, to selectively detect electrons. Figure 9 shows a schematic layout of the detectors. A set of electrodes and coils were placed near the collision point to guide electromagnetically the electrons perpendicular to the antiproton beam. Electrons emitted in the opposite direction away from the MCP-PSD was repelled back toward the detector due to the magnetic mirror effect of the stronger magnetic field there. Adjustment of the voltages to be applied to the electrodes depending on the beam energy allowed efficient (98%) electron transport while minimizing the deflection of the antiproton beam trajectory, as was demonstrated by our simulations [33].

The TOF is shorter for electrons than for antiprotons due to the difference in their velocity. Thus a reaction event can be recognized by an electron signal (i.e. MCP(side) hit and no scintillator hit) followed by an antiproton signal (i.e. coincidence of MCP(downstream) hit and scintillator hit) with an appropriate TOF interval. The signals were processed and recorded by standard NIM and CAMAC systems and a set of computers. We used multichannel TDCs with multi-stop capability. The data with electron-start events were acquired with the common-stop mode of the TDC, in order to suppress the background of electron signals not associated with following antiprotons.

OUTLOOK

Our recent experiment succeeded in observing the first ionization reaction events at the antiproton energy of 250 eV. The results are to be reported in our future publication. We are now planning to take several energy points in the 100–1000 eV region to reveal the cross section curve which is totally unknown for these energies. Even theoretically, no reliable work has been performed. The collision of an antiproton and a helium atom results in a combined nucleus charge of $+1$ and two electrons. The infinitesimal limit of the nuclear distance gives H^-, which is known to be a bound system. Therefore, the ionization cross section goes to zero in the low-energy limit of adiabatic passage approximation (in contrast to the case for antiproton collision with the hydrogen atom). A more realistic calculation must include diabatic effects, whose methodology is not yet established. Therefore, experimental measurement is of importance here.

Atomic capture process will be studied at lower eneries of 10–30 eV. Capture of antiproton by an atom results in formation of antiprotonic atoms. Metastable antiprotonic helium atom has been studied with spectroscopic methods over the decade, but only in media [9, 30]. The cross-beam experiment in vacuo will populate higher-lying states with a principal quantum number $n > 40$ and a much longer lifetime. The overall lifetime of this exotic atom can be observed from the TOF spectra, giving indication on the population distribution of states. Our new measurements are thus expected to clarify the collision dynamics under well-defined single-collision conditions at fixed energies.

SUMMARY

We have succeeded in deceleration, cooling and confinement of a large number of antiprotons provided at the AD facility at CERN, using a combination of the RFQ Decelerator and the multi-ring electrode trap. The antiprotons were then extracted out of the strong magnetic field to form an ultra-slow antiproton beam in the energy range of 10–500 eV. This unique beam opens a new research field of atomic physics. Especially, the cross-beam experiment against our supersonic gas-jet target is expected to reveal the ionization and atomic capture processes involving antiproton projectile.

ACKNOWLEDGMENTS

We are grateful to the AD and PS stuff in the AB department of CERN for their tireless effort in providing us with the antiproton beam. We acknowledge Mr. W. Pirkl and his team for their engagement in the design and operation of the RFQD. This work was supported by the Grant-in-Aid for Creative Scientific Research (10P0101 and 19002004) of the Japanese Ministry of Education, Culture, Sports, Science and Technology (MonbuKagaku-shō), Special Research Projects for Basic Science of RIKEN, and the Hungarian National Science Foundation (OTKA T033079).

REFERENCES

1. J. S. Cohen, Rep. Prog. Phys. **67**, 1769 (2004); J. S. Cohen, AIP Conf. Proc. **793**, 98 (2005).
2. X. M. Tong, K. Hino and N. Toshima, Phys. Rev. Lett. **97**, 243202 (2006); X. M. Tong, T. Shirahama, K. Hino and N. Toshima, Phys. Rev. A **75**, 052711 (2007).
3. K. Sakimoto, Phys. Rev. A **74**, 022709 (2006); Phys. Rev. A **74**, 042711 (2006).
4. L. H. Anderson *et al.*, Phys. Rev. A **41**, 6536 (1990).
5. P. Hvelplund *et al.*, J. Phys. B **27**, 925 (1997).
6. H. Knudsen *et al.*, "Ionization of helium and argon by very slow antiproton impact", submitted to Phys. Rev. Lett.; H. Knudsen *et al.*, Contribution to the Workshop on Cold Antimatter Plasmas and Application to Fundamental Physics, Okinawa, in this book (2008).
7. CERN Antiproton Decelerator web page, http://psdoc.web.cern.ch/PSdoc/acc/ad/index.html
8. ASACUSA collaboration web page, http://cern.ch/ASACUSA
9. T. Yamazaki *et al.*, Phys. Rep. **366**, 183 (2002).
10. E. Widmann *et al.*, Phys. Rev. Lett. **89**, 243402 (2002).
11. H. A. Torii *et al.*, Phys. Rev. A **59**, 223 (1999).
12. M. Hori, A. Dax *et al.*, Phys. Rev. Lett. **96**, 243401 (2006).
13. G. I. Budker and A. N. Skrinski, Sov. Phys. Usp. **21**, 277 (1978).
14. S. van der Meer, IEEE Trans. Nucl. Sci. **NS28**, 1994 (1981).
15. A. M. Lombardi, W. Pirkl, and Y. Bylinsky, Proceedings of the 2001 Particle Accelerator Conference, Chicago (IEEE, Piscataway, NJ) , 585 (2001).
16. M. Hori, private communication. cf. also Nucl. Insturm. Methods. A **522**, 420 (2004).
17. H. A. Torii *et al.*, AIP Conf. Proc. **793**, 293 (2005).
18. A. Mohri, H. Higaki *et al.*, Jpn. J. Appl. Phys. **37**, 664 (1998).
19. H. A. Torii *et al.*, J. Phys. Conf. Ser. **80**, 012022 (2007).
20. GEANT3, CERN Program Library Long Writeup W5013.
21. N. Kuroda, H. A. Torii, K. Yoshiki Franzen, Z. Wang, S. Yoneda, M. Inoue, M. Hori, B. Juhász, D. Horváth, H. Higaki, A. Mohri, J. Eades, K. Komaki, Y. Yamazaki, Phys. Rev. Lett. **94**, 023401 (2005).
22. N. Kuroda, H. A. Torii, M. Shibata, Y. Nagata, D. Barna, M. Hori, D. Horváth, A. Mohri, J. Eades, K. Komaki, and Y. Yamazaki, Phys. Rev. Lett. **100**, 203402 (2008).
23. K. Yoshiki Franzen, N. Kuroda, H. A. Torii, M. Hori, Z. Wang, H. Higaki, S. Yoneda, B. Juhász, D. Horváth, A. Mohri, K. Komaki, and Y. Yamazaki, Rev. Sci. Instrum. **74**, 3305 (2003).
24. Y. Yamazaki, Nucl. Instrum. Methods in Phys. Res. B **154**, 174 (1999).
25. A. Trzcinska *et al.*, Phys. Rev. Lett. **87**, 82501 (2001).
26. J. Jastrzębski *et al.*, Nucl. Phys. A **588**, 405c (1993).
27. M. Wada and Y. Yamazaki, Nucl. Instr. Meth. B **214**, 196 (2004).
28. A. Mohri and Y. Yamazaki, Europhys. Lett. **63**, 207 (2003).
29. Z. Wang, internal report, unpublished, (2002).
30. M. Hori *et al.*, Phys. Rev. A **70**, 012504 (2004).
31. V. L. Varentsov, N. Kuroda, Y. Nagata, H. A. Torii, M. Shibata, Y. Yamazaki. AIP Conf. Proc. **793**, 328 (2005).
32. H. A. Torii *et al.*, Nucl. Instrum. Methods in Phys. Res. A **396**, 257 (1997).
33. Yugo Nagata, "Manipulation of ultraslow antiprotons and first application to atomic collisions", Doctor thesis, Institute of Physics, Komaba, University of Tokyo (2008).

Observation of Ultra-Slow Antiprotons using Micro-channel Plate

H. Imao[*], H.A. Torii[†], Y. Nagata[†,*], H. Toyoda[*], T. Shimoyama[*], Y. Enomoto[†,*], H. Higaki[**], Y. Kanai[*], A. Mohri[*] and Y. Yamazaki[†,*]

[*] *Atomic Physics Laboratory, RIKEN, Wako, Saitama 351-0198, Japan*
[†] *Institute of Physics, University of Tokyo, Komaba 3-8-1, Meguro-ku, Tokyo, Japan*
[**] *Graduate School of Advanced Sciences of Matter, Hiroshima University, Kagamiyama, Higashi-Hiroshima, 739-8530, Japan*

Abstract. Our group ASACUSA-MUSASHI has succeeded in accumulating several million antiprotons and extracting them as monochromatic ultra-slow antiproton beams (10 eV-1 keV) at CERN AD. We have observed ultra-slow antiprotons using micro-channel plates (MCP). The integrated pulse area of the output signals generated when the MCP was irradiated by ultra-slow antiprotons was 6 times higher than that by electrons. As a long-term effect, we also observed an increase in the background rate presumably due to the radioactivation of the MCP surface. Irradiating the antiproton beams on the MCP induces antiproton-nuclear annihilations only on the first layer of the surface. Low-energy and short-range secondary particles like charged nuclear fragments caused by the "surface nuclear reactions" would be the origin of our observed phenomena.

Keywords: antiproton, micro-channel plate
PACS: 25.43.+t, 36.10.Gv

INTRODUCTION

Our group ASACUSA-MUSASHI [1] has successfully developed the system for obtaining monochromatic ultra-slow antiproton beams (10 eV-1 keV) at CERN Antiproton Decelerator (AD) [2, 3]. In this system, the multiring-electrode trap [4] for accumulating antiprotons was placed in a 5-T magnetic field. The trapped antiprotons were effectively cooled by mixing with preloaded electron plasma and radially compressed by applying a rotating electric field. Then, the antiprotons were extracted and delivered to downstream passing through a transport beamline for differential pumping [5].

This novel beam has a lot of applications for studying atomic physics in unexplored energy region. Recently, we have performed atomic collision experiments in unexplored low-energy region by crossing the beam with super-sonic gas-jet helium. One unique feature of the experiments was that this is a low-energy atomic physics experiment under substantial nuclear reactions induced by ultra-slow antiprotons. We observed ultra-slow antiprotons and measured some unique background signals on micro-channel plate (MCP) detectors. Their understanding is important for upcoming precise atomic physics measurements with ultra-slow antiprotons like $\bar{\mathrm{H}}$ spectroscopy.

CP1037, *Cold Antimatter Plasmas and Application to Fundamental Physics,* edited by Y. Kanai and Y. Yamazaki

EXPERIMENT

In the present experiment, ultra-slow antiprotons (approximately 10^5 per spill, with a duty time of $\sim$300 s) were irradiated onto the MCP, which was located in a collision chamber connected to the transport beamline (Fig. 1). The injection energy was controlled to be in the range 30-250 eV. The detection system was designed and constructed to detect antiproton events on the MCP with a high efficiency. The MCP (DLA40; RoentDek) detector consists of a MCP in a chevron configuration and a wire-delay-line anode that enables two-dimensional position readout. Fast pions and kaons generated by antiproton-nuclear annihilations were detected using a set of plastic scintillator plates with lead plates positioned around the MCP. Antiproton events on the MCP were discriminated from other events (e.g., those produced by electrons) on the basis of subsequent scintillator events. Lead plates were inserted to increase the efficiency of the plastic scintillator via $\pi_0 \rightarrow 2\gamma, \gamma \rightarrow 2e^+e^-$ reactions. The design parameters of the scintillator system were optimized using Geant4 simulation code [6] so that it had a high detection efficiency of greater than 99% for an antiproton annihilation event. The electrode inserted in front of the MCP was used for guiding the injected antiprotons and electrons. The entire system was located in the collision chamber that had a base pressure of better than 1×10^{-6} mbar.

An incoming antiproton signal generated by the MCP opens a gate of the data acquisition system. Signals from the MCP were read out on time-to-digital converters with multi-hit capability, charge-sensitive analog-to-digital converters (ADC) and flash ADCs. All data were acquired through a CC/NET CAMAC crate controller (TOYO). The global timing of data acquisition was synchronized using the Labview-based operation system of the ultra-slow antiproton beamline. The CC/NET has a 320 MB system memory that is sufficient to store the data generated by a single spill of the ultra-slow antiprotons ($\sim 10^5$ $\bar{p}$/10 s). The acquired data was transferred to a host computer during the duty time of the beamline ($\sim$5 min) without a significant dead time.

ULTRA-SLOW ANTIPROTONS ON MCP

We observed an unexpectedly large difference between the output signals generated when the MCP was irradiated by ultra-slow antiprotons and those generated when it was irradiated by electrons. An unmistakable signal generated by an antiproton was recorded with an oscilloscope and is shown in Fig. 2(a). Many secondary pulses were generated up to $\sim$600 ns after the first pulse. These secondary pulses were not generated by the pile-up signal of antiprotons, since the antiprotons were extracted as a DC beam at a rate of approximately 10^4 per s during the measurement. For comparison, typical electron signals are shown in Fig. 2(b). The electrons were generated by an ion gauge. There are only small secondary pulses (i.e., ringing) appear in the tail of the signal. The mean pulse height for antiprotons measured by the flush ADC was 1.5 times higher than that for electrons and the integrated pulse area was 6 times higher. This surprising difference indicates that significant surface reactions are induced by ultra-slow antiprotons. A long-term effect that we observed was an increase in the background rate, which is presumably due to the alpha and beta radiation produced by antiproton annihilation. The

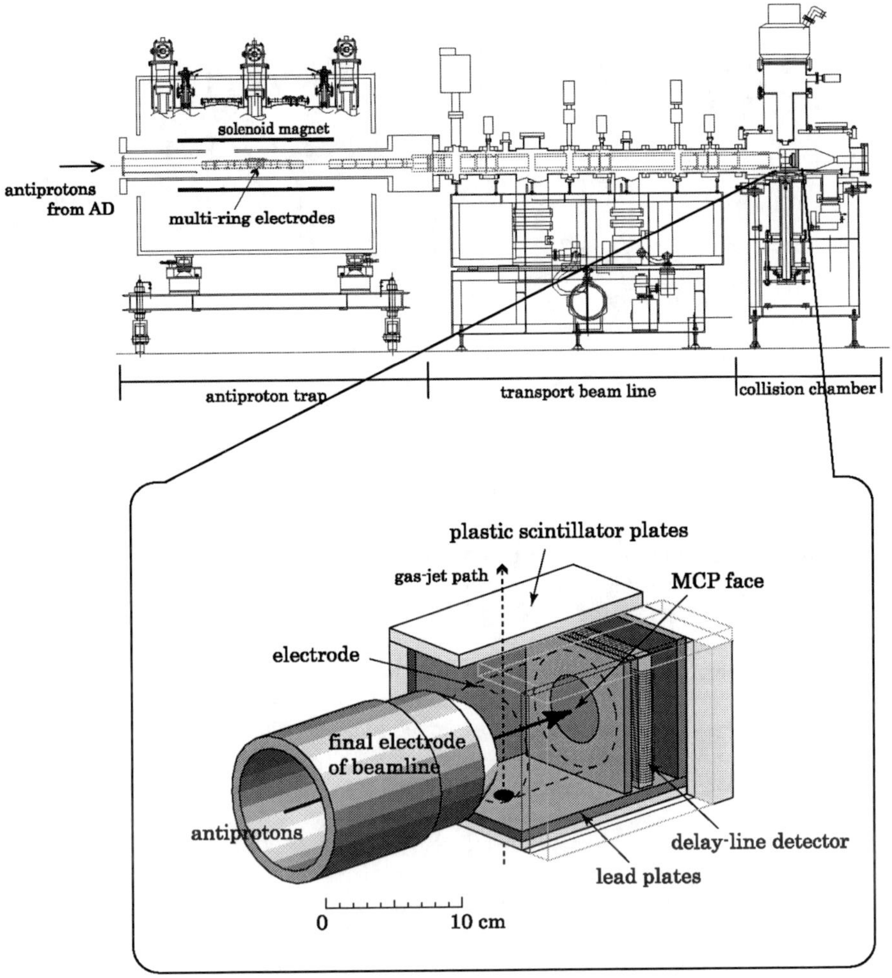

FIGURE 1. Schematic view of the MUSASHI beamline and the detection system for antiprotons. Antiprotons injected from AD were stored in the multi-ring electrodes trap placed in the magnetic field and extracted to the collision chamber passing through the transport beamline. The ultra-slow antiprotons were detected by using the delay-line-anode MCP (DLA40; RoentDek) surrounded by six scintillator plates and four lead plates.

initial background rate of the MCP was around 5 cps before antiproton bombardment; the background rate increased to around 15 cps after irradiation by approximately 10^7 ultra-slow antiprotons over a period of 8 hours.

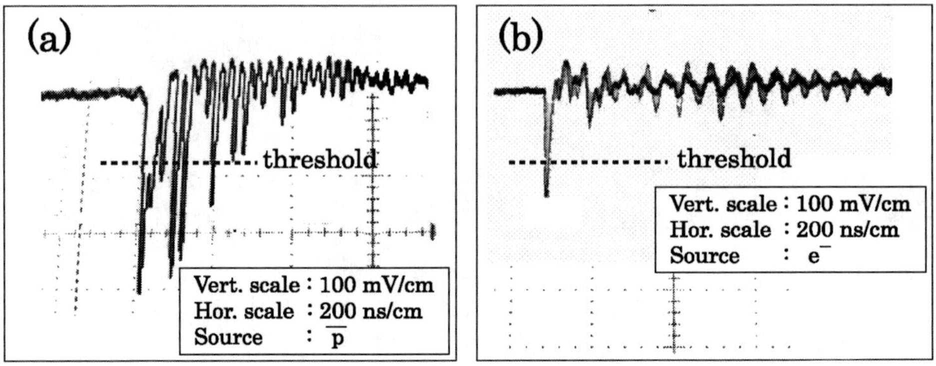

FIGURE 2. Comparison of output signals generated by (a) antiproton irradiation and (b) electron irradiation obtained using an oscilloscope. The trigger threshold was 180 mV.

DISCUSSIONS

In the present measurements, we injected monochromatic antiprotons (30-250 eV) onto the MCP with its front surface biased at +470 V. The initial energy (500-720 eV) of the antiprotons is dissipated on the MCP surface, and secondary electrons are emitted. This emission mechanism, which involves the transfer of kinetic energy followed by kinetic emission of electrons, is similar to that of ordinary particles such as electrons or ions [7, 8]. Further mechanisms that deposit additional energy on the MCP surface are required to explain the huge experimentally observed output of the MCP.

The crucial feature of antiprotons is the large mass energy of 1880 MeV that is released upon annihilation with a nucleon, even when the antiproton is nearly at rest. Injected ultra-slow antiprotons immediately replace atomic electrons and form antiprotonic atoms on the MCP surface. The probability of an atom capturing an antiproton is approximately linearly proportional to the number of electrons the atom possesses [9]. The formed antiprotonic atom is destroyed within picoseconds and the antiproton is absorbed onto the nuclei by the strong interaction. Most of the mass energy released by annihilation is transported away by an average of five pions with mean kinetic energies of 210 MeV. Some of the emitted pions interact with the residual nucleus on the MCP surface. The solid angle for a heavy nucleus, (e.g., lead) with respect to the annihilation location on the nuclear surface is expected to be 30% of 4π [10]. Interactions between the pions and the baryonic system of nucleons induces nuclear excitations with subsequent break up. The nuclear fragments are promptly emitted from the excited nucleus and the recoil of heavy nuclei deposits large energies on the MCP surface, which would cause the high experimentally observed pulse heights. In addition, the persistent secondary pulses imply that the antiproton-induced reactions are followed by the continual generation of unstable nuclei with high probabilities.

In order to evaluate the nuclear fragment emission due to antiproton-nuclear annihilations, we performed a computer simulation by using the chiral invariant phase space (CHIPS) event generator [11] implemented in Geant4 [6]. CHIPS can simulate the emis-

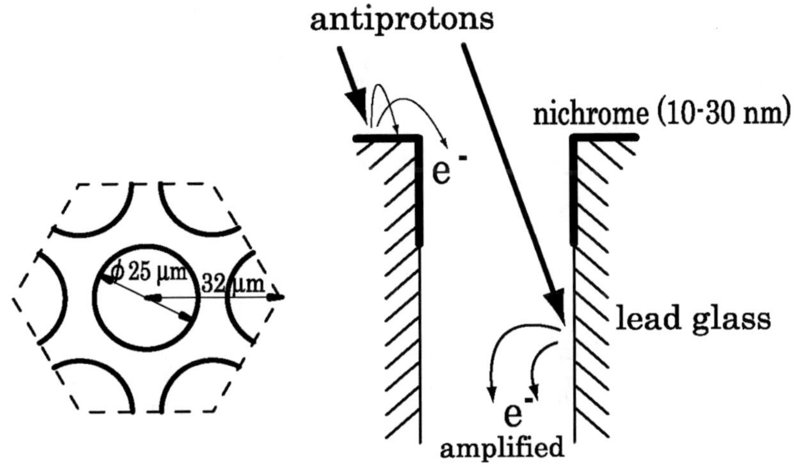

FIGURE 3. Plan and cross-sectional view of the MCP. A nickel-based coating is deposited on the front surface of the MCP.

sion yields and energies of charged nuclear fragments reasonably well [12]. The basic material of MCP is lead silicate glass (Corning 8161 glass [13]) with a nichrome/inconel coating on the front and rear surfaces. In the present setup, approximately 50% of antiprotons injected perpendicular to the MCP face stop on the channel wall, while the others stop on the nickel-based coating (Fig. 3). The escape lengths of secondary electrons for both materials are a few nm. A simulated distribution of the emitted nuclear fragments on the MCP lead silicate glass is plotted in Fig. 4. It shows the generation of a variety of unstable heavy nuclei by the emission of many light nuclei. We note that these considerable nuclear reactions occur only on or near the surface. The emitted energy distributions of the charged nuclear fragments indicate that the energy deposited on the MCP surface by charged nuclear fragments is much greater than the injected energy of 720 eV and this explains the huge MCP signals that were observed. An annihilation reaction predominantly produces light nuclei having energies of the order of several MeVs, which have a stopping range comparable with the distance between neighboring open channels (25 μm). There is a high probability that such light nuclei produce secondary electrons in one or more channels of the MCP. The heavier nuclei also contribute to the local energy deposition. After the equilibration of nuclei by the emission of evaporation particles, some unstable ones undergo radioactive decay. We also simulated the radioactive decay process of unstable nuclei on the MCP glass by using the table of the radioactive decay derived from the Evaluated Nuclear Structure Data File. We found that the observed radioactivation on the MCP surface, which was of the order of 10 cps, can be caused by irradiation of 10^7 antiprotons.

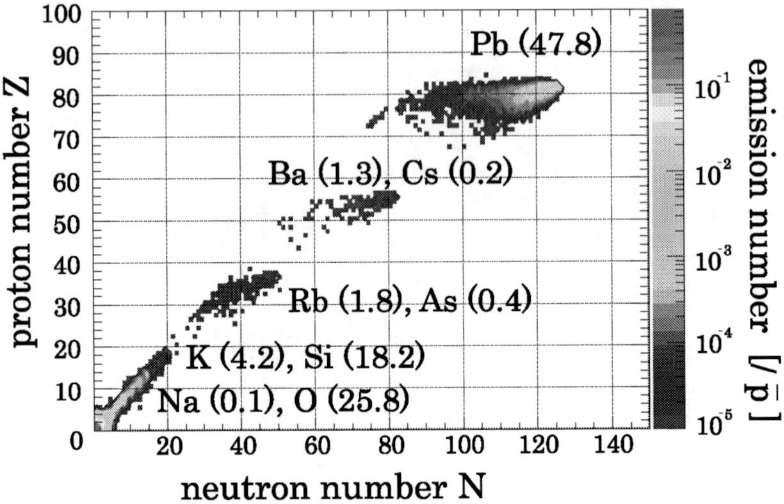

FIGURE 4. Distribution of the emitted nuclear charged fragments on the MCP glass calculated using Geant4 with CHIPS. Original nuclei for each block of the distributions are written next to them along with its weight percent in the MCP glass.

CONCLUSION

In conclusion, we observed significant surface nuclear reactions induced by ultra-slow antiprotons on the MCP for the first time. The pulse shapes and the charge outputs recorded when the MCP was irradiated by ultraslow antiprotons differed drastically from those obtained by electron irradiation. The large energy deposited on the MCP surface is considered to be caused by charged nuclear fragments emitted after antiproton-nucleon annihilation. The present study will be crucial in future sensitive antiparticle experiments with the ultra-slow $\bar{p}$ beam.

ACKNOWLEDGMENTS

We are indebted to CERN staff for their helpful support with the experiment. This research was partially supported by the Japan Society for the Promotion of Science Fellow, by a Grant-in-Aid for Creative Scientific Research (10P0101) from the Japanese Ministry of Education, Science, Sports and Culture, and by the Special Research Projects for Basic Science of RIKEN.

REFERENCES

1. ASACUSA collaboration proposal, CERN/SPSC 307 (1997).
2. N. Kuroda *et al.*, Phys. Rev. Lett. **94**, 023401 (2005).

3. N. Kuroda *et al.*, Phys. Rev. Lett. (2008), in press.
4. A. Mohri *et al.*, Jpn. J. Jppl. Phys. **37**, 664 (1998).
5. K. Y. Franzen *et al.*, Rev. Sci. Inst. **74**, 3305 (2003).
6. S. Agostinelli *et al.*, Nucl. Inst. Meth. **A506**, 250 (2003).
7. G.W. Fraser, Nucl. Inst. Meth. **206**, 445 (1983).
8. G.W. Fraser, Int. J. Mass Spectrom. **215**, 13 (2002).
9. E. Fermi and E. Teller, Phys. Rev. **72**, 399 (1947).
10. D. Polster *et al.*, Phys. Rev. **C51** 1167 (1995).
11. P.V. Degtyarenko *et al.*, Eur. Phys. J. **A9**, 421 (2000).
12. M. Kossov, IEEE Trans. Nucl. Sci. **52**, 2832 (2005).
13. J.L. Wiza, Nucl. Inst. Meth. **162**, 587 (1979).

Sub-Femtosecond Correlated Dynamics Probed with Antiprotons

C.P. Welsch[a-c], K.U. Kühnel[a], C.D. Schröter[a], and J. Ullrich[a]

[a]Max-Planck Institute for Nuclear Physics, D-69117 Heidelberg, Germany
[b]Kirchhoff Institute for Nuclear Physics, University of Heidelberg, D-69120 Heidelberg, Germany
[c]Gesellschaft für Schwerionenforschung, D- 64291 Darmstadt, Germany

Abstract. Low-energy antiprotons are the ideal and perhaps the only tool to study in detail correlated quantum dynamics of few-electron systems in the femto and sub-femtosecond time regime. Unfortunately cooled beams of antiprotons with the necessary beam quality and luminosity are not yet available and cannot be provided with present scientific infrastructures. In order to pave the way for a next-generation low-energy antiproton facility, challenging developments in both, storing and imaging techniques have been launched at MPI-K. A novel ultra-low energy storage ring (USR) to be integrated at the proposed facility for low-energy antiproton and ion research (FLAIR) is being developed to provide electron-cooled beams of antiprotons and possibly highly charged ions in the energy range between 300 and 20 keV/q, maybe even approaching the sub keV regime. To allow for kinematically complete investigations for a variety of different collision processes, a reaction microscope shall be integrated in the ring thus achieving unprecedented luminosities. In this contribution, the present status of experiments in comparison with theory is highlighted and the layout of the USR as well as of the in-ring and an external single-pass reaction microscope is presented.

Keywords: Correlated dynamics, antiproton collisions, few-body problem, electrostatic storage ring, reaction microscope.

PACS: 29.17.+w , 34.10.+x, 34.50.Fa

INTRODUCTION

Whereas the three-body Coulomb problem for single excitation and ionization has claimed to be solved for the first time in a mathematically correct way during 1999 until 2004 for low-energy electron impact on hydrogen and helium [1,2,3], ion-impact ionization still represents a major challenge for theory. Troubling discrepancies have been observed in fully differential cross sections (FDCS) for He single ionization by fast ion impact [4,5,6] and even experimental total cross sections for antiproton impact are in striking disagreement with the predictions of all state-of-the-art theories for low-energy antiproton collisions [7,8], with some recent improvement [9].

For antiprotons at energies below about 300 keV down to 1 keV the interaction time between the projectile passing atoms or molecules is on the order of 80 attoseconds (as) up to 1 femtosecond (fs) and, thus, comparable to the revolution time of outer-shell electrons in atoms or molecules. Since ionization or excitation is the only reaction channel that can occur for antiprotons, and collision energies would be

CP1037, *Cold Antimatter Plasmas and Application to Fundamental Physics,* edited by Y. Kanai and Y. Yamazaki

below the threshold for electron-impact experiments at similarly low velocities and since ultra-short lasers cannot provide strong pulses in this time domain in the foreseeable future, slow antiprotons provide an unsurpassed, precise and probably the only tool to study many-electron dynamics in the strongly correlated, non-linear, sub-femtosecond time regime, the most interesting and, at the same time, most challenging domain for theory.

Exploiting and developing many-particle imaging methods – reaction-microscopes – in combination with novel electrostatic storage rings for slow antiprotons we envision to perform, single and multiple ionization cross section measurements for antiprotons colliding with atoms, molecules and clusters. Total, as well as, for the first time, any differential cross sections up to FDCS including ionization-excitation reactions will become available serving as urgently needed benchmark data for theory. Moreover, the formation of antiprotonic atoms, molecules or of protonium might be explored in kinematically complete experiments at ultra-low energies yielding unprecedented information on (n, l)-distributions of captured antiprotons as well as precise spectroscopic data of the respective energy levels.

MEASUREMENTS WITH LOW-ENERGY ANTIPROTONS

The following figure 1 shows measured and theoretical values of the total cross section for single ionization of Helium by antiproton impact. The corresponding interaction time between the projectile and the target is on the order of 30 as at an energy of $E=1$ *MeV* up to 1 fs at an energy of $E=1$ *keV*. Shorter times, even approaching the zeptosecond regime (1 zs = 10^{-21} s), are easily accessible in fast to relativistic collisions [10]. The main difference in the direct comparison to the situation for ion impact, especially at the low energies of interest here, is that ionization or excitation are the only reaction channels that can occur for antiprotons whereas capture processes dominate in ion collisions.

Moreover, even lowest collision velocities down to 25 eV/u can be realized without falling below the impact ionization threshold in contrast to electron collisions requiring a minimum velocity of about 1836 25 eV/u.

Hence, slow antiprotons provide a precise tool to study many-electron dynamics in the few atto- to femtosecond time regime - comparable to the revolution time of bound electrons.

At the same time, the interaction strength maximizes in this energy regime, which might be important for the possible future control of this motion, and consequently, the experimental [11,12] as well as the theoretical ionization cross sections for helium (see [8,9] and references therein) maximize as illustrated in figure 1.

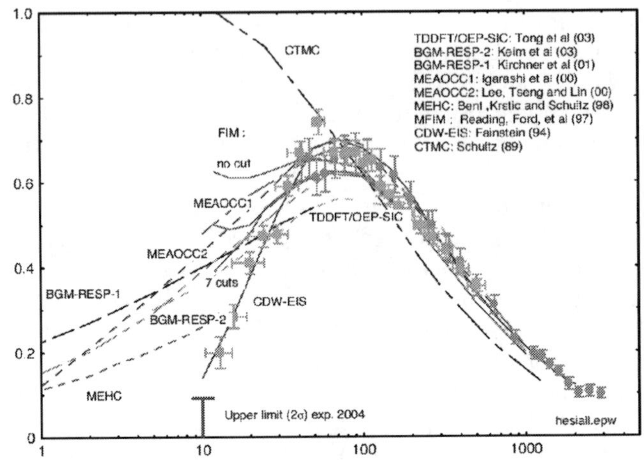

FIGURE 1. Total cross section for single ionization of helium by antiproton impact [7], see new data in [9].

This regime poses utmost challenges to theory: Due to the strong coupling any perturbative approach (valid for $Z/v << 1$) in powers of the interaction strength Z/v must fail (Z, v: projectile charge and velocity in atomic units, a.u., respectively; $v = 1$ a.u. $= 1/137 \cdot c$; c: velocity of light). By reducing the velocity, huge values up to $Z/v = 5$ for the perturbation can be reached at energies of a few keV. Such strong perturbations were so far only achieved in highly-charged heavy-ion collisions with the problem of additional capture channels being open.

Moreover, as demonstrated in numerous approximation calculations (various curves in figure 1, for details see [7]) the correlation between both electrons or at least the time-dependent response of the second electron (not being active in the sense that it remains in a bound state) has to be taken into account.

It can clearly be seen that even on the level of total cross sections pronounced differences between the predictions of different theories arise and all of them are in poor agreement with available experimental data in the low-energy regime. Moreover, even if theoretical predictions are quite close to each other, some of them come to drastically different, even contradictory interpretations. For example, on the basis of the Forced Impulse Method (FIM) results [13] time-dependent electronic correlation, beyond a Hartree-Fock treatment, is claimed to be essential, whereas the equally good predictions within the Basis Generator Model (BGM) [8] including microscopic response lead to the result that an effective one-particle picture is sufficient if the response of the second electron to the time-dependent field is properly taken into account. In the strict sense, this means that dynamical electron-electron correlation is not needed at all!

To conclude, low-energy antiproton collisions are expected to provide the ultimate benchmark for the development of strong-field non-perturbative theories in the presence of correlation. This is especially true if fully differential cross sections are explored for single, double and multiple ionization as will be demonstrated below.

MEASUREMENTS WITH HIGHLY CHARGED IONS

Numerous fully differential cross section measurements, so-called (e,2e) experiments have been performed for single ionization by electron impact since the pioneering work of Erhardt et al [14] (for a review see e.g. [15]), most of them under certain, restricted geometrical conditions in so-called coplanar geometry (see below). Due to immeasurably tiny scattering angles of heavy ionic projectiles, kinematically complete data for ion impact have only become available with the advent of Reaction-Microscopes. Whereas pioneering results were reported a decade ago [16], first FDCS were not published until 2001 (see e.g. [17]). Covering a large part of the final-state phase space they provided, at the same time, first information for electron emission into all three spatial directions, i.e. compared to electron impact for hitherto unexplored geometries.

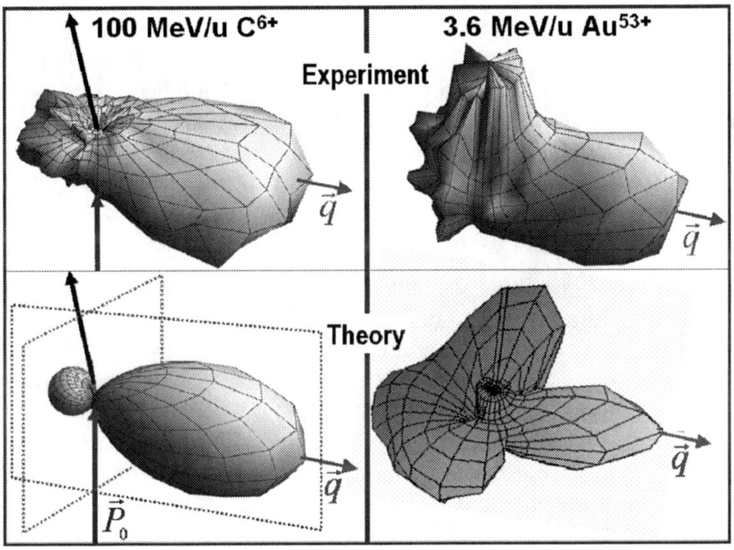

FIGURE 2. Measured (upper row) and calculated (lower row) three-dimensional intensity distributions of low-energy electrons (6.5 eV left and 20 eV right) emitted in singly ionizing 100 MeV/u C^{6+} and 3.6 MeV/u Au^{53+} on helium collisions for fixed momentum transfer along the red arrows (0.75 a.u. left and 1.0 a.u. right) [18].

As pointed out before, one observes in figure 1 that all theoretical predictions merge at high energies on the level of total cross sections. Here, at small perturbations, the first Born approximation (FBA) is expected to be fully applicable, cross sections scale with $(Z/v)^2$ and thus, become independent of the sign of the projectile charge such that electron and ion impact should yield identical results. In view of numerous (e,2e) measurements in this regime, it came as a big surprise that three-dimensional experimental FDCS for ion impact were found to be in significant discrepancy with all state-of-the-art theoretical predictions for single ionization at $Z/v = 0.1$, well in the perturbative area ($Z/v > 1$) [4].

In the above figure 2 (top left), a 100 MeV/u C^{6+} projectile ion impinges with momentum $\vec{P}_0$ along the blue arrow onto a helium atom. Being deflected and emerging along the black arrow, it ionizes the atom, imposing a momentum transfer $\vec{q}$ to the system as illustrated by the red arrow. Electrons of 6.5 eV are emitted into all spatial directions with an intensity represented by the distance of each point on the three-dimensional surface from the origin.

Theoretically, as can be seen from the result of advanced calculations (Figure 4, bottom left), an emission minimum is predicted in the "transverse plane" (red dotted line) perpendicular to the "scattering plane" (blue dotted line, also called coplanar geometry). Whereas this minimum has been consistently observed within the scattering plane in numerous previous ionization experiments using electron projectiles over more than 30 years in agreement with the present results for ion impact, a severe disagreement between our data and all theoretical predictions is found in the transverse plane, never explored in any experiment before. Despite considerable efforts from various sides, and many explanation attempts (see e.g. [19,20] and [21,22] and references therein where the influence of the experimental resolution has been investigated in great detail), the reason for this discrepancy has not yet been uncovered. Interestingly enough, recent experiments at an identical perturbation value for electron impact exploiting a reaction microscope with a factor of three better experimental resolution found the out-of-plane structures as well and theoretical predictions for electron impact trace them back to the contributions of second order amplitudes. Similar calculations for ion-impact, however, fail !

By increasing the projectile charge state while staying at relatively high velocities in order to minimize capture channels (scaling with $\sim v^{-11}$) the regime of strong perturbations up to $Z/v = 4.4$ has been explored using 3.6 MeV/u Au^{53+} projectiles (Figure 2, upper right). As might have been expected, the experimental results deviate dramatically from the theoretical prediction (Figure 2, bottom right) not only in shape, but even more severe in absolute magnitude. It should be mentioned that even in the strongly non-perturbative regime, total as well as differential electron emission cross sections are in excellent agreement with the theoretical predictions (see e.g. [23-25]). Discrepancies arise as soon as FDCS are investigated, i.e. when the quantum four-particle problem is explicitly addressed.

In summary, first measurements for ion impact at various perturbations as well as theoretical studies clearly demonstrate that only FDCS provide a reliable test ground for theory. From the experience gained with positive ions, it is expected that even in situations where theoretical predictions agree with each other on the level of total cross sections for single ionization of helium by antiproton impact, e.g. above the cross section maximum towards higher impact energies as shown in Figure 1, significant differences might arise between different theoretical results and experimental data when FDCS are explored. At lower energies, where the theoretical predictions deviate from each other as well as from available data even for total cross sections, FDCS for antiproton impact will provide the ultimate benchmark for theoretical models to describe sub-femtosecond correlated dynamics.

THE FACILITY FOR LOW-ENERGY ANTIPROTON AND ION RESEARCH (FLAIR)

Currently, the Antiproton Decelerator (AD) at CERN [26] is the only place in the world where physics with low-energy antiprotons can be performed.

This facility has been in operation since 2000, and although the experiments at the AD have produced some widely published and recognized results, such as the first formation of anti hydrogen at rest, they are limited by the relatively low intensity of antiprotons from the AD, approximately 10^5 particles per second, and by the availability of pulsed extraction only. In addition, the particles are delivered from the AD at a kinetic energy of 5 MeV, which is significantly higher than the 100 keV or less which is best suited for most of the experiments.

At the AD, the reduction in kinetic energy from 5 MeV to a few keV is achieved by degrading them in a foil, which causes a rather large increase of the beam divergence and momentum spread, and there is also a high loss of antiprotons in the degrader foil. These effects limit the capture efficiency into traps to about 10^{-4}.

An improvement was achieved by the installation of the RFQ–D used by the ASACUSA collaboration [27] that today provides beams at 100 keV energy. However, the rather large emittance ε = 100 mm mrad and energy spread $\Delta E/E$ = 10% of the output antiproton beam require a large stopping volume for precision measurements on antiprotonic helium for example where a high-power pulsed laser is used to induce transition for high precision spectroscopy. The laser bandwidth today limits the precision of the experiments.

A next generation facility, as evidenced by these examples, clearly needs to overcome the present limitations and must, in addition, pave the way for experiments not possible at present at the AD by providing slow extracted, quasi-DC beams for e.g. nuclear physics type measurements and ultra-show bunches of only a few nanoseconds duration for internal collision experiments with the reaction microscope.

The future FAIR [28] facility on the site of the current GSI laboratory will be an international facility mainly for nuclear and hadron physics, but also for atomic physics, for precision measurements on fundamental interactions and symmetries using atomic physics methods, for plasma research, biophysics and materials research. This will include physics with antiprotons, and the production rate of antiprotons will be at least 10 times higher than what is achieved at the AD today. This is due to the installation of three cooling and decelerating rings, similar to what CERN had at the time of proton-antiproton collisions at the SPS. In the case of FAIR, the antiprotons will be delivered from these rings at 30 MeV kinetic energy. The important new feature at FAIR is that the deceleration from 30 MeV is made at a dedicated facility, FLAIR (Facility for Low-energy Antiproton and Ion Physics), consisting of another two cooling and deceleration rings. In addition to antiprotons, also exotic unstable and highly charged ions will become available at FLAIR.

FLAIR will thus offer the unique possibility for low-energy antiproton and ion research and therefore benefit from maximum synergies between the fields. Details about the proposed research program can be found in the submitted letter of intent [29]. The layout of the facility is shown in figure 3. The building is designed as a complex which includes the experimental areas requested by the experiments presented in the technical proposals submitted by the FLAIR and SPARC [30,31] collaborations, the hall for the low-energy storage ring (LSR) and the additional areas needed for off-line mounting and testing of setups, control and data acquisition rooms, laser labs, power supplies storage rooms, a small workshop and social rooms.

The accelerator structure to decelerate antiprotons and highly charged ions consists of the above mentioned LSR, an electrostatic ultra-low energy storage ring (USR), and finally a universal trap facility (HITRAP) [32]. These components of the facility can provide stored as well as fast and slow extracted cooled beams at energies between 30 MeV and 300 keV (LSR), between 300 keV and 20 keV (USR), and cooled particles at rest or at ultra-low eV energies (HITRAP).

Since one main focus of the FLAIR collaboration is the exploitation of physics with low-energy antiprotons, the deceleration cycle within the facility and maximum particle numbers will be given here in this paper for this particular case.

The particles are injected from the NESR and are slowed down in a first step from 30 MeV to 300 keV in the LSR. Electron cooling will then be applied before transferring the antiprotons to the USR. Here, they are decelerated again to energies as low as 20 keV and can then be used either for in-ring experiments or be transferred via fast or slow extraction to external setups.

Since all the main parameters of the CRYRING facility at the Manne Siegbahn Laboratory are a perfect match with the requirements of the LSR and it was decided to discontinue its funding, a very attractive idea is to move the whole storage ring together with its integrated cooler and a low-energy injector to FLAIR. This would not only provide the LSR as one of the central installations of FLAIR, but would also allow off-line commissioning of the whole FLAIR facility and its experiments without the need of antiprotons or ions from the NESR. Furthermore, training of operators, as well as continuous development of the facility and experiments with ions of other species than those provided from the NESR would become feasible. Details about the storage ring as well as first dedicated experiments towards the machine performance in the FLAIR context can be found in [33].

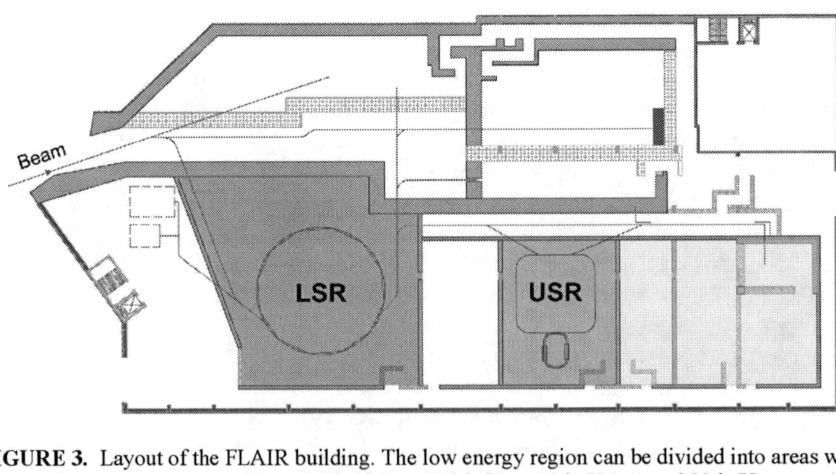

FIGURE 3. Layout of the FLAIR building. The low energy region can be divided into areas with beam energies between 30 MeV → 300 keV (—), 300 keV → 20 keV (—) and 20 keV → rest ( ).

In order to decelerate the beams of antiprotons and highly charged ions that will become available even further, an ultra-low energy electrostatic storage ring shall complement the accelerator infrastructure of FLAIR.

THE ULTRA-LOW ENERGY STORAGE RING (USR)

Electrostatic storage rings have proven to be a valuable tool for molecular and atomic physics in the low energy regime. Around some tens of keV, they allow avoiding problems related to hysteresis effects and remanence of magnetic fields.

Today only three such machines exist, all of them having a comparable, compact racetrack-shape layout and working at a fixed energy of 20 keV [34, 35] or 30 keV [36] with a continuous beam. Two of the rings [34, 36] can be operated at liquid nitrogen temperature and only one of them [35] is equipped with an electron merged beam device, which works at the required low energies but with a rather limited resolution. A double electrostatic ring, operating in a merged beam configuration and at temperatures below 10 K, has been approved at MSL in Stockholm [37], a fixed energy storage ring for energies up to 50 keV is planned at the University of Frankfurt [38] and a cryogenic storage ring (CSR) is presently being built up at the Max-Planck institute for nuclear physics (MPI-K) in Heidelberg [39].

For an ultra-low energy storage ring installed at a next-generation antiproton and ion facility, new and challenging techniques need to be developed to ensure multi-user operation and pave the way for a true multi-purpose facility. Among the particular points of concern are:

- Combination of the electrostatic storage mode with beam deceleration from 300 keV to 20 keV,
- Availability of fast and slow extraction to enable both trap experiments with pulsed beams and nuclear physics type experiments requiring extracted quasi-DC beams,
- Electron cooling at all covered energies for a variety of different ions ranging from antiprotons to bare uranium,
- Layout of a cryogenic vacuum system to ensure final gas pressures below $1 \cdot 10^{-13}$ mbar for reasonable beam life times even for highly charged ions,
- Development of particle detection techniques and beam diagnostic elements, operating at cryogenic temperatures, and designed for lowest beam energies and currents.

The USR is designed to combine these features for the first time and will be one of the essential tools for low-energy physics at FLAIR.

Ring Layout

A preliminary layout of the machine has already been worked out [40,41]. The symmetric machine consists of electrostatic elements only and houses an electron cooler, the in-ring reaction microscope, a combined slow/fast extraction region and the possibility of adding additional experimental sections like e.g. a positron cooler as depicted in the following figure 4.

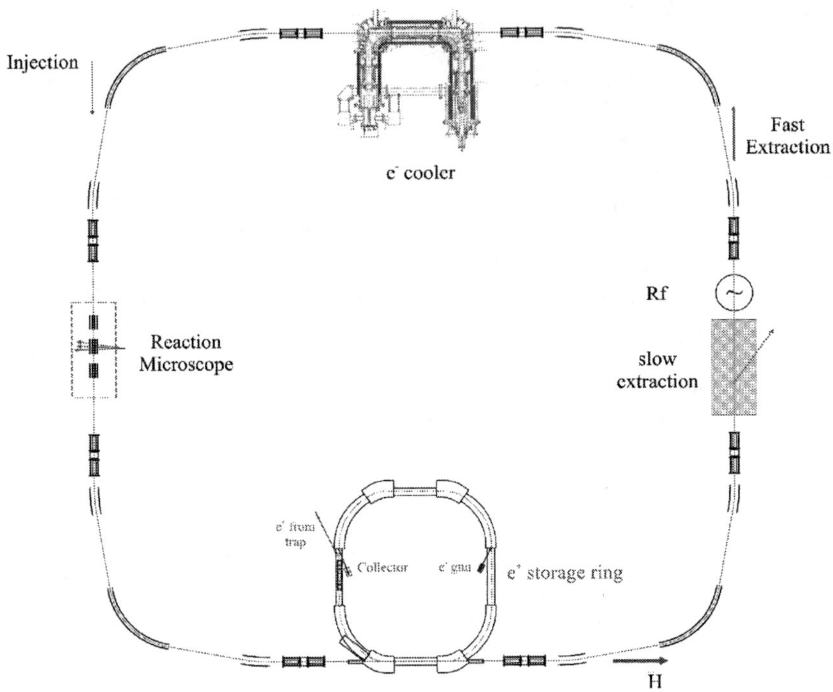

FIGURE 4. Preliminary layout of the ultra-low energy storage ring with the optional merged positron ring [38].

The overall 90° bend in the corner sections is split up into one 70° and two smaller 10° cylinder deflectors to allow for easy injection along one of the straight sections as well as fast extraction of the beam and detection of neutral or any charge-exchanged particles generated by e.g. collisions with the residual gas. Transverse modulation of the beam is realized by pairs of electrostatic quadrupole doublets which might also house small steerers for closed orbit correction [42]. All electrostatic elements are limited by grounded shields placed at a distance of 15 mm on both sides of the respective element. This efficiently reduces the effect of fringe fields on the stored beam and higher order effects.

The main characteristics of the electron cooler are an effective cooler length of roughly one meter, a field homogeneity of $B_x/B_s < 1 \cdot 10^{-3}$ and a maximum solenoid field of B_{max}=300 G. It should be pointed out that electron cooling in this regime, where the electron energy needs to be as low as 10 eV is a new challenge that is presently addressed at the MPI-K [43,44].

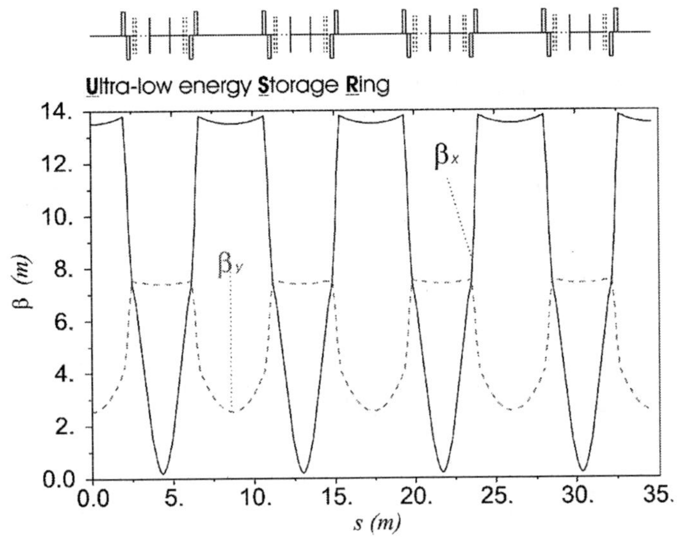

FIGURE 5. Calculated preliminary beta functions in the USR [40].

Depending on the experimental requirements, the machine can be set to different working points with one example given in figure 5. In this case the horizontal beam size is minimized in the 70° bends, where the mechanical aperture is relatively small, while the vertical beam size is at its maximum at this position. An overview of the machine and element parameters is given in table 1.

TABLE 1. Summary of the preliminary parameters of the USR.

Parameter	
Energy Range	300 – 20 keV/q
Circumference	35 m
Base Pressure	$< 5 \cdot 10^{-13}$ mbar
Operating Temperature	4 K
Space Charge Limit at 20 keV/q and $\varepsilon_{x,y}$=1 mm mrad	$2 \cdot 10^{7}$ ions
Estimated cooling time at 20 keV/q	~ 1 sec.

As pointed out before, FLAIR will provide users with beams of both antiprotons and exotic, highly charged ions. To make maximum use of these unique possibilities, the USR shall ideally allow for decelerating all available ion species to lowest energies of 20 keV/q. Since the capture cross sections of highly charged ions are very high at such low energies, a cryogenic vacuum system would be needed to realize ultimate pressures below the 10^{-13} mbar level. Table 1 refers to this cryogenic vacuum system. In case one "only" thinks about decelerating antiprotons, then the vacuum requirements would relax to some $1 \cdot 10^{-11}$ mbar and an operation at room temperature would become feasible.

The collision experiments described earlier in this article require around 10^4 antiprotons per bunch. This number is well below the space charge limit of the USR even at the lowest energies. However, the time structure of the pulses needs to be in the order of a few nanoseconds to allow its use as a trigger signal for the measurements. In this respect, an efficient control of the longitudinal time structure of the stored ion beam is of central importance since it directly determines the quality of the envisaged experiments.

Since many years, it has been a significant challenge for the storage ring accelerator-physics community to develop techniques to reduce the duration of bunches. Up to now, all methods that have been developed go along with various difficulties, which can include reduced stored-beam lifetimes.

This is partially due to the strong space charge limitations that were found in existing electrostatic machines; probably linked to non-linear fields that cannot be avoided in such storage rings. The nature of these effects is not at all understood and it is a part of ongoing studies in the QUASAR group at MPI-K to simulate the motion of a stored ion beam in an electrostatic cooler synchrotron under consideration of non-linear fields as well as space charge effects, together with a particular focus on the needs of both internal and external experiments. Thus, new concepts for the manipulation and control of the longitudinal beam structure are presently being developed:

First, a special operation mode of the storage ring with an optics setting where the phase slippage factor η is negative is being studied. In such a setting the space charge effect would no longer counteract the applied HF voltage. The aim of this work is to develop a thorough understanding of the beam motion and to perform tests with beam in the TSR at the MPI-K.

Second, a scheme where different rf frequencies are subsequently applied to the beam to generate shortest bunches is being investigated. In order to be able to probe sub-femtosecond correlated dynamics, it will be essential to realize ultra-short bunches in the nanosecond regime.

Reaction Microscope

With a luminosity, or in other words, effective collision rates, that are at least six orders of magnitude higher than for a single-pass experiment, the USR provides the ideal tool for studies of the dynamics in ion- (antiproton-) atom (molecule) collisions occurring on the sub-femtosecond timescale. For that purpose, a novel in-ring reaction microscope will be developed for the USR and integrated into the third experimental section. Reaction Microscopes allow for the kinematically complete investigation of the interaction between the stored ion-beam and atomic or molecular gas targets. As illustrated in the following figure 6, the recoil ions as well as the electrons from target ionization are extracted with the help of homogeneous electric and magnetic fields. By measuring the time-of-flight and the detector position of all created particles the vector momenta of multiply emitted electrons and ions become accessible simultaneously.

In combination with the increased luminosity of the beam in the rings as compared to single pass experiments such spectrometers will enable a whole class of novel experiments on many-particle quantum dynamics and spectroscopy in atomic and molecular collisions. Up to now, a fully equipped reaction microscope has only been recently implemented at the Experimental Storage Ring (ESR) [45] for the first time and was demonstrated to be successfully operated but has never been integrated in an electrostatic machine or operated at ultra-low energies of the circulating projectile beam. The spectrometer itself will be oriented perpendicular to the beam axis, requiring a compensation of the electric and magnetic fields by using small correction elements placed along the straight section of the ring.

Since already small field perturbations can yield to beam instabilities at such low beam energies, a careful layout together with detailed numerical simulations is required. For that purpose electric and magnetic fields were calculated using the computer code TOSCA [46]. A quarter section of the storage ring with the reaction microscope and the correction elements is shown in figure 6.

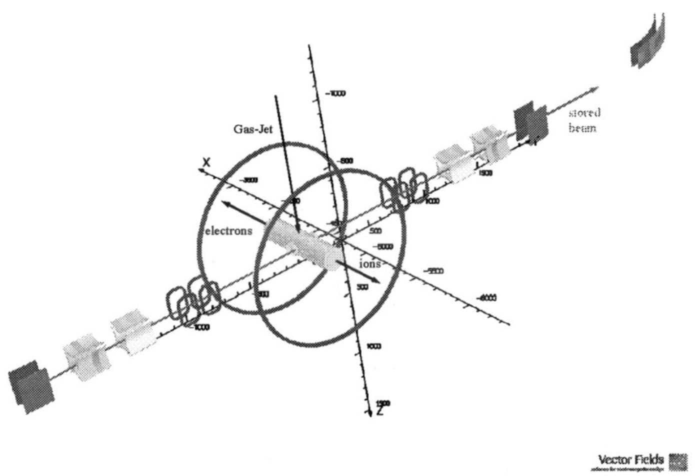

FIGURE 6. Quarter section of the storage ring including the reaction microscope.

The Reaction Microscope has been optimized to increase both resolution and versatility of the spectrometer. The spectrometer is built up asymmetrically featuring a longer recoil-ion detection to increase the time-of-flight resolution for ions. Moreover, the option to implement time-of-flight and spatial focussing optics is foreseen here. In contrast, the electron section is kept much shorter, providing the additional opportunity, when used for positively charged particles to efficiently detect fragments of molecular break-up processes in ion or antiproton collisions with a large solid angle. For the electrodes of the spectrometer, a special design was used to create a setup which is transparent for florescence photons emitted during the collision. This enables us to register these photons in coincidence with all other fragments. Figure 7 shows the geometry of the spectrometer with respect to the beam axis and gas jet.

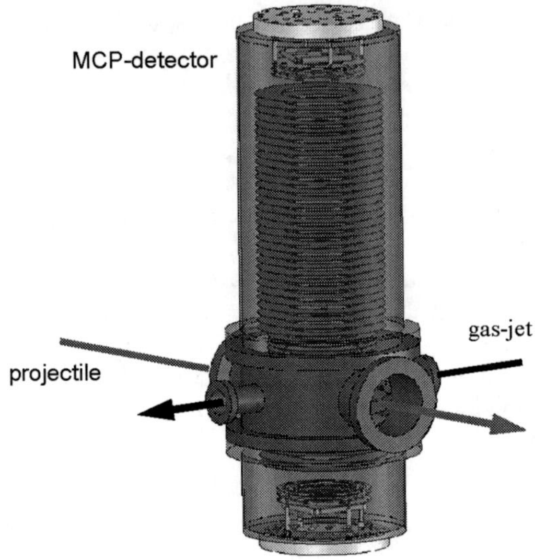

FIGURE 7. Geometry of the asymmetric spectrometer with respect to projectile and gas jet

A critical point in this setup is the question about the functioning of micro channel plate detectors at temperatures of some Kelvin. Only a few measurements were carried out in the past [47,48] to explore this regime and prelimenary results from a recent test in Heidelberg indicate that these detectors can be used at cryogenic temperatures as low as 20 K. It is now planned to extend these tests of the MCP's performance down to temperatures below 10 K.

SUMMARY AND OUTLOOK

The understanding and the envisioned future control of the correlated quantum-motion of few-electron systems on ultra-short time scales are among the most attractive and challenging problems in contemporary atomic, molecular and laser physics. It has been shown in this contribution that slow antiprotons provide a unique, and presently the only, tool to experimentally access this time regime, which is typical for the dynamics of outer-shell electrons in atoms and molecules which, in turn, form chemical bonds and, thus, determine our every-day life. At the same time we have provided striking evidence, that this motion is only poorly understood.

This paper furthermore describes in some detail plans for a next-generation facility for low-energy antiproton and ion research, and in particular the challenges related to the ultra-low energy storage ring project and its integrated reaction microscope. We presently address the many open questions at the MPI-K together with collaborating partners, with the aim to realize fully differential measurements with highest resolution, and with expected rates that are many orders of magnitude larger than what is available today, in the future.

ACKNOWLEDGMENTS

The generous support of the Helmholtz Association of National Research Centers (HGF) under contract number VH-NG-328 and of the Gesellschaft für Schwerionenforschung (GSI) Darmstadt is acknowledged.

REFERENCES

1. T.N. Rescigno, M. Baertschy, W.A. Isaacs, and C.W. McCurdy, Science 286 (1999) 2474; M. Baertschy, T.N. Rescigno, C.W. McCurdy, *Phys. Rev. A* 64 (2001) 022709
2. I. Bray, *Phys. Rev. Lett.* 89 (2002) 273201-1
3. A.S. Kadyrov, A.M. Mukhamedzhanov, A.T. Stelbovics, and I Bray, *Phys. Rev. Lett.* 91 (2003) 253202-1
4. M. Schulz, R. Moshammer, D. Fischer, H. Kollmus, D.H. Madison, S. Jones and J. Ullrich, *Nature*, 422 (2003) 48
5. R. Moshammer, A. Perumal, M. Schulz, V.D. Rodríguez, H. Kollmus, R. Mann, S. Hagmann, and J. Ullrich, *Phys. Rev. Lett.* 87 (2001) 223201
6. M. Schulz, R. Moshammer, A.N. Perumal, and J. Ullrich, *J. Phys. B* 35 (2002) L161
7. ASACUSA Collaboration, CERN/SPSC 2005-001, SPSC-M-728
8. M. Keim, A. Achenbach, H.J. Lüdde and T. Kirchner, *Phys. Rev. A.* 67 (2003) 062711-1
9. H. Knudsen et al., *these proceedings*
10. R. Moshammer et al., *Phys. Rev. Lett.* 79 (1997) 3621
11. L.H. Anderson, P. Hvelplund, H. Knudsen, and S.P. Møller, *Phys. Rev. Lett.* 57 (1986) 2147
12. P Hvelplund, H. Knudsen, U. Mikkelsen, E. Morenzoni, S.P. Møller, E. Uggerhøj, and T. Worm, *J. Phys. B* 27 (1994) 925
13. T. Bronk, J.F. Reading, and A.L. Ford, *J. Phys. B* 31 (1998) 2477
14. H. Erhardt, M. Schulz, T. Tekaat, and K. Willmann, *Phys. Rev. Lett.* 22 (1969) 89
15. A. Lahmam-Bennani, *J. Phys. B.* 24 (1991) 2401
16. R. Moshammer, J. Ullrich, M. Unverzagt, W. Schmidt, P. Jardin, R.E. Olson, R. Mann, R. Dörner, V. Mergel, U. Buck and H. Schmidt-Böcking, *Phys. Rev. Lett.* 73(25) (1994) 3371-3374
17. M. Schulz, R. Moshammer, D.H. Madison, R.E. Olson, P. Marchalant, C.T. Whelan, H.R.J. Walters, S. Jones, M. Foster, H. Kollmus, W. Schmitt, A. Cassimi, and J. Ullrich, *J. Phys. B* 34 (2001) L305
18. J. Ullrich et al., *AIP Conf. Proc.* 793 (2005) 71
19. D. H. Madison, D. Fischer, M. Foster, M. Schulz, R. Moshammer S. Jones, and J. Ullrich, *Phys. Rev. Lett.* 91 (2003) 253201-1
20. Fiol J and Olson R.E., *J. Phys. B* 37, 3947 (2004)
21. M. Dürr, B. Najjari, M. Schulz, A. Dorn, R. Moshammer and J. Ullrich, *J. Phys. C* 88 (2007) 012016
22. M. Dürr, B. Najjari, M. Schulz, A. Dorn, R. Moshammer, A.B. Voitkiv, J. Ullrich, *Phys. Rev. A* 75 (2007) 062708
23. P.D. Fainstein, L. Gulyás, F. Martín, and A. Salin, *Phys. Rev. A* 53 (1996) 3243
24. W. Schmitt, J. Ullrich, R. Moshammer, F. S. C. O'Rourke, H. Kollmus, L. Sarkadi, R. Mann, R.E. Olson, *Phys. Rev. Lett.* 16 (1998) 4337
25. T. Kirchner, L. Gulyás, M. Schulz, R. Moshammer, and J. Ullrich, *Phys. Rev. A.* 65 (2002) 042727
26. S. Baird et al., "The Antiproton Decelerator: AD", *Proc. Part. Acc. Conf.* (1997)
27. W. Pirkl, A. M. Lombardi, and Y. Bylinsky, "First Operating Experience with the CERN Decelerating RFQ for Antiprotons", *Proc. Part. Acc. Conf., Chicago*, USA (2001)
28. W.F. Henning, H.H. Gutbrod, K.D. Groß, and V. Metag (editors), An International Facility for Beams of Ions and Antiprotons (FAIR CDR), GSI Darmstadt (2001)
29. http://www.oeaw.ac.at/smi/flair/LOI.htm
30. http://www.oeaw.ac.at/smi/flair/TechnicalProposal/TP.htm
31. http://www.gsi.de/fair/experiments/sparc/index_e.html
32. W. Quint et al.,"HITRAP: A Facility for Experiments with Trapped Highly Charged Ions", *Hyp. Int.* 132, 457 (2001)

33. H. Danared, A. Källberg, and A. Simonsson, "CRYRING Machine Studies for FLAIR", *Proc. European Part. Acc. Conf.*, Edinburgh, UK (2006)
34. S.P. Møller, "ELISA – an Electrostatic Storage Ring for Atomic Physics", *Proc. European Part. Acc. Conf.*, Stockholm, Schweden (1998)
35. T. Tanabe et al, "An Electrostatic Storage Ring for Atomic and Molecular Science", *Nucl. Instr. and Meth. A* 482 (2002) 595c
36. S. Jinno et al., "TMU electrostatic ion storage ring designed for operation at liquid nitrogen temperature", *Nucl. Instr. and Meth. A* 532 (2004) 477c
37. K.-G. Rensfelt et al.,"DESIREE – a Double Electrostatic Storage Ring", *Proc. European Part. Acc. Conf.*, Lucerne, Switzerland (2004)
38. C.P. Welsch et al.,"An Electrostatic Storage Ring as the Central Maschine of the Frankfurt Ion Storage Experiments", *PRST-AB* 7, 080101 (2004)
39 D Zajfman et al," Physics with colder molecular ions: The Heidelberg Cryogenic Storage Ring CSR", *J. Phys. Conf. Ser.* 4 (2005) 296c
40. C.P. Welsch, M. Grieser, J. Ullrich, A. Wolf,"Layout of the USR at FLAIR", *Proc. European Part. Acc. Conf.*, Edinburgh, Scotland (2006)
41. C.P. Welsch, M. Grieser, J. Ullrich, A. Wolf,"An ultra-low-energy storage ring at FLAIR", *Nucl. Instr. and Meth. A* 546 (2005) 405–417
42. C.P. Welsch, C. Gläßner, M. Grieser, J. Ullrich,"An Electrostatic Quadrupole Doublet with an Integrated Steerer", *Proc. European Part. Acc. Conf.*, Lucerne, Switzerland (2004)
43. H. Fadil, M. Grieser, D. Orlov, A. Wolf," Design of a Low Energy Electron Cooler for the Heidelberg CSR", *Proc. European Part. Acc. Conf.*, Edinburgh, Scotland (2006)
44. C.P. Welsch, A. Smirnov,"Cooling Rates at Ultra-low Energy Storage Rings", *Proc. European Part. Acc. Conf.*, Edinburgh, Scotland (2006)
45. D.Fischer, R. Moshammer, S. Hagmann, T. Ferger, M. Gudmundsson, K. Kühnel, M. Schäfer, M. Nofal, H. Bräuning, C. Brandau, C. Kozhuharov, Th. Stöhlker, G. Lanzanó, E. de Filippo, H. Rothard, and J. Ullrich, *GSI Sci. Rep.* (2006)
46. http://www.vectorfields.com/
47. J.A. Schecker et al., *Nucl. Instr. and Meth. A* 320 (1992) 556-561
48. P. Roth and G.W. Fraser, *Nucl. Instr. and Meth. A* 439 (2000) 134

Low-energy scattering of antihydrogen by helium and molecular hydrogen

E.A.G. Armour*, A.C. Todd*, S. Jonsell†, Y. Liu*, M.R. Gregory* and M. Plummer**

*School of Mathematical Sciences, Nottingham University, Nottingham NG7 2RD, UK
†Department of Physics, University of Wales Swansea, Swansea SA2 8PP, UK
**CLRC Daresbury Laboratory, Warrington WA4 4AD, UK.

Abstract. In this paper, we describe in detail calculations that we have carried out of cross sections for rearrangement processes in very low-energy helium + antihydrogen ($\bar{\mathrm{H}}$) scattering that result in $\mathrm{He}^{+}\bar{p}+\mathrm{Ps}$ or $\mathrm{He}\bar{p}+e^{+}$ or $\alpha\bar{p}+\mathrm{Ps}^{-}$. The interaction between the leptons is taken into account very accurately. Results are presented for all three processes. A description is also given of a preliminary calculation of elastic and antiproton annihilation cross sections for very low-energy $\mathrm{H}_2+\bar{\mathrm{H}}$ scattering.

Keywords: Antihydrogen, Antimatter, Helium, Scattering, Rearrangement, Positron
PACS: 36.10-k; 34.50.-s

INTRODUCTION

Experimentalists in the ATHENA and ATRAP groups at CERN prepared antihydrogen ($\bar{\mathrm{H}}$) in 2002. ATHENA has now been superseded by ALPHA. ALPHA and ATRAP are carrying further work to trap antihydrogen so that its properties can be studied. See, for example, refs. [1,2]. It should be possible to used the trapped antihydrogen to make high-precission tests of fundamental principles of physics such as the CPT invariance of relativistic quantum field theory and the equivalence principle of general relativity. See, for example, ref. [3] and section 4 of this paper.

There is interest in the interaction of $\bar{\mathrm{H}}$ with atoms and molecules for a number of reasons. Firstly, it is a novel area involving the interaction of matter and antimatter. Secondly, to trap antihydrogen, it is necessary to cool it to sub-Kelvin temperatures. In view of this, experimentalists are looking for suitable atoms or molecules that could be used to cool antihydrogen. Thirdly, gases such as He and H_2 may be an impurity in the apparatus used to contain the antihydrogen.

The first calculations to be carried out on the scattering of atoms by antihydrogen were, as might be expected, for $\mathrm{H}\bar{\mathrm{H}}$ scattering at very low energies. The principal causes of loss of $\bar{\mathrm{H}}$ in this process are rearrangement into $p\bar{p}+\mathrm{Ps}$ [4,5] and antiproton annihilation due to the strong interaction [6,7].

In ref. [8] we described preliminary calculations that we had carried out for the rearrangement reaction

$$\mathrm{He}+\bar{\mathrm{H}} \rightarrow \mathrm{He}^{+}\bar{p}+\mathrm{Ps} \tag{1}$$

at very low energies. In these calculations, rearrangement cross sections were obtained from the T-matrix for the reaction which was evaluated using a distorted wave approx-

CP1037, *Cold Antimatter Plasmas and Application to Fundamental Physics,* edited by Y. Kanai and Y. Yamazaki

imation in which the exact scattering wave function was replaced by the wave function for the entrance channel, calculated using the Born–Oppenheimer approximation [9].

As is pointed out in ref. [8], calculations involving rearrangement processes are difficult to carry out as, in most cases, very different coordinates must be used to describe the entrance channel. The entrance channel wave function in (1) is expressed in terms of prolate spheroidal coordinates whereas the positronium (Ps) part of the wave function used to represent the right-hand side is expressed in terms of the coordinates of the Ps centre of mass and the position of the electron relative to the positron. The method used to reduce the numerical integration in the evaluation of the T-matrix elements from nine to five dimensions is described in ref. [8].

The calculation of the leptonic wave function for the entrance channel was carried out using basis functions that contained Hylleraas-type functions that correlate the electron-positron interaction. A defect of the calculation was that no systematic method was used to optimise the non-linear parameters in the trial function used to represent the leptons in the entrance channel. As a consequence of this, the leptonic energy values were considerably higher in value than the very accurate values obtained by Strasburger et al. [10] using a basis set made up of explicitly correlated Gaussian functions. This cast doubt on the accuracy of the leptonic wave functions.

In the present calculation, the accuracy of these wave functions has been very significantly improved by optimising the non-linear parameters in the trial function, by choosing the basis functions used systematically and by including Hylleraas-type functions that correlate the electron-electron interaction in addition to those that correlate the electron-positron interaction.

In addition to reaction (1), we have carried out calculations for two other rearrangement reactions:

$$\mathrm{He}+\bar{\mathrm{H}} \rightarrow \mathrm{He}\bar{p}+e^{+} \tag{2}$$

$$\mathrm{He}+\bar{\mathrm{H}} \rightarrow \alpha\bar{p}+Ps^{-} \tag{3}$$

at very low energies. As our most recent calculations have been for reaction (3), we will focus on this reaction in this paper.

Finally, we will describe a preliminary calculation that we have carried out of elastic and antiproton annihilation cross sections for very low-energy $\mathrm{H}_2+\bar{\mathrm{H}}$ scattering using the generalised Kohn method.

CALCULATION OF THE ENTRANCE CHANNEL WAVE FUNCTIONS

The coordinate system used in our calculations is illustrated in Fig. 1. A and B are the positions of the antiproton ($\bar{p}$) and the He nucleus (α), respectively. A has coordinates $(0,0,-R/2)$ and B, $(0,0,R/2)$, where R is the internuclear distance. The prolate spheroidal coordinates of P are (λ,μ,ϕ), where ϕ is the usual azimuthal angle of spherical polar coordinates,

$$\lambda=\frac{r_A+r_B}{R} \quad \text{and} \quad \mu=\frac{r_A-r_B}{R}.$$

The basis functions that we used to calculate the leptonic part of the entrance channel wave function at a given R value are expressed in terms of these coordinates. They are of the form

$$\begin{aligned}\psi_i &= S_{23} N \lambda_1^{a_i} \lambda_2^{b_i} \lambda_3^{c_i} \mu_1^{d_i} \mu_2^{e_i} \mu_3^{f_i} C(s_i,t_i) \\ &\times \exp[-\alpha_1\lambda, -\alpha_2\lambda_2 - \alpha_3\lambda_3 + \beta_1\mu_1 + \beta_2\mu_2 + \beta_3\mu_3]\end{aligned} \tag{4}$$

The index 1 refers to the positron, 2 and 3 to the electrons and S_{23} is the electron symmetriser. As the He is taken to be in its ground state, the electrons are in a singlet spin state. N is a normalisation constant. $a_i, b_i, c_i, d_i, e_i, f_i$ are non-negative integers. $\alpha_1, \alpha_2, \alpha_3, \beta_1, \beta_2, \beta_3$ are non-linear parameters. The factor $C(s_i,t_i)$ depends on the type of basis function. The forms of this factor that we used in our calculations are shown in Table 1.

Table 1
Types of Basis Function

s_i	t_i	$C(s_i,t_i)$	Type
0	0	1	Product of σ functions
1	1	ρ_{12}	Positron-electron Hylleraas-type
1	2	ρ_{23}	Electron-electron Hylleraas-type
2	1	$\frac{2}{R}(x_1x_2 + y_1y_2)$	Positron-electron configuration interaction (CI) product of π functions
2	2	$\frac{2}{R}(x_2x_3 + y_2y_3)$	Electron-electron CI, product of π functions

The distance between particles i and j, $r_{ij} = \frac{R}{2}\rho_{ij}$.

If Hylleraas-type functions that correlate the interaction between the positron and the electrons are included in the basis set, it is necessary to evaluate matrix elements containing $\frac{r_{12}r_{13}}{r_{23}}$, where r_{ij} is the distance between particles i and j. This is a very difficult task. Factors of the type r_{ij}^2 can easily be expressed in terms of prolate sheroidal coordinates. For $1/r_{ij}$ factors, it is necessary to use the Neumann expansion

$$\frac{1}{r_{ij}} = \frac{2}{R}\sum_{\tau=0}^{\infty}\sum_{\nu=0}^{\tau} D_{\tau\nu}P_\tau^\nu(\lambda_<)Q_\tau^\nu(\lambda_>)P_\tau^\nu(\mu_i)P_\tau^\nu(\mu_j)\cos[\nu(\phi_i-\phi_j)] \tag{5}$$

Here $\lambda_<$ is the lesser and $\lambda_>$ the greater of λ_i and λ_j. $P_\tau^\nu(\lambda)$ and $Q_\tau^\nu(\lambda)$ are the first and second solutions, respectively, to the Legendre equation if $\nu = 0$, or the associated Legendre equation if $\nu \neq 0$.

A triple Neumann expansion was used to evaluate the matrix elements containing $\frac{r_{12}r_{13}}{r_{23}}$. As pointed out in ref. [8], it was possible to evaluate the integrals involved using two dimensional numerical integration. Since carrying out the calculations described in ref. [8], the computer program used to evaluate these integrals has been made very much faster [11]. This made it possible to use it as a subroutine in the overall program that carried out the variational calculations to determine the leptonic energies and wave functions.

At each R value, the variational calculations were carried out using an algorithm that first optimised the non-linear parameters using a set of 320 basis functions made up of all the types that feature in Table 1 with values of $a_i, b_i, c_i, d_i, e_i, f_i$ less than or equal to 1. It then went on to select the 200 basis functions that gave the lowest energy out of a set of 2314, which included the 320 functions used in the above optimization plus all functions with at least one of $a_i, b_i, c_i, d_i, e_i, f_i$ equal to 2 if $s_i = 0$ or 2 [12].

Table 2
Comparison of Results

Internuclear distance, R	Leptonic energy value This calculation†	Leptonic energy value Strasburger *et al.* [10]*	Difference This calculation	Difference for the energy values obtained in ref. [8]§
0.1	−0.793417	−0.802881	0.0095	0.0184
0.5	−1.070414	−1.076846	0.0064	0.0233
1.0	−1.634494	−1.636474	0.0020	0.0152
1.5	−2.099359	−2.100153	0.0008	–

All units are hartree atomic units.
† 200 hundred selected basis functions.
* 768 basis functions.
§ The energy values in ref. [8] were obtained using up to 168 basis functions.

A comparison is made in Table 2 between the energy values we obtain at four R values and the very accurate values obtained by Strasburger et al. [10] using basis functions,

$\{\chi_{ip}\}$, which are explicitly correlated Gaussian functions of the form

$$\chi_{ip} = S_{23} \exp\left[-\sum_{i=1}^{3} \alpha_{ip}(r_i - R_{ip}) - \sum_{i>j}^{3} \beta_{ijp}(r_i - r_j)^2 \right] \tag{6}$$

where α_{ip}, β_{ijp} and R_{ip} are independent non-linear parameters and r_i is the position vector of particle i. S_{23} is the electron symmetriser.

It can be seen from Table 2 that our energy values are always above the values of Strasburger et al. [10] and are thus less accurate. However, the difference is always less than 0.01 a.u., decreases with increasing R and is less than 0.001 a.u. at $R = 1.5a_o$. In view of this, we consider that our leptonic wave functions are likely to be very accurate. Note that our energy values are considerably more accurate than the values for the leptonic wave functions used in our earlier calculation [8].

Our basis sets are much smaller than those used by Strasburger et al. The accuracy of our calculations could be improved by including some basis functions with $s_i = 1$, i.e. Hylleraas-type functions, with at least one of $a_i, b_i, c_i, d_i, e_i, f_i$ equal to 2 or some basis functions with different non-linear parameters in the exponentials. This would substantially increase the time required for the calculation. In any event, it was not practical for us to carry out evaluations of T-matrix elements with leptonic wave functions containing more than about 200 basis functions.

CALCULATION OF REARRANGEMENT CROSS SECTIONS

The rearrangement cross sections for the reactions (1), (2) and (3) are simply related to the T-matrix

$$T_{ji} = \langle \Phi_j \mid V_j \mid \Psi_{k_i}^{(+)} \rangle \quad (j = 1,2,3), \tag{7}$$

where $\Psi_{k_i}^{(+)}$ is the exact scattering wave function for the incident energy under consideration. V_j is the potential that couples the two systems resulting from the scattering, e.g. $He^+\bar{p}$ and Ps. If V_j is not Coulombic asymptotically, Φ_j is the final wave function calculated using the full Hamiltonian for the internal motion of the system with V_j set to zero. The required modification in the Coulombic case will be described later. The value of j is 1, 2 or 3 depending on which reaction is being considered.

T_{ji} was calculated using the form of the distorted wave approximation used by Jonsell et al. [9] in their calculation of rearrangement cross sections for $H + \bar{H}$. In this method, the exact scattering wave function is approximated by the wave function for the entrance channel, calculated using the Born–Oppenheimer (BO) approximation.

The leptonic wave function for the incident channel, $\psi_{BO}(r;R)$, was calculated as described in the previous section. r represents the coordinates of the leptons and R is the internuclear vector. The continuum wave function, $\chi(R)$, for the relative motion of the He and the $\bar{H}$ was calculated using the very accurate BO potential of Strasburger et al. [10].

Calculations for the rearrangement reactions (1) and (2)

$$\mathrm{He}+\bar{\mathrm{H}} \to \mathrm{He}^{+}\bar{p}+\mathrm{Ps} \quad \text{and} \quad \mathrm{He}+\bar{\mathrm{H}} \to \mathrm{He}\bar{p}+e^{+} \tag{8}$$

Details of calculations we have already carried out for these reactions have already been published in ref. [14]. For completeness, the results for these reactions given in this paper are presented in Tables 3 and 4. In the case of the reaction that has Ps as a product, the Ps was taken to be in its ground state. As the collisions are considered to be taking place at the sub-Kelvin temperatures of trapped $\bar{\mathrm{H}}$, only s-wave scattering is considered for these reactions and also for reaction (3). Also, the rearrangement processes that are considered are those to states in which both the products have zero angular momentum.

Table 3
Total cross sections for rearrangement into the $v = 35$ state of $\mathrm{He}^{+}\bar{p}$ and Ps in its groundstate

Incident energy, ε a.u.	Cross section a_0^2	Cross section $\varepsilon^{-\frac{1}{2}}a_0^2$
10^{-10}	14237	0.14237
10^{-9}	4501.5	0.14235
10^{-8}	1421.0	0.14210
10^{-7}	441.9	0.1397
10^{-6}	121.5	0.1215
10^{-5}	20.57	0.0651
10^{-4}	2.30	0.023
10^{-3}	0.27	0.00851

Total cross sections for rearrangement into the Ps in its ground state and $\mathrm{He}^{+}\bar{p}$ in its vibrational state with $v = 35$ are given in Table 3. This is the vibrational state that is closest to threshold. At very low energies, the results in Table 3 are smaller by a factor of about 3 than the cross section for the rearrangement reaction

$$\mathrm{H}+\bar{\mathrm{H}} \to \mathrm{Pn}\ (N=23)+\mathrm{Ps}\ (n=1), \tag{9}$$

where N and n are the principle quantum numbers of $\mathrm{Pn}(p\bar{p})$ and Ps, respectively. This is the largest rearrangement cross section that has been calculated for $\mathrm{H}+\bar{\mathrm{H}}$ [4,5].

It can be shown that if exact Born–Oppenheimer leptonic wave functions were used in the entrance and rearrangement channels, to a very good approximation, the leptonic T-matrix, $t_{ji}(R)$, where

$$t_{ji}(R) = \langle \phi_j^{(\mathrm{lep})} \mid V_j \mid \psi_{\mathrm{BO}} \rangle \tag{10}$$

and $\phi^{(\mathrm{lep})}$ is the leptonic part of Φ_j, can be straightforwardly calculated [13]. A knowledge is required of the energies of these wave functions, as determined by their respective Born–Oppenheimer Hamiltonians, and the overlap of the wave functions on the left and right-hand sides of the matrix element in (10), obtained by setting $V_1 = 1$. Comparison of values of $t_{ji}(R)$, calculated with and without use of this approximation, indicates that it gives accurate results for the BO wave functions that we are using. We have used it to calculate cross sections for the $v = 34$ vibrational state of $\mathrm{He}^+\bar{p}$. We find that they are smaller than those for $v = 35$ by a factor of about 500.

More recently, calculations have been carried out for other states of $\mathrm{He}^+\bar{p}$ and Ps [15]. They were carried out using the method employed to calculate T-matrix elements for reaction (3). It involved eight dimensional numerical integration and will be described in more detail in the section on this reaction. The accuracy of this method was checked by calculating some of the cross sections already obtained as the earlier results were calculated using only five dimensional numerical integration, as described in I. For Ps in its ground state, there was reasonable agreement with the results already obtained for $v = 35$ for the energies less than or equal to 10^{-6} a.u. for which the comparison calculations were carried out. Agreement was less good between the results for $v = 34$, some cross section values differing by factors as large as 4. The origin of these discrepancies is being investigated.

Cross sections for the $v = 33$ and $v = 32$ were obtained using eight dimensional numerical integration. They were found to be smaller than those for $v = 35$ by factors of at least 10 and 20, respectively. If the Ps is taken to be in its first excited s-state, the first state below threshold is the $v = 33$ state of $\mathrm{He}^+\bar{p}$. In this case, cross sections for the $v = 33$ state were found to be a factor of about 1000 smaller than those for $v = 35$ with Ps in its ground state. The cross sections for the $v = 32$ and $v = 31$ states were found to be even smaller.

Table 4
Cross sections for rearrangement into $\mathrm{He}\bar{p} + e^+$

Incident energy, ε a.u.	Cross sections, σ_v, for rearrangement into $\mathrm{He}\bar{p}$ states with $v = 28$ to 33. Units are $\varepsilon^{-\frac{1}{2}} a_0^2$					
	σ_{33}	σ_{32}	σ_{31}	σ_{30}	σ_{29}	σ_{28}
10^{-12}	$1.51 * 10^{-3}$	8.65×10^{-6}	1.84×10^{-5}	1.97×10^{-5}	$9.07 * 10^{-6}$	2.87×10^{-6}
10^{-10}	$1.41 * 10^{-3}$	8.08×10^{-6}	1.72×10^{-5}	1.84×10^{-5}	$8.48 * 10^{-6}$	2.68×10^{-6}
10^{-9}	$1.52 * 10^{-3}$	8.74×10^{-6}	1.86×10^{-5}	1.99×10^{-5}	$9.16 * 10^{-6}$	2.90×10^{-6}
10^{-7}	$1.48 * 10^{-3}$	8.50×10^{-6}	1.81×10^{-5}	1.93×10^{-5}	$8.91 * 10^{-6}$	2.82×10^{-6}
10^{-6}	$1.28 * 10^{-3}$	7.37×10^{-6}	1.57×10^{-5}	1.67×10^{-5}	$7.73 * 10^{-6}$	2.44×10^{-6}
10^{-4}	$2.36 * 10^{-4}$	1.35×10^{-6}	2.90×10^{-6}	3.09×10^{-6}	$1.43 * 10^{-6}$	4.51×10^{-7}
10^{-3}	$8.52 * 10^{-5}$	4.62×10^{-7}	1.09×10^{-6}	1.15×10^{-6}	$5.27 * 10^{-7}$	1.67×10^{-7}

The results obtained for the first six vibrational states of $\mathrm{He}\bar{p}$ below threshold are given in Table 4. It can be seen that the cross sections are all very small. Owing to the complexity of the calculation, it was not possible to include Hylleraas-type basis functions that take into account positron-electron correlation. However, as the cross sections are so small, inclusion of these functions is unlikely to have a significant effect.

Calculations for the rearrangement reaction $\mathrm{He}+\bar{\mathrm{H}} \rightarrow \alpha\bar{p}+\mathrm{Ps}^-(e^+e^-e^-)$

Calculations have been carried out for this rearrangement reaction, using wave functions for the internal motion of Ps^- calculated by Ho [15]. In the T-matrix

$$\langle \psi_3 \mid V_3 \mid \Psi_{\mathrm{BO}} \rangle, \tag{11}$$

the potential V_3 is of the form

$$V_3 = -\frac{1}{r_{A1}} + \frac{2}{r_{B1}} + \frac{1}{r_{A2}} - \frac{2}{r_{B2}} + \frac{1}{r_{A3}} - \frac{2}{r_{B3}}, \tag{12}$$

where

$$\psi_3 = \eta(\rho)\phi_{\mathrm{Ps}^-}(r_{12}, r_{13}, r_{32})\chi_{\alpha\bar{p}}(R). \tag{13}$$

As the potential is Coulombic asymptotically, as in the case of reaction (2), $\eta(\rho)$ is the s-wave Coulomb wave function, of appropriate energy, for a charge of –1 a.u. moving in the field of a charge of 1 a.u., located at the nuclear centre of mass. For a treatment of a simpler rearrangement reaction involving Coulombic scattering, see ref. [17]. ρ is the distance of this charge from the nuclear centre of mass. $\phi_{\mathrm{Ps}^-}(r_{12}, r_{13}, r_{32})$ and $\chi_{\alpha\bar{p}}(R)$ are the wave functions for Ps^- and $\alpha\bar{p}$, respectively.

As in the case of reaction (1), it was possible to use the Σ symmetry of the entrance and rearrangement channel wave functions to integrate directly over one of the nine coordinates involved. However, the presence of the Ps^- in the rearrangement channel makes it necessary to carry out the integrations over the remaining eight coordinates numerically.

Table 5
Cross sections for rearrangement into $\alpha\bar{p}$ and Ps^-

Incident energy, ε a.u.	Cross sections, σ_n, for rearrangement into $\alpha\bar{p}$ with principal quantum number $n = 27$ to 30. Units are $\varepsilon^{-\frac{1}{2}}a_0^2$			
	σ_{30}	σ_{29}	σ_{28}	σ_{27}
10^{-10}	8.1×10^{-4}	3.2×10^{-4}	9.5×10^{-5}	3.5×10^{-5}
10^{-9}	8.5×10^{-4}	1.4×10^{-4}	1.4×10^{-4}	4.3×10^{-5}
10^{-8}	11.9×10^{-4}	1.2×10^{-4}	3.9×10^{-4}	5.7×10^{-6}
10^{-7}	4.9×10^{-4}	9.5×10^{-5}	2.3×10^{-5}	2.0×10^{-5}
10^{-6}	9.8×10^{-4}	9.1×10^{-5}	1.7×10^{-4}	1.1×10^{-5}

The cross section values obtained for the first four states of $\alpha\bar{p}$ below threshold, i.e. those with principal quantum number $n = 30, 29, 28$ and 27, are given in Table 5. These results differ from those in Tables 3 and 4 in that they do not all remain roughly constant at the lower energies and decrease at the upper end. This suggests that there may be inaccuracies in the calculation. One possibility is that these may be due to the use of eight dimensional numerical integration. However, we think that it can be concluded from these results that cross sections for these rearrangement processes are considerably smaller than for the $v = 35$ state of $He^+\bar{p}$ in reaction (1).

$\bar{H} + H_2$ SCATTERING

We have carried out a preliminary calculation for elastic scattering and the antiproton annihilation scattering cross section, in which we treat the strong interaction as a perturbation, [18]. We used prolate spheroidal coordinates as in the $\bar{H} + He$ calculations. The H_2 internuclear distance was taken to be fixed at its equilibrium value, $1.4a_0$.

The method used was a variant of the Kohn variational method. The trial function was of the form

$$\Psi_t = (\cos\tau - a_t \sin\tau)S + (\sin\tau + a_t \cos\tau)C + \sum_{i=1}^{n} c_i\chi_i, \tag{14}$$

where $(a_t, c_1, \ldots, c_n)$ are linear parameters,

$$S = \frac{1}{\sqrt{2\pi}} \frac{2}{R(\lambda - 1)} \sin(c(\lambda - 1))$$

and

$$C = \frac{1}{\sqrt{2\pi}} \frac{2}{R(\lambda - 1)} \cos(c(\lambda - 1))(1 - \exp(-\alpha(\lambda - 1))).$$

S and C were used by Massey and Ridley [19] in a Kohn calculation of $e^- + H_2$ scattering. For applications to $e^+ + H_2$ scattering, see Armour and Humberston [20].

The KE operator matrix elements involving only C and/or S can be calculated *analytically*.

The short-range correlation functions $\{\chi_i\}_{i=1}^{n}$ were chosen to be as follows.

$$\chi_1 = C\exp(-\alpha(\lambda - 1)),$$

$$\chi_i = \frac{1}{\sqrt{2\pi}} \frac{2}{R} \lambda^{n_i} \mu^{m_i} \exp(-\alpha_i \lambda) \exp(\pm\beta_i \mu) \quad (i = 2, 3, \ldots, n),$$

where n_i and m_i are non-negative integers and α_i and β_i are constants. By symmetry, if basis functions with β_i are included, the corresponding function with β_i replaced by $-\beta_i$ must also be included.

The linear parameters are determined by the generalised Kohn method, in which calculations are carried out for a range of τ values. Calculations corresponding to τ values for which the linear inhomogeneous Kohn equations are considered to be nearly ill-conditioned are eliminated and the average taken over the results of the remaining calculations [20]. The method gives an approximation to the phase shift of the lowest spheroidal partial wave. This wave makes the dominant contribution to the scattering cross section at very low energies.

The potential $V(\lambda,\mu)$ is of the form

$$V(\lambda,\mu) = V_N(\lambda,\mu) + V_{\text{lep}}(\lambda,\mu) - E_{\text{H}_2} - E_{\bar{\text{H}}}.$$

$V_N(\lambda,\mu)$ is the Coulomb potential between the protons in the H_2 and the antiproton in $\bar{\text{H}}$. It is of the form

$$V_N(\lambda,\mu) = -\frac{4}{R}\frac{\lambda}{\lambda^2-\mu^2}.$$

$V_{\text{lep}}(\lambda,\mu)$ is the Born–Oppenheimer potential due to the leptons, calculated by Mella [21] using a Quantum Monte Carlo Method. The $-E_{\text{H}_2}$ and $-E_{\bar{\text{H}}}$ terms are included so that

$$V(\lambda,\mu) \longrightarrow 0 \quad \text{as} \quad \lambda \to \infty.$$

The results obtained for elastic cross sections are given in Table 6.

Table 6
Cross sections for $rH_2 + \bar{\text{H}}$ elastic scattering

Incident energy ε a.u.	Cross section for elastic scattering, σ_e (a_0^2)
10^{-10}	4802
5×10^{-10}	4798
10^{-9}	4793
5×10^{-9}	4755
10^{-8}	4708
5×10^{-8}	4258
10^{-7}	2735
5×10^{-7}	2014
10^{-6}	1033
5×10^{-6}	123.3
10^{-5}	8.617
5×10^{-5}	32.48
10^{-4}	11.12

These results were obtained with the H_2 internuclear distance fixed at its equilibrium value, $1.4a_0$. It was found that, on average, the cross sections varied by $\sim 20\%$ if the internuclear distance was increased or decreased by $0.1a_0$.

The antiproton annihilation cross section, $\sigma^{p\bar{p}}$ is determined by first order perturbation theory using the perturbation

$$V = A^{p\bar{p}}\left(\delta(r_A) + \delta(r_B)\right),$$

where r_A and r_B are the position vectors of the antiproton with respect to protons A and B, respectively. $A^{p\bar{p}}$ is an experimentally determined parameter [9]. The antiproton annihilation cross section, $\sigma^{p\bar{p}}$, is given by the relation

$$\sigma^{p\bar{p}} = \frac{\langle\Psi_t|V|\Psi_t\rangle}{F},$$

where F is the incoming flux.

At very low energies,

$$\sigma^{p\bar{p}} = 4.68\varepsilon^{-\frac{1}{2}},$$

where ε is the incident energy.

This value is a factor of ~ 36 larger than in the case of $\mathrm{H}+\bar{\mathrm{H}}$ [7, 22] and a factor of ~ 2 larger than in the case of $\mathrm{He}+\bar{\mathrm{H}}$ [8, 23]. It is surprising that $\sigma^{p\bar{p}}$ for $\mathrm{H}_2+\bar{\mathrm{H}}$ scattering is so much larger than for $\bar{\mathrm{H}}+\mathrm{H}$. The effect on $\sigma^{p\bar{p}}$ of improving the method of calculation will be investigated in the next stage of our calculations.

In this preliminary calculation, we have omitted rearrangement channels such as

$$\mathrm{H}_2+\bar{\mathrm{H}} \longrightarrow \begin{cases} \mathrm{Pn}+\mathrm{PsH} \\ \mathrm{Pn}+\mathrm{H}+\mathrm{Ps}, \\ \mathrm{PnPs}+\mathrm{H} \end{cases}$$

where Pn is protonium $(p\bar{p})$ and $\mathrm{PnPs} = \mathrm{H}\bar{\mathrm{H}}$.

These channels are all open at zero incident energy. They have been included by Cohen [24] using the quasiclassical fermion molecular dynamics method, which is valid for incident energies $\geq \ \sim 2.7$ eV.

His cross section values show that these channels need to be taken into account. This will be the next stage in our treatment of $\mathrm{H}_2+\bar{\mathrm{H}}$ scattering.

TESTS OF FUNDAMENTAL PRINCIPLES OF PHYSICS USING TRAPPED ANTIHYDROGEN

This is another area of research on antihydrogen, different from what has been considered in this paper, in which I am very interested. I have written a short article covering the following topics.

(1) Tests of the CPT and Lorentz invariance of relativistic quantum field theory – phenomenological terms in an extended Dirac equation that might bring about violations of these invariances.

(2) Behaviour of antihydrogen in a gravitational field – tests of Einstein's general theory of relativity including possible violations of the strong equivalence principle through coupling to the curvature tensor.

The article is available from my personal web page [25]. It will be updated, as necessary, to take into account new developments.

CONCLUSION

Since carrying out the calculations described in ref. [8], the method used for calculating the entrance channel leptonic wave function using Hylleraas-type wave functions has been very significantly improved. Amongst other things, it has been possible to optimise the non-linear parameters in the trial function systematically and include Hylleraas-type functions that take into account electron-electron correlation, in addition to those that take into account positron-electron correlation.

At very low energies, the present results for reaction (1) with Ps in its ground state are about a factor of 3 smaller than the largest rearrangement cross section for $H+\bar{H}$. Cross sections for the $v=34$ state of $He^+\bar{p}$ are found to be smaller by a factor of about 500 than for the $v=35$ state. Cross sections for the $v=33$ and $v=32$ states of $He^+\bar{p}$ were found to be smaller than those for $v=35$ by factors of at least 10 and 20, respectively. Cross sections for reaction (1) with Ps in its first excited s-state were found to be even smaller.

Cross sections for rearrangement into the first six vibrational states of $He\bar{p}$ below threshold in reaction (2) are found to be very small in comparison with those for the $v=35$ state in reaction (1).

Results are presented for the first time for reaction (3). It was necessary to use eight-dimensional numerical integration in this case to evaluate the required T-matrix elements. The cross sections for this reaction for the first four $\alpha\bar{p}$ states below threshold, i.e. principal quantum number $n=30$, 29, 28 and 27, were all found to be less than the cross sections for the $v=35$ state of $He^+\bar{p}$ by a factor greater than 100.

The effect of the strong interaction has been omitted in these calculations. As stated in I, when it is included in addition to elastic scattering using a complex potential, the annihilation cross section is found to be $2.62\varepsilon^{-\frac{1}{2}}a_0^2$, which is much larger than the total value for the cross section for the rearrangement reactions resulting from our calculations.

Calculations for $H_2+\bar{H}$ scattering are much more difficult to carry out than for $He+\bar{H}$. However, we have been able to carry out a preliminary calculation of cross sections for elastic scattering and antiproton annihilation for this system.

We find that the elastic cross sections are quite large at very low energies. They are very much larger than in the case of $He+\bar{H}$ and at very low energies are larger than for $\bar{H}+H$ by a factor of about 5. The antiproton annihilation cross section is larger than for $H+\bar{H}$ by a factor of about 36. This result is surprising and needs to be checked in future, more detailed calculations.

Finally, I am very interested in the tests of CPT and Lorentz invariance and of Einstein's principle of equivalence to be carried out using antihydrogen. Thus I have made available on the web a short article that I have written on this subject [25].

ACKNOWLEDGMENTS

We wish to thank Y.K. Ho for sending us his wave functions for Ps^-, M. Mella for making available his potential for $H_2+\bar{H}$ and J.S. Cohen for valuable discussions. We also wish to thank EPSRC(UK) for support for this research through an advanced fellowship grant (EP/D069785/1), two doctoral training grants (GR/P01595/01 and EP/P500389/1) and a research grant (GR/R26672/01).

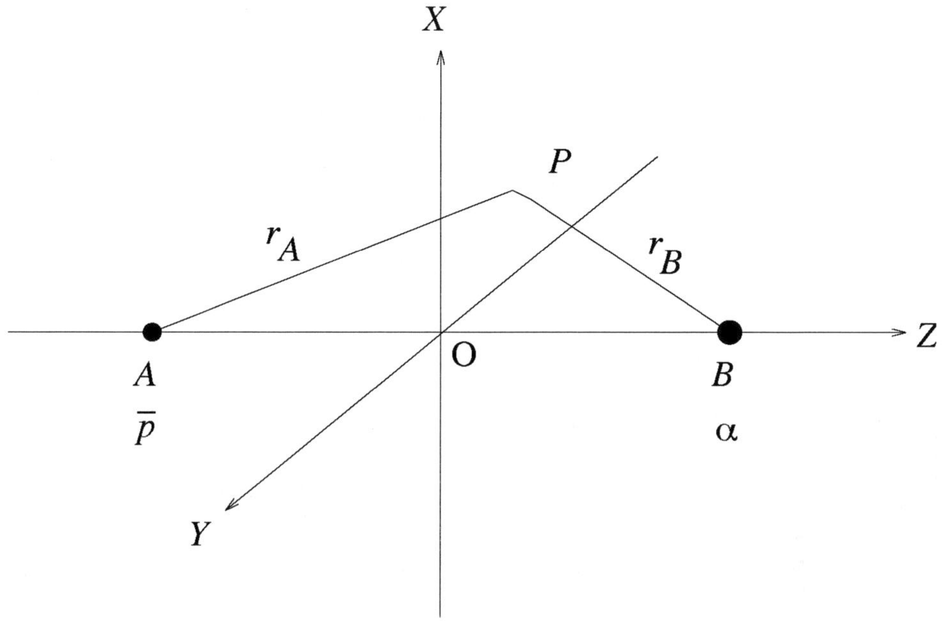

FIGURE 1. Coordinate system

REFERENCES

1. T. Pohl, H.R. Sadeghpour and G. Gabrielse, Phys. Rev. Lett. **97**, 143401 (2006).
2. M. Amoretti et al., Phys. Rev. Lett. **97**, 213401 (2006).
3. G.M. Shore, Nuc. Phys B **717**, 86 (2005).
4. E.A.G. Armour and C.W. Chamberlain, J. Phys. B **35**, L47 (2002).
5. B. Zygelman, A. Saenz, P. Froelich and S. Jonsell, Phys. Rev. A **69**, 042715 (2004).
6. S. Jonsell, A. Saenz, P. Froelich, B. Zygelman and A. Dalgarno, J. Phys. B **37**, 1195 (2004).
7. E.A.G. Armour, Y. Liu and A. Vigier, J. Phys. B **38**, L47 (2005).
8. E.A.G. Armour, S. Jonsell, Y. Liu and A.C. Todd, Nucl. Inst. and Meth. B **247**, 127 (2006).
9. S. Jonsell, A. Saenz, P. Froelich, B. Zygelman and A. Dalgarno, Phys. Rev. A **64**, 052712 (2001).
10. K. Strasburger, H. Chojnacki and A. Sokołowska, J. Phys. B **38**, 3091 (2005).
11. M. Plummer, E.A.G. Armour, A.C. Todd and J.N. Cooper, Frontiers 2007, Computational Science and Engineering Department, Daresbury and Rutherford Laboratories, Science and Technology Facilities Council, UK, 2007, p. 6.
12. A.C. Todd, Ph.D. thesis (University of Nottingham) 2007.
13. E.A.G. Armour and S. Jonsell, J. Phys. A **40**, 701 (2007).
14. E.A.G. Armour, A.C. Todd, S. Jonsell, Y. Liu, M.R. Gregory and M. Plummer, Nucl. Inst. and Meth. B **266**, 363 (2008).
15. M.R. Gregory, Ph.D. thesis (University of Nottingham) 2008.
16. Y.K. Ho, Private communication, 2007.
17. R. Gayet, R. K. Janev and A. Salin, J. Phys. B **6**, 993 (1973).
18. M.R. Gregory and E.A.G. Armour, Nucl. Inst. and Meth. B **266**, 374 (2008).
19. H.S.W. Massey and R.O. Ridley, Proc. Phys. Soc. (London) A**69**, 659 (1956).
20. E.A.G. Armour and J.W. Humberston, Phys. Rep. **204**, 167 (1991).
21. M. Mella, Private communication, 2006.
22. S. Jonsell, A. Saenz, P. Froelich, B. Zygelman and A. Dalgarno, Canadian J. Phys. **83**, 435 (2005).

23. S. Jonsell, Nucl. Inst. and Meth. B **247**, 138 (2006).
24. J.S. Cohen, J. Phys. B: At. Mol. Opt. Phys. **39**, 3561 (2006).
25. http://www.maths.nott.ac.uk/personal/eaga/

Rearrangement and annihilation in antihydrogen-atom scattering

Svante Jonsell

Department of Physics, Swansea University, Singleton Park, Swansea SA2 8PP, U.K.

Abstract. I review some results for annihilation and rearrangement processes in low-energy antihydrogen-hydrogen and antihydrogen-helium scattering. For the strong nuclear force results using a δ-function potential are compared to a scattering length approach. It is found that the δ-function potential does not give correct annihilation cross sections in the case of antihydrogen-helium scattering. Problem associated with the use of the Born-Oppenheimer approximation for rearrangement calculations are reviewed.

Keywords: Antihydrogen, scattering
PACS: 36.10.-k, 34.50.-s

INTRODUCTION

Antihydrogen is now routinely produced by the ALPHA and ATRAP collaborations at CERN, and interesting proposals for new antihydrogen experiments exist [1]. Given the tremendous experimental progress over the past few years there is little doubt that antihydrogen soon will be magnetically trapped in its ground state. This will make experimental studies of atom-antiatom collisions possible. Such collisions are interesting for a number of reasons. Primarily, understanding antihydrogen-atom interaction may aid the ongoing antihydrogen experiments. Relevant issues include: trap loss due to scattering with vacuum impurities, detection of antiatoms through annihilation, and speculations that ordinary atoms may be used for sympathetic cooling of antiatoms. Regardless of its practical implications for these experiments, atom-antiatom scattering is a new fundamental class of processes, qualitatively different from other known scattering processes, which is very interesting in its own right. It may also provide a new way of studying nuclear forces at low energies. Furthermore there may be interest independently of terrestrial experiments. If domains of atomic antimatter exist somewhere in the universe there must also be domain boundaries where matter and antimatter interacts.

In addition to elastic scattering, a number of inelastic channels are open in low-energy antihydrogen-atom (or antihydrogen-molecule) scattering. These include rearrangement processes,

$$\bar{\mathrm{H}} + \mathrm{A} \rightarrow \mathrm{Ps} + \mathrm{A}^{+}\bar{p}, \tag{1}$$

antiproton-nucleus annihilation

$$\bar{\mathrm{H}} + \mathrm{A} \rightarrow e^{-} + e^{+} + \text{decay products} \tag{2}$$

electron-positron annihilation

$$\bar{\mathrm{H}} + \mathrm{A} \rightarrow \mathrm{A}^{+} + \bar{p} + 2\gamma, \tag{3}$$

CP1037, *Cold Antimatter Plasmas and Application to Fundamental Physics,* edited by Y. Kanai and Y. Yamazaki

and atom-antiatom “molecule” formation [2]

$$\bar{\text{H}}+\text{A}\rightarrow(\bar{\text{H}}\text{A})+\gamma. \tag{4}$$

Calculations show that the processes (1) and (2) are the dominating inelastic processes at low temperatures. The former can alternatively be regarded as a resonant annihilation process, as annihilation will ultimately be the fate of the rearrangement products. However, since this annihilation occurs on time scales longer than the time required for the rearrangement products to separate, it is natural to treat the two processes separately.

At very low energies the scattering properties are given by the scattering length a. In order to account both for elastic scattering and inelastic processes we take the scattering length to be complex, $a=\alpha-i\beta$, where $\beta>0$. The elastic scattering cross section in the limit of zero collision energy ε_i then becomes

$$\sigma^{\text{el}}=4\pi(\alpha^2+\beta^2), \tag{5}$$

while the cross section representing the sum of all inelastic processes is

$$\sigma^{\text{inel}}=4\pi\beta/k_i, \tag{6}$$

where $\varepsilon_i=\hbar^2k_i^2/(2\mu)$, and μ is the antiproton-nucleus reduced mass. Note that the elastic cross section is constant at low energies (meaning that the elastic collision rate goes to zero), while the inelastic rate diverges as $1/\sqrt{\varepsilon_i}$ (the inelastic rate is constant). Thus, for low enough energies inelastic processes will always dominate.

THE STRONG NUCLEAR FORCE

The strong nuclear force plays an important role in low-energy atom-antiatom scattering. Since the Coulomb force between the nucleus and antinucleus is attractive, there is a non-vanishing overlap density between them, and hence they interact through nuclear forces. At very short distances ($\sim$ 1 fm) this is the dominating interaction. It should be noted that the nuclear force only affects the s-wave ($J=0$) part of the scattering (which dominates completely at very low energies). For higher partial waves the nuclear interaction is shielded by the angular momentum barrier, as long as the collision energy is small. There is also a possibility that for some atom-antiatom system there may, even for the $J=0$ partial wave, be a barrier in the nucleus-antinucleus interaction, caused by the leptonic part of the interaction. Such a barrier has in fact be calculated (in the Born-Oppenheimer approximation) for the ground state He-$\bar{\text{H}}$ system [3, 4]. However, it turns out that this barrier is too weak to cause any substantial shielding [5].

An obvious consequence of the strong nuclear force is annihilation. However, if the interaction is strong enough, it will also change the elastic cross section. Whereas this shift is moderate for H-$\bar{\text{H}}$ scattering [6, 7] it turns out to have a dramatic effect on He-$\bar{\text{H}}$ elastic scattering [5]. In the latter case the result is also very sensitive to the parameters of the nuclear force. This creates an uncertainty in the calculated cross sections. However, the argument could be turned around. Given that the atomic part of the problem can be solved with sufficient accuracy, any measurement of the atom-antiatom cross section would pin down the low-energy form of the strong force with a greater accuracy. The atomic part of the problem, and in particular the long-range van der Waals interaction, magnifies the effects of the nuclear forces to atomic dimensions.

On an atomic scale, the strong nuclear force can be regarded as having zero range. It is therefore natural to represent it as a delta function potential with a strength given by some constant A [6],

$$V_a(\mathbf{R}) = A\delta(\mathbf{R}), \tag{7}$$

where $\mathbf{R}$ is the internuclear coordinate. The constant A is complex since the strong force causes both annihilation and an energy shift of bound states. Taking this potential in first-order perturbation theory gives the annihilation cross section

$$\sigma^a = \frac{\mathrm{Im}\{A\}}{\hbar F}|\Psi_{\mathbf{k}_i}(0,\mathbf{r}_1,\ldots,\mathbf{r}_N)|^2, \tag{8}$$

where F is the incoming flux and $\Psi_{\mathbf{k}_i}(\mathbf{R},\mathbf{r}_1,\ldots,\mathbf{r}_N)$ is the wavefunction of the atom-antiatom system colliding with relative wave vector $\mathbf{k}_i$, and depending on the internuclear coordinate $\mathbf{R}$ and the coordinates $\mathbf{r}_i$ of the N leptons. (In collisions with antihydrogen we take $\mathbf{r}_1$ to be the coordinate of the positron, and $\mathbf{r}_i$, $i > 1$ the coordinates of the $N-1$ electrons.) Somehow the constant A has to be determined for the nucleus-antinucleus pair at hand. Although it may be possible to calculate A from various model potentials for the strong force, it is most convenient to use experimental data if such exist. For instance, for the antiproton-nucleus systems where the life time τ_{1S} of the $1S$ state has been measured it is possible to extract A from the relation [6]

$$\hbar/\tau_{1S} = \Gamma_{1S} = \mathrm{Im}\{A\}|\phi_{1S}(0)|^2, \tag{9}$$

ϕ_{1S} being the hydrogenic ground-state wavefunction of the appropriate charge and reduced mass. Using the measurements in [8, 9] one then arrives at $\mathrm{Im}\{A^{p\bar{p}}\} = 1.7 \times 10^{-7}$ a.u. $= 6.8 \times 10^8$ eVfm^3 for the proton-antiproton system and $\mathrm{Im}\{A^{\alpha\bar{p}}\} = 3.4 \times 10^{-7}$ a.u. $= 1.4 \times 10^9$ eVfm^3 for the helium-antiproton system. Thus the annihilation constant $\mathrm{Im}\{A\}$ is twice as large for helium as compared to hydrogen. These values are averages over different spin states.

For the case of hydrogen-antihydrogen scattering this simple approach gives relatively good results. A low-energy annihilation cross section $\sigma^a = 0.14/\sqrt{\varepsilon_i}$ a.u. was reported in [6]. This agrees well with a calculation using a Woods-Saxon potential [7] giving a spin-averaged annihilation cross section $\sigma^a = 0.13/\sqrt{\varepsilon_i}$ a.u. and a scattering length approach (see below) giving $\sigma^a = 0.15/\sqrt{\varepsilon_i}$ a.u. [10]. An earlier calculation gave $\sigma^a = 0.26/\sqrt{\varepsilon_i}$ a.u. [11]. Thus three out of four calculations, using very different methods to represent the strong force,give very similar results.

For antihydrogen-helium scattering the situation is quite different. It turns out that in this scattering process the strong force significantly modifies the wavefunction in the entrance channel. This makes it necessary to use non-perturbative methods, i.e. the first-order result (8) is *not* valid. However, it is still true that the strong force is essentially zero range compared to atomic dimensions. Also, the energy scale of the nuclear interaction is MeV, much larger than typical atomic energies ~eV. Thus, even though the nuclei during the collision change their kinetic energy by a few eV due to the interaction with the leptons, on a nuclear scale the antiproton-nucleus collision is essentially a zero-energy collision. Both these facts indicate that it is possible to represent the strong force by its scattering length a_{sf} for the antiproton-nucleus system at hand. In a low-energy collision between an antiproton and a nucleus this scattering length fully determines the wavefunction outside the radius of the nucleus, i.e. for separation $\gtrsim 1$ fm$\sim 10^{-5}$ a.u. For such a small antiproton-nucleus separation the leptons will not see the individual particles, but will effectively interact with a single nucleus with charge decreased one unit due to the antiproton, and likewise with its mass increased by a proton mass. In other words, at distances much less than one atomic unit the antihydrogen-atom wavefunction can be factorized as

$$\Psi(\mathbf{R},\mathbf{r}_1,\ldots,\mathbf{r}_N) \to \chi(\mathbf{R})\psi(\mathbf{r}_1,\ldots,\mathbf{r}_N) \qquad \text{as} \quad R \to 0 \tag{10}$$

Here $\psi(\mathbf{r}_1,\ldots,\mathbf{r}_N)$ is the solution for the N leptons in the Coulomb field $(Z-1)e/(4\pi\varepsilon_0 r_i)$ of an atom with nuclear charge Ze and the antiproton at the origin. The function $\chi(\mathbf{R})$ represents the interaction between the antiproton and the nucleus, and does, at these dimensions, not depend on the leptonic coordinates. For R larger than the nuclear radius, but still $\ll 1$ a.u., this function can be calculated approximating the nuclear interaction to have zero range. Hence the strong force is represented by a boundary condition (in terms of its scattering length a_{sf}) as $R \to 0$.

In determining the form of this boundary condition there is a slight complication. In addition to the strong nuclear force the antiproton and nucleus also interact via the Coulomb interaction. The wavefunction $\chi(\mathbf{R})$ therefore has to be expressed in terms of Coulomb functions rather than plane waves. The details of this "Coulomb corrected" scattering length was worked out long ago in the context of bound states [12], and the theory is readily generalized to continuum states [5, 13]. The form of the function $\chi(\mathbf{R})$ at R just larger than the nuclear radius becomes,

$$\chi(\mathbf{R}) = N[\cot\delta_{sf}^{(0)}(k)F_0(kR) + G_0(kR)]. \tag{11}$$

Here $F_0(kR)$ and $G_0(kR)$ are the zero-angular-momentum regular and irregular Coulomb functions respectively. The wavevector k in their argument is *not* the same as k_i characteristic of the atom-antiatom collision. Instead k is a characteristic of the nucleus-antiproton collision, which has to allow for the amount of the total energy which is stored in the leptonic binding. Balancing the energy at $R = \infty$ with the energy at $R = 0$ gives,

$$\frac{\hbar^2 k^2}{2\mu} + E_{\text{lep}}^{R=0} = \varepsilon_i + E_{\bar{\text{H}}} + E_{\text{Atom}}, \tag{12}$$

where $E_{\text{lep}}^{R=0}$ is the binding energy of the leptons in the limit $R \to 0$, $E_{\bar{\text{H}}}$ and E_{Atom} are the binding energies of the colliding antiatom and atom. The scattering due to the strong nuclear force is determined by the strong-interaction phase shift $\delta_{sf}^{(0)}$ of the antiproton-nucleus collision (not to be confused with the phase shift of the atom-antiatom collision). In terms of the strong-interaction scattering length the zero-energy limit of this phase shift is given by

$$\frac{1}{a_{sf}} = -\frac{2\pi}{b_\mu} \lim_{k\to 0} \cot\delta_{sf}^{(0)}(k), \tag{13}$$

where $b_\mu = a_0 m_e/(Z\mu)$, with a_0 the Bohr radius and m_e the electron mass. From the properties of the Coulomb functions Trueman [12] derived the leading k dependence of $\delta_{\rm sf}^{(0)}(k)$ at low energies,

$$C_0^2(\eta)\cot\delta_{\rm sf}^{(0)}(k) = 2\eta h(\eta) - \frac{1}{ka_{\rm sf}}, \tag{14}$$

where $\eta = 1/(b_\mu k)$,

$$C_0(\eta) = \sqrt{\frac{2\pi\eta}{1-e^{-2\pi\eta}}}, \tag{15}$$

$$h(\eta) = [\Psi(-i\eta)+\Psi(i\eta)]/2 - \ln\eta, \tag{16}$$

and Ψ the digamma function.

Using the equations above we can calculate the $R \to 0$ part of the wavefunction for k much smaller than typical nuclear energies, given that we know the scattering length $a_{\rm sf}$. The scattering length is complex since it accounts both for annihilation and an energy shift. Its value can be determined from measured shifts and lifetimes of bound states, or from low energy scattering data. Bound states correspond to $\cot\delta_{\rm sf}^{(0)} = i$, which gives

$$a_{\rm sf} = b_\mu \left[-Kb_\mu + 2\Psi\left(-\frac{1}{Kb_\mu}\right) + 2\ln Kb_\mu\right], \tag{17}$$

where $K = -\sqrt{-2\mu E_{nl}/\hbar^2}$, and E_{nl} is the (complex) energy of a bound state. For the proton-antiproton system the measured values of the energy shift $\Delta E_{1S} = 721$ eV and lifetime $\Gamma_{1S} = 1097$ eV of the ground state of protonium give $E_{1S} = (-11770.1 - 548.5i)$ eV [8]. Used in Eq. (17) this gives the scattering length $a_{\rm sf} = (0.844 - 0.698i)$ fm. For the alpha particle-antiproton system the ground state energy and life time have not been measured, since in experiments the system annihilates before this state is reached. Based on low-energy scattering data and measurements of the $2P$ and $3D$ states the scattering length $a_{\rm sf} = (1.851 - 0.630i)$ fm has been deduced [9]. Thus the real part of the scattering length is about twice as large for the alpha particle-antiproton system as compared to the proton-antiproton system, while the imaginary part is roughly of the same size.

In Ref. [13] the scattering length approach was compared to calculations using simple model potentials of square-well type. It was verified that as long as the parameters of the potential are tuned to reproduce the same scattering length, the resulting cross sections are completely independent of the extent and depth of the potential. This confirms that the scattering length approach correctly reproduces all the relevant nuclear interaction effects in low-energy atom-antiatom collisions. One should, however, bear in mind that the results are only as accurate as the value of the strong-interaction scattering length used is. There may be a considerable uncertainty in this value, and hence e.g. calculations using different forms for the strong interaction potential may give different results. Only if the strong-force scattering length derived from different strong-force potentials are similar can one expect that they give the same results for the strong-force effects in atom-antiatom scattering.

Using the best potential for hydrogen-antihydrogen available in the literature [14] a low-energy elastic cross section $\sigma^{el} = 765$ a.u. was calculated omitting the effect of the strong nuclear force. When this force was added the elastic cross section shifted to $\sigma^{el} = 892$ a.u. and the annihilation cross section $\sigma^{a} = 0.15/\sqrt{\varepsilon_i}$ a.u. was obtained [13]. It should, however, be noted that there is a considerable uncertainty in these results since they rely on the Born-Oppenheimer approximation which breaks down at a "critical" internuclear separation [15] (see below). On the other hand, a calculation by Armour and Chamberlain used the Kohn variational principle and therefore did not suffer from these problems [16]. They obtained the low-energy elastic cross section $\sigma^{el} = 908$ a.u. in quite good agreement with the result from the Born-Oppenheimer approximation, possibly indicating that the effect on the elastic cross section of this break down is limited. A calculation by Armour et al. using the Born-Oppenheimer approximation for the atomic interaction gave the elastic cross section $\sigma^{el} = 788$ a.u., which was modified to a spin-averaged value $\sigma^{el} = 908$ a.u. and $\sigma^{a} = 0.13/\sqrt{\varepsilon_i}$ a.u. when the strong force was included though a Woods-Saxon potential [7]. Thus both calculations saw an increase in the elastic cross section of about 15%. An earlier calculation by Voronin and Carbonell obtained an increase of almost 50% [11].

For antihydrogen-helium scattering the strong-interaction effect is much more dramatic. Using the Born-Oppenheimer potential (including adiabatic correction) from Ref. [4] an unusually small elastic cross section $\sigma^{el} = 0.65$ a.u. without including the strong force [17]. (A different result was quoted in Ref. [5]. This has been updated as a more accurate potential on a denser grid has become available.) As discussed below the ground-state helium-antihydrogen system does *not* suffer from the break down of the Born-Oppenheimer approximation due to a critical distance. When the strong force is added, using the scattering length approach, the elastic cross section changes to $\sigma^{el} = 1700$ a.u., and the annihilation cross section becomes $\sigma^{a} = 2.62/\sqrt{\varepsilon_i}$ a.u..

How can the strong interaction effect be so dramatic, even though the size of the scattering length is not much larger than for proton-antiproton scattering? Figure 1 shows the real and imaginary parts of the helium-antihydrogen scattering length as a function of a parameter λ scaling a_{sf}. Thus $\lambda = 1$ corresponds to the physical value of a_{sf} whereas $\lambda = 0$ corresponds to no strong interaction. From the graph it is clear that the very small value of the real part of the scattering length $\mathrm{Re}\{a\} = -0.23\,a_0$ when $\lambda = 0$ is untypical, for $\lambda > 0$ its typical order of magnitude is a few times a_0. At the physical value of $\lambda = 1$ the main contribution to the elastic cross section comes from the imaginary part of the helium-antihydrogen scattering length. The scattering length has a "resonant" structure for λ around unity, where its imaginary part attains a maximum, while its real part goes through zero. If the strong-force scattering length is scaled beyond this maximum both the elastic cross section and the annihilation cross section start to decline. Hence, it is not the magnitude of the strong-force scattering length which determines the size of the strong interaction effects, but instead there is a matching condition to the "resonant" value.

For the antihydrogen-hydrogen system the results of a similar scaling is presented in Figure 2. A structure similar to that in Figure 1 is observed, but now the "resonance" occurs when the strong-force scattering length is scaled to more than 20 times its physical value. Analyzing the asymptotic form of the atom-antiatom wavefunction one

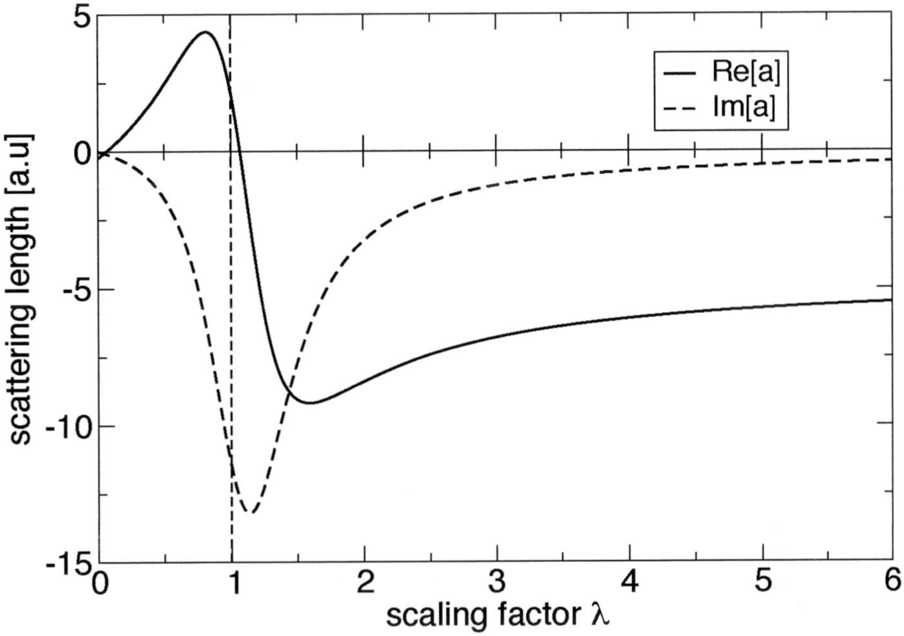

FIGURE 1. Real and imaginary parts of the helium-antihydrogen scattering length as a function of the parameter λ scaling the strong interaction scattering length a_{sf}.

finds that the condition for the resonance in the scattering length is [17]

$$\frac{N_{\rm reg}}{N_{\rm irreg}} = \left| \frac{a_{\rm sf}}{b_\mu} \right| , \tag{18}$$

where $N_{\rm reg}$ and $N_{\rm irreg}$ are the asymptotic normalizations of the atom-antiatom wavefunctions which as $R \rightarrow 0$ become respectively the regular and irregular Coulomb functions. The ratio on the right-hand side is 0.02 for hydrogen-antihydrogen and 0.11 for helium-antihydrogen (the larger value mainly because b_μ is shorter because of the greater charge and the greater reduced mass). The ratio on the right-hand side is ~ 0.5 for hydrogen-antihydrogen, i.e. much larger than the left-hand side, which explains why $a_{\rm sf}$ has to be scaled up a lot to reach the resonant value. For helium-antihydrogen the ratio on the right-hand side is 0.12, thus roughly equal to the left-hand side, giving the resonance condition. Thus two factor contribute a roughly equal amount to the large imaginary part of the helium-antihydrogen scattering length: (*i*) the greater Coulomb focusing and (*ii*) the asymptotic dominance of the irregular solution.

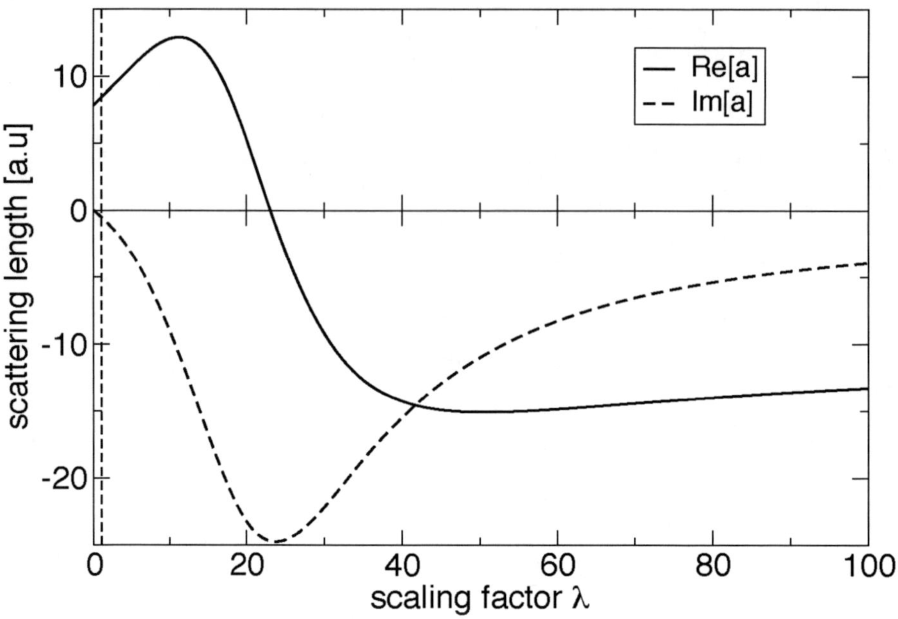

FIGURE 2. Same as Figure 1, but for hydrogen-antihydrogen scattering.

REARRANGEMENT

Another consequence of the antiproton-nucleus attraction is that, even for ultracold ground-state scattering, there are a very large number of rearrangement channels open. Because of the short-range attraction there is a lot of potential energy available for the reaction. For instance, the real ground state of the hydrogen-antihydrogen system is the proton and antiproton bound in the $1S$ state (i.e. protonium $1S$) with binding energy about 12 keV, plus $1S$ positronium with binding energy 6.8 eV. Because of the small radius, and short life time, of ground-state protonium it is not likely to be formed in atom-antiatom collisions. As shown in Figure 3 there is a large number of energetically accessible protonium states. For rearrangement to Ps($1S$) the highest allowed protonium states is $N = 24$ in low-energy collisions. However, several calculations indicate that the cross section of the next lower state, $N = 23$ has a low-energy cross section which is about twice as large as for $N = 24$ [10, 16, 18]. For $N < 23$ the cross section becomes progressively smaller. So far there are no calculations of cross sections for formation of excited positronium. Figure 3 indicates that in low-energy collisions $N \leq 22$ is required for formation of Ps($2S$) or Ps($2P$), while for more highly excited positronium $N \leq 21$.

Calculations of rearrangement cross sections are complicated by the existence of critical distances R_c. For internuclear separations below the critical distance the antiproton-nucleus system does not bind the leptons. It is easy to understand that the hydrogen-antihydrogen system must have a critical distance. At large internuclear separations the leptons are bound by the proton or the antiproton respectively. As $R \rightarrow 0$, on the other

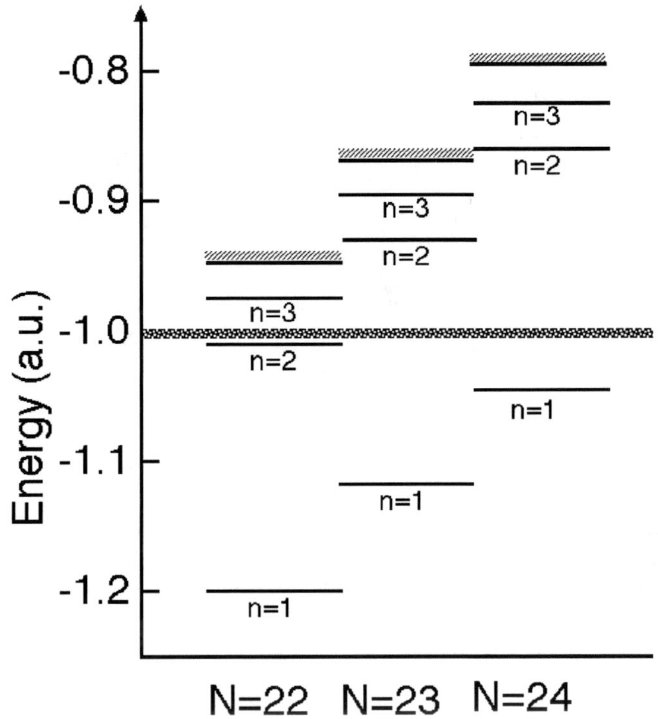

FIGURE 3. Energy levels of protonium with principal quantum number N and positronium with principal quantum number n. The line shows the total energy of hydrogen and antihydrogen in their ground states in the limit of zero collision energy. Rearrangement states below this line are accessible in ultracold collisions. The energies are given in atomic units.

hand, the positive and negative charge are located in the same point, and hence cancel. The leptons then form a free positronium state. At an internuclear separation somewhere between $R = \infty$ and $R = 0$ the leptons cease to be bound by the heavy particles, and this is the critical distance.

Applying the same argument to ground-state helium-antihydrogen scattering shows that for this system there is no critical distance, since the united atom limit of the Helium-antihydrogen system is equivalent to positronium hydride, which is bound by 0.789 a.u.. For excited states of Helium there are, however, critical distances [4]. The argument can be generalized to any element: If an element with atomic number $Z - 1$ binds positronium, then the element with atomic number Z will *not* have a critical distance in scattering with antihydrogen. Using the results from calculations [20] one can deduce that the following atoms do not have critical distances in ground-state collisions with anti-hydrogen: inert gases (He, Ne, Ar, Kr, Xe), at least some alkaline earths (Be, Mg, Ca) and some others (N, F, Zn).

Calculations for the hydrogen-antihydrogen and proton-antihydrogen (antiproton-hydrogen) show that at the critical distance the leading correction to the Born-

Oppenheimer approximation diverges, i.e. the Born-Oppenheimer approximation breaks down [21, 15]. At least for the proton-antihydrogen system this is a consequence of the average separation between the positron and the heavy particles becoming infinitely large. That is, the bound leptonic wavefunction gradually approaches a continuum state, and its energy joins the continuum threshold smoothly. In such a situation it is clear that the assumption behind the Born-Oppenheimer approximation, that the electron/positron rapidly adjusts to changes in the position of the nuclei, breaks down.

Inside the critical distance the system is open for rearrangement in the Born-Oppenheimer approximation. Rearrangement can occur also in systems without a critical distance through non-adiabatic couplings. Still, it is reasonable to expect that systems with a critical distance have larger rearrangement cross sections. Some results for the hydrogen-antihydrogen system are summarized in Table 1. An early calculation using a plane wave approximation of the scattering states gives cross sections which consistently are much lower than the results from other calculations [19]. The distorted wave Born approximation [6, 5] relies on the Born-Oppenheimer approximation, and one would therefore, in view of the problems with the critical distance, expect it to underestimate the rearrangement cross sections. Comparison with calculations using the Kohn variational method [16] and an optical potential method [18], which both go beyond the Born-Oppenheimer approximation, indeed shows this to be the case. For the dominant channel the distorted wave Born approximation gives roughly half the value of more advanced methods. However, for other rearrangement channels also the optical potential method and the Kohn variational method show considerable discrepancies. Preliminary investigations using the simpler antihydrogen-proton system indicate that it is necessary both to include a large number of rearrangement channels, and to include both real and imaginary part of the optical potential (which was not done in Ref. [18]) in order to obtain accurate results [21]. These issues will need further investigation.

Comparing to helium-antihydrogen scattering the dominant rearrangement channel in Table 1 has a roughly three times larger cross section. Thus again it appears that the break-down of the Born-Oppenheimer approximation gives larger rearrangement cross sections. The difference is, however, not as dramatic as one would, perhaps, have expected.

CONCLUSIONS

The strong force is important in matter-antimatter interactions. It is the source for nucleus-antinucleus annihilation, which is one of the dominating inelastic processes. It also has the potential to significantly change elastic cross sections. If the low-energy limit of the strong force is known for a nucleus-antinucleus pair, then it is easy to include it in atom-antiatom calculations through its scattering length. Conversely, experimental results for atom-antiatom scattering could help to pin down the strong force scattering length.

Rearrangement is the other dominating inelastic channel. For a large class of atom-antiatom systems the rearrangement channels cause a break-down of the Born-Oppenheimer approximation, which makes the problem difficult to treat. The most promising methods, the Kohn variational method and the optical potential method, are

TABLE 1. Rearrangement cross sections for low-energy hydrogen-antihydrogen scattering, calculated using various approximations. The low-energy form of the cross section is $\sigma^{\text{rearr}} = C/\sqrt{\varepsilon_i}$, where the constant C is given in atomic units in the table. In all cases only the positronium $1S$ channel is considered, while N stands for different protonium channels.

N	plane wave approx. *	Kohn variational †	Optical pot. **	Distorted Wave Born ‡
24	0.038	0.21	0.32	0.15
23	0.022	0.45	0.48	0.24
22	0.016	0.01	0.14	0.002
21	0.009	–	0.10	0.02
20	0.006	–	0.04	0.001
19	0.003	–	0.03	0.003

* Sinha and Ghosh Ref. [19] (estimated from graph)
† Armour and Chamberlain Ref. [16]
** Zygelman et al. Ref. [18]
‡ Jonsell et al. Ref. [5]

in reasonable agreement for hydrogen-antihydrogen scattering. However, it is to early to definitely assess the accuracy of these methods.

ACKNOWLEDGMENTS

This work was supported by the EPSRC through grant number EP/D069785/1.

REFERENCES

1. See elsewhere in this proceedings.
2. B. Zygelman, A. Saenz, P. Froelich, S. Jonsell, and A. Dalgarno, Phys. Rev. A. **63**, 052722, (2001).
3. K. Strasburger and H. Chojnacki, Phys. Rev. Lett. **88**, 163201 (2002).
4. K. Strasburger, H. Chojnacki and A. Skołowska, J. Phys. B: At. Mol. Opt. Phys. **38**, 3091 (2005).
5. S. Jonsell, P. Froelich, S. Eriksson, and K. Strasburger, Phys. Rev. A **70**, 062708 (2004).
6. S. Jonsell, A. Saenz, P. Froelich, B. Zygelman, and A. Dalgarno, Phys. Rev. A. **64**, 052712 (2001).
7. E. A. G. Armour, Y. Liu and A. Vigier, J. Phys. B: At. Mol. Opt. Phys. **38**, L47 (2005).
8. C. J. Batty, Nucl. Phys. **A601**, 425 (1996).
9. A. Gal, Nucl. Phys. A **699**, 30c (2002).
10. S. Jonsell, A. Saenz, P. Froelich, B. Zygleman and A. Dalgarno, J. Phys. B: At. Mol. Opt. Phys. **37**, 1195 (2004).
11. A. Yu. Voronin and J. Carbonell, Nucl. Phys. A **689**, 529c (2001).
12. T. L. Trueman, Nucl. Phys. **26**, 57 (1961).
13. S. Jonsell, A. Saenz, P. Froelich, B. Zygelman, and A. Dalgarno, Canadian J. Phys. **83**, 435 (2005).
14. K. Strasburger, J. Phys. B: At. Mol. Opt. Phys. **35**, L435 (2002).
15. K. Strasburger, J. Phys. B: At. Mol. Opt. Phys. **37**, 4483 (2004).
16. E. A. G. Armour and C. W. Chamberlain, J. Phys. B: At. Mol. Opt. Phys. **35**, L489 (2002).
17. S. Jonsell, Nucl. Inst. Meth. B **247**, 138 (2006).
18. B. Zygelman, A. Saenz, P. Froelich and S. Jonsell, Phys. Rev. A **69**, 042715 (2004).
19. P. K. Sinha and A. S. Ghosh, J. Phys. B: At. Mol. Opt. Phys. **35**, L281 (2002).
20. J. Mitroy, M. W. J. Bromley and G. G. Ryzhnikh , J. Phys. B: At. Mol. Opt. Phys. **35**, R81 (2002).
21. S. Jonsell, Nucl. Inst. Meth. B **266**, 369 (2008).

Feb. 19 (Tuesday):

18:00-20:00 Reception

Feb. 20 (Wednesday):

Session I — Chair: Y. Yamazaki

8:30-8:40: Welcome

Y. Yamazaki

8:40-9:10: Antimatter Factory

Y. Yamaguchi

9:10-9:40: Testing the validity of the equivalence principle with antimatter: the AEGIS experiment

G. Testera

9:40-10:10: Production of cold antihydrogen for precision studies

A. Kellerbauer

10:10-10:50: Baryogenesis and its implications to fundamental physics

M. Yoshimura

10:50-11:10: coffee break

Session II — Chair: D. Horvath

11:10-11:40: Accelerating universe, WEP violation and antihydrogen atoms

Y. Fujii

11:40-12:10: A new path to measure antimatter free fall

P. Perez

12:10-12:40: SPAN – Spectroscopy by Atomic Neutrinos

N. Sasao

12:40-14:00: Lunch

Session III — Chair: J. Hangst

14:00-14:30: Precision measurements of the positronium decay rate and Energy level

S. Asai

14:30-14:50: Tests of *CP* and *CPT* symmetry with positronium

T. Namba

14:50-15:20: Positronium molecules and many-positron systems
D.B. Cassidy
15:20-15:50: The Quenching of ortho-Positronium
Toshio Hyodo

15:50-16:20: Coffee break

Session IV — Chair: M.Fujiwara

16:20-17:00: Plasma Tools for Antimatter Physics
C. M. Surko
17:00-17:30: Antiproton Compression and Radial Measurements
J. Fajans
17:30-17:50: Electrons Confined with an Axially Symmetric Magnetic Mirror Field
H. Higaki
17:50-18:20: Ionization of Noble Gases by Charged-Particle Impact
K. Bartschat

Feb. 21 (Thursday):

Session V — Chair: S. Asai

8:00-8:30: Antiprotonic helium and CPT invariance
D. Horvath
8:30-8:50: Determination of the antiproton-to-electron mass ratio by high precision spectroscopy of $p\mathrm{He}^+$ atoms
N. Ono
8:50-9:10: The Hyperfine Stuctrure of Antiprotonic Helium and the Antiproton Magnetic Moment
T. Pask
9:10-9:40: Conclusions from recent pionic–atom experiments
D. Gotta
9:40-10:10: Search for double-strangeness production in pbar-p annihilation at CERN/AD
J. Zmeskal

10:10-10:30: coffee break

Session VI — Chair: J. Ullrich

10:30-11:00: Interactions of laser-cooled atoms in a high-magnetic-field atom trap
G. Raithel

11:00-11:30: Rydberg Atom Formation in Cold (Anti)plasmas
T. Pohl

11:30-12:00: Giant Dipole States of Single-Electron and Multi-Electron Systems in Crossed Electric and Magnetic Fields
P. Schmelcher

12:00-12:30: Atomic Physics Research with Highly Charged Ions and Exotic Nuclei at the Future FAIR Facility
T. Stoehlker

12:30-13:00: Particle Physics Techniques in Cold Antimatter Studies
M. C. Fujiwara

13:20-18:30: excursion (with lunch box)

19:30-21:30: banquet
Welcome: K. Kaya

21:30-22:10: A Particle Physicist looks at Global Warming
J. Eades

Feb. 22 (Friday):

Session VII — Chair: J. Walz

8:30-9:00: First attempts at antihydrogen trapping in ALPHA
J. S. Hangst

9:00-9:30: --- title to be announced ---
P. Bowe

9:30-10:00: Antihydrogen Production in Positron Beam Ion Trap
T. Itahashi

10:00-10:30: Measurement of the Ground-State Hyperfine Splitting of Antihydrogen
B.Juhasz

10:30-10:50: coffee

Session VIII — Chair: G. Testera

10:50-11:20: Cryogenic Particle Accumulation for ATRAP Antihydrogen Experiments
C. Storry

11:20-11:50: Solid-state continuous Lyman-alpha source for laser-cooling of antihydrogen and the prospect of antihydrogen gravity measurements
J. Walz

11:50-12:20: Antihydrogen production
E.L. Rizzini

12:20-12:50: Laser ionization of muonium for low-energy muon source
Y. Matsuda

12:50-14:10: lunch

Session IX — Chair: C. M. Surko

14:10-14:40: Ionization of noble gas atoms in slow antiproton collisions
H. Knudsen

14:40-15:10: A new time-dependent scattering theory: application to the capture of antiprotons by hydrogen atoms and helium atoms
X.M. Tong

15:10-15:40: Atomic Collision Experiments with Ultra-Low-Energy Antiprotons
T. A. Torii

15:40-16:00: Observation of Ultra-Slow Antiprotons with Multi-Channel Plate
H. Imao

16:00-16:30: coffee

Session X — Chair: H. Sadeghpour

16:30-17:00: Sub-Femtosecond Correlated Dynamics Probed with Antiprotons
J. Ullrich

17:00-17:30: Low-energy scattering of antihydrogen by helium and molecular hydrogen
E.A.G Armour

17:30-18:00: Annihilation and rearrangement in atom-antihydrogen collisions
S. Jonsell

18:00-18:30: Status and opportunities of FLAIR
E. Widmann

18:30-18:40: closing remark
H. Sadeghpour

(Evening: FLAIR meeting, E. Widmann)

List of Participants

Armour, Edward
University of Nottingham
edward.armour@nottingham.ac.uk

Asai, Shoji
University of Tokyo
Shoji.Asai@cern.ch

Bartchat, Klaus
Drake University
klaus.bartschat@drake.edu

Bowe, Paul D.
University of Aarhus
Paul.Bowe@cern.ch

Cassidy, David B.
University of California Reverside
Cassidy@physics.ucr.edu

Eades, John
University of Tokyo, c/o CERN
John.Eades@cern.ch

Enomoto, Yoshinori
University of Tokyo
enomoto@radphys4.c.u-tokyo.ac.jp

Fajans, Joel
University of California Berkely
JFajans@berkeley.edu

Fujii, Yasunori
Waseda University
fujii@e07.itscom.net

Fujiwara, Makoto C.
TRIUMF
Makoto.Fujiwara@triumf.ca

Gotta, Detlev
Forschngszentrum Jülich
d.gotta@fz-juelich.de

Hangst, Jeffrey S.
University of Aarhus
Jeffrey.Hangst@cern.ch

Higaki, Hiroyuki
Hiroshima University
hhigaki@hiroshima-u.ac.jp

Horváth, Dezsö
KFKI Research Institute for particle
and Nuclear Physics
horvath@mail.dyn.kfki.hu

Hyodo, Toshio
University of Tokyo
hyodo@phys.c.u-tokyo.ac.jp

Imao, Hiroshi
RIKEN
imao@riken.jp

Itahashi, Takahisa
Osaka University
itahashi@kuno-g.phys.sci.osaka-u.ac.jp

Jonsell, Svante
Swansea University
svante.jonsell@btinternet.com

Juhasz, Bertalan
Stefan Meyer Institute for Subatom Physics,
Austrian Academy of Sciences
bertalan.juhasz@oeaw.ac.at

Kanai, Yasuyuki
RIKEN
kanaiyasuyuki@riken.jp

Kaya, Koji
RIKEN
kaya@riken

Kellerbauer, Alban
Max Planck Institute for Nuclear Physics
a.kellerbauer@mpi-hd.mpg.de

Kira, Kentaro
Hiroshima University
Ub044488@hiroshima-u.ac.jp

Knudsen, Helge
University of Aarhus
hk@phys.au.dk

Lodi Rizzini, Evandro
Università di Brescia
lodi@bs.infn.it

Matsuda, Yasuyuki
University of Tokyo
matsuday@phys.c.u-tokyo.ac.jp

Namba, Toshio
University of Tokyo
naniwa@icepp.s.u-tokyo.ac.jp

Ono, Naoya
University of Tokyo
ss67011@mail.ecc.u-tokyo.ac.jp

Pask, Thomas
Stefan Meyer Institute for Subatom Physics,
Austrian Academy of Sciences
thomas.pask@oeaw.ac.at

Perez, Patrice
CEA-Saclay
patrice.perez@cea.fr

Pohl, Thomas
ITAMP,
Harvard-Smithsonian Center for Astrophysics
tpohl@cfa.harvard.edu

Raithel, Georg
University of Michigan
graithel@umich.edu

Sadeghpour, Hossein
ITAMP,
Harvard-Smithsonian Center for Astrophysics
hrs@cfa.harvard.edu

Sasao, Noboru
Kyoto University
sasao@scphys.kyoto-u.ac.jp

Shimamura, Isao
RIKEN
shimamur@ribf.riken.jp

Shimoyama, Takuya
University of Tokyo
shimoyama@radphys4.c.u-tokyo.ac.jp

Stahl, Stefan
Stahl-Electronics
s.stahl@stahl-electronics.com

Storry, Cody
York University
codys@yorku.ca

Surko, Clifford M.
University of California San Diego
csurko@ucsd.edu

Tarek, Mohamed
RIKEN
mohamed@riken.jp

Testera, Gemma
INFN Genova
Gemma.Testera@ge.infn.it

Tong, Xiao-Min
University of Tsukuba
tong@ims.tsukuba.ac.jp

Torii, Hiroyuki A.
University of Tokyo
torii@radphys4.c.u-tokyo.ac.jp

Toyoda, Hiroshi
University of Tokyo
toyoda@radphys4.c.u-tokyo.ac.jp

Ullrich, Joachim
Max Planck Institute for Nuclear Physics
Joachim.Ullrich@mpi-hd.mpg.de

Walz, Jochen
Johannes Gutenberg-Universität Mainz
jwalz@uni-mainz.de

Wiedmann, Eberhard
Stefan Meyer Institute for Subatom Physics,
Austrian Academy of Sciences
eberhard.widmann@oeaw.ac.at

Yamaguchi, Yoshio
RIKEN

Yamazaki, Yasunori
University of Tokyo/RIKEN
yasunori@phys.c.u-tokyo.ac.jp

Yoshimura, Motohiko
Okayama University
yoshim@fphy.hep.okayama-u.ac.jp

Zmeskal, Johann
Stefan Meyer Institute for Subatom Physics,
Austrian Academy of Sciences
johann.zmeskal@oeaw.ac.at

AUTHOR INDEX

H

I

J

K

L

M

N

O

P

Q

R

S

T

U

V

W

Y

Z